Integrating GIS and Remote Sensing in Soil Mapping and Modeling

Integrating GIS and Remote Sensing in Soil Mapping and Modeling

Editors

Dimitris Triantakonstantis
Panagiotis Tziachris

MDPI • Basel • Beijing • Wuhan • Barcelona • Belgrade • Manchester • Tokyo • Cluj • Tianjin

Editors

Dimitris Triantakonstantis
Institute of Soil and Water
Resources
Elgo - Dimitra
Athens
Greece

Panagiotis Tziachris
Institute of Soil and Water
Resources
Elgo - Dimitra
Thessaloniki
Greece

Editorial Office

MDPI
St. Alban-Anlage 66
4052 Basel, Switzerland

This is a reprint of articles from the Special Issue published online in the open access journal *ISPRS International Journal of Geo-Information* (ISSN 2220-9964) (available at: www.mdpi.com/journal/ijgi/special_issues/GIS_remotesensing_soil).

For citation purposes, cite each article independently as indicated on the article page online and as indicated below:

LastName, A.A.; LastName, B.B.; LastName, C.C. Article Title. *Journal Name* **Year**, *Volume Number*, Page Range.

ISBN 978-3-0365-5978-0 (Hbk)
ISBN 978-3-0365-5977-3 (PDF)

Contents

About the Editors

Dimitris Triantakonstantis

Dr. Dimitris Triantakonstantis is an agronomist-soil scientist and Geographical Information Systems/Remote Sensing (GIS/RS) expert. He works as a Research Associate on GIS and RS in soil science at the Institute of Soil and Water Resources (ELGO-DIMITRA, Greece). He is the author of many scientific articles in international journals and conference papers. More specifically, he has published 19 papers in peer-reviewed journals, 6 book chapters, and 22 conference papers. His research interests lie in GIS and remote sensing in soil science, natural resources management, and climate change mitigation and adaptation. He is the Coordinator of the LIFE20 Preparatory Project LIFE GEOCARBON.

Panagiotis Tziachris

Dr. Panagiotis Tziachris is a researcher at the Institute of Soil and Water Resources of the Hellenic Agricultural Organization-DIMITRA (Greece), on the topic of "Geospatial information and new technologies in the agricultural sector". He holds a master's degree in the analysis, design and management of spatial data and a doctoral degree in national spatial data information (NSDI) from the School of Rural and Surveying Engineering of AUTh. He has been working as the head of the GIS department since 2002, and has been a researcher since 2017. He has more than 25 publications in international/national journals and conference proceedings and has participated in or coordinated over 30 national and international programs.

International Journal of *Geo-Information*

MDPI

Article

Hybrid Machine Learning Approach for Gully Erosion Mapping Susceptibility at a Watershed Scale

Sliman Hitouri [1,†], Antonietta Varasano [2,*,†], Meriame Mohajane [2], Safae Ijlil [3], Narjisse Essahlaoui [3], Sk Ajim Ali [4], Ali Essahlaoui [3], Quoc Bao Pham [5], Mirza Waleed [6], Sasi Kiran Palateerdham [7] and Ana Cláudia Teodoro [8]

1 Geosciences Laboratory, Department of Geology, Faculty of Sciences, University Ibn Tofail, BP 133, Kenitra 14000, Morocco; sliman.hitouri@uit.ac.ma

2 ITC-CNR, Construction Technologies Institute, National Research Council, 70124 Bari, Italy; mohajane@itc.cnr.it

3 Geo-Engineering and Environment Laboratory, Water Sciences and Environment Engineering Team, Department of Geology, Faculty of Sciences, Moulay Ismail University, Meknes 50050, Morocco; safae.ijlil@edu.umi.ac.ma (S.I.); essahlaoui.narjisse@gmail.com (N.E.); a.essahlaoui@umi.ac.ma (A.E.)

4 Department of Geography, Faculty of Sciences, Aligarh Muslim University (AMU), Aligarh 202002, UP, India; skajimali.saa@gmail.com

5 Institute of Applied Technology, Thu Dau Mot University, Thu Dau Mot City 820000, Vietnam; phamphambaoquoc@tdmu.edu.vn

6 Department of Geography, Hong Kong Baptist University, Hong Kong, China; mirzawaleed197@gmail.com

7 School of Aerospace Engineering, University of Rome "La Sapienza", 00138 Rome, Italy; sasikiran.palateerdham@uniroma1.it

8 Earth Sciences Institute (ICT) and Department of Geosciences, Environment and Land Planning, Faculty of Sciences, University of Porto, 4169-007 Porto, Portugal; amteodor@fc.up.pt

* Correspondence: varasano@itc.cnr.it

† These authors contributed equally to this work.

Citation: Hitouri, S.; Varasano, A.; Mohajane, M.; Ijlil, S.; Essahlaoui, N.; Ali, S.A.; Essahlaoui, A.; Pham, Q.B.; Waleed, M.; Palateerdham, S.K.; et al. Hybrid Machine Learning Approach for Gully Erosion Mapping Susceptibility at a Watershed Scale. *ISPRS Int. J. Geo-Inf.* **2022**, *11*, 401. https://doi.org/10.3390/ijgi11070401

Academic Editors: Wolfgang Kainz and Dimitris Triantakonstantis

Received: 1 June 2022
Accepted: 12 July 2022
Published: 14 July 2022

Abstract: Gully erosion is a serious threat to the state of ecosystems all around the world. As a result, safeguarding the soil for our own benefit and from our own actions is a must for guaranteeing the long-term viability of a variety of ecosystem services. As a result, developing gully erosion susceptibility maps (GESM) is both suggested and necessary. In this study, we compared the effectiveness of three hybrid machine learning (ML) algorithms with the bivariate statistical index frequency ratio (FR), named random forest-frequency ratio (RF-FR), support vector machine-frequency ratio (SVM-FR), and naïve Bayes-frequency ratio (NB-FR), in mapping gully erosion in the GHISS watershed in the northern part of Morocco. The models were implemented based on the inventory mapping of a total number of 178 gully erosion points randomly divided into 2 groups (70% of points were used for training the models and 30% of points were used for the validation process), and 12 conditioning variables (i.e., elevation, slope, aspect, plane curvature, topographic moisture index (TWI), stream power index (SPI), precipitation, distance to road, distance to stream, drainage density, land use, and lithology). Using the equal interval reclassification method, the spatial distribution of gully erosion was categorized into five different classes, including very high, high, moderate, low, and very low. Our results showed that the very high susceptibility classes derived using RF-FR, SVM-FR, and NB-FR models covered 25.98%, 22.62%, and 27.10% of the total area, respectively. The area under the receiver (AUC) operating characteristic curve, precision, and accuracy were employed to evaluate the performance of these models. Based on the receiver operating characteristic (ROC), the results showed that the RF-FR achieved the best performance (AUC = 0.91), followed by SVM-FR (AUC = 0.87), and then NB-FR (AUC = 0.82), respectively. Our contribution, in line with the Sustainable Development Goals (SDGs), plays a crucial role for understanding and identifying the issue of "where and why" gully erosion occurs, and hence it can serve as a first pathway to reducing gully erosion in this particular area.

Keywords: gully erosion; hybrid machine learning; frequency ratio; Mediterranean area

1. Introduction

Gully erosion is considered the most destructive type of soil erosion, and it is associated with various topographic, climatic, and anthropogenic factors [1], causing serious environmental and human issues across the world [2], especially in arid and semi-arid regions. Gully erosion occurs over a short period of time. Gullies are a common cause of land degradation, as inappropriate land management and land use practices can lead to increased soil erosion, with gullies as the primary landform [3].

The effective functioning of soil has a significant impact on ecosystem services and is linked to the attainment of the Sustainable Development Goals (SDGs). The soil-water system is the most important component in achieving multiple SDGs, with a focus on neutralizing land degradation and restoring land [4,5]. As a result, one of the most significant issues for the long-term development of the environment and economic activity is prevention of land degradation. As a result, extensive planning and erosion protection have always been essential. Therefore, it is a thoughtful environmental issue that loses a considerable quantity of productive soils each year all around the world [6,7]. Hence, mapping soil erosion is very essential for communicating the spatial information risk of gully erosion for managers and decision-makers for its conservation and management planning.

Soil erosion in Morocco has increased dramatically, leading to severe negative effects on crop production, water ecosystems, and the environment. It is estimated that at least 13% of Moroccan lands are affected by soil erosion [8]. However, there is little research on gully erosion in Morocco in the literature [8,9]. Azedou et al. [10] used frequency ratio (FR), logistic regression (LR), and random forest (RF) to project the spatial distribution of gully erosion in the Souss-Massa watershed, Morocco. The results revealed that among the models tested, the RF model had the best prediction performance. Tairi et al. [11] used the revised universal soil loss equation (RUSLE) for estimating soil erosion in the Tifnout Askaoun watershed in Morocco. Such efforts resulted in a vital tool for the local region's long-term land management. It is important to perform soil erosion research in this environment to add to the current literature and assist local governments in developing suitable plans for soil and land management, watershed management, and infrastructure planning.

Recently, expert knowledge methods such as the analytical hierarchy process (AHP) [12–14], bivariate statistical methods (BSMs), such as FR [15,16], certainty factors (CF) [17,18], weight of evidence (WoE) [19], the information value (InfVal) [20], and the evidential belief function (EBF) [16], conditional probability (CP) [21], index of entropy (IOE) [22], multivariate statistical methods (MSMs), such as linear regression (LiR) [23] and logistic regression (LR) [24], and machine learning (ML) methods such as artificial neural networks (ANN) [25,26], support vector machine (SVM) [27,28], RF [29,30], classification and regression trees (CART) [31], and Naive Bayes [32,33], have been applied for soil erosion mapping. ML algorithms are widely employed for a variety of purposes, including soil erosion mapping, due to their superior prediction capacity compared to other traditional methods [34,35]. There are a variety of approaches, each with its own set of pros and cons. ML models, on the other hand, are useful for determining gully erosion and have been utilized for susceptibility mapping [36–38] and ML mode piping erosion susceptibility prediction [35,39]. The RF model and information value approaches are the most often utilized methods in the bivariate model category [36,40]. The RF model has produced positive outcomes in various studies [36,41,42]. Bivariate models can be simply applied within a geographic information system (GIS) due to their straightforward interpretation [43]. These have yielded positive findings in the literature, both in Morocco through studies in the Ourika and Rheraya watersheds [10,20] and elsewhere [38,44–46]. Selecting gully triggering elements, generating susceptibility maps, implementing land management decisions, and establishing future strategies have all been performed using GIS- and ML-based models [47]. Indeed, ML approaches allow for the evaluation of the role of various components and their interactions, which has significant potential and has been increasingly applied in recent years [34]. With the rapid advancement of different ML algorithms in recent years, determining which model is optimal for a certain location has become difficult. It is critical to look into a variety of algorithms and determine which one is best for each situation.

More recently, hybrid/ensemble models are developed in a combined way, via an integration of individual ML models and statistical approaches. The usefulness of hybrid models that have been discussed in previous publications [48,49] lies in their highest accuracies in comparison to individual models [50,51]. Additionally, in this study, the efficiency of different approaches for gully erosion susceptibility, such as FR, RF, SVM, and NB, is investigated. Although these ML techniques have been employed in the past, they have only been used infrequently for gully erosion modeling.

Gullies inflict severe damage in arid and semi-arid areas, for example the GHISS watershed in northern Morocco, and are regarded as a major environmental hazard [52,53]. As a result, executive agencies must continue to identify the reasons for gully erosion development and zoning to build comprehensive management plans. This will be crucial in the development of restoration methods that are based on natural solutions to ensure long-term sustainability. Regardless of the high susceptibility of this study region to gully erosion, no thorough research has been conducted to date to recognize places that are particularly vulnerable to gully erosion; however, the scientific community is working extremely hard and producing intriguing results. In this analysis, a hybrid methodology combining bivariate statistical approaches and ML algorithms was employed to identify locations susceptible to gully erosion. The main objective of this study is to create a gully erosion susceptibility map in an area with identified soil erosion processes, such as the GHISS watershed of northern Morocco. For that objective, the efficacy of the FR as a statistical model as well as other ML techniques were evaluated in terms of their applicability for predicting gully erosion-prone areas.

2. Materials and Methods

2.1. Study Area

The selected study area in this study is the GHISS watershed, located in the province of Al Hoceima, in the northern part of Morocco. The study area lies between longitudes 3°45′ and 4°30′ W and latitudes 34°15′ and 35°17′ N, covering an area of 837.69 km^2 (Figure 1). Elevation ranges from 0 to 2032 m.a.s.l, characterized by mostly steep reliefs with slopes of more than 35%. The climate in this region is described as semi-arid, and the majority of rainfall occurs during September to May, with an average annual rainfall of 300 mm [54], characterized by both seasonal and interannual variabilities [55]. The average temperature ranges from 21 °C, in July, to 10 °C in January [56].

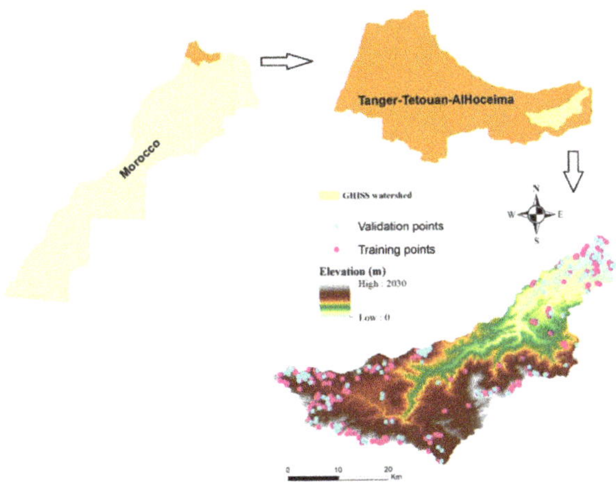

Figure 1. Study area map showing GHISS basin, its elevation profile, and training and validation points used for the gully inventory map.

From a geological point of view, the GHISS basin is characterized by the Ketama unit, which outcrops in the central Rif and is essentially formed of flysch of the Albo-Aptian domain. It belongs to the external domain (intra-Rif) and the flyschs nappes deposits.

Due to its topographical conditions (i.e., slopes higher than 55°) and geological formations (shale, marl, and marl-limestone) [57–59], the watershed has suffered severe soil erosion and decline of forest ecosystems' resources [57]

The study area belongs to the GHISS-Nekkor aquifer, which is an important source of groundwater in that region [60]. However, insufficient and poor sanitation facilities, as well as unsustainable agriculture in the study area, have resulted in deterioration of groundwater quality in this region [60,61]. Previously, a study [62] using the RUSLE method reported that more than 50% of the watershed is considered as moderately eroded. Thus, our research effort is directed towards a deep understanding of gully erosion in this study area.

2.2. Data Used and Methodology

The methodology adopted for this research is divided into three steps, as shown in Figure 2. The first step of the methodology is the data gathering, in which different datasets, including topographic, climatic, and human variables, were preprocessed, and then different geo-environmental variables were generated. Besides geo-environmental variables, a gully erosion inventory map was prepared which was later used in step 2. The second step is the analysis part, in which gully inventory data from step 1 were used to prepare the training and validation data. FR and normalization were performed for training data along with geo-environmental variables, and then all resulting data were fed to three different hybrid ML algorithms (RF-FR, SVM-FR, and NB-FR). In the third step, the validation data prepared in step 2 were used and an accuracy assessment was performed for each of the three hybrid classifiers. Based on the training dataset, the gully erosion maps have been generated using natural breaks classification, available in Arc GIS 10.4. Thus, five categories have been identified: very low, low, moderate, high, and very high gully erosion susceptibility classes. Finally, the gully erosion susceptibility maps were created using the best-suited hybrid classifier. The complete ML implementation was performed in Python utilizing GIS tools and the Jupyter environment.

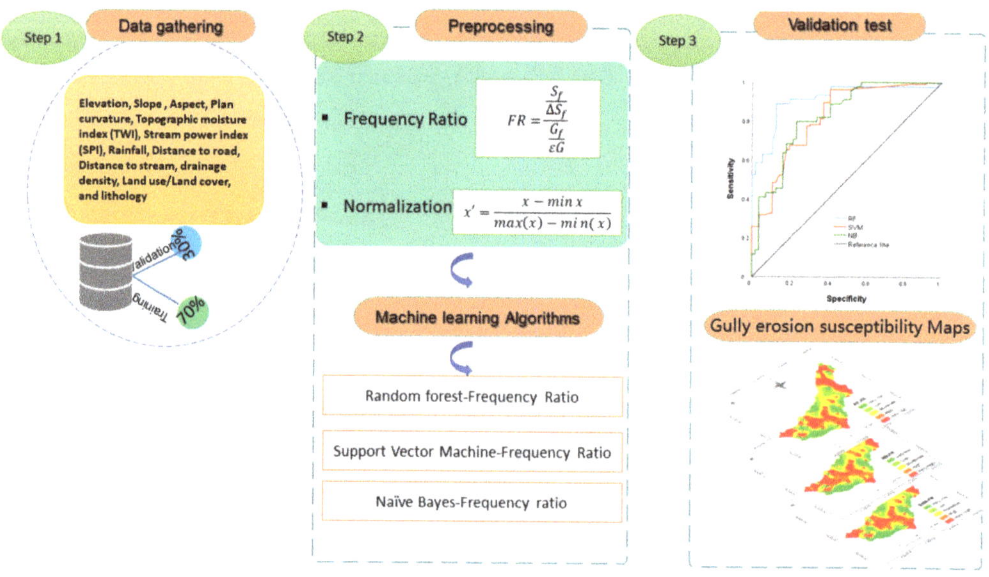

Figure 2. Methodology followed in this study.

2.3. Gully Erosion Inventory Map

The elaboration of the gully erosion inventory map (i.e., target variable) is the first task, and it aims to statistically elucidate the relationship between the distribution of gully erosion (dependent variable) and the conditioning factors (independent variables) of gully erosion hazards [40]. In this study, gully erosion points were identified during field surveys using Global Positioning System (GPS) receivers, and once these locations were recorded, interpretations of high-resolution images from Google Earth were performed. Hence, the 178 points of gully erosion were randomly split into 70% (125) for training, and 30% (53) for validation were kept for the modeling task. An equal total number of non-gully erosion points was randomly selected and split into two sets: 70% (125) for training and 30% (53) for validation. Some examples of point locations and their field photographs are shown in Figure 3.

Figure 3. Gully erosion photos in the GHISS watershed area.

2.4. Parameters' Description

For building binary predictive models, it is necessary to gather both a dependent variable (i.e., target) and a set of independent variables. In this study, 12 variables were selected from different sources (Table 1) due to their importance in gully erosion, as discussed in previous works [37,40]. The selected gully erosion factors classified as topographic, hydrologic, and geologic [37] were used to derive the following variables: elevation, slope, aspect, plan curvature, topographic moisture index (TWI), stream power index (SPI), precipitation, distance to road, distance to stream, drainage density, land use/land cover (LULC), and lithology. All these gully erosion controlling factors were prepared and reclassified based on expert knowledge and statistical analysis using the natural break classification method using GIS tools [63]. To calculate the proportion of gully/non-gully data in each class of each variable, we reclassified each continuous conditioning factor into a set of classes. It should be noted that we adopted an automatic classification for some variables, and for some parameters the classification remains the same as those provided in the source data. The digital elevation model (DEM) with a pixel size of 30×30 m was downloaded from the USGS Earth Explorer website (https://earthexplorer.usgs.gov/), (accessed on 20 August 2021).

It is considered in this study because of its importance in the gully erosion process [37]. Using spatial analysis tools available in ArcGIS 10.4 software, DEM was used to calculate other topographic parameters, including slope, aspect, plan curvature, TWI, and SPI. Due to its effects on vegetation and microclimate [40], elevation plays an important role in gully erosion. It was classified automatically into five classes, including: 0–417, 417–799, 799–1125, 1125–1427, and 1427–2032 m (Figure 4a).

Table 1. Data used in this study.

Conditioning Factor	Unit	Source	Resolution Spatial/Scale
Slope	Degrees (°)	DEM 30 m, from https://earthexplorer.usgs.gov/ (accessed on 20 August 2021)	30 m
Elevation	Meters (m)	DEM 30 m, from https://earthexplorer.usgs.gov/ (accessed on 20 August 2021)	30 m
Plane curvature	-	Morocco DEM 30 m, from https://earthexplorer.usgs.gov/ (accessed on 20 August 2021)	30 m
Aspect	-	DEM 30 m, from https https://earthexplorer.usgs.gov/ (accessedon 20 August 2021)	30 m
Land cover	-	Landsat-8-OLI image, from https://earthexplorer.usgs.gov/ (accessed on 12 July 2021)	30 m
Rainfall	(mm)	ERA-Interim, from https://apps.ecmwf.int/datasets (accessed on 18 July 2021)	30 m
Distance from Road	m	Road map of Morocco	30 m
Distance from stream	m	Stream map of Morocco	30 m
Drainage density	-	DEM 30 m, from https://earthexplorer.usgs.gov/ (accessed on 20 August 2021)	30 m
Lithology	-	Geological map of Morocco 1/1,000,000	30 m
TWI	-	DEM 30 m, from https://earthexplorer.usgs.gov/ (accessed on 20 August 2021)	30 m
SPI	-	DEM 30 m, from https://earthexplorer.usgs.gov/ (accessed on 20 August 2021)	30 m

As reported in previous studies [64,65], slope has an influence on gully erosion, and it becomes more serious in the upslope. In this study, the generated slope factor was classified automatically into five classes: <5, 5–10, 10–20, 20–30, and >30° (Figure 4b).

Aspect has an important effect on gully erosion, as it can influence evapotranspiration, vegetation cover, and incoming solar radiation [66]. In this study, it was classified automatically into nine classes: (1) Flat, (2) North, (3) Northeast, (4) East, (5) Southeast, (6) South, (7) Southwest, (8) West, and (9) Northwest (Figure 4c).

Plan curvature is the curvature of a contour line formed by intersecting a horizontal plane with the surface [67], and it plays an important role in divergence or convergence of water during downslope flow [68]. It was classified automatically into three classes: concave, flat, and convex (Figure 4d).

TWI has been shown to be useful for gully erosion [40], and it was calculated through Equation (1) [69]:

$$\text{TWI} = \ln\left(\text{As}/\tan\beta\right) \quad (1)$$

where A is the watershed area in meters, and β is the slope gradient. This index was classified into five classes (Figure 5b): (18–21), (21–22), (22–23), (23–24), and (24–34).

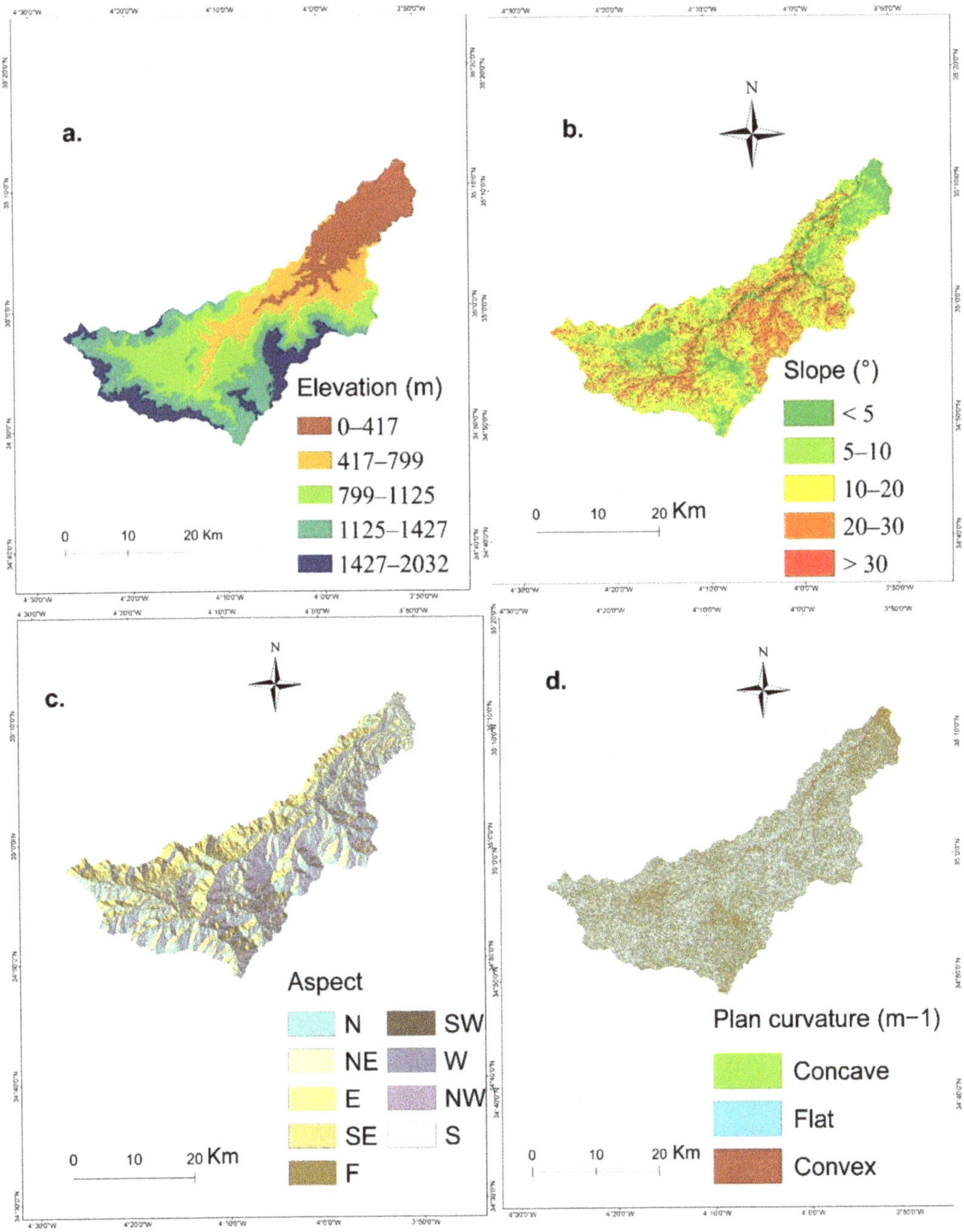

Figure 4. Gully erosion conditioning factors: (**a**) elevation, (**b**) slope, (**c**) aspect, and (**d**) plan curvature.

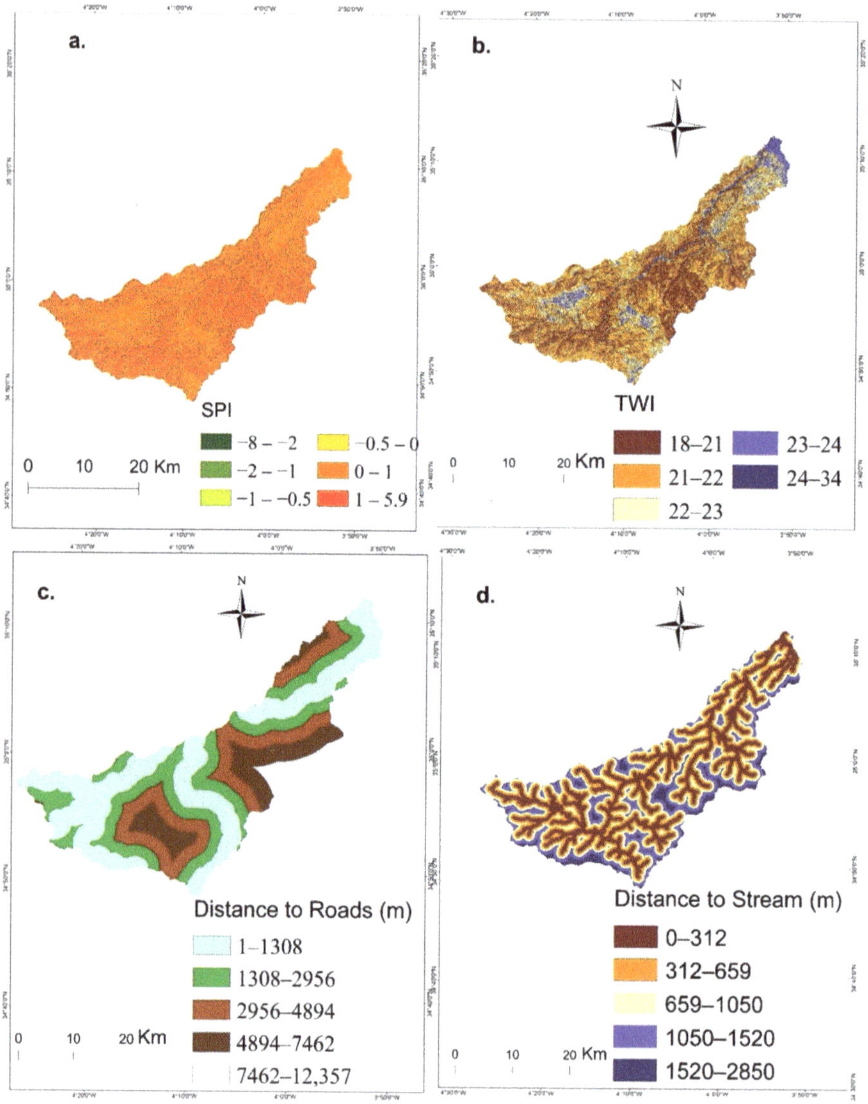

Figure 5. Gully erosion conditioning factors: (**a**) SPI, (**b**) TWI, (**c**) distance to road, and (**d**) distance to stream.

SPI is an index used to measure the capacity and resistance of the soil through surface water flow, runoff, and infiltration, that allows the development of gullies [70]. It was calculated using Equation (2), proposed in [69].

$$SPI = As * \tan\beta \quad (2)$$

where AS is the special area of the basin ($m^2\ m^{-1}$) and β is the slope, in degrees.

It is well-established that gullies occur more in the areas near roads [38]. Hence, distance to road, distance to stream, and drainage density were considered in this study. Using the Euclidean Distance tool available in ArcGIS 10.4, these factors were generated (Figure 5c).

LULC was prepared using the Landsat-8-OLI image acquired on 12 June 2019 downloaded from the United States Geological Survey (USGS) website. First, it was radiometrically and atmospherically calibrated, and then the maximum likelihood supervised classification algorithm using the ENVI software tool was employed. A total of 1079 ground-truth points were randomly selected based on visual interpretation and high-resolution orthorectified Google Earth imagery. Afterwards, based on field investigation, five classes were generated, including: water bodies, forestlands, agricultural lands, buildings/settlements, and bare lands (Figure 6a). The accuracy assessment showed that the generated map had an overall accuracy of 92.4%.

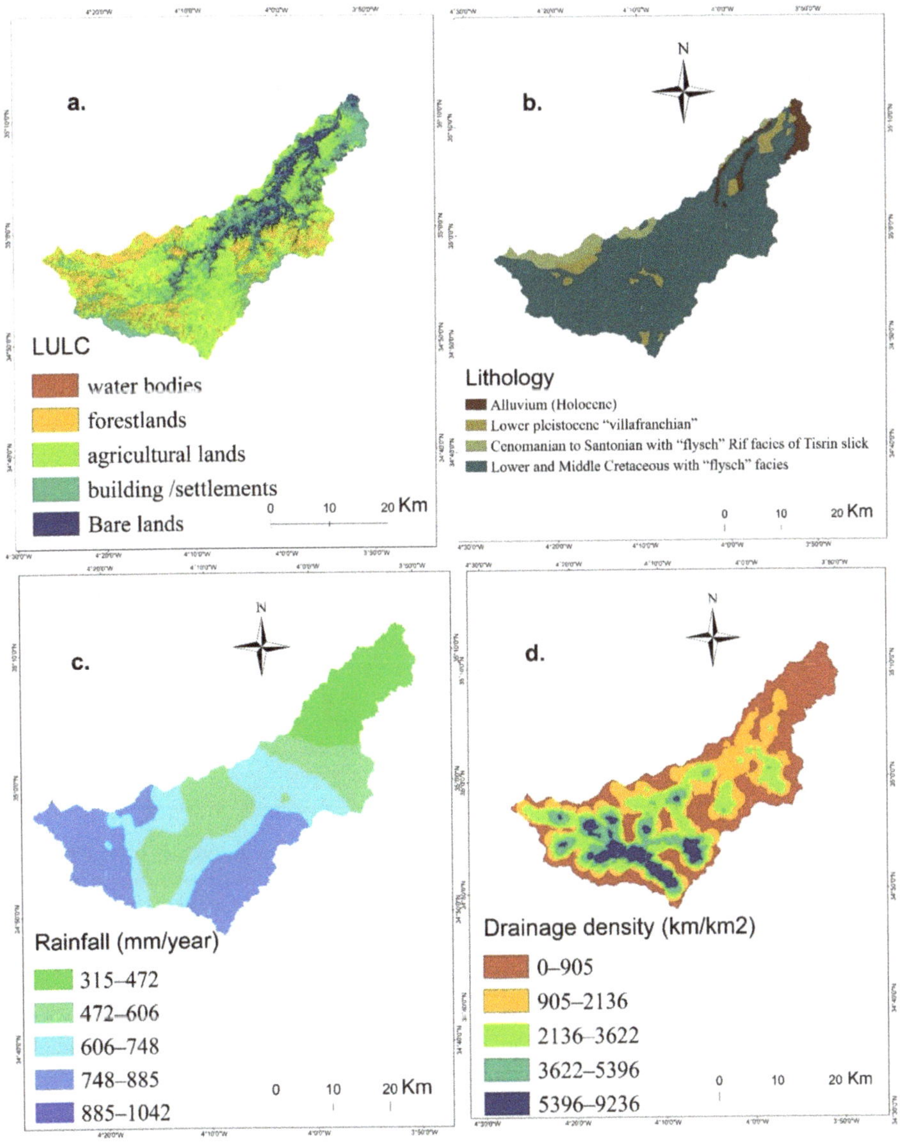

Figure 6. Gully erosion conditioning factors: (**a**) LULC, (**b**) lithology, (**c**) rainfall, and (**d**) drainage density.

This study used monthly precipitation data for a period of 2010 to 2019, downloaded from https://apps.ecmwf.int/datasets, (accessed on 18 August 2021) with a pixel size of $0.25 \times 0.25°$ and resampled to a 30 m pixel size using the nearest neighbor resampling method. A rainfall map was generated using the inverse distance-weighted (IDW) interpolation method, and it was classified into five groups (Figure 6c): (315–461 mm), (461–560 mm), (560–634 mm), (634–778 mm), and (778–1042 mm).

A lithology map of the watershed was digitized from the geological map of Morocco at a scale of 1:000 000. The lithology classes in the study area include four classes, as shown in Figure 6b.

2.5. Multicollinearity Analysis

The correlation of conditioning variables is represented by multicollinearity analysis, which was used to select the optimal controlling factor for gully erosion susceptibility mapping [71]. Many researchers have looked at the value of controlling variables, particularly for gully susceptibility mapping, and virtually all have come to the conclusion that each variable's importance is mostly determined by its surroundings [37,40]. In this study, multicollinearity analysis was performed using variance inflation factors (VIF), which show multiplicative inverse of tolerance (TOL), which is computed as 1 − R2, where R2 is calculated through reverting all resulting variables in multivariate regression [40]. Multiplicative analysis was performed using the 12 geo-environmental variables prepared earlier. From the literature, it is evident that values for TOL = 0.10, and VIF = 0.5 represents issues in overall multicollinearity [72]. For multicollinearity analysis, equations used to derive tolerance and VIF are given as Equations (3) and (4), respectively.

$$\text{Tolerance} = 1 - R^2 j \quad (3)$$

$$\text{VIF} = 1/\text{Tolerance} \quad (4)$$

where R^2j is the coefficient of determination of regression for the variable j.

3. Models and Methods Background

3.1. Frequency Ratio (FR)

The FR model adopts a theory of probability to define the relationship between independent and dependent variables of spatial information using the multi-class mapping approach [73]. It has been used for a variety of environmental hazards, such as landslide [74], flood [75], forest fire [76], and gully erosion susceptibility mapping [77]. It is a bivariate probability statistic index, used to identify the spatial relationship between erosion and different factors contributing towards gully erosion in the region. The FR model can be defined as (Equation (5)):

$$Fr_i = b/a \quad (5)$$

where b is the ratio between erosion pixels by total number of erosion pixels, a is the ratio between no erosion cells by total number of non-erosion cells, while Fr_i donates the importance of the conditioning factor in relation to erosion occurrence. FR > 1 indicates high correlation with the erosion probability, while FR < 1 represents low correlation.

3.2. Random Forest (RF)

The RF model works on the principle of constructing multiple decision trees from different subsets of data. RF is an integrated approach that combines the ideas proposed in [78] with the methods described in [79]. The RF starts growing when the algorithm predicts the variables and targets, leading to a decision tree which can be further pruned [80]. A RF is so large that it is very difficult to explain. It is necessary to summarize its information using quantitative indicators. The famous indicators are the mean decrease Gini index and mean decrease accuracy [80]. RF utilizes mean decrease accuracy and the mean decrease Gini index in the ranking of factors [40,81].

3.3. Support Vector Machine (SVM)

SVM is a typical ML algorithm method which uses statistical learning theory based on the structural risk minimization (SRM) [82]. This algorithm is best-suited to solving regression analysis and classifier problems [64]. Generally, four kinds of computing functions were used in SVM: linear kernel (LN), polynomial kernel (PL), sigmoid kernel, and radial basis function (RBF) [70,83]. The accuracy of the prediction usually depends on the selection of the type of function [84]. The SVM model works well only for linear data; in case of nonlinear datasets, it transforms the nonlinear data into linear by using the so-called "Kernel-trick" [85].

3.4. Naïve Bayes (NB)

The NB model is based on Bayes' theorem, which uses a set of assortment algorithms for classification [86]. This is a family of algorithms, where all explanatory variables are completely independent of each other, which share a common principle [87]. The NB model is well-suited against noise and irreverent models [88]. This model can also be used with a relatively small amount of training data to estimate parameters for classification [89].

3.5. Model Validation

Validation of the developed models is an essential part of any modeling study [90,91]. Thus, several statistical indices were widely used, and among them, accuracy, specificity, sensitivity, and precision were calculated in this research. Overall accuracy (OA) is the probability of occurrence of correctly classified pixels which are computed by the sum of true positive and true negative divided by all available singular tests. The equation form of OA is given in Equation (6). Specificity, also known as the true-positive rate, represents the proportion of gully erosion pixels correctly predicted as gully erosion (Equation (8)). Sensitivity focuses only on correctly classified pixels from the test data and is calculated by dividing the true-negative values by the sum of true negative and false positive (Equation (7)). Precision is the measure of the quality of the results and is calculated by dividing the true positive by the sum of the true positive and the false positive (Equation (9)).

$$\text{Accuracy} = (\text{TP} + \text{TN})/\text{TP} + \text{TN} + \text{FP} + \text{FN} \quad (6)$$

$$\text{Sensitivity} = \text{TP}/\text{TP} + \text{TN} \quad (7)$$

$$\text{Specificity} = \text{TN}/\text{TN} + \text{FP} \quad (8)$$

$$\text{Precision} = \text{TP}/\text{TP} + \text{FP} \quad (9)$$

where TP is true positive, TN is true negative, FP is false positive, and FN is false negative. For the AUC (area under the receiver operating characteristic curve) and the receiver operating characteristic (ROC) curve, on the y-axis, the sensitivity is plotted, and the x-axis shows the specificity in terms of gully erosion probability occurrences.

3.6. Variable Importance Using Information Gain Index

Various statistical indices have been used for feature selection, which include, one-rule attribute elevation (ORAE) [92], forward elimination [93], backward elimination [93], and information gain (IG) [94]. Based on the results of the RF-FR model, IG was used to reveal the importance of each conditioning factor for the modeling process [95].

4. Results

4.1. Results of Frequency Ratio

The FR was calculated to reveal the relationship between gully erosion as the dependent variable and each gully erosion conditioning factor as the independent variables.

Based on findings form this analysis (Table 2), the highest FR value (i.e., 3.528) was found in the class of lithology of Cenomanian to Santonian with "flysch" Rif facies of Tisrin slick, which represents the area more susceptible to gully erosion, mainly composed of the alternation of sandstone marl-limestone flysch and sandstone flysch, more sensitive to erosion, followed by the highest class of rainfall (885–1042) (i.e., 3.058).

The lowest classes of slope, elevation, curvature, plan curvature, distance from roads, distance from stream, drainage density, TWI, and SPI are more susceptible to gully erosion. From the analysis of the LULC parameter, it was observed that this erosion occurred more in the bare lands class, followed by the buildings/settlement class. For the aspect factor, it was also observed that the susceptibility to gully erosion is more pronounced in flat areas.

Table 2. Frequency ratio.

Factors	Classes	No. of Points	% of Points	Classes Area	% of Class Area	FR
Slope (°)	0–7	35,100	22.807	245,253	26.165	0.872
	7–13	45,000	29.240	176,502	18.830	1.553
	13–19	43,200	28.070	256,397	27.354	1.026
	19–27	20,700	13.450	187,187	19.970	0.674
	27–55	9900	6.433	71,995	7.681	0.838
Elevation (m)	3–417	66,600	43.275	228,898	244.194	1.772
	417–799	48,600	31.579	183,996	196.292	1.609
	799–1. 125	11,700	7.602	119,356	127.332	0.597
	1.125–1.427	21,600	14.035	242,315	258.508	0.543
	1. 427–2.032	5400	3.509	162,771	173.648	0.202
Aspect	N	15,300	9.942	122,315	13.049	0.762
	NE	19,800	12.865	107,738	11.494	1.119
	E	17,100	11.111	95,212	10.158	1.094
	SE	20,700	13.450	103,068	10.996	1.223
	F	26,100	16.959	91,816	9.795	1.731
	SW	18,900	12.281	79,477	8.479	1.448
	W	10,800	7.018	86,691	9.249	0.759
	NW	11,700	7.602	112,366	11.988	0.634
	S	13,500	8.772	138,652	14.792	0.593
Plan curvature (100/m)	Concave	50,400	32.749	237,242	25.310	1.294
	Flat	30,600	19.883	194,743	20.776	0.957
	Convex	72,900	47.368	505,350	53.913	0.879
Distance from road (m)	0–1. 308	103,500	65.714	339,677	36.495	1.801
	1. 308–2. 956	5400	3.429	51,948	5.581	0.614
	2.956–4.894	23,400	14.857	239,404	25.722	0.578
	4.894–7.462	6300	4.000	119,729	12.864	0.311
	7.462–12,357	18,900	12.000	179,984	19.338	0.621
Distance from stream (m)	0–312	37,800	24.000	290,993	31.049	0.773
	312–659	53,100	33.714	249,852	26.659	1.265
	659–1.050	36,000	22.857	206,734	22.058	1.036
	1.050–1. 520	24,300	15.429	134,222	14.321	1.077
	1.520–2.850	6300	4.000	55,409	5.912	0.677
Drainage density (km/km^2)	0–905	83,700	53.143	362,876	38.719	1.373
	905–2.136	18,900	12.000	96,881	10.337	1.161
	2.136–3.622	6300	4.000	58,783	6.272	0.638
	3.622–5.396	36,000	22.857	249,700	26.643	0.858
	5.396–9.236	12,600	8.000	168,970	18.029	0.444
TWI	18–21	15,300	9.942	76,047	8.136	1.222
	21–22	72,900	47.368	403,332	43.152	1.098
	22–23	33,300	21.637	192,428	20.588	1.051
	23–24	2700	1.754	18,292	1.957	0.896
	24–34	29,700	19.298	244,575	26.167	0.738

Table 2. *Cont.*

Factors	Classes	No. of Points	% of Points	Classes Area	% of Class Area	FR
SPI	−8–−2	2700	1.754	27,308	2.992	0.586
	−2–−1	16,200	10.526	84,430	9.252	1.138
	−1–−0,5	23,400	15.205	104,777	11.482	1.324
	−0,5–0	60,300	39.181	326,669	35.797	1.095
	0–1	41,400	26.901	294,190	32.238	0.834
	1–5.9	9900	6.433	75,183	8.239	0.781
Rainfall (mm)	315–472	53,100	33.908	231,033	18.601	1.823
	472–606	18,000	11.494	334,596	26.940	0.427
	606–748	10,800	6.897	295,083	23.758	0.290
	748–885	34,200	21.839	276,282	22.245	0.982
	885–1.042	40,500	25.862	105,025	8.456	3.058
Lithology	Alluvium (Holocene)	9000	5.714	40,715	4.344	1.315
	Lower pleistocene "villafranchian"	22,500	14.286	52,106	5.560	2.569
	Cenomanian to Santonian with "flysch" Rif facies of Tisrin slick	34,200	21.714	57,680	6.155	3.528
	Lower and Middle Cretaceous with "flysch" facies	91,800	58.286	786,673	83.941	0.694
LULC	Water bodies	0	0.000	134	0.014	0.000
	Forestlands	900	0.571	55,437	5.960	0.096
	Agricultural lands	10,800	6.857	124,566	13.392	0.512
	Buildings/settlements	57,600	36.571	322,690	34.692	1.054
	Bare lands	88,200	56.000	427,319	45.941	1.219

4.2. Results of Multicollinearity Assessment

Multicollinearity analysis represents the correlation of conditioning variables, correlated or interrelated [46]. It was applied in this study to analyze the correlation among the gully erosion factors (independent variables). To perform this, we used two indexes: TOL and VIF [64]. If the value of TOL is less than 0.1 and the value of VIF is greater than 10 [96], collinearity exists amongst the variables. Our results (Table 3) showed that LULC had the lowest tolerance value of the gully erosion conditioning factors (0.454), while aspect had the highest tolerance value (0.893). Regarding the variance inflation factor (VIF), the highest value was 2.203 (LULC), and the lowest value was 1.120 (aspect). These gully erosion conditioning factors had tolerance values greater than 0.1, and the VIF values were less than 0.1 and 10, indicating that no collinearity exists amongst these factors. Therefore, all 12 conditioning factors were kept in this research.

Table 3. Multicollinearity analysis.

Factors	Collinearity Statistics	
	TOL	VIF
Aspect	0.893	1.120
Slope	0.691	1.447
Plan curvature	0.747	1.338
Distance to stream	0.797	1.254
Distance to road	0.782	1.279
Drainage density	0.757	1.321
Elevation	0.756	1.323
Rainfall	0.748	1.337
Lithology	0.778	1.285
LULC	0.454	2.203
SPI	0.766	1.306
TWI	0.577	1.732

4.3. Identification of Gully Zones

Based on the training dataset, the gully erosion maps have been generated using natural breaks classification, available in Arc GIS 10.4. Thus, five categories have been identified: very low, low, moderate, high, and very high gully erosion susceptibility classes. Figure 7 and Table 4 present the spatial distribution of the susceptibility classes in the gully erosion susceptibility maps. In the gully erosion susceptibility map constructed using the RF model, 25.98% of the study area had a very high susceptibility to erosion, while 17.88%, 19.30%, 19.64%, and 17.20% of the area was classified as very low, low, moderate, and high susceptibilities, respectively. In the case of the SVM model, 22.62% of the area was classified as very high susceptibility, while 15.01%, 14.4%, 20.00%, and 27.95% had very low, low, moderate, and high susceptibilities, respectively. For the NB model, 27.10% of the study area was classified to the very high gully hazard category, while 17.40%, 17.67%, 20.37%, and 17.44% had very low, low, moderate, and high susceptibilities, respectively. Figure 7 shows the spatial distribution of gully erosion in the watershed. By comparing the spatial distribution of gully erosion obtained by all the used models, a quiet homogeneity of the most erosion-prone areas throughout the watershed was clearly observed. The most eroded areas were located in different parts of the watershed, which were characterized mainly by variations in slope, rapidly increasing the transport of sediments. The characteristic lithology of this watershed might be another reason [97]. In addition, inappropriate agriculture practices and overgrazing also acted as other driving forces in gully erosion in this study area [62].

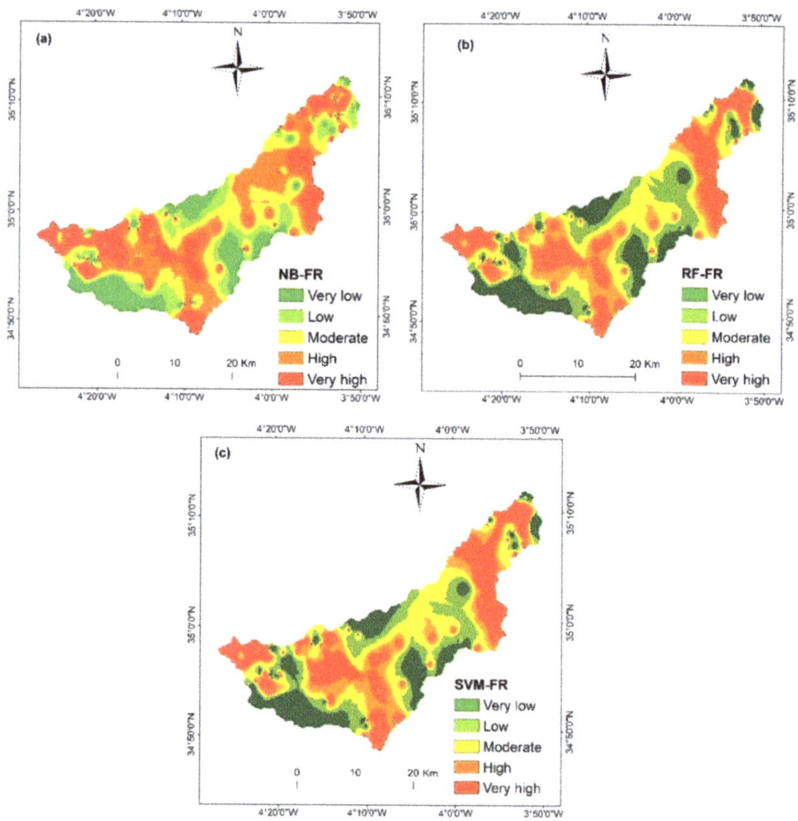

Figure 7. Gully erosion susceptibility mapping using: (**a**) NB-FR, (**b**) RF-FR, and (**c**) SVM-FR.

Table 4. Percentages of gully erosion susceptibility classes.

Susceptibility Class	RF		SVM		NB	
	Class	% of Area	Class	% of Area	Class	% of Area
Very low	5954	17.88	4996	15.01	5791	17.40
Low	6423	19.30	4791	14.4	5880	17.67
Moderate	6538	19.64	6658	20.00	6779	20.37
High	5723	17.20	9302	27.95	5805	17.44
Very high	8648	25.98	7529	22.62	9021	27.10

4.4. Variable Importance

The importance of variables for gully erosion mapping was performed based on the RF model. As can be seen in Figure 8, TWI (1.78), LULC (1.73), distance from stream (1.47), drainage density (1.45), slope (1.42), aspect (1.34), rainfall (1.28), SPI (1.06), and distance from road (1.02) were the most importance factors for gully erosion susceptibility mapping, whereas elevation (0.79), plan curvature (0.76), and lithology (0.74) were of the least importance.

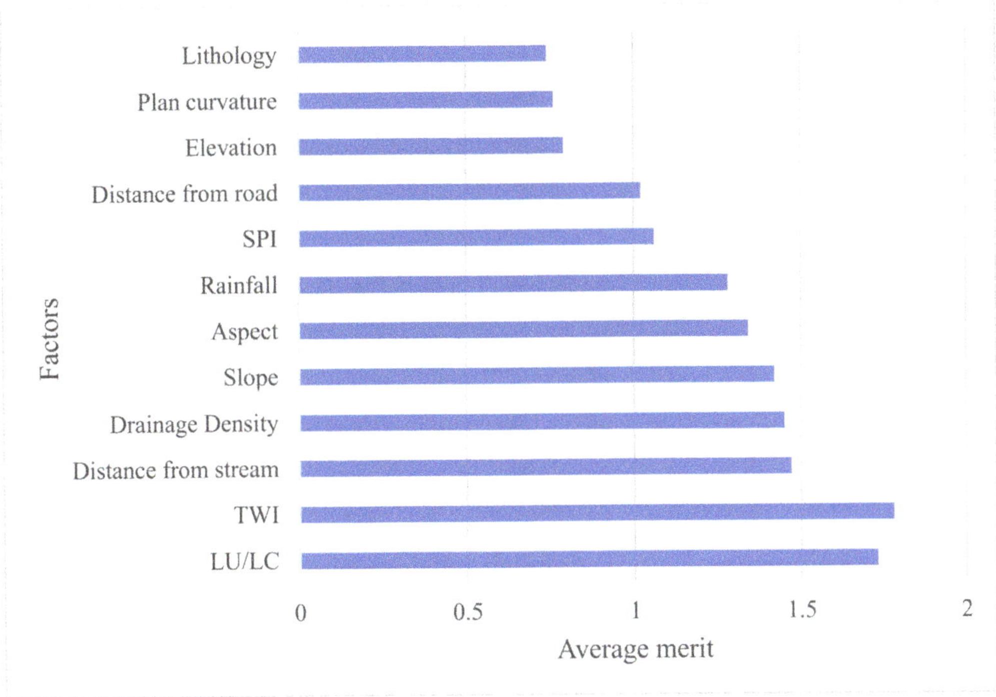

Figure 8. The importance of conditioning factors.

4.5. Validation of Gully Erosion Models

The validation results for both training and validation datasets using accuracy, precision, and AUC are presented in Table 5 and Figure 9. The statistical parameters for each model were almost the same for both training and validation datasets, with a slight difference in favor of the AUC for the FR-NB model.

Table 5. Model statistical measures assigned to the training and validation datasets.

	FR-RF		SVM-FR		NB-FR	
	Training	**Validation**	**Training**	**Validation**	**Training**	**Validation**
Accuracy	86.29	86.11	80.64	80.55	65.72	65.74
Precision	83.58	83.05	78.35	77.96	67.89	68.08
AUC	0.83	0.83	0.78	0.79	0.69	0.79

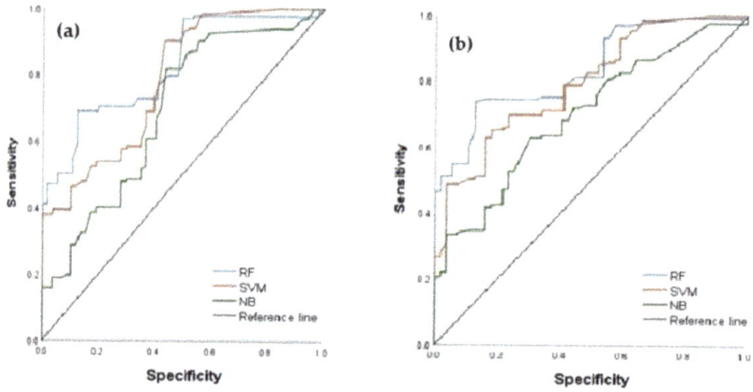

Figure 9. (**a**) ROC curves of success rate. (**b**) ROC curves of prediction rate.

5. Discussion

The effective functioning of soil has a significant impact on ecosystem services and is linked to the attainment of the SDGs. The soil-water system is the most important component in achieving multiple SDGs, with a focus on neutralizing land degradation and restoring land [4,5]. With the shifting land-use trends and compaction of productive soil, excessive degradation in productive land is observed globally due to gully erosion. As a result, one of the most significant issues for the long-term development of the environment and economic activity is preventing land degradation. Therefore, extensive planning and erosion protection have always been essential.

Machine learning methods are reliable tools for mitigating and controlling the influence of gully erosion in different regions all over the world. Based on the Web of Science (WoS) database and using the common keywords "gully erosion susceptibility" and "machine learning algorithms", nine papers published between 2019 and 2021 were selected from different parts of the world, and their results are reported in Table 6. RF generates models with high accuracy in comparison to the different approaches, and this is due to its ability to handle large datasets and produce fast classifications, based on multiple features. Additionally, RF is widely used to assess the importance of each variable used in order to calculate a multi-classifier and evaluates its own accuracy and its suitability for the modeling process [98].

Table 6. A comparison of machine learning models in gully erosion susceptibility.

Region	ML Model	Performances Based on Accuracy/AUC	Paper Reference
Brazil (Rio das Velhas watershed)	RF	0.996	[99]
	LR	0.935	
	NB	0.947	
	ANN	0.987	

Table 6. *Cont.*

Region	ML Model	Performances Based on Accuracy/AUC	Paper Reference
Iran (Robat Turk Watershed)	RF	0.893	[34]
	CDTree	0.808	
	KLR	0.825	
	BFTree	0.789	
India	RF	90.38	[100]
	BRT	88.29	
	Naïve bayes	86.37	
Brazil (South Mato Grosso)	MDA	78.47	[101]
	LR	77.62	
	CART	82.81	
	RF	86.09	
India	MARS	91.4	[36]
	FDA	84.2	
	RF	96.2	
	SVM	88.3	
China	RF	0.944	[46]
	GBDT	0.938	
	XGBoost	0.947	
India (Hinglo River basin)	RF	0.87	[35]
	GBRT	0.80	
	NBT	0.81	
	TE	0.82	
Iran (Bastam watershed)	ADTree	0.922	[102]
	NBTree	0.939	
	LMT	0.944	
Iran (Fars province)	RF	0.958	[64]
	BRT	0.991	
	SVM	0.914	

Abbreviations: random forest (RF), logistic regression (LR), naïve Bayes (NB), artificial neural network (ANN), credal decision trees (CDTree), kernel logistic regression (KLR), best first decision tree (BFTree), boosted regression tree (BRT), multivariate discriminant analysis (MDA), classification and regression tree (CART), gradient boosted decision trees (GBDT), extreme gradient boosting (XGBoost), multivariate additive regression splines (MARS), flexible discriminant analysis (FDA), support vector machine (SVM), gradient boosted regression tree (GBRT), naïve Bayes tree (NBT), tree ensemble (TE), alternating decision tree (ADTree), logistic model tree (LMT).

In this research, the relationship between gully erosion occurrence and various environmental factors was investigated for the GHISS watershed. We used three hybrid ML models (i.e., RF-FR, SVM-FR, and NB-FR) for gully erosion susceptibility mapping. We found that the FR-RF model achieved better performance results compared to the other models. Our findings are consistent with previous studies, for example [36,40]. In one study [103], the authors argued that RF's better performance is because it is less prone to both over-fitting and outliers in the training dataset.

In terms of accuracy, the RF model is followed by the SVM-FR model due to its capacity to handle non-linear data, and it has yielded good results for both classification and regression problems in many applications [104].

The performance of NB-FR was slightly weaker than the other models. It assumes conditional independence between features [105]. It has been used with great results in previous papers [106].

Based on our results, among model variables, LULC and topographic moisture index (TMI) showed the maximum importance factors in enhancing the performance of hybrid models. Land-use change affects gully erosion by altering the hydrological and physicochemical properties of the soil. Other factors, such as distance from stream, drainage density, and slope, also showed reasonable importance output after LULC and TWI. Vegetation stabilizes gullies because of the role of plant roots. Thus, areas with no and sparse vegetation are most affected by gully erosion and widely exposed to rainfall and runoff. The land resources in the northern part of Morocco that have been found to be influenced by environmental and anthropogenic activities mainly include: rainfall irregularity [107,108], steep slopes, and weak geologic units, i.e., the type of geological formation that forms the northern parts make the lands more prone to erosion. Moreover, Cannabis cultivation has become a complex and challenging practice to control, leading to the loss of forestlands and accelerating soil erosion processes [76]. Thereby, the approach developed in this study could be effective in gully erosion prediction. The maps generated here can be a good reference to reduce the phenomenon of gully erosion and can be used as a valuable tool for the establishment of sustainable strategies and actions. It should be highlighted that there is not one perfect algorithm in comparison to others, because each one has its specific advantages and drawbacks, and each algorithm highlights its own usefulness for each study case.

6. Conclusions

The use of ML algorithms for environmental hazard modeling is an emerging focus in many studies, thanks to the technological advancements of Internet of Things (IoT). The relationship between gully erosion occurrence and various environmental factors was investigated for the study area. As investigated in this study, three hybrid models (FR-RF, FR-SVM, and FR-NB) have been elaborated for gully erosion susceptibility mapping. The results of this study showed that the FR-RF hybrid model outperformed the other developed models for gully erosion susceptibility mapping. In summary, this study showed that the use of hybrid ML models for gully erosion is better than single ML models, which is consistent with previous studies. The methodology proposed in this study can be applied to areas influenced by identical environmental and anthropogenic activities, which includes, for instance, rainfall irregularity, steep slopes, and weak geologic units, for mapping gully erosion. For the elaboration of new studies, researchers are encouraged to use the above three models to address new questions and research directions. Further research may consider the application of deep learning approaches in gully erosion mapping from local to regional scale areas.

Author Contributions: Conceptualization, Sliman Hitouri and Antonietta Varasano; Formal analysis, Antonietta Varasano; Funding acquisition, Antonietta Varasano; Investigation, Meriame Mohajane, Mirza Waleed and Sasi Kiran Palateerdham; Methodology, Antonietta Varasano; Resources, Meriame Mohajane and Ali Essahlaoui; Software, Antonietta Varasano, Meriame Mohajane, Safae Ijlil and Narjisse Essahlaoui; Supervision, Ali Essahlaoui; Visualization, Narjisse Essahlaoui; Writing—original draft, Sliman Hitouri, Antonietta Varasano, Meriame Mohajane, Safae Ijlil, Sk Ajim Ali and Quoc Bao Pham; Writing—review & editing, Ana Cláudia Teodoro. Sliman Hitouri and Antonietta Varasano contributed equally to this paper by performing analyses and designing the study. All the co-authors drafted and revised the article together. All authors have read and agreed to the published version of the manuscript.

Funding: This research received no external funding.

Institutional Review Board Statement: Not applicable.

Informed Consent Statement: Not applicable.

Data Availability Statement: Not applicable.

Conflicts of Interest: The authors declare no conflict of interest.

References

1. Domazetović, F.; Šiljeg, A.; Lončar, N.; Marić, I. Development of Automated Multicriteria GIS Analysis of Gully Erosion Susceptibility. *Appl. Geogr.* **2019**, *112*, 102083. [CrossRef]
2. Fadul, H.M.; Salih, A.A.; Ali, I.A.; Inanaga, S. Use of Remote Sensing to Map Gully Erosion along the Atbara River, Sudan. *Int. J. Appl. Earth Obs. Geoinf.* **1999**, *1*, 175–180. [CrossRef]

3. Magliulo, P. Assessing the Susceptibility to Water-Induced Soil Erosion Using a Geomorphological, Bivariate Statistics-Based Approach. *Environ. Earth Sci.* **2012**, *67*, 1801–1820. [CrossRef]
4. Dooley, E.; Roberts, E.; Wunder, S. Land Degradation Neutrality under the SDGs: National and International Implementation of the Land Degradation Neutral World Target. *Elni. Rev.* **2015**, *1*, 2–9. [CrossRef]
5. Safriel, U. Land Degradation Neutrality (LDN) in Drylands and beyond—Where Has It Come from and Where Does It Go. *Silva Fenn.* **2017**, *51*, 20–24. [CrossRef]
6. Durán Zuazo, V.H.; Rodríguez Pleguezuelo, C.R. Soil-Erosion and Runoff Prevention by Plant Covers. A Review. *Agron. Sustain. Dev.* **2008**, *28*, 65–86. [CrossRef]
7. Lal, R. Soil Erosion Impact on Agronomic Productivity and Environment Quality. *Crit. Rev. Plant Sci.* **1998**, *17*, 319–464. [CrossRef]
8. Peter, K.D.; d'Oleire-Oltmanns, S.; Ries, J.B.; Marzolff, I.; Ait Hssaine, A. Soil Erosion in Gully Catchments Affected by Land-Levelling Measures in the Souss Basin, Morocco, Analysed by Rainfall Simulation and UAV Remote Sensing Data. *CATENA* **2014**, *113*, 24–40. [CrossRef]
9. Simonneaux, V.; Cheggour, A.; Deschamps, C.; Mouillot, F.; Cerdan, O.; Le Bissonnais, Y. Land Use and Climate Change Effects on Soil Erosion in a Semi-Arid Mountainous Watershed (High Atlas, Morocco). *J. Arid Environ.* **2015**, *122*, 64–75. [CrossRef]
10. Azedou, A.; Lahssini, S.; Khattabi, A.; Meliho, M.; Rifai, N. A Methodological Comparison of Three Models for Gully Erosion Susceptibility Mapping in the Rural Municipality of El Faid (Morocco). *Sustainability* **2021**, *13*, 682. [CrossRef]
11. Tairi, A.; Elmouden, A.; Bouchaou, L.; Aboulouafa, M. Mapping Soil Erosion–Prone Sites through GIS and Remote Sensing for the Tifnout Askaoun Watershed, Southern Morocco. *Arab. J. Geosci.* **2021**, *14*, 811. [CrossRef]
12. Kachouri, S.; Achour, H.; Abida, H.; Bouaziz, S. Soil Erosion Hazard Mapping Using Analytic Hierarchy Process and Logistic Regression: A Case Study of Haffouz Watershed, Central Tunisia. *Arab. J. Geosci.* **2015**, *8*, 4257–4268. [CrossRef]
13. Saha, S.; Gayen, A.; Pourghasemi, H.R.; Tiefenbacher, J.P. Identification of Soil Erosion-Susceptible Areas Using Fuzzy Logic and Analytical Hierarchy Process Modeling in an Agricultural Watershed of Burdwan District, India. *Environ. Earth Sci.* **2019**, *78*, 649. [CrossRef]
14. Senouci, R.; Taibi, N.-E.; Teodoro, A.C.; Duarte, L.; Mansour, H.; Yahia Meddah, R. GIS-Based Expert Knowledge for Landslide Susceptibility Mapping (LSM): Case of Mostaganem Coast District, West of Algeria. *Sustainability* **2021**, *13*, 630. [CrossRef]
15. Senanayake, S.; Pradhan, B.; Huete, A.; Brennan, J. Assessing Soil Erosion Hazards Using Land-Use Change and Landslide Frequency Ratio Method: A Case Study of Sabaragamuwa Province, Sri Lanka. *Remote Sens.* **2020**, *12*, 1483. [CrossRef]
16. Tehrany, M.S.; Shabani, F.; Javier, D.N.; Kumar, L. Soil Erosion Susceptibility Mapping for Current and 2100 Climate Conditions Using Evidential Belief Function and Frequency Ratio. *Geomat. Nat. Hazards Risk* **2017**, *8*, 1695–1714. [CrossRef]
17. Arabameri, A.; Pradhan, B.; Rezaei, K. Gully Erosion Zonation Mapping Using Integrated Geographically Weighted Regression with Certainty Factor and Random Forest Models in GIS. *J. Environ. Manage.* **2019**, *232*, 928–942. [CrossRef]
18. Azareh, A.; Rahmati, O.; Rafiei-Sardooi, E.; Sankey, J.B.; Lee, S.; Shahabi, H.; Ahmad, B.B. Modelling Gully-Erosion Susceptibility in a Semi-Arid Region, Iran: Investigation of Applicability of Certainty Factor and Maximum Entropy Models. *Sci. Total Environ.* **2019**, *655*, 684–696. [CrossRef]
19. Hembram, T.K.; Paul, G.C.; Saha, S. Comparative Analysis between Morphometry and Geo-Environmental Factor Based Soil Erosion Risk Assessment Using Weight of Evidence Model: A Study on Jainti River Basin, Eastern India. *Environ. Process.* **2019**, *6*, 883–913. [CrossRef]
20. Meliho, M.; Khattabi, A.; Mhammdi, N. A GIS-Based Approach for Gully Erosion Susceptibility Modelling Using Bivariate Statistics Methods in the Ourika Watershed, Morocco. *Environ. Earth Sci.* **2018**, *77*, 655. [CrossRef]
21. Pourghasemi, H.R.; Mohammady, M.; Pradhan, B. Landslide Susceptibility Mapping Using Index of Entropy and Conditional Probability Models in GIS: Safarood Basin, Iran. *CATENA* **2012**, *97*, 71–84. [CrossRef]
22. Pournader, M.; Ahmadi, H.; Feiznia, S.; Karimi, H.; Peirovan, H.R. Spatial Prediction of Soil Erosion Susceptibility: An Evaluation of the Maximum Entropy Model. *Earth Sci. Inform.* **2018**, *11*, 389–401. [CrossRef]
23. Nosrati, K. Assessing Soil Quality Indicator under Different Land Use and Soil Erosion Using Multivariate Statistical Techniques. *Environ. Monit. Assess.* **2013**, *185*, 2895–2907. [CrossRef] [PubMed]
24. Sarkar, T.; Mishra, M. Soil Erosion Susceptibility Mapping with the Application of Logistic Regression and Artificial Neural Network. *J. Geovisualization Spat. Anal.* **2018**, *2*, 8. [CrossRef]
25. Gholami, V.; Sahour, H.; Hadian Amri, M.A. Soil Erosion Modeling Using Erosion Pins and Artificial Neural Networks. *CATENA* **2021**, *196*, 104902. [CrossRef]
26. Gholami, V.; Booij, M.J.; Nikzad Tehrani, E.; Hadian, M.A. Spatial Soil Erosion Estimation Using an Artificial Neural Network (ANN) and Field Plot Data. *CATENA* **2018**, *163*, 210–218. [CrossRef]
27. Dinh, T.V.; Nguyen, H.; Tran, X.-L.; Hoang, N.-D. Predicting Rainfall-Induced Soil Erosion Based on a Hybridization of Adaptive Differential Evolution and Support Vector Machine Classification. *Math. Probl. Eng.* **2021**, *2021*, 6647829. [CrossRef]
28. Arabameri, A.; Asadi Nalivan, O.; Chandra Pal, S.; Chakrabortty, R.; Saha, A.; Lee, S.; Pradhan, B.; Tien Bui, D. Novel Machine Learning Approaches for Modelling the Gully Erosion Susceptibility. *Remote Sens.* **2020**, *12*, 2833. [CrossRef]
29. Ghosh, A.; Maiti, R. Soil Erosion Susceptibility Assessment Using Logistic Regression, Decision Tree and Random Forest: Study on the Mayurakshi River Basin of Eastern India. *Environ. Earth Sci.* **2021**, *80*, 328. [CrossRef]
30. Phinzi, K.; Ngetar, N.S.; Ebhuoma, O. Soil Erosion Risk Assessment in the Umzintlava Catchment (T32E), Eastern Cape, South Africa, Using RUSLE and Random Forest Algorithm. *S. Afr. Geogr. J.* **2021**, *103*, 139–162. [CrossRef]

31. Choubin, B.; Moradi, E.; Golshan, M.; Adamowski, J.; Sajedi-Hosseini, F.; Mosavi, A. An Ensemble Prediction of Flood Susceptibility Using Multivariate Discriminant Analysis, Classification and Regression Trees, and Support Vector Machines. *Sci. Total Environ.* **2019**, *651*, 2087–2096. [CrossRef]
32. Lee, S.; Lee, M.-J.; Jung, H.-S.; Lee, S. Landslide Susceptibility Mapping Using Naïve Bayes and Bayesian Network Models in Umyeonsan, Korea. *Geocarto Int.* **2020**, *35*, 1665–1679. [CrossRef]
33. Mosavi, A.; Sajedi-Hosseini, F.; Choubin, B.; Taromideh, F.; Rahi, G.; Dineva, A. Susceptibility Mapping of Soil Water Erosion Using Machine Learning Models. *Water* **2020**, *12*, 1995. [CrossRef]
34. Lei, X.; Chen, W.; Avand, M.; Janizadeh, S.; Kariminejad, N.; Shahabi, H.; Costache, R.; Shahabi, H.; Shirzadi, A.; Mosavi, A. GIS-Based Machine Learning Algorithms for Gully Erosion Susceptibility Mapping in a Semi-Arid Region of Iran. *Remote Sens.* **2020**, *12*, 2478. [CrossRef]
35. Saha, S.; Roy, J.; Arabameri, A.; Blaschke, T.; Tien Bui, D. Machine Learning-Based Gully Erosion Susceptibility Mapping: A Case Study of Eastern India. *Sensors* **2020**, *20*, 1313. [CrossRef]
36. Gayen, A.; Pourghasemi, H.R.; Saha, S.; Keesstra, S.; Bai, S. Gully Erosion Susceptibility Assessment and Management of Hazard-Prone Areas in India Using Different Machine Learning Algorithms. *Sci. Total Environ.* **2019**, *668*, 124–138. [CrossRef]
37. Pal, S.C.; Arabameri, A.; Blaschke, T.; Chowdhuri, I.; Saha, A.; Chakrabortty, R.; Lee, S.; Band, S.S. Ensemble of Machine-Learning Methods for Predicting Gully Erosion Susceptibility. *Remote Sens.* **2020**, *12*, 3675. [CrossRef]
38. Soleimanpour, S.M.; Pourghasemi, H.R.; Zare, M. A Comparative Assessment of Gully Erosion Spatial Predictive Modeling Using Statistical and Machine Learning Models. *CATENA* **2021**, *207*, 105679. [CrossRef]
39. Razavi-Termeh, S.V.; Sadeghi-Niaraki, A.; Choi, S.-M. Gully Erosion Susceptibility Mapping Using Artificial Intelligence and Statistical Models. *Geomat. Nat. Hazards Risk* **2020**, *11*, 821–844. [CrossRef]
40. Pourghasemi, H.R.; Sadhasivam, N.; Kariminejad, N.; Collins, A.L. Gully Erosion Spatial Modelling: Role of Machine Learning Algorithms in Selection of the Best Controlling Factors and Modelling Process. *Geosci. Front.* **2020**, *11*, 2207–2219. [CrossRef]
41. Avand, J.; Janizadeh, S.; Naghibi, S.A.; Pourghasemi, H.R.; Bozchaloei, S.K.; Blaschke, T. A Comparative Assessment of Random Forest and K-Nearest Neighbor Classifiers for Gully Erosion Susceptibility Mapping. *Water* **2019**, *11*, 2076. [CrossRef]
42. Pham, Q.B.; Mukherjee, K.; Norouzi, A.; Linh, N.T.T.; Janizadeh, S.; Ahmadi, K.; Cerdà, A.; Doan, T.N.C.; Anh, D.T. Head-Cut Gully Erosion Susceptibility Modelling Based on Ensemble Random Forest with Oblique Decision Trees in Fareghan Watershed, Iran. *Geomat. Nat. Hazards Risk* **2020**, *11*, 2385–2410. [CrossRef]
43. Arabameri, A.; Chen, W.; Lombardo, L.; Blaschke, T.; Tien Bui, D. Hybrid Computational Intelligence Models for Improvement Gully Erosion Assessment. *Remote Sens.* **2020**, *12*, 140. [CrossRef]
44. Ahmadpour, H.; Bazrafshan, O.; Rafiei-Sardooi, E.; Zamani, H.; Panagopoulos, T. Gully Erosion Susceptibility Assessment in the Kondoran Watershed Using Machine Learning Algorithms and the Boruta Feature Selection. *Sustainability* **2021**, *13*, 10110. [CrossRef]
45. Chen, Y.; Chen, W.; Janizadeh, S.; Bhunia, G.S.; Bera, A.; Pham, Q.B.; Linh, N.T.T.; Balogun, A.-L.; Wang, X. Deep Learning and Boosting Framework for Piping Erosion Susceptibility Modeling: Spatial Evaluation of Agricultural Areas in the Semi-Arid Region. *Geocarto. Int.* **2021**, *12*, 1–27. [CrossRef]
46. Yang, A.; Wang, C.; Pang, G.; Long, Y.; Wang, L.; Cruse, R.M.; Yang, Q. Gully Erosion Susceptibility Mapping in Highly Complex Terrain Using Machine Learning Models. *ISPRS Int. J. Geo-Inf.* **2021**, *10*, 680. [CrossRef]
47. Saha, S.; Sarkar, R.; Thapa, G.; Roy, J. Modeling Gully Erosion Susceptibility in Phuentsholing, Bhutan Using Deep Learning and Basic Machine Learning Algorithms. *Environ. Earth Sci.* **2021**, *80*, 295. [CrossRef]
48. Kadavi, P.; Lee, C.-W.; Lee, S. Application of Ensemble-Based Machine Learning Models to Landslide Susceptibility Mapping. *Remote Sens.* **2018**, *10*, 1252. [CrossRef]
49. Lee, S.; Pradhan, B. Probabilistic Landslide Hazards and Risk Mapping on Penang Island, Malaysia. *J. Earth Syst. Sci.* **2006**, *115*, 661–672. [CrossRef]
50. Costache, R.; Ngo, P.T.T.; Bui, D.T. Novel Ensembles of Deep Learning Neural Network and Statistical Learning for Flash-Flood Susceptibility Mapping. *Water* **2020**, *12*, 1549. [CrossRef]
51. Ijlil, S.; Essahlaoui, A.; Mohajane, M.; Essahlaoui, N.; Mili, E.M.; Van Rompaey, A. Machine Learning Algorithms for Modeling and Mapping of Groundwater Pollution Risk: A Study to Reach Water Security and Sustainable Development (Sdg) Goals in a Mediterranean Aquifer System. *Remote Sens.* **2022**, *14*, 2379. [CrossRef]
52. Arabameri, A.; Cerda, A.; Rodrigo-Comino, J.; Pradhan, B.; Sohrabi, M.; Blaschke, T.; Bui, D.T. Proposing a Novel Predictive Technique for Gully Erosion Susceptibility Mapping in Arid and Semi-Arid Regions (Iran). *Remote Sens.* **2019**, *11*, 2577. [CrossRef]
53. Bouamrane, A.; Bouamrane, A.; Abida, H. Water Erosion Hazard Distribution under a Semi-Arid Climate Condition: Case of Mellah Watershed, North-Eastern Algeria. *Geoderma* **2021**, *403*, 115381. [CrossRef]
54. Nouayti, N.; Cherif, E.K.; Algarra, M.; Pola, M.L.; Fernández, S.; Nouayti, A.; Esteves da Silva, J.C.G.; Driss, K.; Samlani, N.; Mohamed, H.; et al. Determination of Physicochemical Water Quality of the Ghis-Nekor Aquifer (Al Hoceima, Morocco) Using Hydrochemistry, Multiple Isotopic Tracers, and the Geographical Information System (GIS). *Water* **2022**, *14*, 606. [CrossRef]
55. Bouhout, S.; Haboubi, K.; Zian, A.; Elyoubi, M.S.; Elabdouni, A. Evaluation of Two Linear Kriging Methods for Piezometric Levels Interpolation and a Framework for Upgrading Groundwater Level Monitoring Network in Ghiss-Nekor Plain, North-Eastern Morocco. *Arab. J. Geosci.* **2022**, *15*, 1016. [CrossRef]

56. Benabdelouahab, S.; Salhi, A.; Stitou, J.; Himi, M.; Draoui, M.; Casas, A. Application Des SIG et de La Tomographie Électrique Pour Contribuer à La Protection de l'aquifère de Martil-Alila (Maroc). In Proceedings of the Euromediterranean Scientific Congress on Engineering, Algeciras, Spain, 19–20 May 2011.
57. Bourjila, A.; Dimane, F.; EL Ouarghi, H.; Nouayti, N.; Taher, M.; EL Hammoudani, Y.; Saadi, O.; Bensiali, A. Groundwater Potential Zones Mapping by Applying GIS, Remote Sensing and Multi-Criteria Decision Analysis in the Ghiss Basin, Northern Morocco. *Groundw. Sustain. Dev.* **2021**, *15*, 100693. [CrossRef]
58. El Motaki, H.; El-Fengour, A.; Aissa, E.; Madureira, H.; Monteiro, A. The Global Change Impacts on Forest Natural Resources in Central Rif Mountains in Northern Morocco: Extensive Exploration and Planning Perspective. *GOT—J. Geogr. Spat. Plan.* **2019**, *17*, 75–92. [CrossRef]
59. Leikine, M.; Asebriy, L.; Bourgois, J. About the Age of the Ketama Unit's Anchi-Epizonal Metamorphism, Central Rif, Morocco. *Comptes Rendus—Acad. Sci. Ser. II* **1991**, *313*, 787–793.
60. Mansour, S.; Kouz, T.; Thaiki, M.; Ouhadi, A.; Mesmoudi, H.; Hassani Zerrouk, M.; Mourabit, T.; Dakak, H.; Cherkaoui Dekkaki, H. Spatial Assessment of the Vulnerability of Water Resources against Anthropogenic Pollution Using the DKPR Model: A Case of Ghiss-Nekkour Basin, Morocco. *Arab. J. Geosci.* **2021**, *14*, 699. [CrossRef]
61. Benyoussef, S.; Arabi, M.; El Ouarghi, H.; Ghalit, M.; Azirar, M.; El Midaoui, A.; Ait Boughrous, A. Impact of Anthropic Activities on the Quality of Groundwater in the Central Rif (North Morocco). *Ecol. Eng. Environ. Technol.* **2021**, *22*, 69–78. [CrossRef]
62. Taher, M.; Mourabit, T.; Bourjila, A.; Saadi, O.; Errahmouni, A.; El Marzkioui, F.; El Mousaoui, M. An Estimation of Soil Erosion Rate Hot Spots by Integrated USLE and GIS Methods: A Case Study of the Ghiss Dam and Basin in Northeastern Morocco. *Geomat. Environ. Eng.* **2022**, *16*, 95–110. [CrossRef]
63. Tsangaratos, P.; Ilia, I.; Hong, H.; Chen, W.; Xu, C. Applying Information Theory and GIS-Based Quantitative Methods to Produce Landslide Susceptibility Maps in Nancheng County, China. *Landslides* **2017**, *14*, 1091–1111. [CrossRef]
64. Amiri, M.; Pourghasemi, H.R.; Ghanbarian, G.A.; Afzali, S.F. Assessment of the Importance of Gully Erosion Effective Factors Using Boruta Algorithm and Its Spatial Modeling and Mapping Using Three Machine Learning Algorithms. *Geoderma* **2019**, *340*, 55–69. [CrossRef]
65. Valentin, C.; Poesen, J.; Li, Y. Gully Erosion: Impacts, Factors and Control. *CATENA* **2005**, *63*, 132–153. [CrossRef]
66. Hembram, T.K.; Paul, G.C.; Saha, S. Modelling of Gully Erosion Risk Using New Ensemble of Conditional Probability and Index of Entropy in Jainti River Basin of Chotanagpur Plateau Fringe Area, India. *Appl. Geomat.* **2020**, *12*, 337–360. [CrossRef]
67. Conforti, M.; Aucelli, P.P.C.; Robustelli, G.; Scarciglia, F. Geomorphology and GIS Analysis for Mapping Gully Erosion Susceptibility in the Turbolo Stream Catchment (Northern Calabria, Italy). *Nat. Hazards* **2011**, *56*, 881–898. [CrossRef]
68. Arabameri, A.; Chandra Pal, S.; Costache, R.; Saha, A.; Rezaie, F.; Seyed Danesh, A.; Pradhan, B.; Lee, S.; Hoang, N.-D. Prediction of Gully Erosion Susceptibility Mapping Using Novel Ensemble Machine Learning Algorithms. *Geomat. Nat. Hazards Risk* **2021**, *12*, 469–498. [CrossRef]
69. Moore, I.D.; Grayson, R.B.; Ladson, A.R. Digital Terrain Modelling: A Review of Hydrological, Geomorphological, and Biological Applications. *Hydrol. Process.* **1991**, *5*, 3–30. [CrossRef]
70. Band, S.S.; Janizadeh, S.; Chandra Pal, S.; Saha, A.; Chakrabortty, R.; Shokri, M.; Mosavi, A. Novel Ensemble Approach of Deep Learning Neural Network (DLNN) Model and Particle Swarm Optimization (PSO) Algorithm for Prediction of Gully Erosion Susceptibility. *Sensors* **2020**, *20*, 5609. [CrossRef]
71. Chowdhuri, I.; Pal, S.C.; Saha, A.; Chakrabortty, R.; Roy, P. Evaluation of Different DEMs for Gully Erosion Susceptibility Mapping Using In-Situ Field Measurement and Validation. *Ecol. Inform.* **2021**, *65*, 101425. [CrossRef]
72. Bui, D.T.; Lofman, O.; Revhaug, I.; Dick, O. Landslide Susceptibility Analysis in the Hoa Binh Province of Vietnam Using Statistical Index and Logistic Regression. *Nat. Hazards* **2011**, *59*, 1413–1444. [CrossRef]
73. Rahmati, O.; Pourghasemi, H.R.; Zeinivand, H. Flood Susceptibility Mapping Using Frequency Ratio and Weights-of-Evidence Models in the Golastan Province, Iran. *Geocarto Int.* **2016**, *31*, 42–70. [CrossRef]
74. Lee, S.; Sambath, T. Landslide Susceptibility Mapping in the Damrei Romel Area, Cambodia Using Frequency Ratio and Logistic Regression Models. *Environ. Geol.* **2006**, *50*, 847–855. [CrossRef]
75. Samanta, R.K.; Bhunia, G.S.; Shit, P.K.; Pourghasemi, H.R. Flood Susceptibility Mapping Using Geospatial Frequency Ratio Technique: A Case Study of Subarnarekha River Basin, India. *Model. Earth Syst. Environ.* **2018**, *4*, 395–408. [CrossRef]
76. Mohajane, M.; Costache, R.; Karimi, F.; Bao Pham, Q.; Essahlaoui, A.; Nguyen, H.; Laneve, G.; Oudija, F. Application of Remote Sensing and Machine Learning Algorithms for Forest Fire Mapping in a Mediterranean Area. *Ecol. Indic.* **2021**, *129*, 107869. [CrossRef]
77. Boroughani, M.; Pourhashemi, S.; Hashemi, H.; Salehi, M.; Amirahmadi, A.; Asadi, M.A.Z.; Berndtsson, R. Application of Remote Sensing Techniques and Machine Learning Algorithms in Dust Source Detection and Dust Source Susceptibility Mapping. *Ecol. Inform.* **2020**, *56*, 101059. [CrossRef]
78. Breiman, L. Random Forests Machine Learning. Kluwer Academic Publishers: Dordrecht, The Netherlands, 2001; Volume 45.
79. Ho, T.K. The Random Subspace Method for Constructing Decision Forests. *IEEE Trans. Pattern Anal. Mach. Intell.* **1998**, *20*, 832–844. [CrossRef]
80. Rodriguez-Galiano, V.F.; Chica-Olmo, M.; Chica-Rivas, M. Predictive Modelling of Gold Potential with the Integration of Multisource Information Based on Random Forest: A Case Study on the Rodalquilar Area, Southern Spain. *Int. J. Geogr. Inf. Sci.* **2014**, *28*, 1336–1354. [CrossRef]

81. Arabameri, A.; Pradhan, B.; Rezaei, K. Spatial Prediction of Gully Erosion Using ALOS PALSAR Data and Ensemble Bivariate and Data Mining Models. *Geosci. J.* **2019**, *23*, 669–686. [CrossRef]
82. Pai, P.-F.; Hsu, M.-F. An Enhanced Support Vector Machines Model for Classification and Rule Generation. In *Computational Optimization, Methods and Algorithms*; Koziel, S., Yang, X.-S., Eds.; Studies in Computational Intelligence; Springer: Berlin/Heidelberg, Germany, 2011; Volume 356, pp. 241–258, ISBN 978-3-642-20858-4.
83. Rahmati, O.; Tahmasebipour, N.; Haghizadeh, A.; Pourghasemi, H.R.; Feizizadeh, B. Evaluation of Different Machine Learning Models for Predicting and Mapping the Susceptibility of Gully Erosion. *Geomorphology* **2017**, *298*, 118–137. [CrossRef]
84. Garosi, Y.; Sheklabadi, M.; Conoscenti, C.; Pourghasemi, H.R.; Van Oost, K. Assessing the Performance of GIS- Based Machine Learning Models with Different Accuracy Measures for Determining Susceptibility to Gully Erosion. *Sci. Total Environ.* **2019**, *664*, 1117–1132. [CrossRef]
85. Phinzi, K.; Abriha, D.; Bertalan, L.; Holb, I.; Szabó, S. Machine Learning for Gully Feature Extraction Based on a Pan-Sharpened Multispectral Image: Multiclass vs. Binary Approach. *ISPRS Int. J. Geo-Inf.* **2020**, *9*, 252. [CrossRef]
86. Ranjitha, K.V. Classification and Optimization Scheme for Text Data Using Machine Learning Naïve Bayes Classifier. In Proceedings of the 2018 IEEE World Symposium on Communication Engineering (WSCE), Singapore, 28–30 December 2018; IEEE: Singapore, 2018; pp. 33–36.
87. Serrano-Cinca, C.; Gutiérrez-Nieto, B. Partial Least Square Discriminant Analysis for Bankruptcy Prediction. *Decis. Support Syst.* **2013**, *54*, 1245–1255. [CrossRef]
88. Pourghasemi, H.; Gayen, A.; Park, S.; Lee, C.-W.; Lee, S. Assessment of Landslide-Prone Areas and Their Zonation Using Logistic Regression, LogitBoost, and NaïveBayes Machine-Learning Algorithms. *Sustainability* **2018**, *10*, 3697. [CrossRef]
89. Bhargavi, P.; Tech, M.; Jyothi, D.S. Applying Naive Bayes Data Mining Technique for Classification of Agricultural Land Soils. *Int. J. Comput. Sci. Netw. Secur.* **2009**, *6*, 189–193.
90. Pham, B.T.; Jaafari, A.; Avand, M.; Al-Ansari, N.; Dinh Du, T.; Yen, H.P.H.; Phong, T.V.; Nguyen, D.H.; Le, H.V.; Mafi-Gholami, D.; et al. Performance Evaluation of Machine Learning Methods for Forest Fire Modeling and Prediction. *Symmetry* **2020**, *12*, 1022. [CrossRef]
91. Domazetović, F.; Šiljeg, A.; Lončar, N.; Marić, I. GIS Automated Multicriteria Analysis (GAMA) Method for Susceptibility Modelling. *MethodsX* **2019**, *6*, 2553–2561. [CrossRef]
92. Yildirim, P. Filter Based Feature Selection Methods for Prediction of Risks in Hepatitis Disease. *Int. J. Mach. Learn. Comput.* **2015**, *5*, 258–263. [CrossRef]
93. Mao, K.Z. Orthogonal Forward Selection and Backward Elimination Algorithms for Feature Subset Selection. *IEEE Trans. Syst. Man Cybern. Part B Cybern.* **2004**, *34*, 629–634. [CrossRef]
94. Lee, C.; Lee, G.G. Information Gain and Divergence-Based Feature Selection for Machine Learning-Based Text Categorization. *Inf. Process. Manag.* **2006**, *42*, 155–165. [CrossRef]
95. Dash, M.; Liu, H. Feature Selection for Classification. *Intell. Data Anal.* **1997**, *1*, 131–156. [CrossRef]
96. Imdadullah, M.; Aslam, M.; Altaf, S. Mctest: An R Package for Detection of Collinearity among Regressors. *R J.* **2016**, *8*, 495–505. [CrossRef]
97. Salhi, A.; Benabdelouahab, T.; Martin-Vide, J.; Okacha, A.; El Hasnaoui, Y.; El Mousaoui, M.; El Morabit, A.; Himi, M.; Benabdelouahab, S.; Lebrini, Y.; et al. Bridging the Gap of Perception Is the Only Way to Align Soil Protection Actions. *Sci. Total Environ.* **2020**, *718*, 137421. [CrossRef]
98. Phinzi, K.; Holb, I.; Szabó, S. Mapping Permanent Gullies in an Agricultural Area Using Satellite Images: Efficacy of Machine Learning Algorithms. *Agronomy* **2021**, *11*, 333. [CrossRef]
99. Lana, J.C.; Castro, P.D.; Lana, C.E. Assessing Gully Erosion Susceptibility and Its Conditioning Factors in Southeastern Brazil Using Machine Learning Algorithms and Bivariate Statistical Methods: A Regional Approach. *Geomorphology* **2022**, *402*, 108159. [CrossRef]
100. Hembram, T.K.; Saha, S.; Pradhan, B.; Abdul Maulud, K.N.; Alamri, A.M. Robustness Analysis of Machine Learning Classifiers in Predicting Spatial Gully Erosion Susceptibility with Altered Training Samples. *Geomat. Nat. Hazards Risk* **2021**, *12*, 794–828. [CrossRef]
101. Bouramtane, T.; Hilal, H.; Rezende-Filho, A.T.; Bouramtane, K.; Barbiero, L.; Abraham, S.; Valles, V.; Kacimi, I.; Sanhaji, H.; Torres-Rondon, L.; et al. Mapping Gully Erosion Variability and Susceptibility Using Remote Sensing, Multivariate Statistical Analysis, and Machine Learning in South Mato Grosso, Brazil. *Geosciences* **2022**, *12*, 235. [CrossRef]
102. Arabameri, A.; Chen, W.; Loche, M.; Zhao, X.; Li, Y.; Lombardo, L.; Cerda, A.; Pradhan, B.; Bui, D.T. Comparison of Machine Learning Models for Gully Erosion Susceptibility Mapping. *Geosci. Front.* **2020**, *11*, 1609–1620. [CrossRef]
103. Costache, R.; Popa, M.C.; Tien Bui, D.; Diaconu, D.C.; Ciubotaru, N.; Minea, G.; Pham, Q.B. Spatial Predicting of Flood Potential Areas Using Novel Hybridizations of Fuzzy Decision-Making, Bivariate Statistics, and Machine Learning. *J. Hydrol.* **2020**, *585*, 124808. [CrossRef]
104. Tien Bui, D.; Bui, Q.-T.; Nguyen, Q.-P.; Pradhan, B.; Nampak, H.; Trinh, P.T. A Hybrid Artificial Intelligence Approach Using GIS-Based Neural-Fuzzy Inference System and Particle Swarm Optimization for Forest Fire Susceptibility Modeling at a Tropical Area. *Agric. For. Meteorol.* **2017**, *233*, 32–44. [CrossRef]
105. Ng, S.S.Y.; Xing, Y.; Tsui, K.L. A Naive Bayes Model for Robust Remaining Useful Life Prediction of Lithium-Ion Battery. *Appl. Energy* **2014**, *118*, 114–123. [CrossRef]

106. Nguyen, P.T.; Tuyen, T.T.; Shirzadi, A.; Pham, B.T.; Shahabi, H.; Omidvar, E.; Amini, A.; Entezami, H.; Prakash, I.; Phong, T.V.; et al. Development of a Novel Hybrid Intelligence Approach for Landslide Spatial Prediction. *Appl. Sci.* **2019**, *9*, 2824. [CrossRef]
107. Gourfi, A.; Daoudi, L.; Shi, Z. The Assessment of Soil Erosion Risk, Sediment Yield and Their Controlling Factors on a Large Scale: Example of Morocco. *J. Afr. Earth Sci.* **2018**, *147*, 281–299. [CrossRef]
108. Salhi, A.; Martin-Vide, J.; Benhamrouche, A.; Benabdelouahab, S.; Himi, M.; Benabdelouahab, T.; Casas Ponsati, A. Rainfall Distribution and Trends of the Daily Precipitation Concentration Index in Northern Morocco: A Need for an Adaptive Environmental Policy. *SN Appl. Sci.* **2019**, *1*, 277. [CrossRef]

International Journal of *Geo-Information*

Article

Enhancing the Accuracy of Land Cover Classification by Airborne LiDAR Data and WorldView-2 Satellite Imagery

Chun-Ta Wei [1], Ming-Da Tsai [2,*], Yu-Lung Chang [2] and Ming-Chih Jason Wang [3]

1 School of Defense Science, Chung Cheng Institute of Technology, National Defense University, Taoyuan 33551, Taiwan; 1120512301@ccit.ndu.edu.tw
2 Department of Environmental Information and Engineering, Chung Cheng Institute of Technology, National Defense University, Taoyuan 33551, Taiwan; chang72118@gmail.com
3 Department of History and Geography, University of Taipei, Taipei 100234, Taiwan; wmcsm@go.utaipei.edu.tw
* Correspondence: tsai.md5478@ccit.ndu.edu.tw; Tel.: +886-3-380-0364 (ext. 307)

Abstract: The Full Waveform LiDAR system has been developed and used commercially all over the world. It acts to record the complete time of a laser pulse and has a high-resolution sampling interval compared to the traditional multiple-echo LiDAR, which only provides signals within a single target range. This study area mainly collects data from Riegl LMS-Q680i Full Waveform LiDAR and WorldView-2 satellite imagery, which focuses on buildings, vegetation, grassland, asphalt roads and other ground types as the surface objects. The amplitude and pulse width are selected as waveform basic parameters. The parameter of topography is slope, and the height classification parameters of the test ground are 0–0.5 m, 0.5–2.5 m, and 2.5 m. To eliminate noise, the neighborhood average is applied on the LiDAR parameter values and analyzed as the classification accuracy comparison. This survey uses Decision Tree as the classification method. Comparing the data between neighborhood average and non-neighborhood average, the data classification accuracy improves by 7%, and Kappa improves by 5.92%. NDVI image data are utilized to distinguish the artificial from natural ground. The results show that the neighborhood average with previous data can improve the classification accuracy by 5%, and Kappa improves by 4.25%. By adding NIR-2 of WorldView-2 satellite imagery to the neighborhood average analysis, the overall classification accuracy is improved by 2%, and the Kappa value by 1.21%. This article shows that utilizing the analysis of neighborhood average and image parameters can effectively improve the classification accuracy of land covers.

Keywords: LiDAR; full waveform; decision tree; accuracy

Citation: Wei, C.-T.; Tsai, M.-D.; Chang, Y.-L.; Wang, M.-C.J. Enhancing the Accuracy of Land Cover Classification by Airborne LiDAR Data and WorldView-2 Satellite Imagery. *ISPRS Int. J. Geo-Inf.* **2022**, *11*, 391. https://doi.org/10.3390/ijgi11070391

Academic Editors: Dimitris Triantakonstantis and Wolfgang Kainz

Received: 20 April 2022
Accepted: 8 July 2022
Published: 12 July 2022

1. Introduction

This study aims to use WorldView-2 satellite imagery and LMS-Q680i airborne LiDAR data to produce research on classification and interpretation of features. In the section concerning satellite imagery, the majority of the image data are from the WorldView-2 satellite, which provides eight different wavebands. Each of these has different radiation characteristics on the same surface objects. The paper will use the 8th band, the near-infrared band-2 (NIR-2), to collect image data. Then, the 5th band, red and 7th band, and near-infrared band-1 (NIR-1) will be used to calculate the Normalized Difference Vegetation Index (NDVI) as a significant parameter of the image classification. Next, the decision tree is applied to analyze the LiDAR data classification parameters (echo width, amplitude, slope, custom height values, after neighborhood averages' echo width, amplitude, slope) to define the land covers, which could be called the "LiDAR decision tree process".

This survey is divided into two experimental stages. In the first stage, the LiDAR decision-tree process mentioned above is applied. In the second phase of the experiment, the poor classification results are chosen, and then the image NDVI and image NIR-2 band parameters are added to verify whether the addition of LiDAR data to the image data can improve the overall classification accuracy.

Guo et al. [1] used the random forest classification method to classify urban landmarks by traditional multiple-echo, full-waveform LiDAR and multispectral image parameters. In their research, four parameters, which are elevation, amplitude, width, and backward reflection waveform, are applied to define the building class. The waveform parameters of the artificial ground class have relatively low importance.

To detect vegetation in urban areas, Höfle et al. [2] use a neural network, decision-tree classifier, and full waveform LiDAR as parameters to collect data. The effects of multiple echoes, gridded segmentation, radiometric correction, and classification on vegetation detection in urban areas are analyzed, compared, and discussed. The results show that the accuracy of classification reaches up to 98% with the above classification methods and point-cloud density can be effectively reduced to 10 (pts./m^2). Full waveform LiDAR data definitely enhance the characterization conditions for describing objects, and can effectively help to distinguish objects. Wu et al. [3] selected the Random Forest (RF) as a classifier. The spectral bands of GF-2, NDVI, and normalized digital surface model derived from LiDAR data, and their grey-level co-occurrence matrix (GLCM) textures including mean, variance, homogeneity, contrast, dissimilarity, entropy, second moment, and correlation, were generated to create seven scenarios with different combinations of RF input variables. This study aims to provide a reference for the efficient improvement of land-cover classification, and to offer support for extending the applications of classification algorithms and data sources.

2. Theory

2.1. LiDAR

The full waveform LiDAR system records the echo waveform by dense sampling, recording the continuous digital number (DN) value, recording the echo waveform reflected by different substances, and obtaining the point-cloud information in the full waveform through waveform post-processing. The point-cloud information of the full waveform is more complete than that of the traditional multi-echo point-cloud information [4]; the complete waveform information is recorded, making the point-cloud data denser. It is more valuable than the traditional multiple echo in the study of classification and application [5].

The full waveform LiDAR point cloud will affect the complexity of the return waveform due to the size of the footprint when recording the surface. The large footprint covers a large area of the ground. The waveform complexity recorded is relatively high, which is difficult to analyze and process in post-processing analysis. The advantage of a large footprint is that the research area can be quickly scanned. The small-footprint point cloud covers a small area and the waveform information recorded is relatively simple. It is necessary to increase the scanning frequency to achieve an increase in point-cloud density. At present, most commercial LiDARs are mainly small footprints [6].

2.2. NDVI

The more vigorous the growth of colorful plants, the more red light chlorophyll absorbs, and the stronger the reflection of infrared light. The difference between the two is used: proportion of subtraction and addition. NDVI has no unit, and its value ranges from −1.0 to 1.0. The larger the value, the greater the growth of the plant. The formula is as follows:

$$NDVI = \frac{(NIR - Red)}{(NIR + Red)} \tag{1}$$

In the formula, *NIR* and *Red* respectively represent the near infrared light band and the red-light band, and the output value of *NDVI* is between −1.0 and 1.0. A positive increase in the value of *NDVI* represents the degree of increase in vegetation. A value close to 0 or negative, indicates the absence of vegetation. For places with a vegetation distribution, there will be a higher *NDVI* value. With higher reflection of near-infrared light band and lower reflection of red-light band, a place with dense vegetation distribution yields an NDVI value between 0.1 and 0.6, and its value depends on the density and greening at the top of the plant. Soil and rock materials will produce values close to 0, because the two have similar values in the red and near-infrared bands, and water, clouds, and snow will have higher red-band values than in the near-infrared band, so they will produce negative values. Therefore, to use the *NDVI* method to extract vegetation information, the threshold must be manually adjusted before it can be applied to different satellite images.

2.3. Decision Tree Classification

The decision-tree mining technology is an algorithm that divides and predicts the structure of the tree and the diagram. Multistage classification techniques are also possible, in which a series of decisions are taken to determine the correct label for a pixel. The more common multistage classifiers are called decision trees [7]. They consist of a number of connected classifiers (or decision nodes), none of which is expected to perform the complete segmentation of the image data set.

Frequently, decision tree strategies can be designed manually, particularly when they are required to perform quite specific labelling tasks [8]. However, as with single stage classifier and neural network training, it would be of value to have automated design procedures available. Since the number of possible tree structures, even for a moderately small number of classes, is astronomical, it is very difficult to design an optimal classifier [9]. Classification accuracy and efficiency, however, rely heavily on the tree chosen. Therefore, various heuristic methods for decision-tree design have been developed, details of which can be found in Safavian and Landgrebe [10].

To make the design task easier, binary decision trees are often adopted. Discrimination ability is not necessarily weakened by choosing a binary approach, since a general decision tree can be uniquely transformed into an equivalent binary tree [11]. The decision tree is also called the classification tree; the decision tree is one of the common functions of classification in the data mining processing technology. This study uses classification and regression trees (CART) in MATLAB to classify point-cloud data. CART is one of the commonly used data detection and data analysis tools, which can automatically detect the potential structure, important patterns, and relationships of highly complex data. The basic principle of CART is to use entropy calculation to divide the sample data into two parts, and each time the data will be divided into two subsets. Then, according to different training samples, classification is carried out layer by layer. The tree node at the top of the classification tree usually has the most influence on the classified data, and the influence of the classification tree node gradually decreases according to the layering [12].

2.4. Accuracy Evaluation

Olofsson et al. [13] discussed the basic protocols needed to produce scientifically rigorous and transparent accuracy and area estimates. This set of good practice recommendations provides guidelines to assist scientists and practitioners in designing and implementing precision assessment and area estimation methods for land change assessment using remote sensing.

For the evaluation and validation of the classification results, the accuracy of the classification results will be evaluated according to the real point-cloud data drawn manually. The method used to assess the classification accuracy is the error matrix. As shown in Table 1, this method uses geo-authentic references to verify the quality of classified data, further validated by user accuracy (UA), producer accuracy (PA), and overall accuracy (OA) [14,15].

It is assumed that the categories classified by the ground truth data are N categories from A to N in total. The decision-tree classification results used in this study are a to n in a total of n categories. They are listed on the abscissa and ordinate of the error matrix, respectively. The value of each coordinate from $X(1, 1)$ to $X(n, N)$ in the matrix is filled in with the number of classified point clouds.

Table 1. Error Matrix.

Classification Results	Ground Truth					
	Class A	Class B	…	Class N	Total Category Pixels	PA
Class a	$X(1,1)$	$X(1,2)$	…	$X(1,N)$	$\sum_{i=1}^{N} X(1,i)$	$\frac{X(1,1)}{\sum_{i=1}^{N} X(1,i)}$
Class b	$X(2,1)$	$X(2,2)$	…	$X(2,N)$	$\sum_{i=1}^{N} X(2,i)$	$\frac{X(2,2)}{\sum_{i=1}^{N} X(2,i)}$
⋮	⋮	⋮	⋮	⋮	⋮	⋮
Class n	$X(n,1)$	$X(n,2)$	…	$X(n,N)$	$\sum_{i=1}^{N} X(n,i)$	$\frac{X(n,N)}{\sum_{i=1}^{N} X(n,i)}$
Total Ground truth Category	$\sum_{i=1}^{N} X(i,1)$	$\sum_{i=1}^{N} X(i,2)$	…	$\sum_{i=1}^{N} X(i,N)$		
UA	$\frac{X(1,1)}{\sum_{i=1}^{N} X(i,1)}$	$\frac{X(2,2)}{\sum_{i=1}^{N} X(i,2)}$	…	$\frac{X(n,N)}{\sum_{i=1}^{N} X(i,N)}$		

User Accuracy (UA): The meaning of the row coordinates of Category A in the error matrix is the same as the method used in this study to classify Category A objects in the coordinates classified for real data in light. The number of point clouds in Category A is filled in $X(1, 1)$. In Category A coordinates, the number of pixels in Category B is filled in $X(2, 1)$. In order to determine if point clouds in class A are classified as N by this study, or if they are filled in sequentially, after completing the coordinates of class A row, it can be determined that in the coordinates of real data class A objects, classes B to n from $X(2, 1)$ to $X(n, 1)$ are the number of misclassified point clouds, and classes a from $X(1, 1)$ are the correct number of point clouds that can be completely classified into class A coordinates by this study. Therefore, user accuracy is defined as the true classified objects. This method can be used to classify the ratio to the true objects, and then the error matrix is completed.

$$UA = \frac{X(1,1)}{\sum_{i=1}^{n} X(i,1)} \times 100\% \tag{2}$$

The number of misclassified point clouds is defined as Commission Error (*CE*).

$$CE = \frac{\sum_{i=1}^{n} X(i,1) - X(1,1)}{\sum_{i=1}^{n} X(i,1)} \times 100\% = 1 - UA \tag{3}$$

Producer Accuracy (*PA*): The meaning of coordinates listed in Category A in the error matrix represents how much of the category A in this study in LiDAR does belong to genuine category A. The remaining $X(1, 2)$ to $X(1, N)$ actually belong to genuine category B to N. Therefore, the producer accuracy is defined as the point cloud that this study classifies as category A and the ratio of the genuine local category A.

$$PA = \frac{X(1,1)}{\sum_{i=1}^{N} X(1,i)} \times 100\% \tag{4}$$

Except for $X(1, 1)$ point clouds, which are classified as category A in this study and are truly category A, the remaining point clouds are defined as Omission Error (*OE*) because they are missing and are not classified as category A.

$$OE = \frac{\sum_{i=1}^{N} X(1,i) - X(1,1)}{\sum_{i=1}^{N} X(1,i)} \times 100\% = 1 - PA \tag{5}$$

After completing the error matrix, the diagonal elements $X(1, 1)$, $X(2, 2)$, $X(3, 3)$ ..., $X(n, N)$ are the number of point clouds that are fully classified correctly, so the number of point clouds that are fully classified correctly is defined, and the ratio to all cloud numbers is the overall accuracy (OA).

$$OA = \frac{\sum_{i=1}^{N} X(i,i)}{\sum_{i=1}^{n} \sum_{j=1}^{N} X(i,j)} \times 100\% \tag{6}$$

The Kappa statistics, which are generated by the mutual operation of error matrices, can show the error of the whole point-cloud classification, and take into account the factors of misjudgment and omission, so as to provide an indicator of how good the classification results are compared with random classification.

$$Kappa = \frac{Total\ Overall\ Accuracy - Expected\ Accuracy}{1 - Expected\ Accuracy} = \frac{N\sum_{i=0}^{n} X_{ii} - \sum_{i=0}^{n}(X_{i+} \cdot X_{+i})}{N^2 - \sum_{i=0}^{n}(X_{i+} \cdot X_{+i})} \tag{7}$$

3. Research Process and Methods

3.1. Research Area

This article takes the campus located in Taoyuan city as the study area (as shown in Figure 1). The image is the orthophoto taken by LiDAR during the same period. Though its range is only 650 m × 600 m this area is appropriate to conduct verification of classification results. The average altitude in research area is about 200 m, and the features covered include: different kinds of trees, buildings, asphalt roads, PU ground, cement floor, grassland, and other features. The floor height of buildings in the area is less than three floors, and the grassland is mainly short turf. The collected LiDAR data are the LMS-Q680i full waveform airborne LiDAR data. Considering that the ground features (vegetation) will show different scenes depending on the season, the satellite images collected are mainly WorldView-2 satellite images.

Figure 1. Orthophoto of the study area. (650 m × 600 m).

3.2. Work Flow

This research focuses on the analysis of LiDAR data and satellite images. We used LiDAR parameters and satellite images to classify objects based on LiDAR point-cloud data and added image parameters to the point-cloud data to verify whether the classification accuracy can be improved. The work flow chart is shown in Figure 2 below.

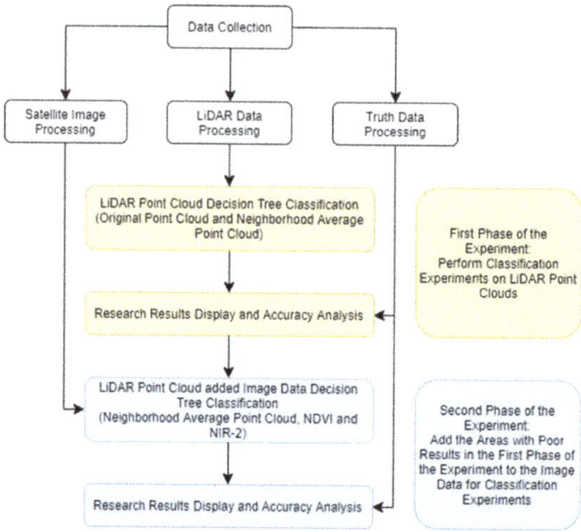

Figure 2. Work Flow Chart.

3.3. Data Collection and Processing of Satellite Image

3.3.1. WorldView-2 Satellite Imagery

WorldView-2 is the third high-resolution optical satellite owned by American Digital Globe. It has a high optical resolution of 50 cm and a commercial high-resolution optical remote sensing satellite that provides eight bands. The multi-spectral image data of eight bands provide more parameters for remote sensing research. In addition to the traditional multi-spectral bands red, green, blue, and near infrared-1 (NIR-1), four bands such as red-edge, coastal, yellow, and near infrared-2 (NIR-2) are added. The NIR-2 band has a spectral range between 860 and 1040 nm. Compared to NIR-1, this band is less affected by interference from the Earth's atmosphere and is suitable for research and application of plants or biological quality.

3.3.2. Image Fusion and Processing

Image Fusion technology can effectively combine high-resolution full-color images and low-resolution multi-spectral color images to fuse together into a quantity of data that preserves both high-resolution images and multi-spectral color images. The purpose is to obtain more information than a single image. Improving image quality for photointerpretative data fusion is discussed by Gross and Schott [16] and van der Meer [17].

In this study, ERADS IMAGINE software was used to process image fusion, and the method used was Principal Component Analysis (PCA) for computation fusion. In the World View-2 images collected by the institute, the resolution of the multi-spectral image in Figure 3a,b is 2.2 m, and the resolution of the full-color image in Figure 3c,d is 0.5. m. Using ERDAS IMAGINE software, the main components of the two types of images are fused. After the fusion, the resolution of the color image is increased to 0.6 m, as shown in Figure 3e,f.

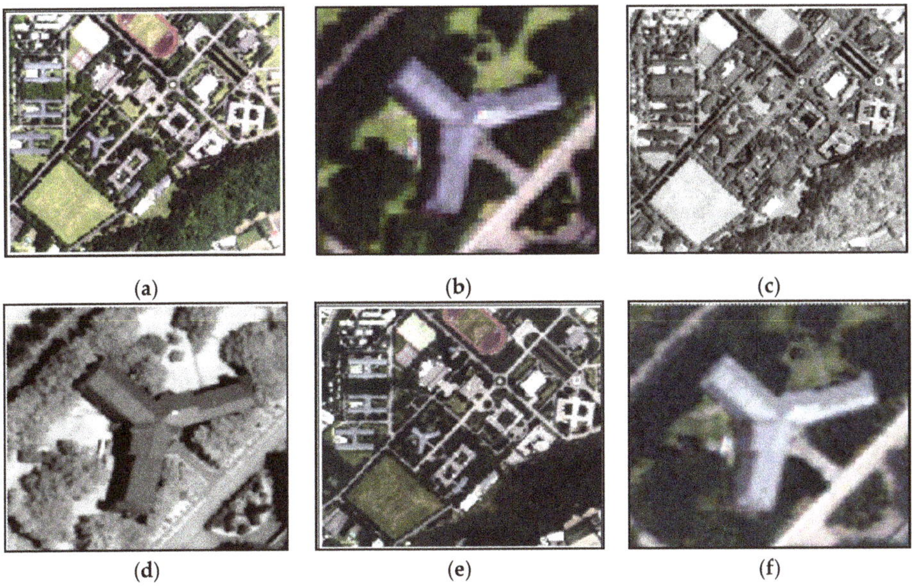

Figure 3. Fusion processing. (**a**) Multispectral image; (**b**) Multispectral zoom image; (**c**) Full color image; (**d**) Full color zoom image; (**e**) Image after fusion; (**f**) Fusion zoom image.

3.3.3. NDVI Calculation

We use the NIR−1 band and the red band in the satellite image to calculate the NDVI value and use the Matlab software to calculate the NDVI image (as shown in Figure 4).

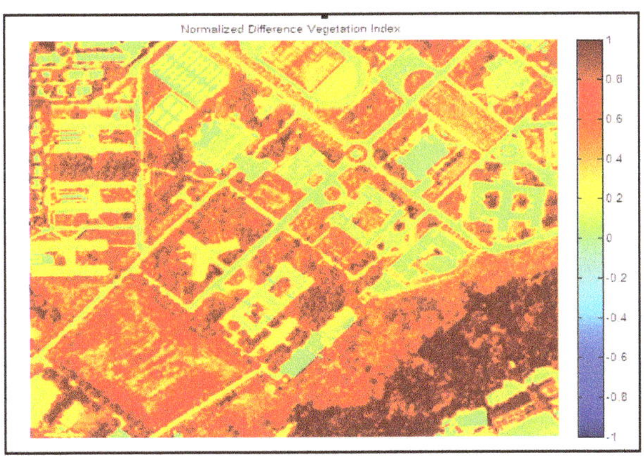

Figure 4. NDVI image.

3.4. Data Collection and Processing of LiDAR Airborne Image

3.4.1. LiDAR Point Cloud Data Collection and Processing

The LiDAR data collected in this article are obtained by scanning the Taoyuan city with LMSQ680i airborne LiDAR. LMS-Q680i is a full waveform no-load LiDAR system manufactured by Riegl Company. It has a good linear scan. The scanning speed of the point arrangement method reaches 266,000 points/s, and the scanning line is 10–200 lines/s. In the study, the positioning and orientation system on the LiDAR and the author's campus armaments building GPS ground tracking station were simultaneously tested. After the

LiDAR data were downloaded, the simultaneous joint measurement ground tracking station data was imported into the LiDAR data for processing. We obtained the correct coordinate point-cloud data, and the relevant information of the positioning and orientation system is shown in Figure 5.

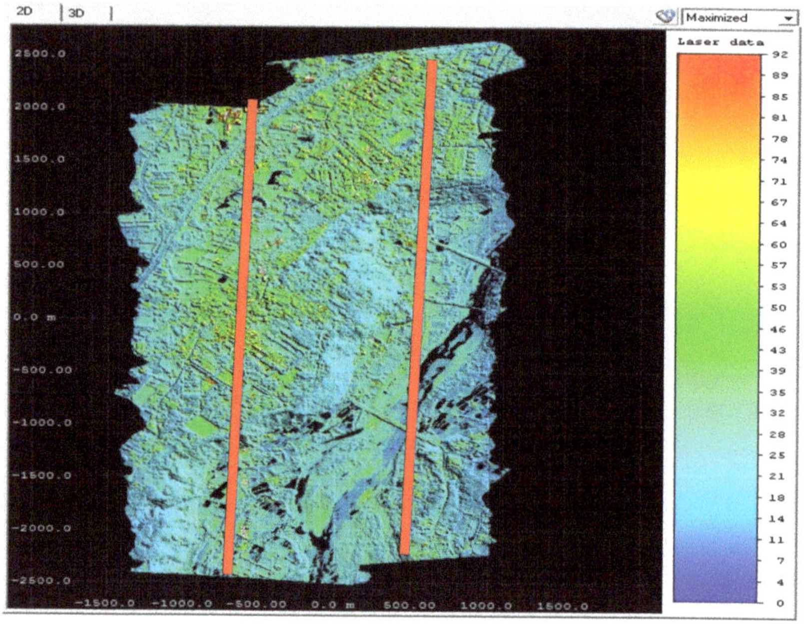

Figure 5. Point cloud data of LiDAR's two flight belts.

The LiDAR data used in this study area were taken for the LiDAR data of the two flight bands (as shown in Figure 5). The average measurement height is about 1500 m, and the average point-cloud density is 2 points/m.

Since the data format of each company's LiDAR system is different, in order to allow the data of different companies to be interchangeable, the American Society for Photogrammetry and Remote Sensing (APSRS) issued the "ASPRS LIDAR Data Exchange Format Standard (LAS1.0)" in May 2003. LAS successively released LAS1.1, 1.2, 1.3, 1.4 and 2.0 versions.

This research mainly uses the LAS1.2 data format. In the research, the Riegl program was also used to convert the ASCII file data format and the ASCII output parameters of the pulse width. After obtaining the point-cloud data of the study area, the first stage of the study is to classify and compare the accuracy between the original point-cloud data and the point-cloud data after the neighborhood average.

The point-cloud neighborhood averaging in this research mainly uses the Neighborhood Toolset command in the ArcGIS software neighborhood analysis tool set to perform point-cloud averaging analysis. Uses the width, amplitude, and slope values in the light data to enter the decision-tree classification operation. In the original data, it was found that the distribution of the point-cloud eigenvalues was too discrete, resulting in too much noise, so neighborhood analysis was used to analyze the space. The characteristics of the point cloud are averaged to filter out excessive noise values, which obtains a more average feature distribution. The analysis results are shown in Figures 6–8, where Figure 6a is the image map presented by the original data, and Figure 6b is the image averaged by the neighborhood. It can be seen from the figure that the average color distribution of the image after the neighborhood is smoother, indicating that too much noise has been filtered out. Too much noise will cause misjudgment of decision-tree classification. The research expects to improve the classification accuracy after filtering to remove noise.

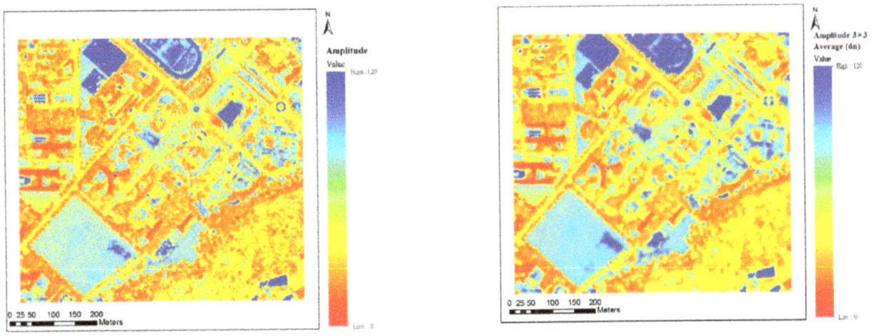

(**a**) (**b**)

Figure 6. Amplitude image. (**a**) Raw amplitude image; (**b**) Amplitude value image after neighborhood average.

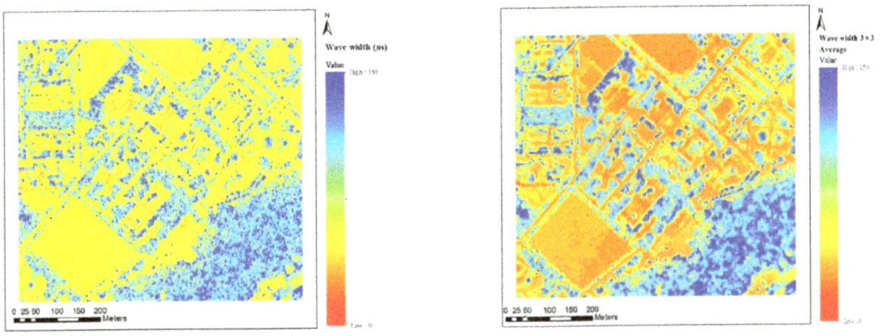

(**a**) (**b**)

Figure 7. Pulse width image. (**a**) Original pulse width image; (**b**) Pulse width value image after neighborhood average.

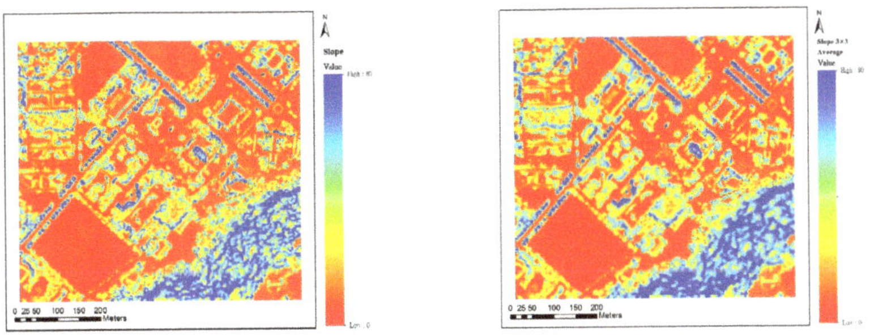

(**a**) (**b**)

Figure 8. Slope value image. (**a**) Original slope value image; (**b**) Slope value image after neighborhood average.

3.4.2. LiDAR Data Feature Value Extraction

The LiDAR parameters mainly used in this study include amplitude value, pulse width value, slope value, and custom height value. The amplitude value and wave width value are the values calculated by Gaussian fitting of the echo waveform by Riegl program. After fitting, the waveform presents a Gaussian distribution curve [18]. The amplitude value is the highest value of the waveform signal intensity, the full width half maximum (FWHM) of the waveform is the wave width value, and FWHM is defined as the wave width value of the waveform when the standard deviation of Gaussian distribution is $2\sqrt{2ln2}\sigma \approx 2.3548\sigma$. The amplitude and slope width of point cloud can be obtained from the waveform fitted by Riegl program [19]. The slope value is the point-cloud slope value calculated by using the slope analysis in ArcGIS software. The slope represents the steepest position on the surface inclination. The LiDAR data are three-dimensional spatial point data. The features in the study area can be classified according to their heights, which can be used as classification parameters. In the study area, according to the actual observation of the terrain by people, most of them are grassland and 0.5 m in height. The ground man-made structures are the main types, 0.5 to 2.5 m are mainly the middle canopy and short buildings, and 2.5 m or more are the high buildings and the tall canopy. Therefore, the study set these three levels as the custom height parameter values. The first layer is 0 to 0.5 m, the second layer is 0.5 to 2.5 m, the third layer is more than 2.5 m (as shown in Figure 9), and the self-defined height value is used as the classification parameter of the decision tree.

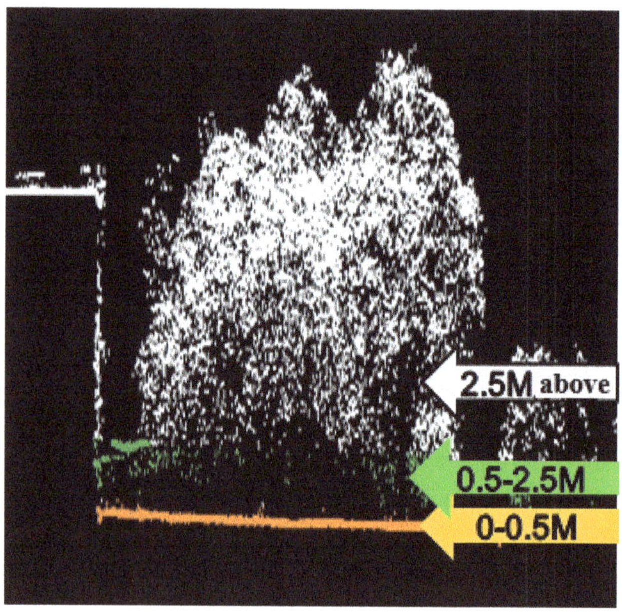

Figure 9. Schematic diagram of custom height value.

3.5. Ground Truth Data Search

The true value data in this study area are divided into canopy, building, grassland, road, and other ground (PU, cement land, bare mud land) through actual exploration and comparison of aerial photograph data. After the completion of the classifications, the classified images are used as the reference for truth value, and then the true value of the point cloud is plotted. The true value of the point cloud is shown in Figure 10. The truth data are mainly used in the verification of the classification results.

Figure 10. Point cloud truth data.

4. Results

This study is divided into two experimental stages. The first stage experiment is to compare and analyze the original point cloud with the point cloud after neighborhood average. The second stage experiment reanalyzes the areas with poor classification in the first stage experiment, reduces the study area, and resamples. At the same time, the NDVI and NIR-2 band values in the data and the point-cloud data are added to re-calculate the decision tree. After completion, the results are compared and analyzed the results, and determined whether adding image data can improve the classification accuracy of point clouds.

The pruning levels of the untrimmed classification decision tree are different. In order to have a unified comparison standard, the decision tree is pruned into 10 pruning levels as the basis for accuracy comparison, and then the error and fitting status of the 10 pruning levels are observed through the fitting diagram, so as to judge the characteristics of the classification data. The operation formula of each node of the decision tree is the result judged by the sampling data of the decision-tree operation. The closer the root node operation formula is, the more important its parameters are.

4.1. The First Stage Experiment: Original Value and Neighborhood Mean Point Cloud Classification

The first phase of the experiment mainly used LiDAR data for classification. In this paper, the features in the study area are divided into five categories, namely building, canopy, grassland, road, and other ground categories, while other ground categories include cement ground, and PU ground, and bare mud land. The original values of the point cloud and the point cloud averaged by the neighborhood are classified into decision trees, and the accuracy of the results is evaluated after the classification is completed.

First, we picked training samples in the study area and circled the training areas for the five feature categories. The circled training area is formed by overlaying LiDAR point-cloud data with aerial photograph orthophoto images taken at the same time to help understand the point. According to the actual distribution of cloud features, and after the actual feature survey of the researchers, the scope of the training area was selected for the study area, as shown in Figure 11. The colors selected in the training area are: red for buildings, blue for crowns, green for grasslands, orange for roads, and purple for other ground types. Sampling is to sample the point cloud according to the self-defined height of the study. The point-cloud samples of the second and third floors of the self-defined height values are taken for buildings and tree crowns, and the point-cloud samples of the second and first floors are taken for roads, grasslands and other ground classes. There are overlapping areas of blue and purple in the sampling area. Blue is the point cloud on the crown, and purple is the point cloud under the crown. We used the amplitude, width, and slope values in the LiDAR data to draw a histogram of the sampled training area data and recorded the histogram statistics in Tables 2–4. The statistical table shows the ratio of the average amplitude value of the neighborhood. The overall standard deviation value of the original is reduced by 21.61%, the neighborhood average pulse width value is lower than the original standard value by 55.07%, and the neighborhood average slope value is lower than the original standard value by 25.45%, so this means that the point cloud is averaged in the neighborhood, which can filter out the excessive noise of the point cloud and improve the average distribution of the point cloud's eigenvalues.

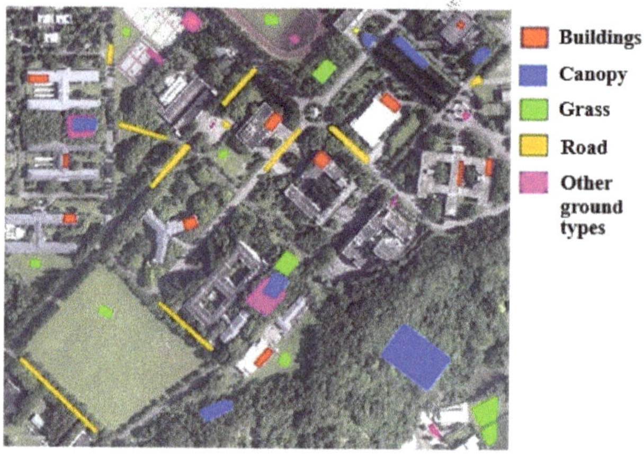

Figure 11. Sampling diagram of training samples in the first phase of the experimental study area.

Table 2. Training Amplitude Value–Comparison Table of Original Value and Neighborhood Standard Deviation.

Amplitude Value (Unit: DN)					
Category \ Method	Original Mean	Original Value Standard Deviation	Neighborhood Average	Neighborhood Standard Deviation	The Average Standard Deviation of Various Neighborhoods Decreases
Buildings	32.72	17.7	35.18	16.60	6.32%
Tree	21.92	8.02	21.42	3.61	55.04%
Other ground	2.49	3.41	1.72	2.37	30.40%
Grass	32.67	13.69	32.30	11.50	16.00%
Road	28.21	7.45	28.13	5.3	28.94%
Overall standard deviation		50.27		39.38	
Neighborhood average overall standard deviation decreases: 21.61%					

Table 3. Comparison table of training pulse width value–original value and neighborhood mean standard deviation.

Amplitude Value (Unit: DN)					
Category \ Method	Original Mean	Original Value Standard Deviation	Neighborhood Average	Neighborhood Standard Deviation	The Average Standard Deviation of Various Neighborhoods Decreases
Buildings	45.19	6.11	44.84	2.71	55.60%
Tree	54.96	9.69	54.45	2.26	76.68%
Other ground	43.64	2.46	43.17	1.15	53.21%
Grass	47.21	7.46	46.82	5.74	23.04%
Road	43.31	1.95	42.96	0.57	70.92%
Overall standard deviation		27.68		12.43	
Neighborhood average overall standard deviation decreases: 55.07%					

Table 4. Training slope value–original value and neighborhood standard deviation comparison table.

Category \ Method	Original Mean	Original Value Standard Deviation	Neighborhood Average	Neighborhood Standard Deviation	The Average Standard Deviation of Various Neighborhoods Decreases
	Amplitude Value (Unit: DN)				
Buildings	16.39	17.13	15.62	12.60	26.44%
Tree	33.56	15.83	33.09	11.99	24.26%
Other ground	2.49	3.41	1.723	2.37	30.40%
Grass	1.22	1.45	0.74	1.19	18.16%
Road	1.02	0.62	0.64	0.507	17.83%
Overall standard deviation		38.43		28.65	
Neighborhood average overall standard deviation decreases: 25.45%					

4.1.1. Decision Tree Classification Based on Original Values

The original amplitude value, wave width value, slope value, and the point-cloud value in the sampling area of self-determined height value are classified by MATLAB. The pruning level of the original decision tree without pruning is 83 levels, and the decision tree is pruned to 10 levels, as shown in Figure 12. After completing the decision-tree calculation, we extract all the LiDAR point clouds in the study area, and extract the point-cloud data that conform to the decision-tree classification operation, as shown in Figure 13. Figure 13a–c respectively represent the results of local point-cloud profile and the results of point-cloud classification in the study area, as well as the comparison and explanation drawings. Figure 13a shows the point-cloud profile area shown in the light blue box of Figure 13c, from which the distribution of point-cloud results can be observed. Figure 13b shows the classification results of the whole study area, and Figure 13c shows the color representative significance in the comparison aerial photos and explanation results.

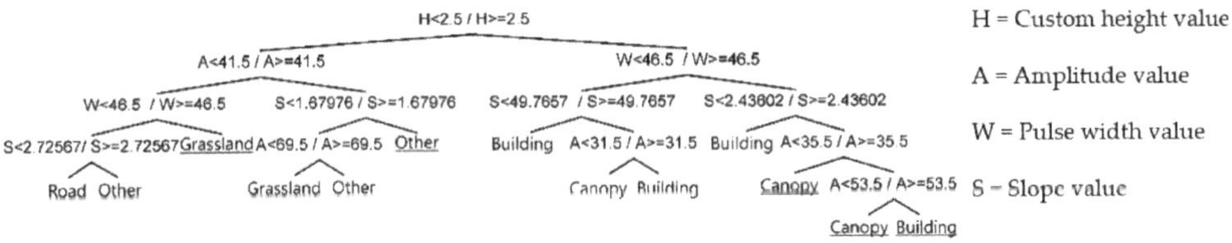

Figure 12. Pruning the decision tree with the original value of the first stage experiment.

4.1.2. Decision Tree Classification Based on Neighborhood Average

The amplitude value, wave width value, slope value, and self-determined height value point-cloud data in the neighborhood average point-cloud data are classified by the MATLAB program. The pruning level of the original decision tree without pruning is 58 levels, and the branches are pruned to 10 levels, as shown in Figure 14. After completing the decision-tree calculation, we classify and extract all the LiDAR point clouds in the study area and extract the point-cloud data that conform to the decision-tree classification calculation, as shown in Figure 15. Figure 15a–c respectively represent the results of the local point-cloud profile and the results of point-cloud classification in the study area, as well as the comparison and explanation drawings. Figure 15a shows the point-cloud profile area shown in the light blue box of Figure 15c, from which the distribution of point-cloud results can be observed. Figure 15b shows the classification results of the whole study area, and Figure 15c shows the color representative significance in the comparison aerial photos and explanation results.

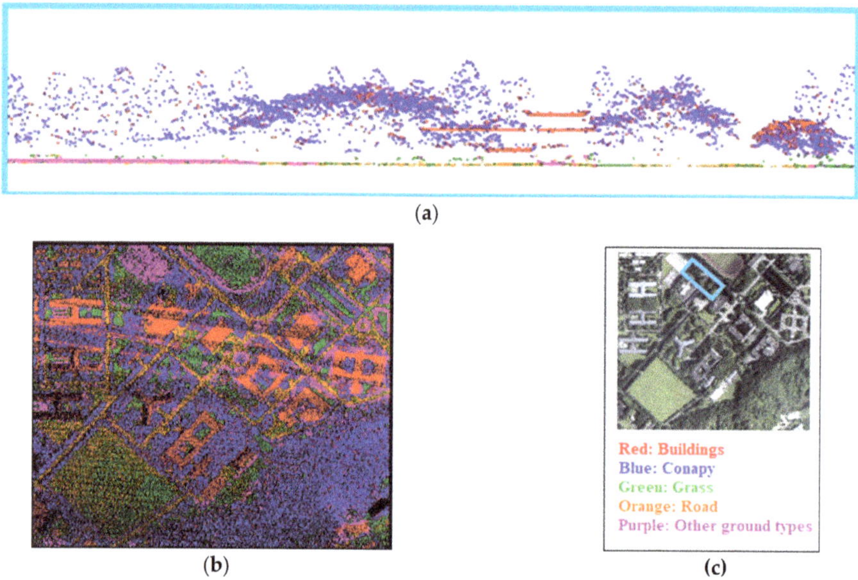

Figure 13. Classification results of the original value of the first stage experiment. (**a**) Local point cloud profile results; (**b**) Point cloud classification results in the study area; (**c**) Comparison diagram.

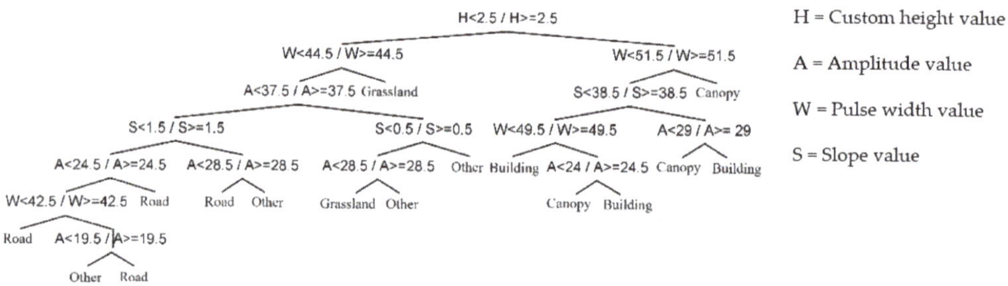

Figure 14. Decision tree trimmed by the mean value of neighborhood in the first stage experiment.

4.1.3. Research Results Display and Accuracy Analysis

We calculated the error matrix of the truth data and the classified results (as shown in Tables 5 and 6), found the classification accuracy of the original value and the average of the neighborhood, and the overall accuracy of the average of the neighborhood in the selected verification area of 83% was higher than the original value of 7%; Kappa is higher than the original value of 5.92%. The accuracy evaluation results verify that the point-cloud value classification after the neighborhood average can improve the classification results of the original value. From the error matrix, the point-cloud classification can be observed on the ground. The classification accuracy of objects is relatively poor. The average ground category of neighborhoods is 87%, 58% and 48% for roads, grasses and other grounds, respectively. Compared with 96% for buildings and 97% for canopy, the classification accuracy is relatively low.

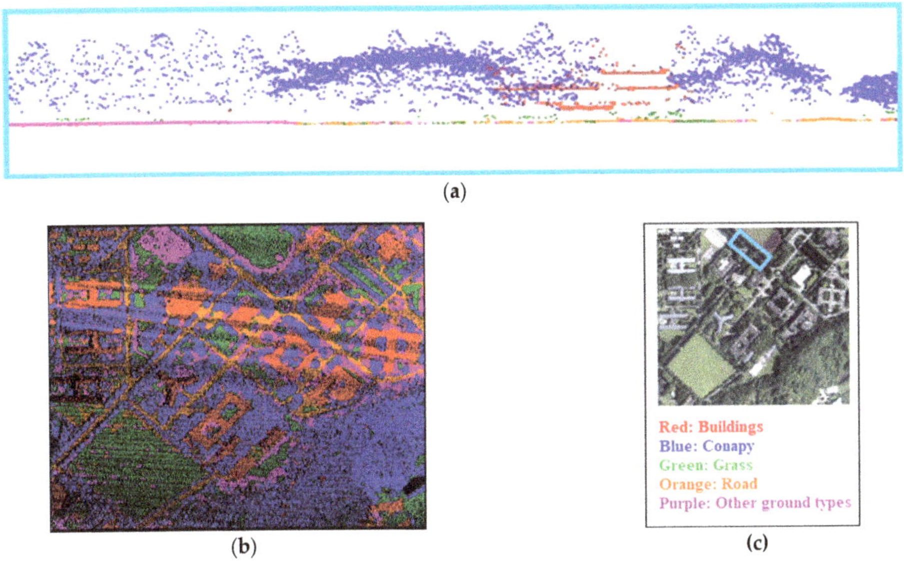

Figure 15. Results of the neighborhood average classification in the first phase of the experiment. (**a**) Local point cloud profile results; (**b**) Point cloud classification results in the study area; (**c**) Comparison diagram.

Table 5. Error matrix of original value in the first stage experiment.

Classification Results	Buildings	Canopy	Road	Grassland	Other Ground	Classification Total	Producer Accuracy (PA)
Buildings	22,153	6073	0	0	42	28,268	78%
Canopy	2362	38,226	0	0	8	40,596	94%
Road	0	34	26,471	7081	2044	35,630	74%
Grassland	4	419	3679	10,129	7738	21,969	46%
Other ground	2	63	2537	1659	7224	11,485	63%
Ground truth category total	24,521	44,815	32,687	18,869	17,056	137,948	
User Accuracy (UA)	90%	85%	81%	54%	42%		
Overall Accuracy (OA): 76%, Kappa: 43.26%							

4.2. Second Stage Experiment: Adding Image Data to the Classification of the Neighborhood Average

The results of the first-stage experimental research show that the decision-tree classification averaged in the neighborhood has better results, but the classification results in the ground category (grassland, road, other ground categories) are less satisfactory, so the second stage experiment takes the first in the stage where the experimental classification is not good, and the NDVI value and satellite image NIR-2 band data are added for classification and accuracy evaluation. The question is whether the addition of image information in the experiment can improve the classification accuracy of the point cloud.

Table 6. The mean error matrix of the neighborhood in the first stage experiment.

Classification Results	Buildings	Canopy	Road	Grassland	Other Ground	Classification Total	Producer Accuracy (PA)
Buildings	23,648	988	0	0	31	24,667	96%
Canopy	866	43,328	0	0	19	44,213	98%
Road	0	18	28,377	4536	687	33,618	84%
Grassland	7	477	1133	11,038	8070	20,725	53%
Other ground	0	4	3177	3295	8249	14,725	56%
Ground truth category total	24,521	44,815	32,687	18,869	17,056	137,948	
User Accuracy (UA)	96%	97%	87%	58%	48%		
Overall Accuracy (OA): 83%, Kappa: 49.18%							

The second phase of the experiment mainly carried out five classifications of LiDAR data, namely building, canopy, grassland, road, and other ground categories, among which other ground categories include cement ground, PU ground, and bare mud land. After that, NDVI and NIR-2 point clouds were added to the neighborhood mean value to classify the decision tree, and the accuracy of the results was evaluated after the classification was completed.

4.2.1. Circle Selection in Decision Tree Training Area

We re-calculated the data of the study area. Since the results of the first-stage experiment classification found that it is easier to misjudge in the ground category, the area with more misjudgments in the study area was taken as the study area of the second-stage experiment (as shown in Figure 16), the NDVI value and the image NIR-2 value were added to the LiDAR data attribute data, and the point cloud was sampled according to the customized height studied during the sampling. The colors selected in the training area are: red for buildings, blue for canopy, green for grassland, orange for roads, and purple for other ground classes. The point-cloud samples of the second and third floors of self-defined height values are taken for buildings and canopy, and the point-cloud samples of the second and first floors are taken for roads, grasslands, and other ground classes.

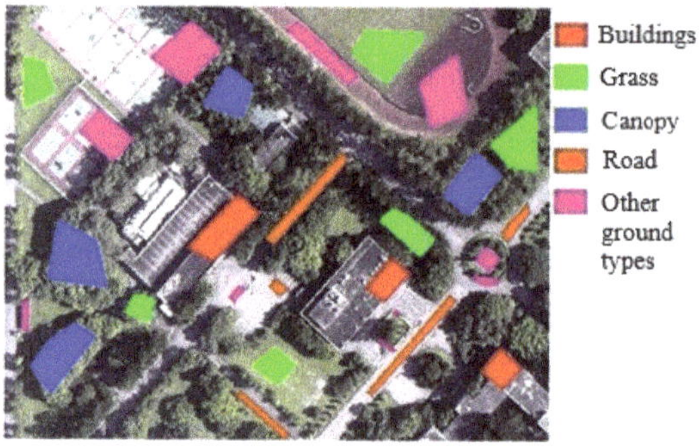

Figure 16. Sampling diagram of training samples in the second phase experimental study area.

4.2.2. Decision Tree Classification Based on Neighborhood Average

The amplitude value, wave width value, slope value and self-determined height value point-cloud data in the neighborhood average point-cloud data are classified by the MATLAB program. The pruning level of the original decision tree without pruning is 38 levels, and the branches are pruned to 10 levels, as shown in Figure 17. We extract the point-cloud data that conform to the classification operation of the decision tree, as shown in Figure 18. Figure 18a–c respectively represent the results of local point-cloud profile and the results of point-cloud classification in the study area, as well as the comparison and explanation drawings. Figure 18a shows the point-cloud profile area shown in the light blue box of Figure 18c, from which the distribution of point-cloud results can be observed. Figure 18b shows the classification results of the whole study area, and Figure 18c shows the color representative significance in the comparison aerial photos and explanation results.

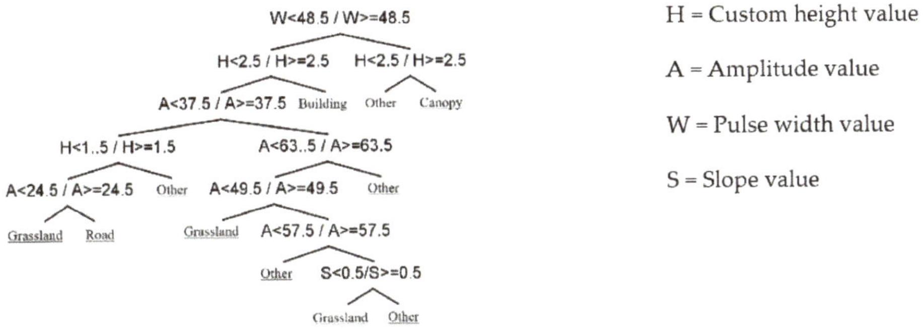

Figure 17. Pruning the decision tree by the average value of the experimental neighborhood in the second stage.

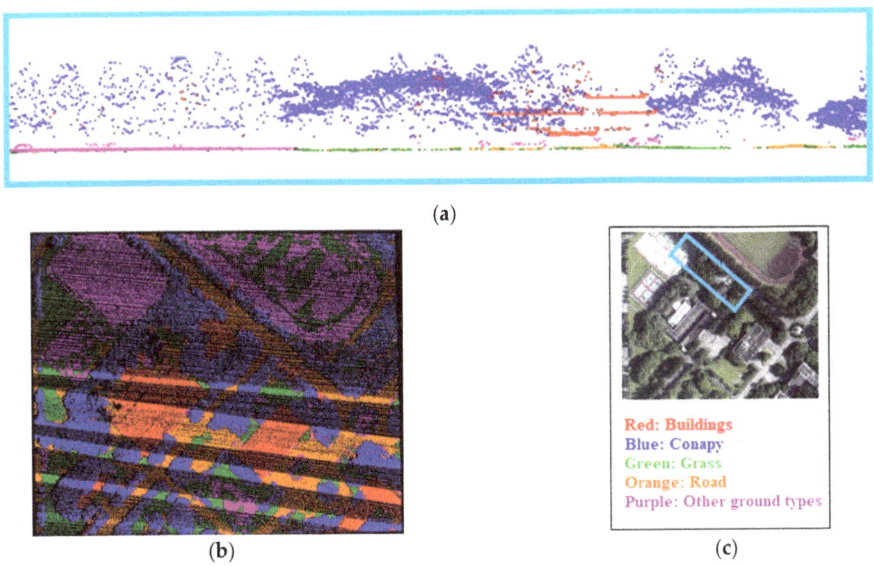

Figure 18. Results of the neighborhood average classification in the second experiment. (**a**) Local point cloud profile results; (**b**) Point cloud classification results in the study area; (**c**) Comparison diagram.

4.2.3. Decision Tree Classification Based on the Neighborhood Average Value and NDVI Value

The amplitude value, wave width value, slope value, self-determined height value and NDVI value in the neighborhood average point-cloud data are classified by the MATLAB program. The pruning level of the original decision tree without pruning is 48 levels, and the branches are pruned to 10 levels, as shown in Figure 19. After completing the decision-tree calculation, we extract all the LiDAR point clouds in the study area and extract the point-cloud data that conform to the decision-tree classification operation, as shown in

Figure 20. Figure 20a–c respectively represent the results of local point-cloud profile and the results of point-cloud classification in the study area, as well as the comparison and explanation drawings. Figure 20a shows the point-cloud profile area shown in the light blue box of Figure 20c, from which the distribution of point-cloud results can be observed. Figure 20b shows the classification results of the whole study area, and Figure 20c shows the color representative significance in the comparison aerial photos and explanation results.

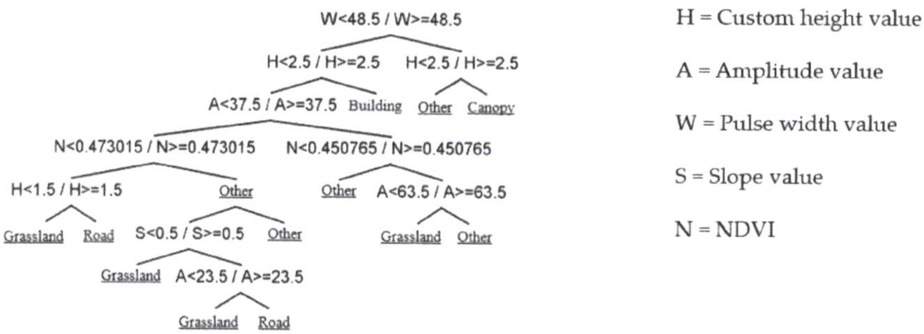

Figure 19. The second stage experiment adds NDVI value to trim the decision tree.

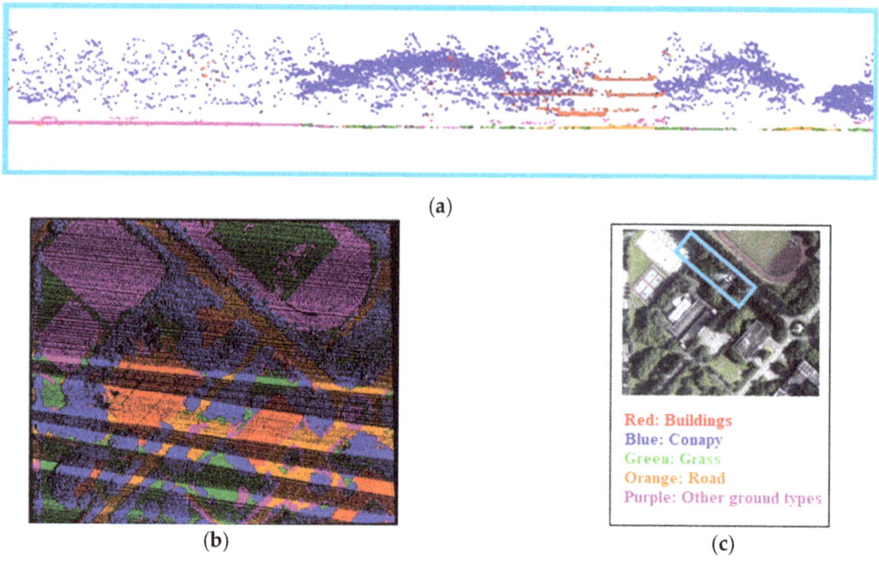

Figure 20. The second stage experiment added NDVI value classification results. (**a**) Local point cloud profile results; (**b**) Point cloud classification results in the study area; (**c**) Comparison diagram.

4.2.4. Adding NIR-2 Band Value Decision Tree Classification by Neighborhood Average

The amplitude value, wave width value, slope value, self-determined height value and NIR-2 value in the neighborhood average point-cloud data are classified by the MATLAB program. The pruning level of the original decision tree without pruning is 43 levels, and the branches are pruned to 10 levels, as shown in Figure 21. After completing the decision-tree calculation, we extract all the LiDAR point clouds in the study area, and extract the point-cloud data that conform to the decision-tree classification operation, as shown in Figure 22. Figure 22a–c respectively represent the results of local point-cloud profile and the results of point-cloud classification in the study area, as well as the comparison and explanation drawings. Figure 22a shows the point-cloud profile area shown in the light blue box of Figure 22c, from which the distribution of point-cloud results can be observed. Figure 22b shows the classification results of the whole study area, and Figure 22c shows the color representative significance in the comparison aerial photos and explanation results.

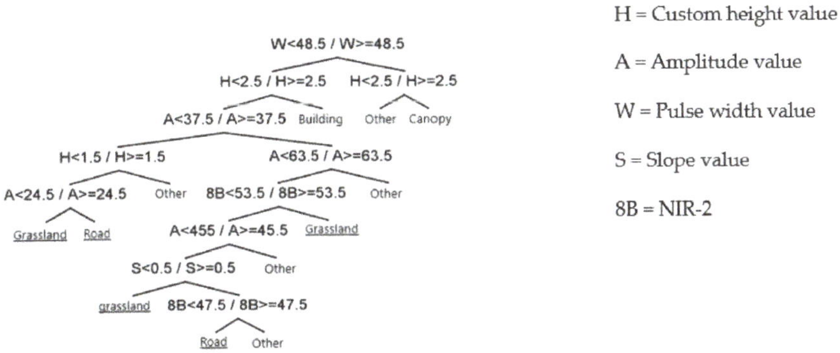

Figure 21. The second stage experiment adds image NIR-2 value to trim the decision tree.

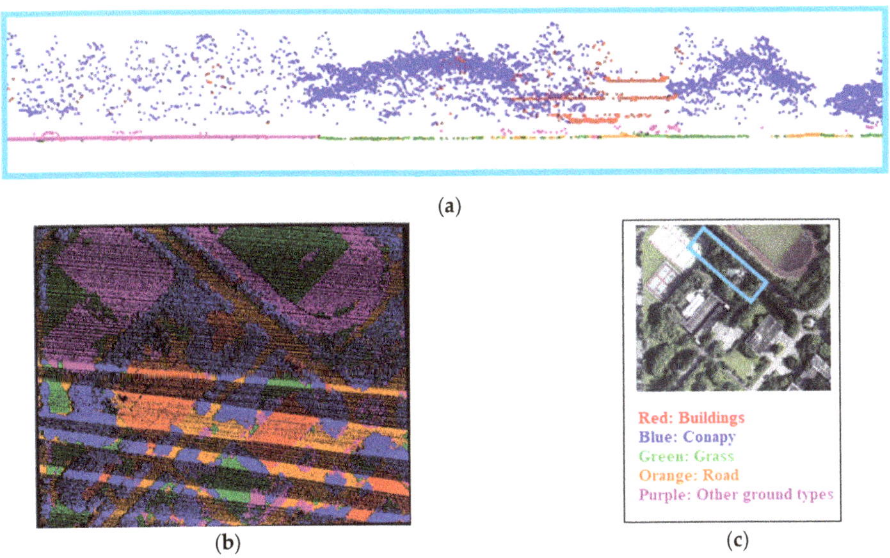

Figure 22. The second stage experiment added image NIR-2 value classification results. (**a**) Local point cloud profile results; (**b**) Point cloud classification results in the study area; (**c**) Comparison diagram.

4.2.5. Achievement Display and Accuracy Analysis

We calculated the error matrix of the truth data and the classified results (as shown in Tables 7–9), found the neighborhood average, then added the neighborhood average to NDVI and NIR-2 classification accuracy, using the error matrix to compare the differences between its three accuracies.

Table 7. Neighbor mean error matrix of the second stage experiment.

Classification Results	Buildings	Canopy	Road	Grassland	Other Ground	Classification Total	Producer Accuracy (PA)
Buildings	24,029	1355	0	0	31	25,415	95%
Canopy	485	42,961	0	0	19	43,465	99%
Road	0	12	28,395	2965	1055	32,427	88%
Grassland	2	0	3985	13,464	4712	22,163	61%
Other ground	5	487	339	2440	11,239	14,510	77%
Ground truth category total	24,521	44,815	32,687	18,869	17,056	137,948	
User Accuracy (UA)	98%	96%	87%	71%	66%		
Overall Accuracy (OA): 87%, Kappa: 52.60%							

Table 8. Error matrix of neighborhood mean value plus NDVI value in the second stage experiment.

Classification Results	Buildings	Canopy	Road	Grassland	Other Ground	Classification Total	Producer Accuracy (PA)
Buildings	24,029	1355	0	0	30	25,414	95%
Canopy	485	42,961	0	0	19	43,465	99%
Road	0	12	27,555	1001	591	29,159	94%
Grassland	2	0	3185	16,345	651	20,183	81%
Other ground	5	487	1947	1523	15,765	19,727	80%
Ground truth category total	24,521	44,815	32,687	18,869	17,056	137,948	
User Accuracy (UA)	98%	96%	84%	87%	92%		
Overall Accuracy (OA): 92%, Kappa: 56.85%							

Table 9. Neighbor mean value plus NIR-2 value error matrix in the second stage experiment.

Classification Results	Buildings	Canopy	Road	Grassland	Other Ground	Classification Total	Producer Accuracy (PA)
Buildings	24,029	1355	0	0	31	25,415	95%
Canopy	485	42,961	0	0	19	43,465	99%
Road	0	12	28,874	4200	1365	34,451	84%
Grassland	2	0	2677	12,514	1725	16,918	74%
Other ground	5	47	1136	2155	13,916	17,259	81%
Ground truth category total	24,521	44,815	32,687	18,869	17,056	137,948	
User Accuracy (UA)	98%	96%	88%	66%	82%		
Overall Accuracy (OA): 89%, Kappa: 53.81%							

Compared with the average value of the NDVI added to the neighborhood, the overall accuracy of the NDVI value increased by 5%, the Kappa value increased by 4.25%, other ground and grass types in the UA increased by 26% and 16%, and the road type decreased by 3% In the PA part, other ground categories, grass categories and road categories were increased by 9%, 20% and 6%, respectively. The result shows that the ground category classification has become better.

Compared with the NIR-2 single band image value, the nir-2 single band image value improves the overall accuracy by 2%, and the Kappa value increases by 1.21%. In UA, other ground and Road classes increased by 16% and 1%, respectively, and in grassland class decreased by 5%. In PA, other ground classes and grassland classes increased by 10% and 13%, respectively, and Road classes decreased by 3%. The results show that the distribution of NIR-2 single band image values is better in grassland and other ground classes, so it is proved that adding image information to point-cloud data can effectively improve the classification accuracy of the point cloud. Because the decision tree takes 10 pruning levels in classification, and the decision leaf nodes of building in progress and crown are the same, the classification accuracy of building and crown is the same.

5. Conclusions and Suggestions

5.1. Conclusions

In this study, the parameters of the full waveform light were used for the classification and accuracy evaluation of the ground features. After verification, the averaged point-cloud results had better classification accuracy, but the classification results were poorer on grasslands and other artificial structures. If it is supplemented with image parameters, the poor part of the point-cloud classification can be improved.

The first phase of the experiment mainly classifies and evaluates the original point cloud and the point cloud averaged by the neighborhood. Through decision-tree classification, the following conclusions are drawn after analysis and research:

(1) The point-cloud data are averaged by neighborhood analysis, and the standard deviation of each parameter value is reduced, which means that the point cloud can effectively filter out noise after neighborhood average, especially for the bandwidth value. In other words, there is the best average distribution.

(2) Comparison was made of the accuracy of the point-cloud decision-tree division between the original point cloud and the neighborhood average. The overall accuracy of the point cloud classification after neighborhood averaging was 83% higher than the original value of 7%, and Kappa was higher than the original value of 5.92%, so the average of the point-cloud neighborhood can filter out excessive noise and improve the classification accuracy.

(3) Since the neighborhood average converts the point cloud into grid data, and then averages the point clouds in the grid data in a rectangular manner, the overlapping of the point clouds with different values will result in point clouds with different values. After averaging the neighborhood, the average becomes the same value. As shown in Figure 23, the vertically overlapping parts will average the parameter values of buildings and crowns to the same value during averaging, as shown in the yellow box of Figure 23a. After averaging, the entire vertical surface is averaged, so that the building parameter values are averaged to become the canopy class, and the neighborhood average will cause the point cloud to misjudge where different types of features overlap. The neighborhood average point cloud is suitable for use in an open surface area without overlap; if it is used in an area with much shadow, it will cause a false judgment of the overlapping part of the shadow.

In the second phase of the experiment, the areas with poor classification results in the first phase of the experimental research were taken, the scope of the study area was narrowed, the second decision-tree classification was conducted, and re-sampling was conducted. After analysis and research, the following conclusions were drawn:

(1) After averaging the neighborhood point-cloud value for the decision tree operation, the image parameter NDVI value and image NIR-2 value were added, respectively, for the decision-tree operation. Research results show that adding NDVI and NIR-2 can improve the overall accuracy; especially for the ground object categories that are more difficult to distinguish in the first phase of the experiment there are better classification results. The overall accuracy of the NDVI value is increased by 5%, the Kappa value is increased by 4.25%, and the NIR-2 value is added to increase the overall accuracy by 2%. The Kappa value is increased by 1.21%, so LiDAR data combined with satellite image data can effectively improve the classification accuracy of LiDAR point clouds.

(2) Because the difference between the shooting time of the satellite image and the LiDAR image is 1 year, the vegetation has different extent, which will cause errors in the classification. For example, in as shown in Figure 24, the turf part of the playground in the satellite image is bare soil, and the turf on the aerial photo image taken by LiDAR at the same time is the part without bare soil. Therefore, when the decision tree is added to the satellite image parameters for classification, the playground turf will be classified incorrectly.

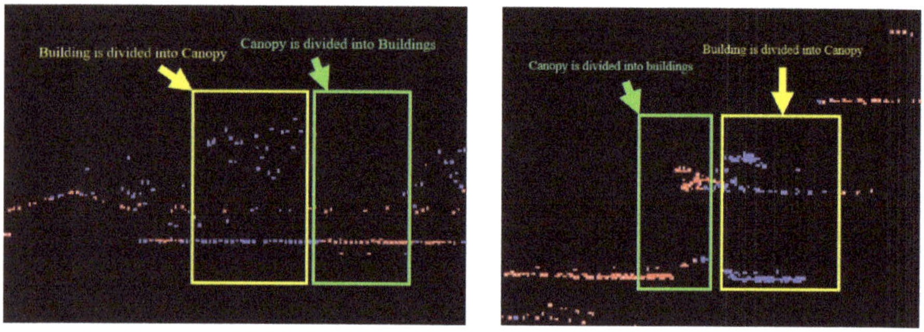

(a) (b)

Figure 23. (**a**) Neighborhood average limit graph I; (**b**) Neighborhood average limit graph II.

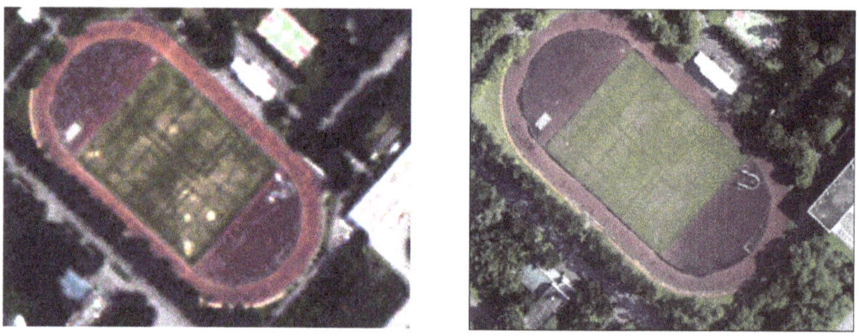

(a) (b)

Figure 24. Vegetation Difference Image. (**a**) Satellite imagery; (**b**) Aerial photograph.

5.2. Suggest

After this study, the following suggestions are summarized:

(1) Since this research mainly hopes to obtain the classification results of point clouds simply and quickly and provide the application of disaster prevention and relief layers, the selected LiDAR parameters are the parameter data (amplitude value, pulse width value, and slope value) that can be obtained quickly. The classification accuracy of the LiDAR parameters selected in the study is poor in the ground class; it is recommended that other LiDAR parameters (number of echoes, back reflection, elevation difference, multiple echo percentage, etc.) be added for classification research to improve the ground features classification accuracy.

(2) In this study, the neighborhood average point cloud is performed on the entire study area. In order to avoid the misjudgment of overlapping point clouds caused by the neighborhood average, it is recommended that the neighborhood average point cloud can be performed only on the open and uncovered areas to avoid misjudgment. The situation arises, and the classification accuracy can also be improved.

(3) This study uses the CART decision-tree algorithm, which is mainly due to the shortcomings of the decision tree without information hiding, and the analysis results can be presented in the form of a decision tree, which can directly see the important changes and their segmentation effects, which can make the research results more clear and understandable. On the other hand, the decision tree is simple and fast. If it is used in image interpretation, it can reduce the labor and time of the interpreter at work and can understand the correlation of the measurement area in a short time. Geographic information provides various scientific and practical applications. At present, many full-waveform classification researches use Random Forest, Support Vector Machine (SVM), Neural Network (NN) and other classification methods for classification. It is recommended that different algorithms can be used for classification to achieve better classification results.
(4) In terms of obtaining LiDAR and images, it is recommended to use LiDAR point-cloud data and satellite images in the same year and season to reduce the differences in ground features and vegetation and avoid lowering the classification accuracy.
(5) The decision-tree classification results in the study are only applicable in this study area. This study has not been verified and compared in different study areas. It is recommended that LiDAR data at different altitudes can be used for classification studies to compare between the different flight zones. The parameter difference of the research, research and analysis of the general LiDAR classification parameter data, to improve the versatility of point-cloud data in different research areas.
(6) This study mainly studies the Red, NIR-1 and near-NIR-2 band information in the World View-2 satellite image, using NDVI and NIR-2 parameters and adding LiDAR point-cloud data for classification. For the advantages and disadvantages and limitations of image bands, it is recommended that each band image be analyzed and studied. After calculating the best classification value between bands, it can be added to the LiDAR point cloud for classification to effectively improve the classification results and accuracy.

Author Contributions: Conceptualization, Ming-Da Tsai; Data curation, Chun-Ta Wei; Formal analysis, Yu-Lung Chang; Investigation, Yu-Lung Chang; Methodology, Ming-Da Tsai and Yu-Lung Chang; Project administration, Chun-Ta Wei and Ming-Da Tsai; Resources, Ming-Da Tsai; Software, Ming-Da Tsai; Supervision, Ming-Da Tsai; Validation, Chun-Ta Wei and Yu-Lung Chang; Visualization, Chun-Ta Wei, Writing—original draft, Chun-Ta Wei and Yu-Lung Chang; Writing—review & editing, Chun-Ta Wei and Ming-Chih Jason Wang. All authors have read and agreed to the published version of the manuscript.

Funding: This research received no external funding.

Acknowledgments: Thanks to Strong Engineering Consulting Co., Ltd. for assisting LiDAR data collection and data preprocessing and other related technical support operations.

Conflicts of Interest: The authors declare no conflict of interest.

References

1. Guo, L.; Chehata, N.; Mallet, C.; Boukir, S. Relevance of airborne LiDAR and multispectral image data for urban sceneclassification using Random Forests. *ISPRS J. Photogramm. Remote Sens.* **2011**, *66*, 56–66. [CrossRef]
2. Höfle, B.; Hollaus, M.; Hagenauer, J. Urban vegetation detection using radiometrically calibrated small-footprint full-waveform airborne LiDAR data. *ISPRS J. Photogramm. Remote Sens.* **2012**, *67*, 134–147. [CrossRef]
3. Wu, Q.; Zhong, R.; Zhao, W.; Song, K.; Du, L. Land-cover classification using GF-2 images and airborne lidar data based on Random Forest. *Int. J. Remote Sens.* **2019**, *40*, 2410–2426. [CrossRef]
4. Diaz, J.C.F. Lifting the Canopy Veil Airborne LiDAR for Archeology of Forested Areas. *Imagingnotes* **2011**, *26*, 31–34.
5. Teo, T.A.; Huang, C.H. Object-based Land Cover Classification Using Airborne Lidar and Different Spectral Images. *Terr. Atmos. Ocean. Sci.* **2016**, *27*, 491–504. [CrossRef]
6. Mallet, C.; Bretar, F. Full-waveform topographic LiDAR:State-of-the-art. *ISPRS J. Photogramm. Remote Sens.* **2009**, *64*, 1–16. [CrossRef]
7. Breiman, L.; Friedman, J.H.; Olshen, R.A.; Stone, C.J. *Classification and Regression Trees*, 1st ed.; Routledge: London, UK, 1984.

8. Swain, P.H.; Hauska, H. The Decision Tree Classifier: Design and Potential. *IEEE Trans. Geosci. Electron.* **1977**, *15*, 142–147. [CrossRef]
9. Mui, J.K.; Fu, K.S. Automated Classification of Nucleated Blood Cells Using a Binary Tree Classifier. *IEEE Trans. Pattern Anal. Mach. Intell.* **1980**, *PAMI-2*, 429–443. [CrossRef]
10. Safavian, S.R.; Landgrebe, D.A. A Survey of Decision Tree Classifier Methodology. *IEEE Trans. Syst. Man Cybern.* **1991**, *21*, 660–674. [CrossRef]
11. Rounds, E.M. A Combined Nonparametric Approach to Feature Selection and Binary De cision Tree Design. *Pattern Recognit.* **1980**, *12*, 313–317. [CrossRef]
12. Breiman, L.; Friedman, J.; Stone, C.J.; Olshen, R.A. *Classification and Regression Trees*; Wadsworth & Brooks/Cole Advanced Books & Software: Monterey, CA, USA, 1984.
13. Olofsson, P.; Foody, G.M.; Herold, M.; Stehman, S.V.; Woodcock, C.E.; Wulder, M.A. Good practices for estimating area and assessing accuracy of land change. *Remote Sens. Environ.* **2014**, *148*, 42–57. [CrossRef]
14. Congalton, R.G. A review of assessing the accuracy of classifications of remotely sensed data. *Remote Sens. Environ.* **1991**, *37*, 35–46. [CrossRef]
15. Lin, Y.C. Digital Terrain Modelling from Small-footprint, Full-Waveform Airborne Laser Scanning Data. Doctoral Dissertation, School of Civil Engineering and Geosciences, Newcastle University, Newcastle, UK, 2009.
16. Gross, H.N.; Schott, J.R. Application of Spectral Mixture Analysis and Image Fusion Techniques for Image Sharpening. *Remote Sens. Environ.* **1998**, *63*, 85–94. [CrossRef]
17. Van Der Meer, E. What Does Multisensor Image Fusion Add in Terms of Information Content for Visual Interpretation? *Int. J. Remote Sens.* **1997**, *18*, 445–452. [CrossRef]
18. Congalton, R.G.; Green, K. *Assessing the Accuracy of Remotely Sensed Data: Principles and Practices*; CRC Press: Boca Raton, FL, USA, 2008.
19. Heinzel, J.; Koch, B. Exploring Full-Waveform LiDAR Parameters for Tree Species Classification. *Int. J. Appl. Earth Obs. Geoinf.* **2011**, *13*, 152–160. [CrossRef]

International Journal of Geo-Information

MDPI

Article

Crop Identification Based on Multi-Temporal Active and Passive Remote Sensing Images

Hebing Zhang, Hongyi Yuan *, Weibing Du and Xiaoxuan Lyu

School of Surveying and Land Information Engineering, Henan Polytechnic University, Jiaozuo 454000, China; jzitzhb@hpu.edu.cn (H.Z.); dwb@hpu.edu.cn (W.D.); 112004010003@home.hpu.edu.cn (X.L.)
* Correspondence: yhy@home.hpu.edu.cn

Abstract: Although vegetation index time series from optical images are widely used for crop mapping, it remains difficult to obtain sufficient time-series data because of satellite revisit time and weather in some areas. To address this situation, this paper considered Wen County, Henan Province, Central China as the research area and fused multi-source features such as backscatter coefficient, vegetation index, and time series based on Sentinel-1 and -2 data to identify crops. Through comparative experiments, this paper studied the feasibility of identifying crops with multi-temporal data and fused data. The results showed that the accuracy of multi-temporal Sentinel-2 data increased by 9.2% compared with single-temporal Sentinel-2 data, and the accuracy of multi-temporal fusion data improved by 17.1% and 2.9%, respectively, compared with multi-temporal Sentinel-1 and Sentinel-2 data. Multi-temporal data well-characterizes the phenological stages of crop growth, thereby improving the classification accuracy. The fusion of Sentinel-1 synthetic aperture radar data and Sentinel-2 optical data provide sufficient time-series data for crop identification. This research can provide a reference for crop recognition in precision agriculture.

Keywords: time series; Sentinel-1; Sentinel-2; random forest; feature-level fusion

Citation: Zhang, H.; Yuan, H.; Du, W.; Lyu, X. Crop Identification Based on Multi-Temporal Active and Passive Remote Sensing Images. *ISPRS Int. J. Geo-Inf.* **2022**, *11*, 388. https://doi.org/10.3390/ijgi11070388

Academic Editors: Dimitris Triantakonstantis, Panagiotis Tziachris and Wolfgang Kainz

Received: 17 April 2022
Accepted: 8 July 2022
Published: 11 July 2022

Publisher's Note: MDPI stays neutral with regard to jurisdictional claims in published maps and institutional affiliations.

1. Introduction

As the world's population continues to grow and the COVID-19 pandemic brings uncertainties, food security has gained increasing attention [1–3]. At the same time, driven by the digital revolution in agriculture, two major changes have taken place in world agriculture: the development of precision agriculture, and the development of the agricultural digital economy [4–6]. Integrating GIS and remote sensing to obtain and analyze crop planting information is the key to realizing intelligent agriculture [7]. Remote sensing, with its rapidity, accuracy, and numerous other advantages, is now widely used to extract and classify crops, which is vital for intelligent agriculture [8].

Multi-temporal passive remote sensing plays an important role in agriculture [9]. Since the growth of crops follows seasonal rhythms and regular phenological variations, a time series of multi-temporal remote-sensing data can characterize crops as a function of time [10–12]. Sonobe et al. (2017) used data from the Landsat 8 Operational Land Imager to classify crops in Hokkaido, Japan and achieved an overall accuracy of 94.5% [13]. Yi et al. (2019) extracted various features from multi-temporal Sentinel-2 data for crop classification and obtained an overall accuracy of 94% [14].

The above Landsat 8 and Sentinel-2 optical data were obtained by passive remote sensing platforms [15]. Sentinel-2 is a multispectral high-resolution imaging satellite that carries a Multispectral Instrument (MSI) [16]. It has two satellites 2A and 2B, and each satellite conducts an Earth observation every 10 days under constant-observation conditions [16]. The complementarity of the two satellites can achieve a temporal resolution of 5 days. The MSI has 13 bands that cover from 442 to 2202 nm, and the highest spatial resolution is 10 m [16]. With its three bands in the red edge, Sentinel-2 can provide rich information for crop detection and thereby greatly improve the estimation accuracy for chlorophyll content, the fractional cover of forest canopies, and leaf area index [17]. In addition, Sentinel-2 data have high temporal resolution, so their use can provide rich temporal information for short-term crop identification over large areas [18]. Although the vegetation indices (VIs) feature of optical images provides an efficient method for crop identification, they are easily affected by cloud cover and are difficult to obtain as complete time series [19].

Since microwaves can penetrate clouds, active remote sensing can be done independently of weather conditions, which greatly facilitates the acquisition of continuous time-series images [20]. Sentinel-1 is a radar satellite launched by the European Space Agency with high spatial resolution and a short revisit period. It has four working modes: wave (WV), interferometric wide swath (IW), extra-wide swath (EW), and stripmap (SM) [21], and can provide single-polarized (HH or VV), dual-polarized (HH + HV or VV + VH), multi-temporal, high-resolution (down to 5 m), and C-band synthetic aperture radar (SAR) imaging data. Researchers have already applied Sentinel-1 data to crop identification [22–25]. Teja et al. used Sentinel-1 temporal SAR data to estimate Kharif rice planting areas and obtained an overall accuracy of 91% [25]. To better identify crops, many scholars used both Sentinel-1 and -2 data [26–28]. However, few studies have focused on feature-level fusion, which is helpful for dealing with information redundancy [29].

The random forest (RF) classifier can classify fused remote sensing data [30]. RF is an ensemble learning method used to solve classification and regression problems [31]. Ensemble learning is a machine learning scheme that improves prediction accuracy by integrating multiple models to solve a given problem [32]. Multiple classifiers participating in ensemble classification produce more accurate results than a single classifier [32]. In addition, RF can extract multisource features through multiple decision trees, thus making the most of the features used in fused data. In particular, for time-series features, RF uses the statistical characteristics of time-series data to extract and analyze the time-series features of different samples [33]. Finally, due to the simple and efficient decision-making method of majority voting, RF classifiers can quickly classify large quantities of remote-sensing data [34].

This paper focuses on the autumn crops in Wen County, Jiaozuo City. Using multi-temporal Sentinel-1 and -2 data, this study fused the backscattering coefficient of Sentinel-1 and the normalized difference vegetation index (NDVI) of Sentinel-2 data at the feature-level, then used a GIS platform to make samples, and finally used RF for classification. The classification results of fused data, single-temporal Sentinel-2 data, multi-temporal Sentinel-1 data, and multi-temporal Sentinel-2 data were compared to elucidate the advantages of multi-temporal data and fused data for crop identification.

2. Materials and Methods

2.1. Study Area

Wen County, Jiaozuo City is situated in northwest Henan Province, China (112°57′39″–113°02′43″ E, 34°50′15″–34°57′37″ N) and has a warm temperate continental monsoon climate. The average temperature is between 14 and 15 °C, and the annual rainfall is between 550 and 700 mm. Most of Wen County is plain, and the average elevation (relative to the average sea level) is between 102.3 and 116.1 m. The multiple historical floodings of the Yellow River and the Qin River formed the unique landforms of the south beach, the north depression, and the middle hill in Wen County. The inland rivers belong to the Yellow River system. The Yellow River, Qin River, and the drainage river system flow through the area, providing sufficient water resources and convenient irrigation. The main types of soil in this area are yellow fluvo-aquic soil and cinnamon fluvo-aquic soil, and long-term farming gives this area significant production potential. This paper focuses mainly on the autumn crops in Wen County, mainly including corn, peanut, and Chinese yam, accounting for about 96% of the total crops. Other non-major crops, mainly including sweet potatoes, oil crops, vegetables and fruits, fall outside the scope of this paper. Figure 1 shows the geography of Wen County.

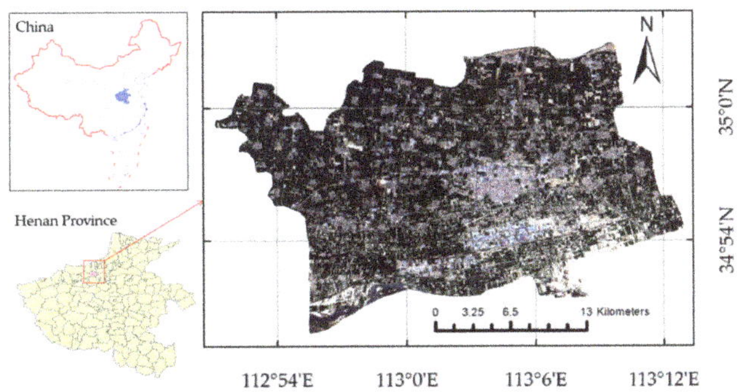

Figure 1. Geographical location of study area.

2.2. Sentinel-1, -2 Data

For this research, ground range detected (GRD) Sentinel-1 C-band (5.405 GHz) level-1 images in IW mode were downloaded from the website of ASF Data Search. IW mode supports the merging of wider strip widths (250 km), and the downloaded images have medium resolution (10 m). GRD products include two polarization modes, VV and VH. The range and azimuth resolutions are 20 and 22 m, respectively, and the pixel spacing is 10 m. Table 1 shows the data from Sentinel-1 imagery products.

Table 1. Sentinel-1 SAR data.

Product	Acquisition Date	Characteristics
Sentinel-1	1 July 2020	Data product: Level-1 GRD Imaging mode: IW Imaging frequency: C band (5.405 GHz) Polarization: VV and VH
Sentinel-1	25 July 2020	
Sentinel-1	6 August 2020	
Sentinel-1	18 August 2020	
Sentinel-1	30 August 2020	
Sentinel-1	11 September 2020	
Sentinel-1	23 September 2020	
Sentinel-1	5 October 2020	
Sentinel-1	17 October 2020	
Sentinel-1	29 October 2020	

This study uses multi-temporal Sentinel-2 data for crop identification. Table 2 lists the dates of Sentinel-2 image acquisition. These Level-2A data cover different plant growth periods from June 2020 to October 2020, and the data were obtained from the PIE (Pixel Information Expert) Engine platform.

Table 2. Sentinel-2 multispectral data.

Product	Acquisition Date	Characteristics
Sentinel-2	25 June 2020	Product level: Level-2A Imaging instrument: MSI
Sentinel-2	1 July 2020	
Sentinel-2	10 July 2020	
Sentinel-2	5 August 2020	
Sentinel-2	20 August 2020	
Sentinel-2	1 September 2020	
Sentinel-2	25 September 2020	
Sentinel-2	1 October 2020	
Sentinel-2	10 October 2020	

2.3. Field Data

To understand the local crop-planting situation in Wen County and obtain the training set and testing set labels for crop classification, this study conducted field visits at three different sites in the study area in September 2020. This study consulted the local agricultural production departments and farmers in detail about the planting phenology of autumn crops. The statistical results are given in Table 3, and Figure 2 shows representative crop images.

Table 3. Phenological of autumn crops in the study area.

Month	June			July			August			September			October		
Ten-Day	**E**	**M**	**L**	**E**	**M**	**L**	**E**	**M**	**L**	**E**	**M**	**L**	**E**	**M**	**L**
Peanut		sowing			germination		flowering			pod setting		maturity			
Maize			sowing			jointing			tasseling		maturity				
Chinese yam				flowering						fruiting				maturity	

Figure 2. (**a.1–a.4**) Images of peanuts at sowing, germination, flowering, and harvest, respectively. (**b.1–b.4**) Images of maize at seedling, jointing, tasseling, and maturity, respectively. (**c.1–c.4**) Images of Chinese yam at sowing, climbing pole, fruiting, and maturity, respectively.

In the field visits, in addition to recording the crop attributes for each parcel, the center latitude and longitude coordinates of each parcel were recorded with a handheld differential GPS positioning tool using the WGS1984-UTM coordinate system (zone 49N), with less than 2 m positioning error. Using the ArcGIS10.6 (Esri, Redlands, CA, USA) software platform, the data generated by GPS positioning and attribute data were merged and converted into ESRI shapefile format and matched to Google's high-resolution remote sensing images. This study used visual interpretation to label polygons on Google Image with the shapefile as the center. In terms of the ground sample distance of Sentinel-1 and -2 data, each polygon was at least 10 m from the boundary to avoid pixel mixing at the boundary. Some of the labels are shown in Figure 3. "Others" includes non-major crops, established areas, water, roads, and trees.

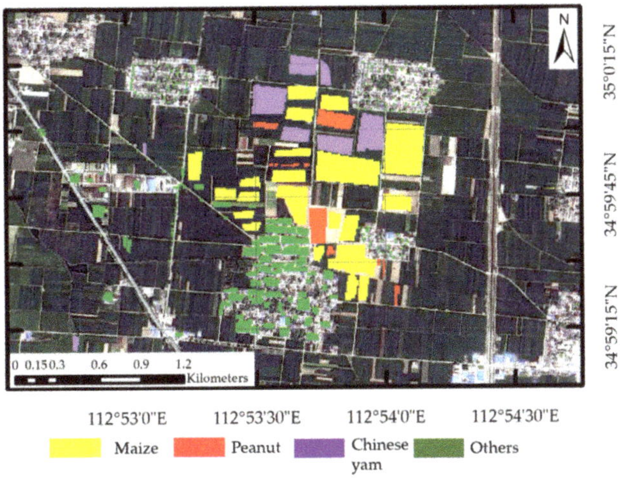

Figure 3. Examples of labels from field visits.

Table 4 lists the ratio of labeled area to corresponding crop area. The crop area data came from the 2020 Jiaozuo City Statistical Yearbook data.

Table 4. Ratio of labeled area to corresponding crop area.

Category	Government Statistics Area (hm^2)	Sample Label Area (hm^2)	Ratio (%)
Maize	17,281	85.24	0.49
Chinese yam	5228	20.71	0.40
Peanut	3497	15.53	0.44
Others	22,124	24.91	0.11

2.4. Methods

The research method used herein mainly focused on the fusion of Sentinel-1 and -2 data. This paper not only studied the role of multi-temporal data in crop identification but also discusses the advantages of fusing active and passive remote-sensing images for crop classification. Figure 4 shows a flowchart of the process.

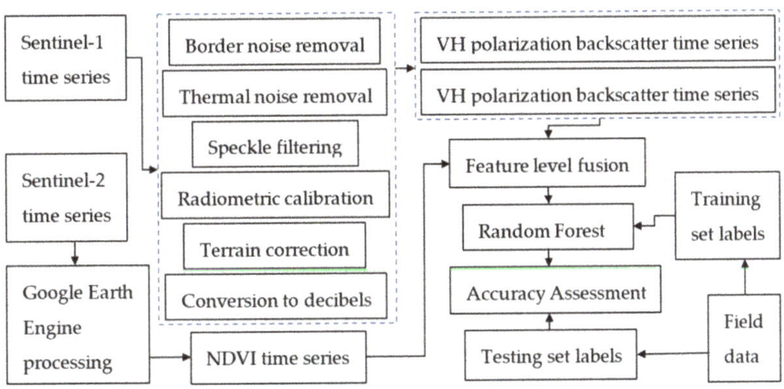

Figure 4. Flow chart for of crop identification.

2.4.1. Time-Series Datasets

Sentinel-1 was preprocessed using the open-source remote-sensing processing software SeNtinel Applications Platform (SNAP) of the European Space Agency and ENVI5.3 (Esri, Redlands, CA, USA). The main processes included removal of border noise and thermal noise, speckle filtering, radiometric calibration, terrain correction, conversion to decibels, and image clipping [35]. The time-series datasets with VV and VH polarization were obtained by preprocessing, which includes the backscattering coefficient. This study normalized features in both datasets: the pixel values of each image were transformed to a common scale, and the inherent similarities and variations were preserved. Feature normalization can improve the performance of machine learning algorithms. The normalization uses the min-max normalization type, with a minimum value of 2% and a maximum value of 98%, to reduce sensitivity to outlier.

Sentinel-2 data were preprocessed using PIE Engine, the optical features extracted, and a normalized difference vegetation index (NDVI) time-series dataset was constructed. Extracting the NDVI time-series features of Sentinel-2 is the key to studying crop mapping [14]. The NDVI correlates strongly with leaf area index and plant chlorophyll, and it is an important tool to study vegetation growth status and vegetation coverage and eliminate radiation errors [36]. Its time-series curve reflects the growth cycle of crops, including sowing, germination, heading, maturity, and harvest [37–39]. NDVI is calculated according to the normalized transformation of near-infrared and red reflectance, which is given by Equation (1).

$$\mathrm{NDVI} = \frac{\rho_{NIR} - \rho_{RED}}{\rho_{NIR} + \rho_{RED}}, \quad (1)$$

where ρ_{NIR} is the reflectance in the near-infrared band and ρ_{NIR} is the reflectance in the red band.

To extract the study area, the vector boundaries of Wen County were read as feature classes in the PIE Engine and overlaid onto Sentinel-2 images. Then, this study performed operations such as cloud removal, NDVI band math, filtering, and batch exporting. In particular, the quality assessment band was used to detect clouds and cloud shadows for simple and efficient cloud detection and removal. To facilitate the calculations involved in machine learning, feature normalization was performed on the acquired NDVI time-series dataset.

2.4.2. Fusion of Active and Passive Remote Sensing

The fusion of different sensor data generally requires georeferencing and image fusion [40]. In this study, both active and passive remote sensing time-series datasets were projected to the WGS1984-UTM coordinate system (zone 49N), so georeferencing is not required before fusion.

Most of the fusion of optical and radar images adopts an early fusion strategy, in which the optical image time series and the radar image time series are stacked together in the form of a data cube [41–44]. This fusion method does not extract vegetation index and backscattering coefficient, so it is simple and easy to implement. However, the fusion data have too much information, most of which is not important for crop identification, so information redundancy is a problem [29]. Therefore, this study adopted here the feature-level fusion method, whereby this study selected the backscattering coefficient feature from Sentinel-1 and the VI feature from Sentinel-2, and obtained the fused data by stacking the corresponding time-series dataset. In more detail, the fusion data contain 9 NDVI, 10 VV polarization backscattering coefficients and 10 VH polarization backscattering coefficients.

2.4.3. Random Forest Classifier

This study used a semantic segmentation algorithm based on the RF classifier to classify Sentinel-1 and -2 fused data. Unlike instance segmentation, which treats a single object as a distinct entity regardless of its category, semantic segmentation treats all objects of the same category as belonging to one entity [45,46]. Due to the complexity of remote-sensing image applications and the similar structure of crops, this study used a classification method of semantic segmentation. RF is a supervised machine learning algorithm that meets the needs of semantic segmentation [47]; it consists of multiple decision trees, and the category of each pixel of its output depends on the maximum number of votes in the set of trees. RF is faster than other classifiers, easier to parameterize, and robust [48], which supports the analysis of various features used herein. Figure 5 shows the structure of RF, and the RF classifier involved four main steps:

(1) A sample set with capacity N was extracted N times with one-at-a-time replacement until N samples were formed, which were then used as the samples at the root node of the decision tree to train the decision tree;
(2) Each sample has M features. When the decision tree needed to be split, m << M features were selected at random from these M features. The feature with the best classification ability of these m features was selected as the splitting feature of the node;
(3) To form the decision tree, each node was split as per step 2 until the feature selected by the child node was the feature used when the parent node was split; that is, the child node was a leaf node. At this point, the splitting stopped. Note that each tree grew to the maximum extent, and no pruning was done during the formation of the decision tree; and
(4) This study followed steps 1–3 to build k decision trees to form a RF. Assuming that the set of categories was $\{c_1, c_2, \ldots, c_N\}$, the prediction output of h_i in sample x was expressed as an N-dimensional vector $\left(h_i^1(x), h_i^2(x), \ldots, h_i^N(x)\right)^T$, where $h_i^j(x)$ represented the output of h_i in category c_j, and the decision was made by the majority voting (Equation (2)).

$$H(x) = \begin{cases} c_j, \sum_{i=1}^{k} h_i^j(x) > 0.5 \sum_{m=1}^{N} \sum_{i=1}^{k} h_i^m(x) \\ \text{reject, others} \end{cases}, \quad (2)$$

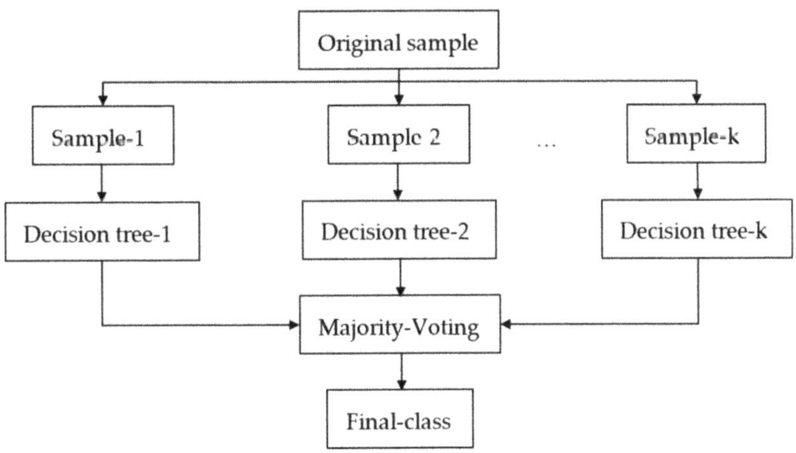

Figure 5. The structure of Random Forest classifier.

That was, if a category got more than half of the votes, the prediction would be that category, otherwise the prediction would be rejected.

The RF classifier had two parameters to set: the number k of decision trees in the classifier and the number m of features to consider when finding the best split [49]. The more decision trees in the RF classifier, the better the classification, but the longer the calculation time; the smaller the number of features, the smaller the variance, although the bias would increase. Therefore, the classification accuracy and time efficiency were comprehensively considered. This study used 300 decision trees, and the root of the total number of features was taken as the number of features.

2.4.4. Training and Prediction

For classification training and testing, the sample labels were overlaid on the fused images to obtain the samples. To avoid the influence of spatial autocorrelation [50], samples from two sites in the study area were used as the training set, and the other samples were used as the testing set. Table 5 lists the number of pixels and parcels for the training set and the testing set.

Table 5. Number of parcels and number of pixels for the training set and the testing set.

Sequence	Category	No. of Parcels		No. of Pixels	
		Training Set	Testing Set	Training Set	Testing Set
1	Maize	22	11	4867	5848
2	Peanut	19	9	1224	670
3	Chinese yam	13	6	1862	437
4	Others	250	305	1404	1714
-	All	304	331	9357	8669

The training set data were input into the parameterized RF classifier for training, and then the trained RF model was used to make predictions from the testing-set images. To evaluate the advantages of fused data and multi-temporal data, the time series of backscatter coefficients of Sentinel-1, the NDVI time series of Sentinel-2, and the NDVI images of single-temporal Sentinel-2 (on this day, all categories are easily distinguished) were also used to train the classifier and to make predictions with the same parameters.

2.4.5. Accuracy

Four schemes were used to evaluate the accuracy of the predictions. The most commonly used technique to assess the accuracy of crop classification involves the confusion matrix [51]. The confusion matrix compares the predicted images of the testing set with the labels of the testing set and produces the overall accuracy (OA%) and Kappa coefficient (K) (Equations (3) and (4)), which are the probability that the pixels are correctly classified and measure the consistency between the classification result and the actual result [48]. The OA was given by

$$OA = \frac{\sum_{i=1}^{n} p_{i,j}}{\sum_{j=1}^{n} \sum_{i=1}^{n} p_{i,j}}, \quad (3)$$

where $p_{i,j}$ is the total number of pixels belonging to category i and assigned to category j and n is the number of categories. The Kappa coefficient was given by

$$K = \frac{N^2 \times OA - \sum_{i=1}^{n} a_i b_i}{N^2 - \sum_{i=1}^{n} a_i b_i}, \quad (4)$$

where N is the total number of pixels, $a_1, a_2, \ldots, a_n$ are the numbers of real pixels in each type, and $b_1, b_2, \ldots, b_n$ are the numbers of pixels predicted for each type [52]. In addition, to fully verify the classification results obtained, this study compared the classification results of the fused data with the 2020 Jiaozuo City Statistical Yearbook data.

3. Results

3.1. Time Series Curve

Figure 6 shows the time-varying curves of VV and VH polarization backscatter coefficients for different categories after feature normalization. The pixel value of this backscatter coefficient is the average value of each class sample. It can be seen from Figure 6 that the pixel values changed with time; the pixel values of VV changed little, while the pixel values of VH changed greatly.

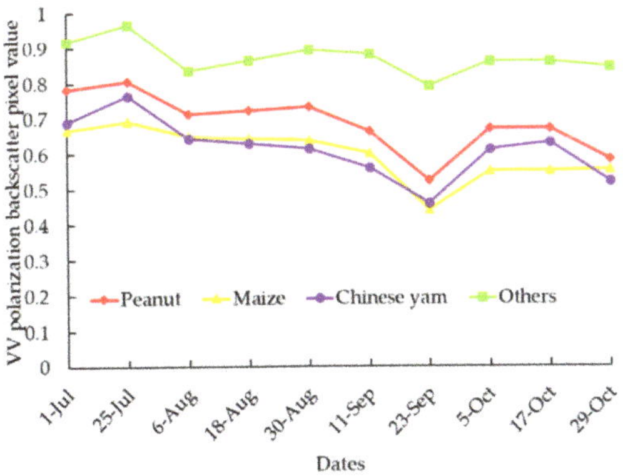

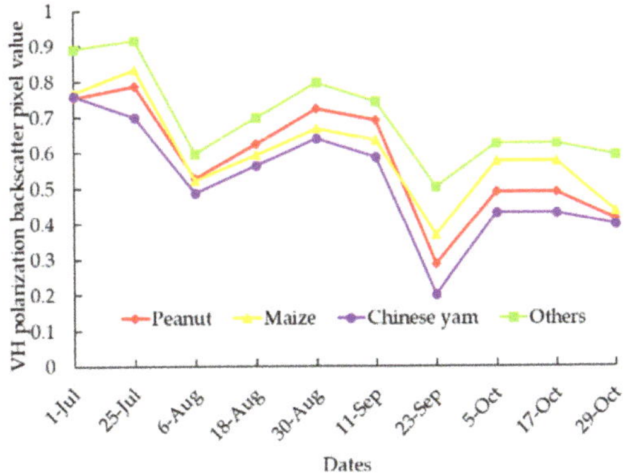

Figure 6. Temporal profiles of VV (**left**) and VH (**right**) polarization backscatter coefficients.

Figure 7 shows the NDVI pixel values after feature normalization and as a function of time for different categories. The NDVI pixel value is the average of the NDVI pixel values for all samples in that category. The results showed that the NDVI value of crops changed with time, which was consistent with the growth process of crops [11]. On August 5, the categories were easy to distinguished, so this study selected the Sentinel-2 data on this day as the single-temporal data.

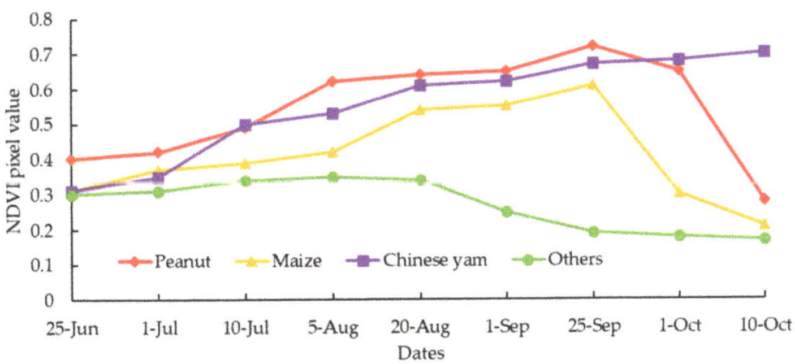

Figure 7. NDVI pixel values as a function of time.

3.2. Accuracy

The OAs and Kappa coefficients are given in Table 6. Compared with the 5 August 2020 Sentinel-2 data, the OA of the multi-temporal Sentinel-2 data increased by 6.3%, and the Kappa coefficient increased by 0.047. The fused multi-temporal Sentinel-1 and -2 data achieved the highest OA of 90.5% compared with multi-temporal Sentinel-2 (87.6%), 5 August 2020 Sentinel-2 (81.3%), and multi-temporal Sentinel-1 (73.4%).

Table 6. Results of accuracy assessment.

Data Category	Overall Accuracy (%)	Kappa Coefficient
5 August 2020 Sentinel-2	81.3	0.814
Multi-temporal Sentinel-1	73.4	0.732
Multi-temporal Sentinel-2	87.6	0.861
Fused multi-temporal Sentinel-1 and -2	90.5	0.881

For further comparison, the confusion matrix of the fused data is shown in Table 7. Maize achieved the highest producer's accuracy at 91.5%, followed by other land cover (89.6%), yams (88.6%), and peanuts (85.2%). In terms of user's accuracy, maize obtained the highest accuracy (96.4%), followed by other land cover (81.4%), peanuts (79.5%), and yams (75.1%).

Table 7. Confusion matrix of fused multi-temporal Sentinel-1 and -2 data. The value displayed for each class is the number of pixels.

		Truth Data				
		Maize	**Peanut**	**Chinese Yam**	**Others**	**PA (%)**
Classifier Results	Maize	5351	114	68	315	91.5
	Peanut	22	571	54	23	85.2
	Chinese yam	8	30	387	12	88.6
	Others	169	3	6	1536	89.6
	UA (%)	96.4	79.5	75.1	81.4	
		OA = 90.5%		Kappa = 0.881		

Abbreviations: User's accuracy (UA, %), producer's accuracy (PA, %), overall accuracy (OA, %), and Kappa coefficient (Kappa).

3.3. Comparison of Details of Prediction

Figure 8 visually compares different predictions and testing set labels. On the one hand, multi-temporal Sentinel-2 data produced less image noise than single-temporal Sentinel-2 data and greater internal uniformity of the parcels. On the other hand, the fusion of multi-temporal Sentinel-1 and -2 data reduced the image noise of multi-temporal Sentinel-2 data and improves the prediction accuracy of multi-temporal Sentinel-1 data. With multi-temporal Sentinel-1 data, the edges of autumn crop parcels are the main areas of poor prediction, with minor linear objects, such as paths and streams, being incorrectly predicted as autumn crops.

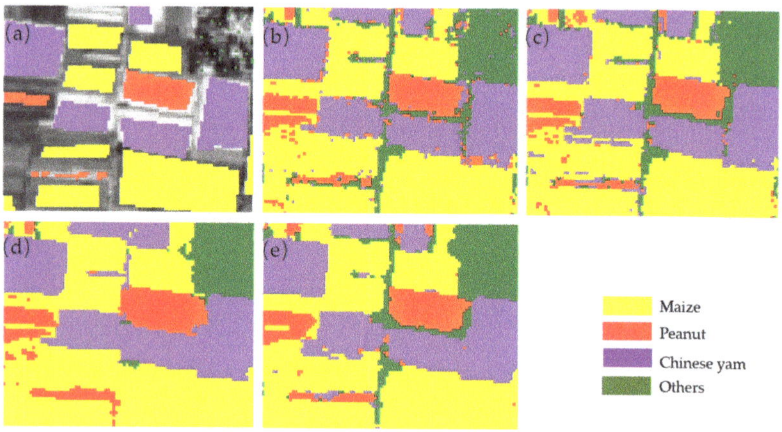

Figure 8. Comparison of different predicted results and partial testing set labels: (**a**) Partial testing set la-bels; (**b**) Single-temporal Sentinel-2 data; (**c**) Multi-temporal Sentinel-2 data; (**d**) Multi-temporal Sentinel-1 data; (**e**) Fusion of multi-temporal Sentinel-1 and -2 data.

3.4. Crop Mapping

Figure 9 shows the spatial distribution of autumn crops in the study area, as determined by the fusion of multi-temporal Sentinel-1 and -2 data. From a spatial point of view, peanut, maize, and Chinese yam are mainly distributed in areas outside the southeastern part of Wen County. In terms of area, maize covers the largest fraction of crop area (74.95%), followed by Chinese yam (15.27%) and peanut (9.48%).

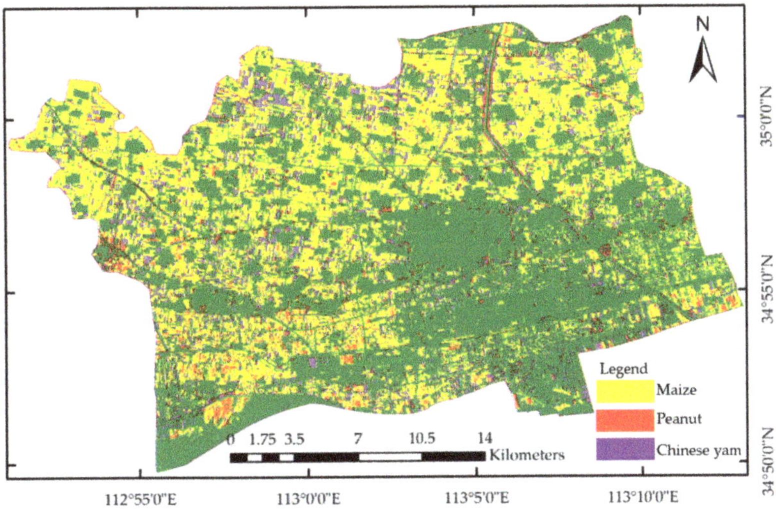

Figure 9. Crop classification map produced by fusion of multi-temporal Sentinel-1 and -2 data.

3.5. Comparison with Government Data

As shown in Table 8, the areas of peanut, maize, and Chinese yam reported by the government are all within the range predicted herein based on the fused data. Therefore, the classification results obtained herein correlate significantly with the 2020 Jiaozuo Statistical Yearbook data.

Table 8. Comparison of predicted area with statistical yearbook data.

Category	Predicted Area (hm^2)	Government Statistics Area (hm^2)
Peanut	3188 ± 335	3497
Maize	15,947 ± 1674	17,281
Chinese yam	4756 ± 499	5228

4. Discussion of Results

This study focuses on crop classification based on multi-temporal remote sensing data. The results showed that the use of the multi-temporal Sentinel-2 data significantly improved the classification accuracy. This result is attributed to the fact that crops have certain similarities during growth due to their physicochemical properties such as moisture and chlorophyll content and thereby have highly similar spectral characteristics [53]. This means that different crops may have the same spectral characteristics. The use of multi-temporal remote sensing imagery is thus crucial to clarify these ambiguities.

The time series of multi-temporal remote-sensing data characterized crops on a temporal basis, which provided a broader basis for identifying crop types, as demonstrated by the following four points: (1) The temporal dimension provided by the time-series data thus resolved the problem of the different crops having the same spectral characteristics or different spectral characteristics corresponding to the same crop, thus allowing the accurate, time-resolved determination of crops characteristics [54]. (2) Crop identification based on time-series data is not limited by seasons or crop phenology. By reconstructing or decomposing time-series curves, the effects of crop phenological changes in different periods can be eliminated and more general growth characteristics of crops can be explored, thereby eliminating the spurious variations caused by seasonal factors [55]. (3) The variations detected based on time-series data reflected multi-year changes in crops [56], which allows people to analyze how crops evolve over time. (4) The high temporal resolution of the time-series data allowed the accurate extraction of the temporal variations in crops.

The results show that the Sentinel-2 data lead to much greater accuracy than the Sentinel-1 data [57,58]. The NDVI of Sentinel-2 is an important index for crop area inversion and is widely used to extract crops and other vegetation [14,36]. For Sentinel-1 data, crops can only be extracted according to the change of backscattering coefficient, which is difficult to detect and distinguish crops.

The fusion of Sentinel-1 and -2 data is the key to this study. The results show that, compared with other data, the fused Sentinel-1 and -2 data provided the highest classification accuracy. This study used a feature-level fusion method to construct a multi-source remote-sensing fusion model that integrates multi-source features such as spectrum, time series, and backscatter coefficient. On the one hand, the combination of Sentinel-1 and -2 data provided more features for identifying crops. On the other hand, SAR data has strong anti-interference ability and is not affected by cloud cover. It can be acquired during the day or night under different weather conditions. Using Sentinel-1 data to supplement Sentinel-2 data solves the problem of cloud cover and insufficient time series data.

Some identification errors occurred in this study. A comparison of the classification details revealed that the pixels that should be classified as peanuts are instead classified as Chinese yams, and vice versa. Considering the growth phenology and vegetation indexes of the two crops, it can be found that their growth states are similar, which explains why it is difficult to distinguish Chinese yam from peanuts using remote sensing. However, the classification of maize is better because the phenology and VIs of maize differ from those of other crops. In addition, maize covers a large area so the samples are abundant, which is conducive to training an RF classifier.

The crop classification map and government data show the distribution of crops in Wen County. Maize is most widely distributed, much more so than Chinese yam or peanuts, because maize is the main food crop and has strong market demand, whereas peanuts and Chinese yam are local-specialty cash crops with small market demand. Wen is a typical agricultural county in China, so the good classification obtained in this work means that the method proposed herein can be transferred to other regions of China.

Future research should consider the following three aspects: (1) In feature selection, the multi-source remote sensing fusion model should be studied by integrating the spectral band, spectral index, texture feature, spectral index variation, and their combinations. (2) A variety of machine learning algorithms can be tested in comparative research. (3) Finally, to improve classification accuracy, multi-source remote-sensing fused data can be further enriched by, for example, airborne hyperspectral data, point cloud data, elevation data, and high-resolution data.

5. Conclusions

Given the population growth of recent decades, the rational use of land resources for crop planting has gained importance, and remote sensing provides an effective method to monitor crops. Radar satellites can be used to monitor the Earth's surface on cloudy and rainy days, and optical data can be used to obtain vegetation indices to monitor crops. Therefore, compared with traditional remote-sensing data, the combination of radar and optical data can improve the accuracy with which crops are classified. The multi-sensor developed by the Copernicus program offers the fusion of data to improve crop identification. In this study, multi-source features such as backscatter coefficient, VIs, and time series were extracted from Sentinel-1 and Sentinel-2 data, and these features were fused at the feature-level. By using a RF classifier, this study obtained a Kappa coefficient of 0.881 and an overall accuracy of 90.5%, which shows that the fusion of multi-temporal active and passive remote sensing data can improve the accuracy of crop classification.

Author Contributions: Conceptualization, Hebing Zhang and Hongyi Yuan; methodology, Hongyi Yuan; software, Hongyi Yuan; validation, Hongyi Yuan, Hebing Zhang, Xiaoxuan Lyu and Weibing Du; formal analysis, Hongyi Yuan; investigation, Hebing Zhang; resources, Hongyi Yuan; data curation, Hongyi Yuan; writing—original draft preparation, Hongyi Yuan; writing—review and editing, Hebing Zhang, Hongyi Yuan, Xiaoxuan Lyu and Weibing Du; visualization, Hongyi Yuan; supervision, Hebing Zhang; project administration, Hebing Zhang; funding acquisition, Hebing Zhang. All authors have read and agreed to the published version of the manuscript.

Funding: This research was funded by the National Natural Science Foundation of China (grant number U21A20108), the Ministry of Education Cooperation in Production and Education (grant number 202102245019), the Joint Fund of Collaborative Innovation Center of Geo-Information Technology for Smart Central Plains, Henan Province and Key Laboratory of Spatiotemporal Perception and Intelligent processing, Ministry of Natural Resources (grant number No.211102), the PI project of Collaborative Innovation Center of Geo-Information Technology for Smart Central Plains, Henan Province (grant number 2020C002), the Doctoral Fund of Henan Polytechnic University(grant number B2022-8), the Fundamental Research Funds for the Universities of Henan Province(grant number NSFRF220424) and the project of Provincial Key Technologies R & D Program of Henan (grant number 222102320306).

Institutional Review Board Statement: Not applicable.

Informed Consent Statement: Not applicable.

Data Availability Statement: Not applicable.

Acknowledgments: We thank Piesat Information Technology Co., Ltd. for providing classification codes and related remote sensing data (website: https://www.piesat.cn/, accessed on 16 April 2022).

Conflicts of Interest: The authors declare no conflict of interest.

References

1. Ayanlade, A.; Radeny, M. COVID-19 and food security in Sub-Saharan Africa: Implications of lockdown during agricultural planting seasons. *NPJ Sci. Food* **2020**, *4*, 13. [CrossRef] [PubMed]
2. Stark, J.C. Food production, human health and planet health amid COVID-19. *Explor.-J. Sci. Health* **2021**, *17*, 179–180. [CrossRef] [PubMed]
3. Wang, Y.; Peng, D.; Yu, L.; Zhang, Y.; Yin, J.; Zhou, L.; Zheng, S.; Wang, F.; Li, C. Monitoring Crop Growth During the Period of the Rapid Spread of COVID-19 in China by Remote Sensing. *IEEE J. Sel. Top. Appl. Earth Obs. Remote Sens.* **2020**, *13*, 6195–6205. [CrossRef] [PubMed]
4. Chen, J.; Yang, A. Intelligent Agriculture and Its Key Technologies Based on Internet of Things Architecture. *IEEE Access* **2019**, *7*, 77134–77141. [CrossRef]
5. Cheng, L.; Zhang, Y. Analysis of intelligent agricultural system and control mode based on fuzzy control and sensor network. *J. Intell. Fuzzy Syst.* **2019**, *37*, 6325–6336.
6. Tseng, F.-H.; Cho, H.-H.; Wu, H.-T. Applying Big Data for Intelligent Agriculture-Based Crop Selection Analysis. *IEEE Access* **2019**, *7*, 116965–116974. [CrossRef]
7. Han, C.; Zhang, B.; Chen, H.; Wei, Z.; Liu, Y. Spatially distributed crop model based on remote sensing. *Agric. Water Manag.* **2019**, *218*, 165–173. [CrossRef]
8. Lin, F.; Weng, Y.; Chen, H.; Zhuang, P. Intelligent greenhouse system based on remote sensing images and machine learning promotes the efficiency of agricultural economic growth. *Environ. Technol. Innov.* **2021**, *24*, 101758. [CrossRef]
9. Zhang, H.; Huang, Q.; Zhai, H.; Zhang, L. Multi-temporal cloud detection based on robust PCA for optical remote sensing imagery. *Comput. Electron. Agric.* **2021**, *188*, 106342. [CrossRef]
10. Murthy, C.S.; Raju, P.V.; Badrinath, K.V.S. Classification of wheat crop with multi-temporal images: Performance of maximum likelihood and artificial neural networks. *Int. J. Remote Sens.* **2003**, *24*, 4871–4890. [CrossRef]
11. Vuolo, F.; Neuwirth, M.; Immitzer, M.; Atzberger, C.; Ng, W.-T. How much does multi-temporal Sentinel-2 data improve crop type classification? *Int. J. Appl. Earth Obs. Geoinf.* **2018**, *72*, 122–130. [CrossRef]
12. Wang, H.; Lin, H.; Munroe, D.K.; Zhang, X.; Liu, P. Reconstructing rice phenology curves with frequency-based analysis and multi-temporal NDVI in double-cropping area in Jiangsu, China. *Front. Earth Sci.* **2016**, *10*, 292–302. [CrossRef]
13. Sonobe, R.; Yamaya, Y.; Tani, H.; Wang, X.F.; Kobayashi, N.; Mochizuki, K. Mapping crop cover using multi-temporal Landsat 8 OLI imagery. *Int. J. Remote Sens.* **2017**, *38*, 4348–4361. [CrossRef]
14. Yi, Z.; Jia, L.; Chen, Q. Crop Classification Using Multi-Temporal Sentinel-2 Data in the Shiyang River Basin of China. *Remote Sens.* **2020**, *12*, 4052. [CrossRef]
15. Zhou, Y.; Flynn, K.C.; Gowda, P.H.; Wagle, P.; Ma, S.; Kakani, V.G.; Steiner, J.L. The potential of active and passive remote sensing to detect frequent harvesting of alfalfa. *Int. J. Appl. Earth Obs. Geoinf.* **2021**, *104*, 102539. [CrossRef]

16. Vuolo, F.; Zoltak, M.; Pipitone, C.; Zappa, L.; Wenng, H.; Immitzer, M.; Weiss, M.; Baret, F.; Atzberger, C. Data Service Platform for Sentinel-2 Surface Reflectance and Value-Added Products: System Use and Examples. *Remote Sens.* **2016**, *8*, 938. [CrossRef]
17. Ramoelo, A.; Dzikiti, S.; Deventer, H.V.; Maherry, A.; Cho, M.A.; Gush, M. Potential to monitor plant stress using remote sensing tools. *J. Arid Environ.* **2015**, *113*, 134–144. [CrossRef]
18. Kattenborn, T.; Lopatin, J.; Förster, M.; Braun, A.C.; Fassnacht, F.E. UAV data as alternative to field sampling to map woody invasive species based on combined Sentinel-1 and Sentinel-2 data. *Remote Sens. Environ.* **2019**, *227*, 61–73. [CrossRef]
19. Ebel, P.; Xu, Y.; Schmitt, M.; Zhu, X.X. SEN12MS-CR-TS: A Remote-Sensing Data Set for Multimodal Multitemporal Cloud Removal. *IEEE Trans. Geosci. Remote Sens.* **2022**, *60*, 1–14. [CrossRef]
20. Park, N.-W.; Lee, H.; Chi, K. Feature Extraction and Fusion for Land-Cover Discrimination with Multi-Temporal SAR Data. *Korean J. Remote Sens.* **2005**, *21*, 145–162.
21. Bhogapurapu, N.; Dey, S.; Bhattacharya, A.; Mandal, D.; Lopez-Sanchez, J.M.; McNairn, H.; Lopez-Martinez, C.; Rao, Y.S. Dual-polarimetric descriptors from Sentinel-1 GRD SAR data for crop growth assessment. *ISPRS J. Photogramm. Remote Sens.* **2021**, *178*, 20–35. [CrossRef]
22. Clauss, K.; Ottinger, M.; Kuenzer, C. Mapping rice areas with Sentinel-1 time series and superpixel segmentation. *Int. J. Remote Sens.* **2018**, *39*, 1399–1420. [CrossRef]
23. Kussul, N.; Mykola, L.; Shelestov, A.; Skakun, S. Crop inventory at regional scale in Ukraine: Developing in season and end of season crop maps with multi-temporal optical and SAR satellite imagery. *Eur. J. Remote Sens.* **2018**, *51*, 627–636. [CrossRef]
24. Singha, M.; Dong, J.; Sarmah, S.; You, N.; Zhou, Y.; Zhang, G.; Doughty, R.; Xiao, X. Identifying floods and flood-affected paddy rice fields in Bangladesh based on Sentinel-1 imagery and Google Earth Engine. *ISPRS J. Photogramm. Remote Sens.* **2020**, *166*, 278–293. [CrossRef]
25. Subbarao, N.V.; Mani, J.K.; Shrivastava, A.; Srinivas, K.; Varghese, A.O. Acreage estimation of kharif rice crop using Sentinel-1 temporal SAR data. *Spat. Inf. Res.* **2021**, *29*, 495–505. [CrossRef]
26. Chong, L.U.; Liu, H.J.; Lu, L.P.; Liu, Z.R.; Kong, F.C.; Zhang, X.L. Monthly composites from Sentinel-1 and Sentinel-2 images for regional major crop mapping with Google Earth Engine. *J. Integr. Agric.* **2021**, *20*, 1944–1957.
27. Steinhausen, M.J.; Wagner, P.D.; Narasimhan, B.; Waske, B. Combining Sentinel-1 and Sentinel-2 data for improved land use and land cover mapping of monsoon regions. *Int. J. Appl. Earth Obs. Geoinf.* **2018**, *73*, 595–604. [CrossRef]
28. Veloso, A.; Mermoz, S.; Bouvet, A.; Thuy Le, T.; Planells, M.; Dejoux, J.-F.; Ceschia, E. Understanding the temporal behavior of crops using Sentinel-1 and Sentinel-2-like data for agricultural applications. *Remote Sens. Environ.* **2017**, *199*, 415–426. [CrossRef]
29. Chen, Y.J.; Tian, S.F. Feature-Level Fusion between Gaofen-5 and Sentinel-1A Data for Tea Plantation Mapping. *Forests* **2020**, *11*, 1357. [CrossRef]
30. Veerabhadraswamy, N.; Devagiri, G.M.; Khaple, A.K. Fusion of complementary information of SAR and optical data for forest cover mapping using random forest algorithm. *Curr. Sci.* **2021**, *120*, 193–199. [CrossRef]
31. Belgiu, M.; Dragut, L. Random forest in remote sensing: A review of applications and future directions. *Isprs J. Photogramm. Remote Sens.* **2016**, *114*, 24–31. [CrossRef]
32. Biedrzycki, J.; Burduk, R. Decision Tree Integration Using Dynamic Regions of Competence. *Entropy* **2020**, *22*, 1129. [CrossRef]
33. Wang, X.; Zhang, J.; Xun, L.; Wang, J.; Wu, Z.; Henchiri, M.; Zhang, S.; Zhang, S.; Bai, Y.; Yang, S.; et al. Evaluating the Effectiveness of Machine Learning and Deep Learning Models Combined Time-Series Satellite Data for Multiple Crop Types Classification over a Large-Scale Region. *Remote Sens.* **2022**, *14*, 2341. [CrossRef]
34. Li, Y.; Chang, C.; Wang, Z.; Li, T.; Li, J.; Zhao, G. Identification of Cultivated Land Quality Grade Using Fused Multi-Source Data and Multi-Temporal Crop Remote Sensing Information. *Remote Sens.* **2022**, *14*, 2109. [CrossRef]
35. Suresh, G.; Gehrke, R.; Wiatr, T.; Hovenbitzer, M. Synthetic aperture radar (SAR) based classifiers for land applications in Germany. *ISPRS—Int. Arch. Photogramm. Remote Sens. Spat. Inf. Sci.* **2016**, *41*, 1187. [CrossRef]
36. Zhang, H.K.; Roy, D.P.; Yan, L.; Li, Z.; Huang, H.; Vermote, E.; Skakun, S.; Roger, J.C. Characterization of Sentinel-2A and Landsat-8 top of atmosphere, surface, and nadir BRDF adjusted reflectance and NDVI differences. *Remote Sens. Environ.* **2018**, *215*, 482–494. [CrossRef]
37. Fan, J.; Huang, J.; Zhang, M. Retrieval of Cropping Index in China Using Time Series of SPOT Vegetation NDVI. *Sens. Lett.* **2013**, *11*, 1134–1140. [CrossRef]
38. Pan, Z.; Huang, J.; Zhou, Q.; Wang, L.; Cheng, Y.; Zhang, H.; Blackburn, G.A.; Yan, J.; Liu, J. Mapping crop phenology using NDVI time-series derived from HJ-1 A/B data. *Int. J. Appl. Earth Obs. Geoinf.* **2015**, *34*, 188–197. [CrossRef]
39. Wei, W.; Wu, W.; Li, Z.; Yang, P.; Zhou, Q. Selecting the Optimal NDVI Time-Series Reconstruction Technique for Crop Phenology Detection. *Intell. Autom. Soft Comput.* **2016**, *22*, 237–247. [CrossRef]
40. Zhao, Q.; Huett, C.; Lenz-Wiedemann, V.I.S.; Miao, Y.; Yuan, F.; Zhang, F.; Bareth, G. Georeferencing Multi-source Geospatial Data Using Multi-temporal TerraSAR-X Imagery: A Case Study in Qixing Farm, Northeast China. *Photogramm. Fernerkund. Geoinf.* **2015**, *2*, 173–185. [CrossRef]
41. Arkhipkin, O.P.; Sagatdinova, G.N. The application of optical and radar data fusion in space monitoring of water objects. *Sovrem. Probl. Distantsionnogo Zondirovaniya Zemli Iz Kosm.* **2020**, *17*, 91–102. [CrossRef]
42. Moskvitin, A.E.; Ushenkin, V.A. Fusion of radar and optical images from the Earth remote sensing systems. *Radiotekhnika* **2019**, *83*, 183–191.

43. Xu, Z.; Zhao, J.; Zhang, F.; Zhang, L.; Yang, T.; Li, Q.; Pan, S. Photonics-Based Radar-Lidar Integrated System for Multi-Sensor Fusion Applications. *IEEE Sens. J.* **2020**, *20*, 15068–15074. [CrossRef]
44. Zhou, Y.; Zhang, L.; Cao, Y.; Huang, Y. Optical-and-Radar Image Fusion for Dynamic Estimation of Spin Satellites. *IEEE Trans. Image Process.* **2020**, *29*, 2963–2976. [CrossRef]
45. Chu, H.; Ma, H.; Li, X. Pedestrian instance segmentation with prior structure of semantic parts. *Pattern Recognit. Lett.* **2021**, *149*, 9–16. [CrossRef]
46. Hafiz, A.M.; Bhat, G.M. A survey on instance segmentation: State of the art. *Int. J. Multimed. Inf. Retr.* **2020**, *9*, 171–189. [CrossRef]
47. Kang, B.; Nguyen, T.Q. Random Forest With Learned Representations for Semantic Segmentation. *IEEE Trans. Image Process.* **2019**, *28*, 3542–3555. [CrossRef]
48. Rodriguez-Galiano, V.F.; Ghimire, B.; Rogan, J.; Chica-Olmo, M.; Rigol-Sanchez, J.P. An assessment of the effectiveness of a random forest classifier for land-cover classification. *ISPRS J. Photogramm. Remote Sens.* **2012**, *67*, 93–104. [CrossRef]
49. Xu, J.; Zhu, Y.; Zhong, R.; Lin, Z.; Lin, T. DeepCropMapping: A multi-temporal deep learning approach with improved spatial generalizability for dynamic corn and soybean mapping. *Remote Sens. Environ.* **2020**, *247*, 111946. [CrossRef]
50. Wang, D.; Zhou, Q.-B.; Yang, P.; Chen, Z.-X. Design of a spatial sampling scheme considering the spatial autocorrelation of crop acreage included in the sampling units. *J. Integr. Agric.* **2018**, *17*, 2096–2106. [CrossRef]
51. Pena-Barragan, J.M.; Ngugi, M.K.; Plant, R.E.; Six, J. Object-based crop identification using multiple vegetation indices, textural features and crop phenology. *Remote Sens. Environ.* **2011**, *115*, 1301–1316. [CrossRef]
52. Yang, S.; Gu, L.; Li, X.; Jiang, T.; Ren, R. Crop Classification Method Based on Optimal Feature Selection and Hybrid CNN-RF Networks for Multi-Temporal Remote Sensing Imagery. *Remote Sens.* **2020**, *12*, 3119. [CrossRef]
53. Zarco-Tejada, P.J.; Rueda, C.A.; Ustin, S.L. Water content estimation in vegetation with MODIS reflectance data and model inversion methods. *Remote Sens. Environ.* **2003**, *85*, 109–124. [CrossRef]
54. Wardlow, B.D.; Egbert, S.L.; Kastens, J.H. Analysis of time-series MODIS 250 m vegetation index data for crop classification in the U.S. Central Great Plains. *Remote Sens. Environ.* **2007**, *108*, 290–310. [CrossRef]
55. Zhu, Z.; Woodcock, C.E.; Olofsson, P.; Zhu, Z.; Woodcock, C.E.; Olofsson, P. Continuous monitoring of forest disturbance using all available Landsat imagery. *Remote Sens. Environ.* **2012**, *122*, 75–91. [CrossRef]
56. Verbesselt, J.; Hyndman, R.; Zeileis, A.; Culvenor, D. Phenological change detection while accounting for abrupt and gradual trends in satellite image time series. *Remote Sens. Environ.* **2010**, *114*, 2970–2980. [CrossRef]
57. Meroni, M.; D'Andrimont, R.; Vrieling, A.; Fasbender, D.; Lemoine, G.; Rembold, F.; Seguini, L.; Verhegghen, A. Comparing land surface phenology of major European crops as derived from SAR and multispectral data of Sentinel-1 and-2. *Remote Sens. Environ.* **2021**, *253*, 112232. [CrossRef]
58. Valero, S.; Arnaud, L.; Planells, M.; Ceschia, E. Synergy of Sentinel-1 and Sentinel-2 Imagery for Early Seasonal Agricultural Crop Mapping. *Remote Sens.* **2021**, *13*, 4891. [CrossRef]

International Journal of *Geo-Information*

MDPI

Article

Sub-Surface Geotechnical Data Visualization of Inaccessible Sites Using GIS

Tariq Ahmed Awan [1,*], Muhammad Usman Arshid [2], Malik Sarmad Riaz [1], Moustafa Houda [3], Mirvat Abdallah [3], Muhammad Shahkar [4], Mirsina Mousavi Aghdam [5] and Marc Azab [3]

1 Civil Engineering Department - National University of Technology (NUTECH), Islamabad, Pakistan; sarmadriaz@nutech.edu.pk
2 University of Engineering and Technology, Taxila, Pakistan; usman.arshid@uettaxila.edu.pk
3 College of Engineering and Technology, American University of the Middle East, Kuwait; moustafa.houda@aum.edu.kw (M.H.); mirvat.abdallah@aum.edu.kw (M.A.); marc.azab@aum.edu.kw (M.A.)
4 Ghulam Ishaq Khan Institute of Engineering Sciences and Technology, Topi, Pakistan; m.shahkar@giki.edu.pk
5 Department of Civil and Environmental Engineering and Architecture, University of Cagliari, Via Marengo, 2, 09123 Cagliari, Italy; smousavi@gdgeo.com
* Correspondence: tariqahmed@nutech.edu.pk

Citation: Awan, T.A.; Arshid, M.U.; Riaz, M.S.; Houda, M.; Abdallah, M.; Shahkar, M.; Aghdam, M.M.; Azab, M. Sub-Surface Geotechnical Data Visualization of Inaccessible Sites Using GIS. *ISPRS Int. J. Geo-Inf.* **2022**, *11*, 368. https://doi.org/10.3390/ijgi11070368

Academic Editors: Wolfgang Kainz and Dimitris Triantakonstantis

Received: 18 April 2022
Accepted: 24 June 2022
Published: 29 June 2022

Publisher's Note: MDPI stays neutral with regard to jurisdictional claims in published maps and institutional affiliations.

Abstract: Geotechnical investigation, in hilly areas, for high-rise projects, becomes a problematic issue and costly process due to difficulties in mobilization and assembling the drilling equipment on mountainous terrains. The objective of this study is to map soil properties of study areas, especially at inaccessible sites, for reconnaissance. Digital soil maps for Tehsil Murree, Pakistan, have been developed using the emerging Geographical Information System (GIS). The research work involved the creation of an exhaustive database, by collecting and rectifying geotechnical data, followed by the digitization of the acquired data through integration with GIS, in an attempt to visualize, analyze and interpret the collected geotechnical information spatially. The soil data of 205 explanatory holes were collected from the available geotechnical investigation (GI) reports. The collection depth of soil samples, which were initially used for the design of deep and shallow foundations by different soil consultancies in the Murree area, was approximately 50 ft. below ground level. Appropriate spatial interpolation methods (i.e., the Kriging) were applied for the preparation of smooth surface maps of soil standard penetration tester number values, soil type and plasticity index. The accuracy of developed SPT N value and plasticity maps were then evaluated using the linear regression method, in which the predicted values of soil characteristics from developed maps and actual values were compared. SPT N value maps were developed up to a depth of 9.14 m below ground level and at every 1.52 m interval. The depth of refusal was considered in the developed maps. Soil type and plasticity maps were generated up to 15.24 m depth, again at every 1.52 m intervals, using color contours, considering the maximum predicted foundation depth for high-rise projects. The study has implications for academics and practitioners to map the soil properties for inaccessible sites using GIS, as the resulting maps have high accuracy.

Keywords: Geographical Information System (GIS); spatial interpolation; SPT N; plasticity index; soil maps

1. Introduction

The characterization of sub-surface conditions and properties is one of the most essential activities required for the successful planning, design, and construction of engineering structures. Risks related to improper ground conditions commonly represent the greatest concern in many infrastructure projects [1]. Most of these risks are directly, or indirectly, affected by the quantity and quality of subsurface investigations [1,2]. Therefore, in order to eliminate these risks, it is of great importance to increase the density of ground investigation data and improve the site sub-surface characterization to reduce the probability of encountering unexpected soil conditions during the construction phase and the likelihood of an excessive over-design which often leads to increased costs and delays. Sufficient site investigations also increase the reliability of estimates of characteristic soil and rock properties and geotechnical design parameters which, in turn, can reduce the likelihood of performance failure and/or allow designers to achieve the target reliability for a lesser cost [3]. The application of geotechnical data is not limited to structures and civil engineering projects. It is widely used in other disciplines, including infrastructure, environmental and risk analysis [4].

In most cases, geotechnical data are displayed and managed using traditional and ineffective methods, unlike geographic and geomorphologic data that are available in spatial data infrastructures (SDIs). As a result, geotechnical data are not readily available on a decent spatial resolution for many developed areas.

Subsurface soil data exploration can be done by geophysical methods, such as Ground-penetrating radar (GPR) and SPT. However, geotechnical investigation, especially in hilly areas for high-rise projects, becomes a problematic issue and costly process, due to the difficulties in mobilization and assembling the drilling equipment on mountainous terrains. For this purpose, soil mapping can be done with readily available geotechnical data using GIS for inaccessible sites, such as the top of a hill or the bottom of a valley. However, the data explored by these conventional techniques are often unorganized, heterogeneous, and not properly archived. Geographic Information System (GIS) technology can be used as an efficient tool to store, organize, analyze and display this type of data in the form of digital layers and maps [5]. In geotechnical practice, GIS can be used in at least four ways: data integration, data visualization and analysis, planning and summarizing site activities, and data presentation [5]. GIS provides promising tools for integrating multiple data types, such as raster format data (e.g., photos and scanned maps), vector format data (e.g., Shapefiles or computer-aided drafting (CAD) files) and it may consist of open-access information, such as soil surveys, topographic maps and borehole log data [6]. In addition, the exploration of subsurface soil is often an expensive task and geotechnical engineers deal with uncertainty by interpolating data from limited sources [7]. The problems associated with this poor data management have been recognized by many engineering agencies and organizations, and have prompted researchers to explore recent advancements in information technology, such as Web GIS platforms, in order to utilize them for exchanging geotechnical data. These data have to be collected, validated and stored within a unified spatial database to establish stratigraphy and pertinent properties of the soil at a specific site [8].

GIS is well-known for its capacity to integrate the capabilities of spatial analysis, database management and graphic visualization. The detailed geotechnical investigations can be pared down to only sensitive projects with access to preliminary data. By paring down the most suitable options, based on examples of existing surrounding structures, early soil type judgments can help plan risk mitigation strategies and improve design capabilities. These geotechnical soil type maps can be used to guide future infrastructure development. Examples of applications of GIS in the geotechnical field can be soil survey planning to soil management in addition to hydrologic response unit generation. Orhan and Tosun developed maps for Turkey to check the feasibility of foundations in residential areas, However, at site-specific scale, where local geological features alter, they may be less useful. [9]. Local authorities in South Africa, Palestine, Turkey and Thailand have all developed geotechnical mapping and they continue to update and improve their databases, but all these studies are restricted to shallow depth [10–13]. As another example, geotechnical characteristics were assessed using spatial modeling of geotechnical parameters by developing 3D models for the development of the Lavrion Technological [14]. A study was conducted to develop interpolated soil maps for Najaf, Iraq, using Kriging and Ordinary Kriging. The study is limited to soil USCS classification of Najaf, Iraq. [15]. ArcGIS was also used to map the organic composition of the soil in Najaf city [16]. For Baghdad, a GIS-based database of soil parameters has been created [17]. Another study was conducted for Wasit city, to map shear strength, consolidation parameters, and groundwater level to aid in the preliminary design phase, but this study is also limited to a depth of 10 m [18].

Since soil properties are highly variable spatially [19], and in order to provide an accurate estimate of characteristic values, continuous variability should be considered. The traditional method of soil analysis and interpretation is laborious, and time-consuming, hence, becoming expensive [20].

Geostatistical techniques (Kriging) are widely recognized as an important spatial interpolation technique in land resource inventories [21,22]. Several research studies have been conducted focusing on different interpolation methods of soil characteristics [23–26]. For example, the study done by (Srivastava et al. 2019) revealed that, for soil moisture data, kriging with external drift (KED) is performing better than ordinary kriging (OK), inverse distance weighting (IDW) and Spline methods, in terms of model accuracy and performance using cross-validation techniques. However, the performance of spatial interpolation techniques depends on several parameters, such as sample size, sampling design, sample spatial distribution, data (density and variation), surface type, data quality, input data uncertainty, and poor choice or implementation techniques. Thus, it is difficult to assign spatial interpolation methods as the "best practices" based on a given dataset without a prior investigation of their accuracy [27]. In our study, the KED interpolation method was adopted in order to develop continuous geotechnical spatial maps for the studied area. Kriging was the interpolation technique used, in which a linear combination of observations from close sampled sites were used to produce predictions at unsampled sites of Tehsil Murree, Pakistan. Kriging addresses several of the limitations of standard interpolation techniques used in GIS.

The aim of this study was to fill a gap by providing GIS soil characterization spatial maps for the area of Tehsil Murree in Pakistan. In fact, for a hilly area such as Tehsil Murree, it was not easy to collect the soil samples from the top of the mountain and difficult to set up the soil sampling equipment. To address this, spatial interpolation of the available geotechnical data was used to develop continuous soil maps. The created maps were used to show the variation in soil type, SPT N values, and Plasticity index.

The contribution of this study is to add to the geotechnical database of the country by making soil maps of one of Teshil Muree, which is the most developed tourist site of Pakistan. Geotechnical mapping of inaccessible sites is done using GIS so that these maps can be used in reconnaissance survey and table study for projected development in the future.

2. Materials and Methods

2.1. Study Area

The Tehsil Murree is one of the renowned hill stations of the Rawalpindi district with coordinates of 33.9078 N and 73.3915 E at an elevation of about 7500 feet. The Digital Elevation Model (DEM) of the study area was used to determine the elevation required at any given location. Figure 1a shows the DEM of the study region, whereas Figure 1b shows the study area's location.

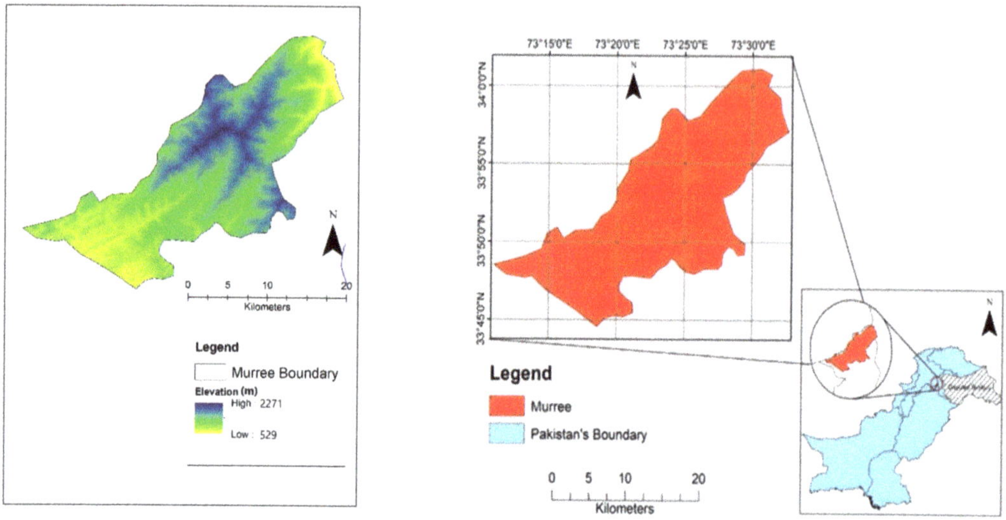

(**a**) (**b**)

Figure 1. (**a**) Digital elevation model of the study area (**b**) Map showing the location of the study area in Pakistan.

The neighboring districts of the study area are Haripur and Abbottabad. Both of these districts are located in Khyber Pakhtunkhwa, on the northern and western sides, and the Jhelum River on the eastern side. Kotli Satian, on the other hand, is found on the southern side of Murree. The soil formation of Muree Tehsil is characterized by lower Miocene sandstone and claystone. The sandstone is often reddish grey to purple-grey in color, thick-bedded, fine to medium-grained, micaceous, cross-bedded, and calcareous [24]. The location of the study area was in the sub-Himalayan mountainous region. According to the Ministry of Housing and Works of Pakistan, Murree was classified to be in earthquake zone 3, which falls in the class of moderate hazard. The hill slopes in the Murree area are rather steep, ranging from 17° to 30°. As the slope gets steeper, the shear force increases, which may result in landslides. One of the main reasons that may cause landslides in Muree Tehsil is rapid development, such as the construction of road networks and immature geology with disturbed slopes [25]. Therefore, land sliding is a common occurrence, especially with steep slopes, such as along highways, streams and disturbed slopes.

2.2. Methodology

Geotechnical reports of 112 sampling sites for deep and shallow foundation designs from four soil investigation firms in the Muree Tehsil area were collected. The available data were used to develop soil maps for the study area covering a maximum depth of 15.24 m below ground level. The methodology adopted consisted of six steps as shown in Figure 2.

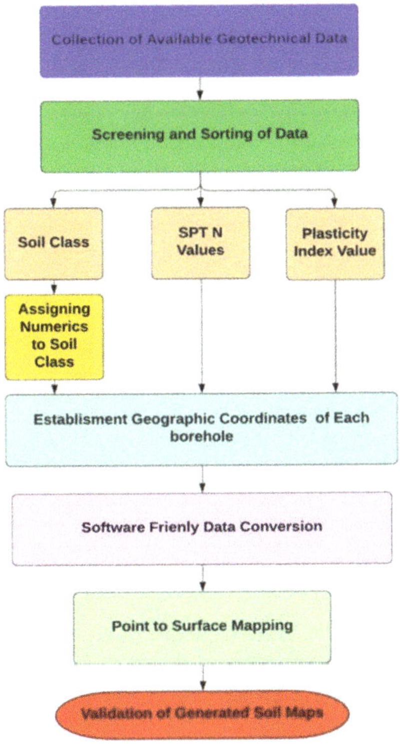

Figure 2. Flowchart of methodology.

From collected geotechnical data soil classification, borehole coordinates, plasticity index and SPT N values were extracted at 1.52 m intervals and up to the level of 15.24 m below ground level. The classification of the soil type was performed, based on the USCS soil classification system [28]. Plasticity Index of soil samples collected from a borehole was determined by Atterberg's limit test [29].

The extracted soil classification data was used to develop soil type maps by assigning an ascending wise numeric value to each soil class according to their gradation size, as listed in Table 1.

Table 1. Soil sample classes and assigned values to be used in the GIS environment.

Soil Class	Symbol	Assigned Values	Soil Class	Symbol	Assigned Values
Heavy Clay	MH	1	Silty Clayey Gravel	GC-GM	9
Clay	CL	2	Silty Gravel	GM	10
Silty Clay	CL-ML	3	Well Graded Gravel	GW	11
Silt	ML	4	Poorly Graded Silty Gravel	GP-GM	12
Clayey Sand	SC	5	Well Graded Silty Gravel	GW-GM	13
Clayey Silty Sand	SC-SM	6	Shale		14
Silty Sand	SM	7	Sandstone		15
Clayey Gravel	GC	8			

Regarding soil investigation reports in which borehole coordinates were not given, the coordinates of these boreholes were established by Google Earth by identifying the site on which the soil investigation was held. The extracted data, such as soil classification, SPT N value and plasticity index, were transformed into MS excel to make the data compatible with the ArcGIS format. The goal was to develop 1.52 m. interval soil maps using the retrieved data for Tehsil Murree. For this purpose, ArcGIS 10.3 was used to develop continuous interpolated maps. Ordinary Kriging interpolation technique was used to develop the continuous maps for soil classification, SPT N and plasticity index data. Kriging is a multiphase procedure that comprises interpretive statistical data analysis, variogram modeling, surface creation, and (optionally) variance surface exploration. Kriging creates an approximated surface from a scattering of z-valued points. Kriging is based on the assumption that the direction or distance between sample points implies a spatial correlation that may be utilized to explain surface variation using Equation (1).

$$Z(x.) = \sum_{i=1}^{N} \lambda_i z(x_i) \quad (1)$$

Kriging interpolation can reliably be used for interpolation of geotechnical databases into continuous maps [30]. The validation of developed soil maps was done at 52 points using 8 boreholes in which generated values of maps were compared with the actual value of soil investigation reports that were not used in initial map making. The regression method, using MATLAB, was used for determining the accuracy of SPT N maps.

3. Results

The geotechnical data collection points and location of the study area are shown in Figure 3.

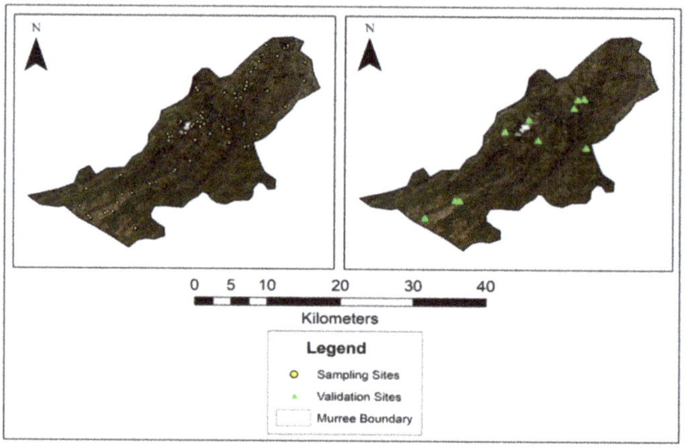

Figure 3. Sampling sites map and validation sites map (locations of the boreholes).

In this study, interpolated maps of SPT N values were elaborated for Tehsil Murree at an interval of 1.52 m to a depth up to 9.14 m below ground level. Since the study area was hilly, there was mostly refusal after 9.14 m depth. The SPT N was represented by the color range in the legends and the red color represents the refusal. Figure 4 shows the SPT N value maps. It can be noticed that with an increase in depth, N values increased throughout the region. Refusal in the southern region was evident throughout the depth. At shallow depth, the N value was low because of the presence of peat and clayey soil and at a greater depth the N value increased because of the presence of Murree formation, which consists of consolidated sandstone and foothills of Margalla lie [26].

The accuracy of interpolated N value maps was determined by 52 points obtained from field investigations which were compared with predicted SPT N values at the same point in the study area. These additional 52 points were not used in the generation of SPT N value maps. Figure 5 depicts a statistical comparison of created and measured cell values.

Regression analysis was done in which the R^2 coefficient of determination was calculated. The acquired results by regression analysis demonstrated that the kriging approach maps the geotechnical parameters, and that these values were comparable to the original data. The coefficient of determination (R^2) was 0.8233. The coefficient of correlation being greater than 0.8 indicated a strong correlation between the two variables [31].

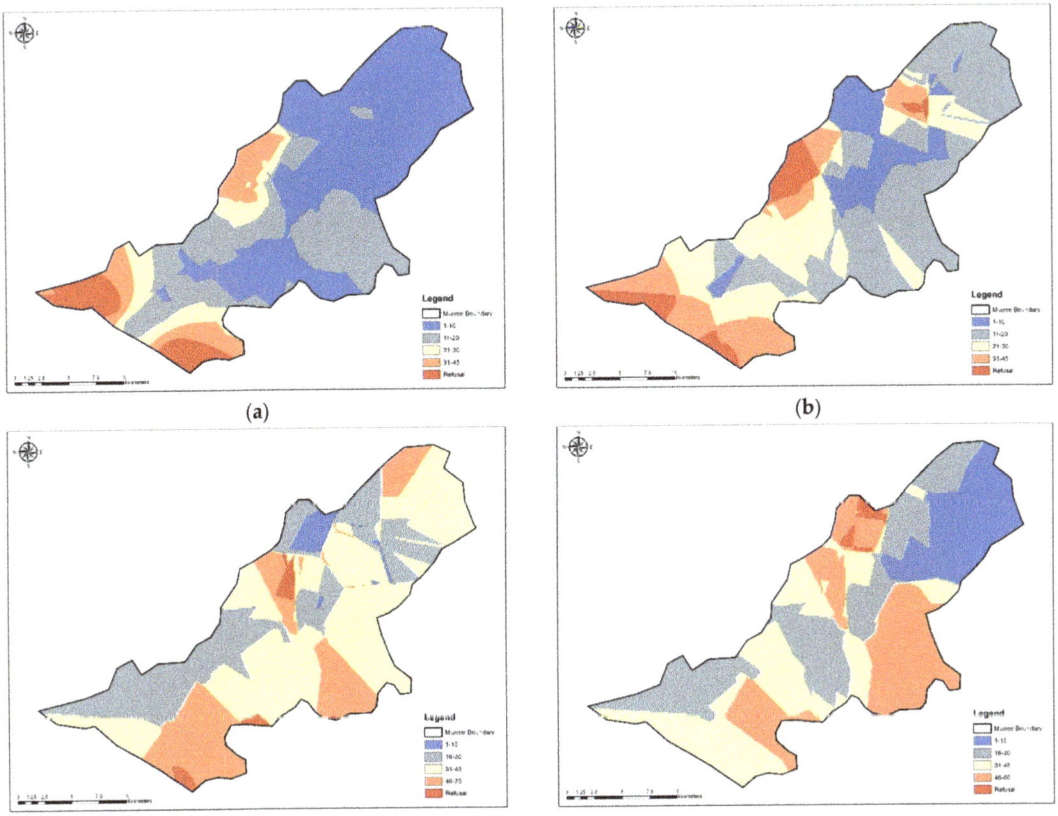

(**a**) (**b**) (**c**) (**d**)

Figure 4. *Cont.*

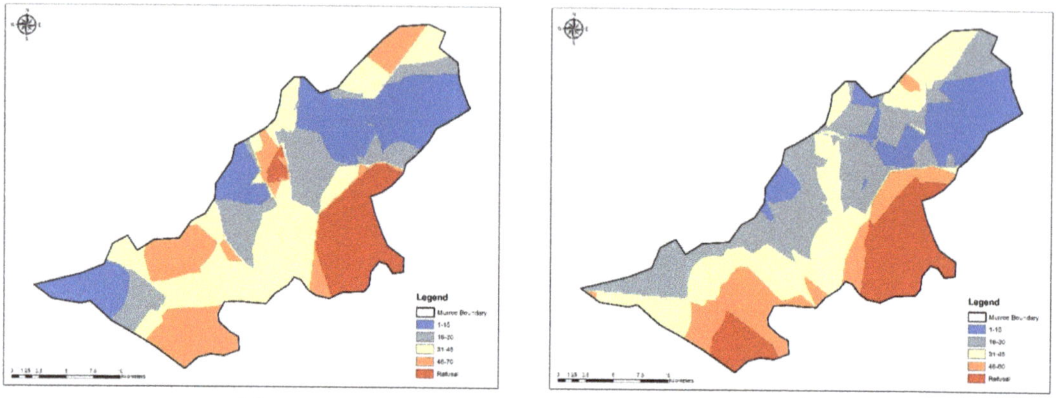

(e) (f)

Figure 4. SPT N maps at (**a**) 1.52 m depth, (**b**) 3.05 m depth, (**c**) 4.57 m depth, (**d**) 6.10 m depth, (**e**) 7.62 m depth and (**f**) 15.2 mdepth.

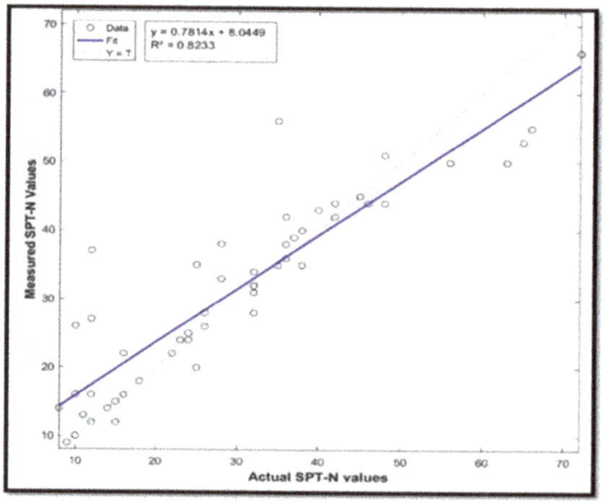

Figure 5. Validation of generated soil maps.

The plasticity Index is the range of moisture content in which soil remains in a plastic state while in transition from a semi-solid state to a liquid state. The plasticity index indicates the degree of plasticity of the soil. Thematic maps were used to show the variation of plasticity index up to 50 ft. depth of soil strata at different locations of Tehsil Murree. Interpolated maps of the plasticity index were developed at the interval of 5 ft. The variation in plasticity index is represented by the color range in legends, as shown in Figure 5. Interpolated maps show that the majority of the areas of Murree represent medium-range plastic soils, due to the presence of clay and silty clay.

For the preparation of soil type maps, soil in the bore sites was classified using the Unified soil classification system. The soil types were digitized according to the gradation size listed in Table 1. At the shallow depths, a higher density and detailed maps were available because there was a large amount of data available at shallow depths. Also, most of the residential buildings' bore logs were limited to a shallow depth which was almost 6.1–9.14 m depth. At shallow depth, the types of soil found were mainly MH, CL, CL-ML, ML, SC, SC-SM, SM, GC, GC-GM, GM, GW-GM, Shale and Sandstone. However, the dominant soil types were mostly CL, CL-ML, ML, and GC and, in some areas, MH,

SC, SC-SM, shale, and sandstone were also found. As the depth increased the variation in soil types decreased. The major reason for this was that as the depth increased mostly uniform soil was found and the other reason was that a lesser amount of data was available at greater depth. The data acquired at greater depth were for large-scale projects, such as high-rise buildings.

The soils found at greater depths of the study area were usually clayey gravels, shale, and sandstone. However, clay and silty clay were also found in some specific areas. The following maps were developed by using the Kriging interpolation technique at intervals of 1.52 m up to 15.24 m depth, showing the identified soil types.

The generated soil maps could be used for feasibility studies at the preliminary investigation level, and rough cost estimation, especially at the inaccessible site. The accuracy of generated soil maps was checked by validating the SPT N value maps. The technique was used to assess the validity of the created maps for the predicted data by collecting samples from ten field survey sites and comparing the actual test outcomes in relation to the developed maps. Figure 5 depicts a statistical comparison of generated and measured cell values. The acquired results indicated that the maps developed by the Kriging technique were able to predict the geotechnical parameters accurately and that these values were comparable to the original data. Using generated maps, the cost of ground investigations for the development projects could be reduced. This research aimed to create soil maps, especially for inaccessible sites, and ready-to-use maps that bring useful ground information for geotechnical engineering projects. It is recommended to prepare similar maps for additional geotechnical properties, such as uniaxial compressive strength, effective cohesion and the angle of internal friction using the accurate interpolation techniques.

4. Discussions

In this study, the significance of GIS in geotechnical engineering was shown, specifically in the field of soil mapping. SPT N value has a direct relation with allowable bearing capacity. So, the higher the SPT N value, the higher will be the bearing capacity. The generated SPT N maps depicted the lower SPT N value in the northeastern area which indicated lower bearing capacity in that area than the remaining part of the study area. Similarly, in soil type maps the presence of clay gravels, gravels, shale and sandstone indicated harder strata, as there was the refusal of SPT. Figure 6 shows the generated maps of P.I which are linked with soil type maps of Figure 7. The plasticity index was mostly determined by the quantity of clay present. It signified the soil's fineness and ability to change form without changing volume. A high PI implied that the soil contained too much clay or colloids. So, for the areas where fine soil was present, the observed value of P.I was higher and vice versa. When the PL was more than, or equal to, the LL, it had no value. So, for the areas where fine soil was present, the observed value of P.I was higher and vice versa. GIS software was used to analyze the study work, demonstrating the potential of GIS software in the geotechnical engineering area. The primary goal of this study was to ascertain the scope of GIS and its adaptability to map the soil properties at inaccessible sites. The produced maps to estimate the soil characteristics were easy to understand as they were built upon basic statistical principles. The end-user community of these maps could be engineers, employers and even researchers who are interested in exploring the economic and deliberate value of GIS, due to the benefits of this rapidly growing technology. Further to the economic aspects, the other fundamental advantages of GIS to produce such geotechnical maps can be generalized as (1) easy to visualize the spatial information by GIS output maps with clear legends and different groups of coloring and patterns (2) GIS can be applied to site selection, zoning, planning, and conservation measures (3) any other available data, like geophysical parameters on the subsurface conditions, can be incorporated with spatial data to validate the prediction and increase the precision of maps (4) it is possible to update/modify and refine datasets, data models, and the relation between attributes to produce better future maps if needed. For example, this study did not include data about the southeastern part of the Tehsil Murree, as there is no development in this area. However, the maps could be updated in the future when more sampling data becomes available. In the current research effort, we only used the ordinary Kriging interpolation method. Future research could compare the accuracy of different Kriging methods in the ArcGIS software platform.

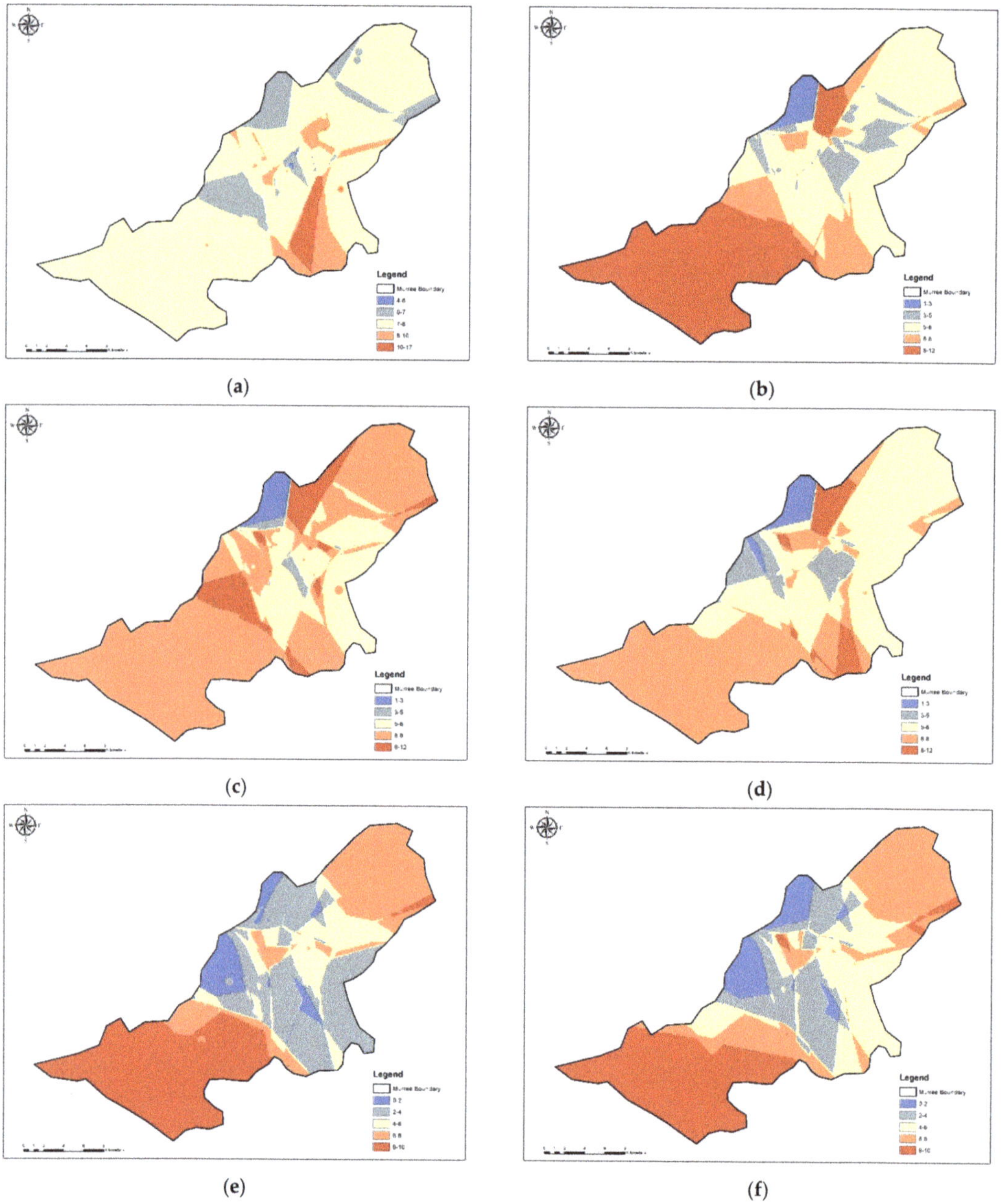

Figure 6. *Cont.*

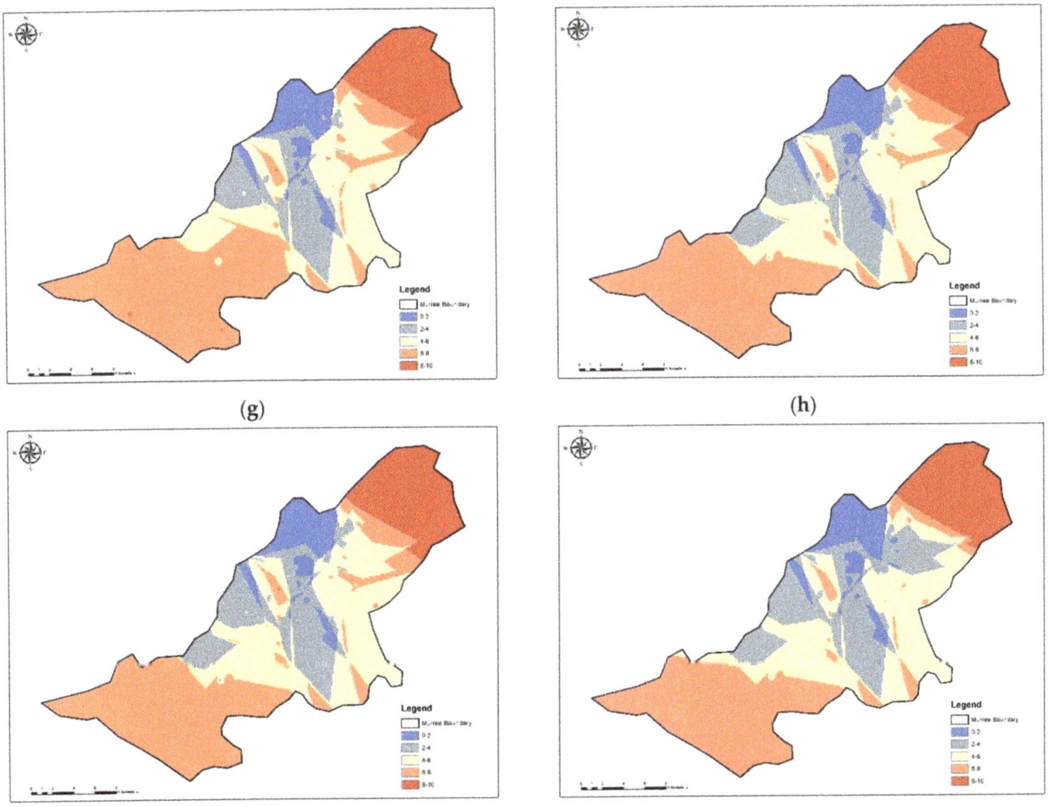

(**g**) (**h**) (**i**) (**j**)

Figure 6. Plasticity maps at (**a**) 1.52 m depth, (**b**) 3.05 m depth, (**c**) 4.57 m depth, (**d**) 6.10 m depth, (**e**) 7.62 m depth and (**f**) 9.14 m depth, (**g**) 10.67 m depth, (**h**) 12.19 m depth and (**i**) 13.72 m depth, (**j**) 15.24 m depth.

Although the use of GIS-based methods for the prediction of geotechnical parameters can generally be beneficial, there are disadvantages that might increase the risk of over- or under-estimation of the strength parameters of the soil. Accuracy and precision, as the main challenging points, are the functions of the scale at which a map (paper or digital) is created. Inaccuracies may result from mistakes of many sorts, for example, regression models cannot work properly if the input data has errors (that is poor quality data). If the data preprocessing is not performed well to remove missing values or redundant data, outliers or imbalanced data distribution, the validity of the regression model suffers. Regression models are susceptible to collinear and nonlinearity problems and as the number of variables increases, the reliability of the regression models decreases. To assure the workability and prediction power of the regression models, several checks should be required. Some diagnostic tests and techniques make it possible to examine the data and evaluate the model assumptions and investigate whether or not there are observations with a large, undue influence on the analysis. Testing the assumptions of linear regression was not included in this research. However, it would be beneficial to test the four principal assumptions of linear regression, which would justify the use of linear regression models for purposes of prediction of geotechnical parameters in future research: (i) linearity and additivity of the relationship between dependent and independent variables, (ii) statistical independence of the errors, (iii) homoscedasticity (constant variance) of the errors and (iv) normality of the error distribution).

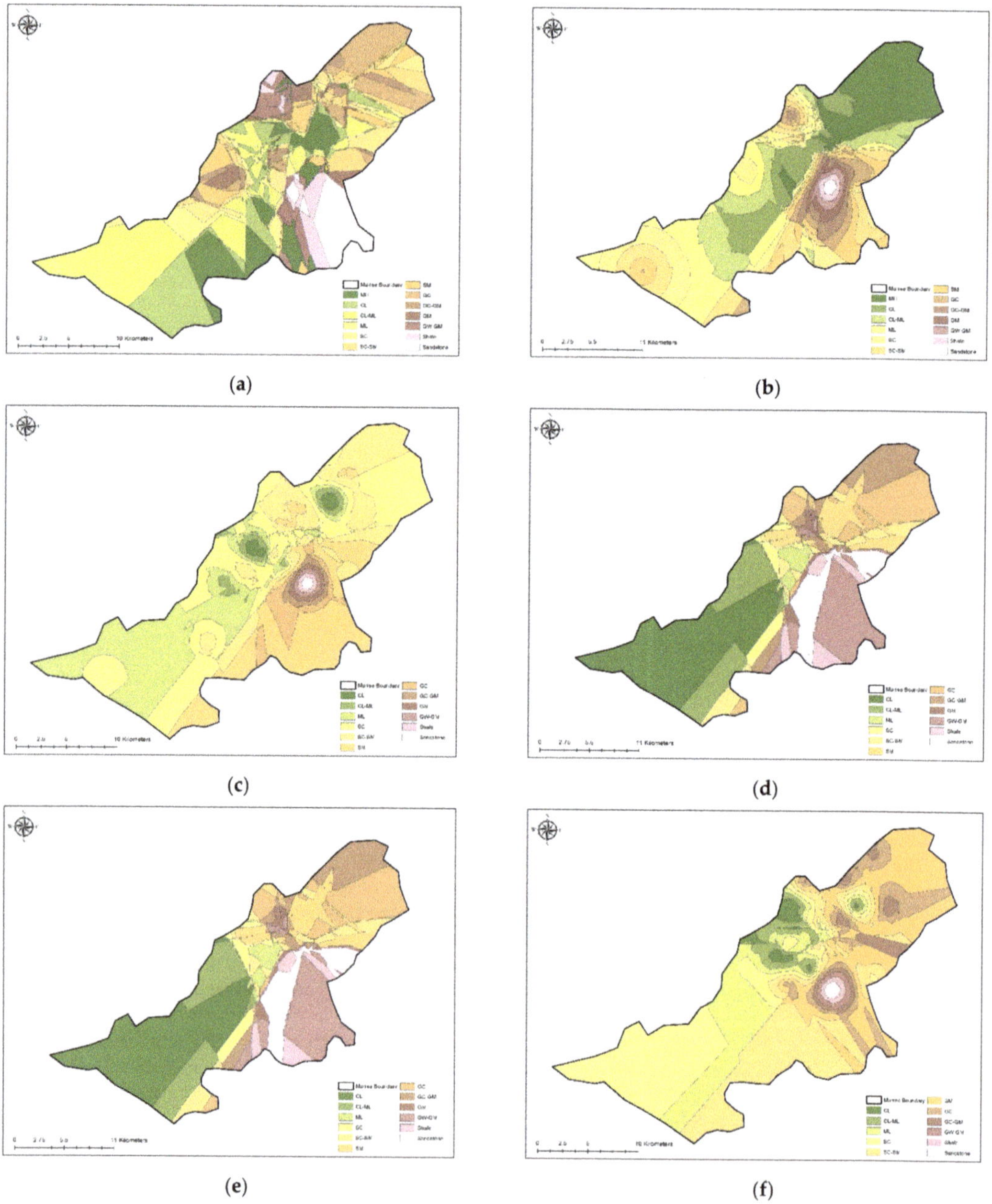

Figure 7. *Cont.*

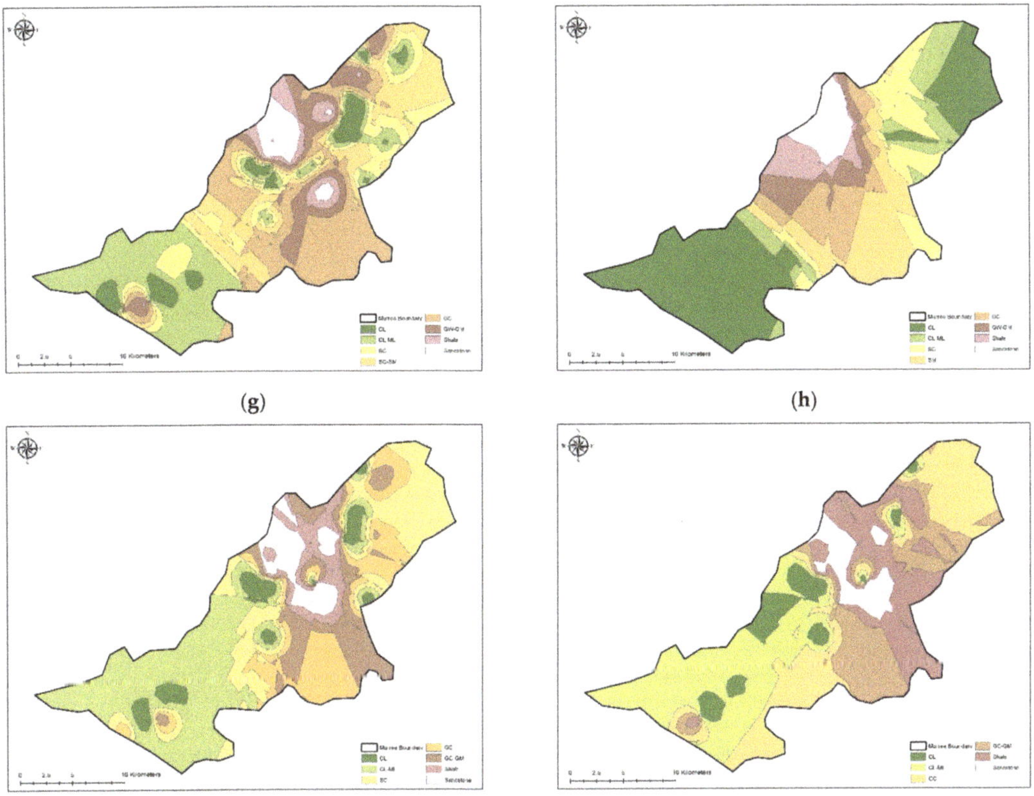

(g) (h) (i) (j)

Figure 7. Plasticity maps at (**a**) 1.52 m depth, (**b**) 3.05 m depth, (**c**) 4.57 m depth, (**d**) 6.10 m depth, (**e**) 7.62 m depth and (**f**) 9.14 m depth, (**g**) 10.68 m depth, (**h**) 12.19 m depth and (**i**) 13.72 m depth, (**j**) 15.24 depth.

Empirical Bayesian Kriging (EBK) is an alternative method that could be used to achieve a straightforward and robust method of data interpolation and prediction. EBK automatically configures the construction process of Kriging to help reduce the uncertainties. The method allows modeling of the relationship between geotechnical properties and co-variables and to capture the nonstationary aspects of these parameters. The biggest advantage is that EBK takes into consideration the semi-variogram error. This leads to the best spatial distribution performance among the most common interpolation techniques. It is recommended to run an EBK analysis and compare the results with the estimated spatial distribution of SPTs over the study area. It is also very important for the end-users to know the uncertainty of the modeled spatial distribution of geotechnical maps in order to determine the reliability of maps that might be used in different applications. An uncertainty map is inherently produced in the spatial distribution generated by EBK.

5. Conclusions

Based on the previous use of GIS in geotechnical engineering, this paper proposed a promising method to predict soil properties for inaccessible sites using GIS. The resulted interpolated soil maps showed that geotechnical properties of soil, such as SPT N values, plasticity index, and soil types, could be mapped accurately using GIS. This would save a considerable amount of money spent on initial ground investigation steps.

The SPT N values, plasticity index and soil types were mapped using the Kriging interpolation technique with ArcGIS 10.3. The software was sourced from University of Tartu, Estonia. The SPT N value-interpolated maps for Tehsil showed that the N values were directly related to the depth. As the depth increased the larger were the N values observed. Usually, at the greater depths, the bedrock was found to consist of shale and sandstone. Similarly, interpolated plasticity index maps showed the range of plasticity index of subsurface soil, which represented medium range plastic soils with a degree of plasticity at shallow depth. The map of interpolated soil types exhibited the variation of soil strata for depths up to 15.24 m below ground level.

Using these maps, soil characteristics for inaccessible sites could be estimated. The developed soil maps are important for preliminary design and hazard identification, such as landslides, especially at the preliminary investigation level. The SPT N value maps and plasticity index maps play an important role in the determination of the bearing capacity of foundations. The SPT N values can also be used to evaluate soil strength and designers can use these maps for the selection of suitable types of foundation and depths of excavation. These developed soil maps represent the continuous and smooth surface of geotechnical parameter values using the Kriging technique, which reduces the probability of negligence of geotechnical hazards.

Author Contributions: Conceptaulization, Tariq Ahmed Awan, Muhammad Usman Arshid and Malik Sarmad Riaz; methodology, Tariq Ahmed Awan, Muhammad Usman Arshid and Muhammad Shahkar; validation, Malik Sarmad Riaz; data collection, Tariq Ahmed Awan, Muhammad Usman Arshid and Muhammad Shahkar; data curation Tariq Ahmed Awan, Muhammad Usman Arshid; writing—original draft prepration, Tariq Ahmed Awan, Malik Sarmad Riaz; writing—reviewing and editing, Malik Sarmad Riaz, Moustafa Houda, Mirvat Abdallah, Mirsina Mousavi Aghdam and Marc Azab; visualization Tariq Ahmed Awan, Muhammad Usman Arshid, Malik Sarmad Riaz and Muhammad Shahkar; supervision Moustafa Houda, Mirvat Abdallah, Mirsina Mousavi Aghdam and Marc Azab; project administration, Moustafa Houda, Mirvat Abdallah, Mirsina Mousavi Aghdam and Marc Azab. All authors have read and agreed to the published version of the manuscript.

Funding: This research received no external funding.

Institutional Review Board Statement: Not applicable.

Informed Consent Statement: Not applicable.

Data Availability Statement: Data will be available on demand.

Acknowledgments: The authors thank Engr Abdullah Tauqeer (University of Tartu) for helping with the ArcGIS software provision to the authors.

Conflicts of Interest: The authors declare no conflict of interest.

References

1. Clayton, C.R. *Managing Geotechnical Risk: Improving Productivity in UK Building and Construction*; Thomas Telford: Southampton, UK, 2001.
2. Hoek, E.; Palmieri, A. Geotechnical Risks on Large Civil Engineering Projects. 1998. Available online: https://www.rocscience.com/assets/resources/learning/hoek/Geotechnical-Risks-on-Large-Civil-Engineering-Projects-1998.pdf (accessed on 21 January 2022).
3. Loehr, J.E.; Ding, D.; Likos, W.J. Effect of Number of Soil Strength Measurements on Reliability of Spread Footing Designs. *Transp. Res. Rec.* **2015**, *2511*, 37–44. [CrossRef]
4. Zand, A.G. *Enabling Geotechnical Data for Broader Use by the Spatial Data Infrastructures*; University of Southern California: Los Angeles, CA, USA, 2011.
5. Player, R.S.V. *Geographic Information System (GIS) Use in Geotechnical Engineering*; American Society of Civil Engineers: Reston, VA, USA, 2006.
6. Singh, A.; Noor, S.; Chitra, R.; Gupta, M. Applications of GIS in Geotechnical Engineering: Some Case Studies. *Int. J. Sci. Eng. Sci.* **2018**, *2*, 34–38.
7. Bardet, J.-P.; Zand, A. Spatial Modeling of Geotechnical Information Using GML. *Trans. GIS* **2009**, *13*, 125–165. [CrossRef]
8. Dodagoudar, G.R. An integrated geotechnical database and GIS for 3D subsurface modelling: Application to Chennai City, India. *Appl. Geomat.* **2018**, *10*, 47–64.

9. Orhan, A.; Tosun, H. Visualization of geotechnical data by means of geographic information system: A case study in Eskisehir city (NW Turkey). *Environ. Earth Sci.* **2010**, *61*, 455–465. [CrossRef]
10. Jardaneh, I. Geotechnical Map for the City of Nablus-Palestine. *An-Najah Univ. J. Res. A (Nat. Sci.)* **2007**, *21*, 201–219.
11. Kolat, Ç.; Doyuran, V.; Ayday, C.; Süzen, M.L. Preparation of a geotechnical microzonation model using Geographical Information Systems based on Multicriteria Decision Analysis. *Eng. Geol.* **2006**, *87*, 241–255. [CrossRef]
12. Suwanwiwattana, P.; Chantawarangul, K.; Mairaing, W.; Apaphant, P. *The Development of Geotechnical Database of Bangkok Subsoil Using GRASS-GIS*; Citeseer: Princeton, NJ, USA, 2001.
13. Kleinhans, I. *A Critical Appraisal of Regional Geotechnical Mapping in South Africa*; University of Pretoria: Pretoria, South Africa, 2006.
14. Sotiropoulos, N.; Benardos, A.; Mavrikos, A. Spatial modelling for the assessment of geotechnical parameters. *Procedia Eng.* **2016**, *165*, 334–342. [CrossRef]
15. Al-Mamoori, S.K.; Al-Maliki, L.A.J.; Al-Sulttani, A.; El-Tawil, K.; Hussain, H.M.; Al-Ansari, N. Horizontal and vertical geotechnical variations of soils according to USCS classification for the city of An-Najaf, Iraq using GIS. *Geotech. Geol. Eng.* **2020**, *38*, 1919–1938. [CrossRef]
16. Al-Mamoori, S.K.; Al-Maliki, L.A.; Hussain, H.M.; Al-Ali, M.J. Distribution of sulfate content and organic matter in AN-najaf and al-kufa cities'soil using gis. *Kufa J. Eng.* **2018**, *9*, 92–111. [CrossRef]
17. Ahmed, B.A.; Yahya, A.Y.; Mohammed, A.H. Database for Baghdad Soil Using GIS Techniques. *J. Eng.* **2012**, *18*, 1307–1324.
18. Aldefae, A.H.; Mohammed, J.; Saleem, H.D. Digital maps of mechanical geotechnical parameters using GIS. *Cogent Eng.* **2020**, *7*, 1779563. [CrossRef]
19. Burrough, P.A. Soil variability: A late 20th century view. *Soils Fert.* **1993**, *56*, 529–562.
20. Shit, P.K.; Bhunia, G.S.; Maiti, R. Spatial analysis of soil properties using GIS based geostatistics models. *Modeling Earth Syst. Environ.* **2016**, *2*, 107. [CrossRef]
21. Bhunia, G.S.; Shit, P.K.; Maiti, R. Comparison of GIS-based interpolation methods for spatial distribution of soil organic carbon (SOC). *J. Saudi Soc. Agric. Sci.* **2018**, *17*, 114–126. [CrossRef]
22. Hengl, T.; Rossiter, D.G.; Stein, A. Soil sampling strategies for spatial prediction by correlation with auxiliary maps. *Soil Res.* **2003**, *41*, 1403–1422. [CrossRef]
23. Arshid, M.U.; Shabbir, F.; Hussain, J.; Elahi, A. Assessment of variation in soil parameters, for design of lightly loaded structural foundations. *Life Sci. J.* **2013**, *10*, 217–220.
24. Critelli, S.; Garzanti, E. Provenance of the lower Tertiary Murree redbeds (Hazara-Kashmir Syntaxis, Pakistan) and initial rising of the Himalayas. *Sediment. Geol.* **1994**, *89*, 265–284. [CrossRef]
25. Khan, A.N.; Collins, A.E.; Qazi, F. Causes and extent of environmental impacts of landslide hazard in the Himalayan region: A case study of Murree, Pakistan. *Nat. Hazards* **2011**, *57*, 413–434.
26. Williams, V.S.; Pasha, M.K.; Sheikh, I.M. Geologic Map of the Islamabad-Rawalpindi Area, Punjab, Northern Pakistan. The Survey. 1999. Available online: https://pubs.usgs.gov/of/1999/0047/report.pdf (accessed on 10 January 2022).
27. Yao, X.; Fu, B.; Lü, Y.; Sun, F.; Wang, S.; Liu, M. Comparison of four spatial interpolation methods for estimating soil moisture in a complex terrain catchment. *PloS ONE* **2013**, *8*, e54660. [CrossRef]
28. *Standard, A.J.A.I. D2487-17*; Standard Practice for Classification of Soils for Engineering Purposes (Unified Soil Classification System). ASTM: West Conshohocken, PA, USA, 2018.
29. *4318–00, (2003)*; Standard Test Methods for Liquid Limit, Plastic Limit, and Plasticity Index of Soils. American Society For Testing: West Conshohocken, PA, USA, 2018; Volume 4, pp. 582–595.
30. Arshid, M.U.; Kamal, M.A. Regional geotechnical mapping employing kriging on electronic geodatabase. *Appl. Sci.* **2020**, *10*, 7625. [CrossRef]
31. Smith, M. *Neural Networks for Statistical Modeling*; Thomson Learning: New York, NY, USA, 1993.

International Journal of
Geo-Information

Article

Fine Crop Classification Based on UAV Hyperspectral Images and Random Forest

Zhihua Wang [1], Zhan Zhao [2,*] and Chenglong Yin [3]

1 College of Geomatics and Geoinformation, Guilin University of Technology, Guilin 541006, China; 1020211836@glut.edu.cn
2 School of Surveying and Geo-Informatics, Shandong Jianzhu University, Jinan 250101, China
3 School of Geographic Sciences, East China Normal University, Shanghai 200241, China; 52213901009@stu.ecnu.edu.cn
* Correspondence: zhaozhan19@sdjzu.edu.cn

Abstract: The classification of unmanned aerial vehicle hyperspectral images is of great significance in agricultural monitoring. This paper studied a fine classification method for crops based on feature transform combined with random forest (RF). Aiming at the problem of a large number of spectra and a large amount of calculation, three feature transform methods for dimensionality reduction, minimum noise fraction (MNF), independent component analysis (ICA), and principal component analysis (PCA), were studied. Then, RF was used to finely classify a variety of crops in hyperspectral images. The results showed: (1) The MNF–RF combination was the best ideal classification combination in this study. The best classification accuracies of the MNF–RF random sample set in the Longkou and Honghu areas were 97.18% and 80.43%, respectively; compared with the original image, the RF classification accuracy was improved by 6.43% and 8.81%, respectively. (2) For this study, the overall classification accuracy of RF in the two regions was positively correlated with the number of random sample points. (3) The image after feature transform was less affected by the number of sample points than the original image. The MNF transform curve of the overall RF classification accuracy in the two regions varied with the number of random sample points but was the smoothest and least affected by the number of sample points, followed by the PCA transform and ICA transform curves. The overall classification accuracies of MNF–RF in the Longkou and Honghu areas did not exceed 0.50% and 3.25%, respectively, with the fluctuation of the number of sample points. This research can provide reference for the fine classification of crops based on UAV-borne hyperspectral images.

Keywords: unmanned aerial vehicle hyperspectral image; fine classification of crops; characteristic transform; random forest

Citation: Wang, Z.; Zhao, Z.; Yin, C. Fine Crop Classification Based on UAV Hyperspectral Images and Random Forest. *ISPRS Int. J. Geo-Inf.* **2022**, *11*, 252. https://doi.org/10.3390/ijgi11040252

Academic Editors: Dimitris Triantakonstantis, Panagiotis Tziachris and Wolfgang Kainz

Received: 24 February 2022
Accepted: 7 April 2022
Published: 12 April 2022

1. Introduction

The classification of crops supported by remote sensing technology is one of the core means of realizing the promotion of agricultural development [1]. Compared with spaceborne and manned aviation hyperspectral systems, the UAV-borne hyperspectral remote sensing system can simultaneously obtain remote sensing images with nanometer-level hyperspectral resolution and centimeter-level high spatial resolution, which is more suitable for the current situation of the Chinese agricultural structure [2–5]. The continuous development of hyperspectral remote sensing technology has promoted the development of agricultural science and technology [6]. UAV hyperspectral imagery has a wide range of applications in crop yield estimation, single species classification, and disease monitoring, but many studies have focused on hyperspectral images of low spatial resolution [7–10].

To date, a classification algorithm based on deep learning has been used to classify crops [11,12], but the deep learning method has high requirements on the number of samples. Under the conditions of a small image area and a small number of samples, the application of the deep learning method in the fine classification of crops is limited to a certain extent [13]. Compared with deep learning algorithms, random forests have the characteristics of high efficiency, fewer samples required, and high classification accuracy, and their applications in environmental monitoring, land use, etc. have achieved impressive results [14–17]. Hongyan Yang et al. used the random forest classification algorithm to classify crops. The obtained classification accuracy was high, but the research categories were few, so it was difficult to determine the performance of the random forest classification algorithm in the case of more complex categories and feature information [18].

Hyperspectral images have many bands and a large amount of data, and large information redundancy affects the efficiency of classification operations [19,20]. Redundant bands also have an impact on classification accuracy when the number of classification samples is small [13]. Therefore, dimensionality reduction aiming at the problems of a large number of spectra and the large amount of calculation using feature transform is very important. Zahra Dabiri et al. systematically performed a comparison of independent component analysis (ICA), principal component analysis (PCA), and minimum noise fraction transformation (MNF) for classification of six tree species using APEX hyperspectral imagery [21]. However, there have yet been few targeted studies on the classification of various crops by different transforms.

Therefore, this study used high-spatial resolution hyperspectral image data from a UAV and adopted three dimensionality reduction transformation methods, MNF, ICA, and PCA. The random forest (RF) classification algorithm was then used to classify all kinds of crops in Honghu City, Hubei Province. The purposes of this study were as follows: first, to explore the optimal classification combination and study the effect of different feature transforms on the classification accuracy of crop hyperspectral images; second, to study the stability of the classification accuracy of different classification combinations in the face of different random sample points. It was expected that this research would achieve ideal results in the above aspects and provide a reference for the application of UAV-borne hyperspectral image crop fine classification.

2. Study Area and Data Source

The two regions selected for this study are both located in Honghu City, Jingzhou City, Hubei Province, China. Honghu City is located at the southeastern end of the Jianghan Plain, spanning from 113°07′ to 114°05′ east of the Greenwich Meridian and 29°39′ to 30°12′ north of the Equator. It has a subtropical humid monsoon climate. The crop planting scale is small, and land fragmentation is common, in the two study areas [22]. The location of the study area is shown in Figure 1.

The original dataset used in this study is the UAV-borne Hyperspectral Image (WHU-HI) dataset, which contains two separate UAV-borne hyperspectral datasets: WHU-HI-Long Kou and WHU-HI-HongHu [23,24]. All datasets were obtained in the agricultural areas of different crop types in Honghu City through the Headwall Nano-Hyperspec sensor installed on the UAV platform, and all preprocessing processes, such as radiometric calibration and geometric correction, were completed in the HyperSpec software. The image sizes of the two datasets were 254.650 m × 185.200 m and 40.420 m × 20.425 m, respectively; there were 270 bands between 400 and 1000 nm. The spatial resolutions of the two sets of UAV hyperspectral images were 0.463 m and 0.043 m, respectively. The Longkou and Honghu areas each contained 9 types of ground objects and 22 types of ground objects, among which crops accounted for 6 types and 18 types, respectively. To avoid the interference of other ground objects in crop classification, the classification of other ground objects was carried out at the same time. A real land type survey of the study area is shown in Figure 2.

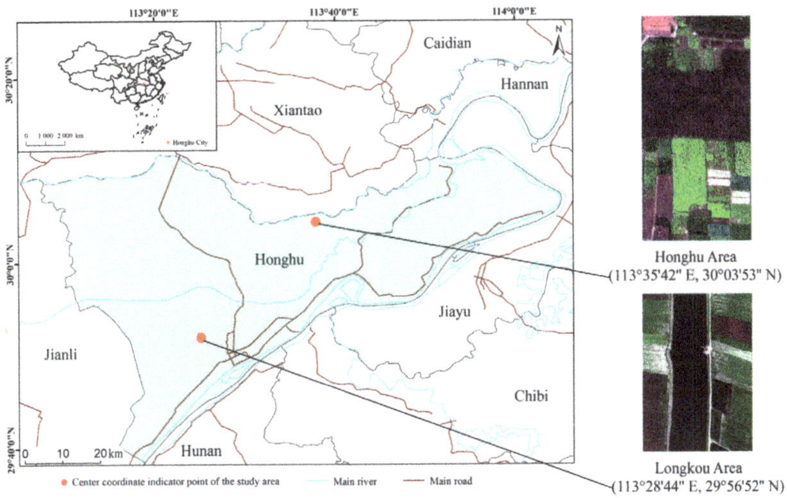

Figure 1. Location of the study area.

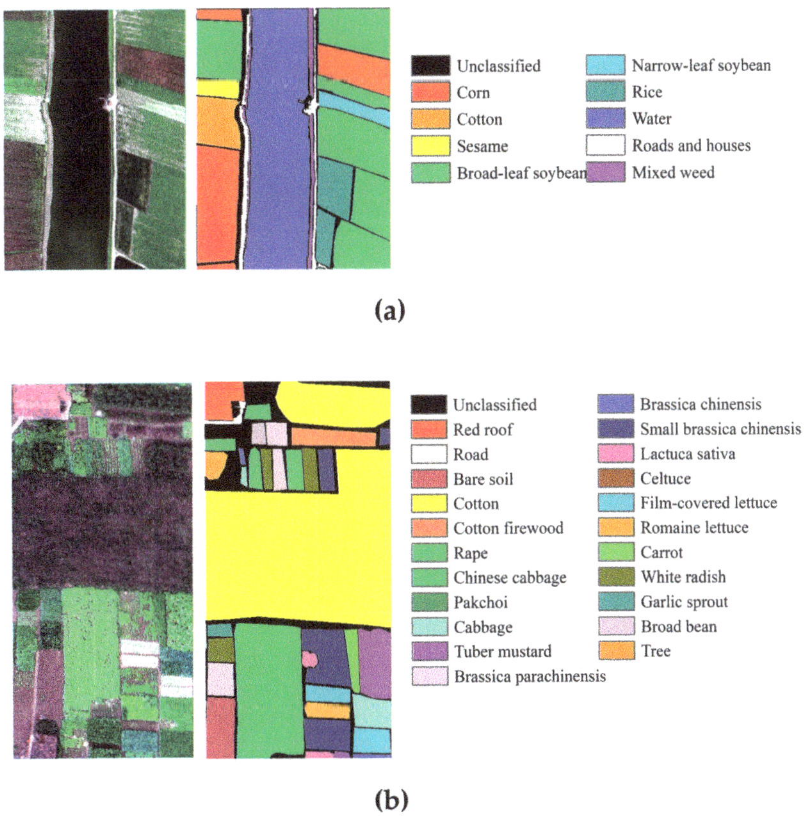

Figure 2. (**a**) Survey map of real land types in the Longkou area; (**b**) survey map of real land types in the Honghu area.

3. Research Methods

The main research route of this study is as Figure 3: MNF, PCA, and ICA feature transforms were carried out using WHU-HI data sets, and then random sample sets were selected using the results of the three feature transforms. Seven random sample sets of Train25–Train300 were selected, and then random forest classification was carried out. Finally, accuracy evaluation and stability comparison were carried out on the classification results.

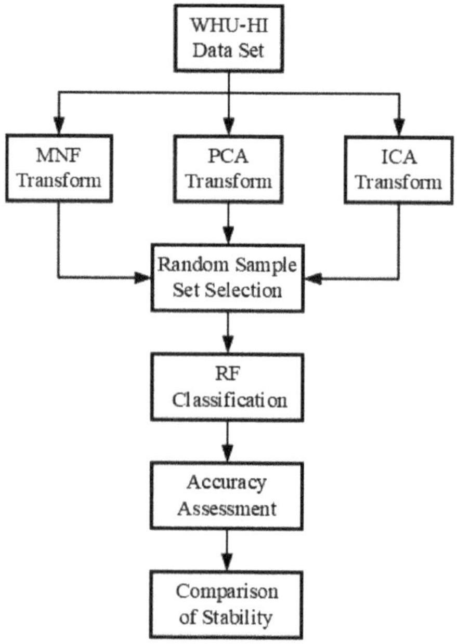

Figure 3. Flow chart of study.

3.1. Sample Acquisition Strategy and Accuracy Assessment Method

Samples can provide numerical feature information about categories for the supervised classification process, and the number and distribution of random sample points can have a great impact on the classification results. Therefore, sample selection is a key step in the subsequent classification process. The sample acquisition strategy of this study was to select a certain number of random points—that is, randomly select different numbers of random sample points, increasing from 25 to 300 for each type of ground object (25, 50, 100, 150, 200, 250, 300)—to generate seven random sample sets (referred to as Train25–Train300). In this study, a confusion matrix was used to evaluate the accuracy of the classification results, and the accuracy of each classification result was presented in the confusion matrix to compare the classification results with the true ground information.

3.2. Feature Transform

In this study, feature transform was performed first, and most of the feature information of the original image was concentrated into a few bands, thereby significantly reducing the processing time of the subsequent classification process [25]. This study compared the effects of different feature transforms, including the PCA transform, MNF transform, and ICA transform. The basic principle of the PCA transform is to concentrate the main information in the image data into a few bands to filter out redundant information so that the subsequent classification processing can be carried out smoothly [26]. MNF transform is a linear transform created by superimposing two PCA transforms [27]. The MNF transform adjusts and distinguishes noise through the covariance matrix in the PCA transform and then transforms the noise-whitening data. Considering the influence of noise, a better transform effect can be achieved when there is noise in the data [28–30]. The ICA transform can convert hyperspectral data into independent components to better complete the discovery and separation of hidden noise in the image, dimensionality reduction, noise reduction, anomaly detection, classification and spectral endmember extraction, and data fusion [31].

In the eigenvalue curve corresponding to the transformed image, as the number of eigenvalues increases, each eigenvalue gradually decreases until it converges to a certain constant. At this time, the transformed image reaches the best state, that which contains as much information as possible under the number of eigenvalues. Because of this, the band images after the eigenvalues corresponding to the convergent constants can be screened out to avoid too much redundant band information affecting the subsequent classification process.

MNF, PCA, and ICA transforms were carried out on the original data. Combined with the display status of band information after feature transform, the effective feature information was further screened, and the best ideal transform image was explored. To ensure the accuracy of the ideal band selection work, first, the derivative analysis method was used to screen the curve. When the derivative of the curve was 0, its eigenvalues converged to a stable value, and the corresponding number of bands at this time was the preliminarily determined ideal band. The number was set to N here. Second, verification work was carried out, and the image information at the (N + 1) band was presented separately for verification. If the corresponding information amount was small, it was finally determined that the N band was ideal for the transformed band.

3.3. Principle of the RF Classifier

The RF classifier, first proposed by Breiman, is a combined classifier obtained by taking the decision tree as the basic unit and performing ensemble learning through multiple combinations [32]. In the application of remote sensing image classification and change detection, it has obtained better experimental results than other classification algorithms because of its advantages of high accuracy, few parameters, and strong robustness [33].

The basic principle is to extract K sample subsets from the training sample set by the Bootstrap method. The value of K usually takes the number of sample subsets corresponding to when the out of bag error begins to converge to obtain the model with the best classification accuracy and generalization ability. The number of random features for node splitting m is usually taken as $\sqrt{\mathrm{M}}$ (M is the dimension of the feature vector). According to the value of m, a decision tree model is established for the K sample subsets. By classifying all the samples to be classified, the classification result sequence $\{h_1(x), h_2(x), \ldots, h_k(x)\}$ is obtained. The final classification result is determined by majority voting [32,34]. The basic formula of the final classification decision process is as follows:

$$\mathrm{H(x)} = \mathrm{arg}max_Y \sum_{i=1}^{K} I(h_i(x) = Y) \tag{1}$$

In the formula, H(x) represents the final classification decision of the random forest, $h_i(x)$ represents the classification result of a single decision tree model, Y represents the output variable (target variable), and $I(.)$ represents the indicative function.

The basic steps of RF algorithm classification in this study were as follows:

(1) First, m feature variables were selected from all features using the square root method.
(2) Second, m characteristic variables were used to establish a decision tree for the random sample (Train25–Train300).
(3) The above two steps were repeated K times, that is, K decision trees were generated to form a random forest.
(4) Finally, using the decision results of each decision tree, the category of final prediction was determined through the voting method.

In the actual classification process using an RF classifier, the number of trees N and the number of randomly selected feature variables M are the main factors determining the classification accuracy [35,36]. Through multiple parameter debugging and comparative analysis of classification experiments, the number of decision trees was determined to be 100, and the number of randomly selected feature variables was the square root of the number of bands of each image to be classified.

4. Results and Analysis

4.1. Feature Transform

Figure 4 shows the characteristic curve of the MNF transform. The derivative values of the curves in the Longkou and Honghu areas were 0 when the number of eigenvalues is 18. The corresponding eigenvalues were 2.45 and 1.52, respectively, and the eigenvalue changes tended to be stable.

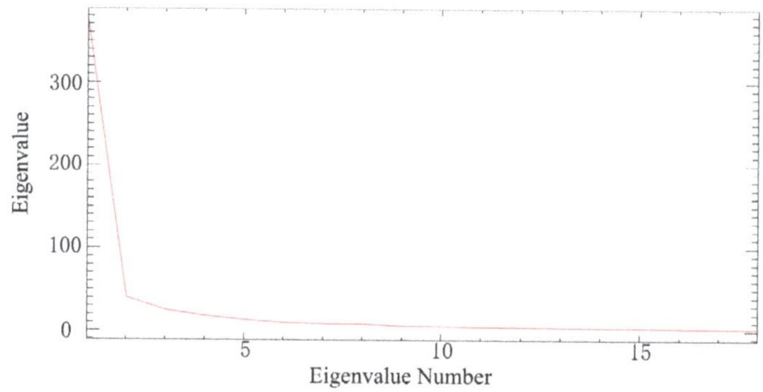

(a)

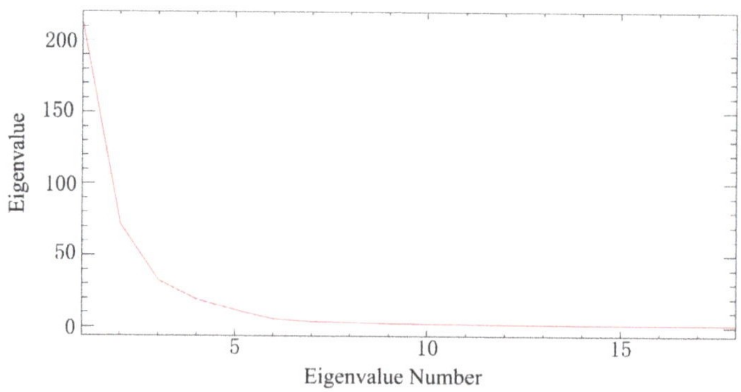

(b)

Figure 4. (**a**) The MNF transform eigenvalue curve of the Longkou area; (**b**) the MNF transform eigenvalue curve of the Honghu area.

As shown in the curves in Figure 5, the PCA transform eigenvalue curves of the two regions converged at the 12th and 8th bands, respectively. Figure 6 shows the display status of the image information corresponding to each transformed (N + 1) band. There was little information after the 18 bands of MNF transformation in these two regions, so the optimal ideal bands for the MNF transform in the two regions were both 18 bands. Similarly, the optimal ideal bands for the PCA transform in Longkou and Honghu regions were 12 bands and 8 bands, respectively, and the optimal ideal bands for the ICA transform were 18 bands and 44 bands, respectively. Figure 7 is a color image composed of the first three components of each transform.

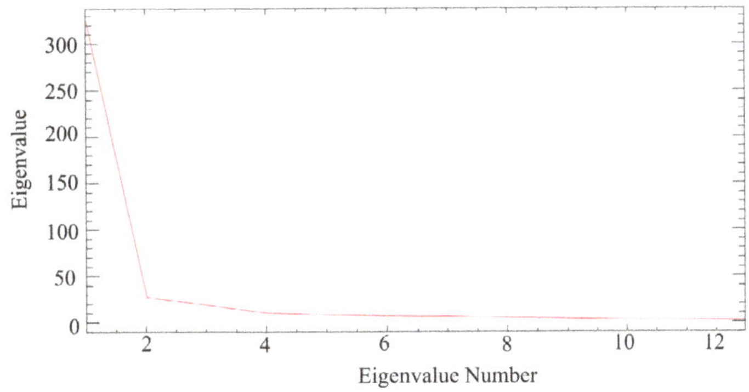

(a)

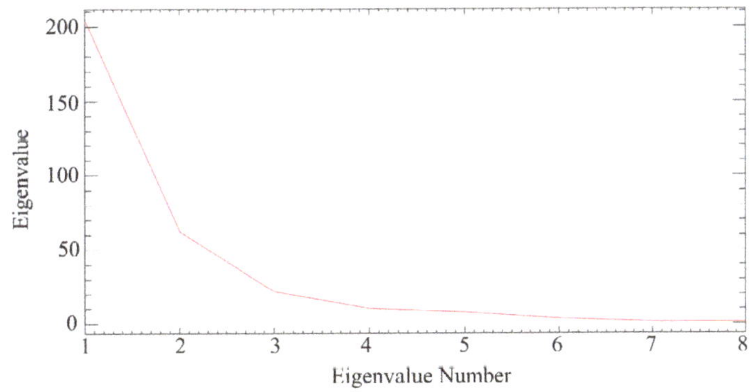

(b)

Figure 5. (**a**) The PCA transform eigenvalue curve of the Longkou area; (**b**) the PCA transform eigenvalue curve of the Honghu area.

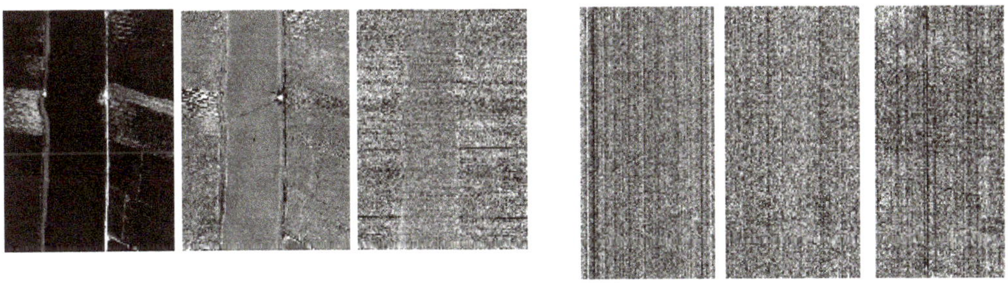

(a) (b) (c) (d) (e) (f)

Figure 6. (**a**–**c**) The MNF transform at 19 bands, ICA transform at 19 bands, and PCA transform at 13 bands, respectively, in the Longkou area; (**d**–**f**) the MNF transform at 19 bands, ICA transform at 45 bands, and PCA transform at 9 bands, respectively, in the Honghu area.

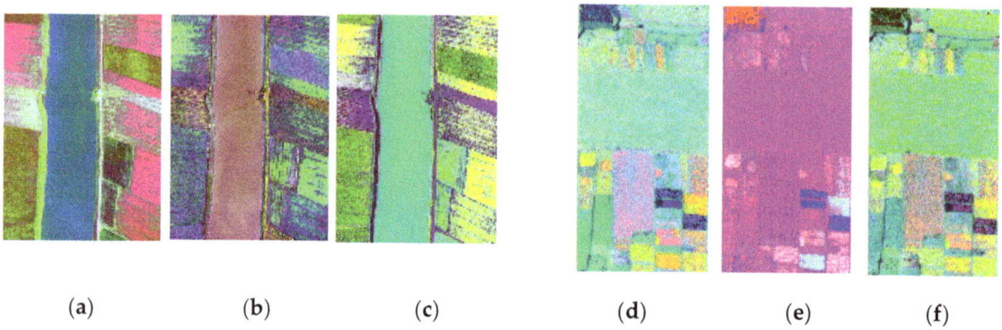

Figure 7. (a–c) Images of the Longkou area after MNF, ICA, and PCA transforms, respectively; (d–f) images of the Honghu area after MNF, ICA, and PCA transforms, respectively.

4.2. Accuracy Assessment

The final accuracy evaluation is shown in Tables 1 and 2, which show that the classification accuracy of the two regions was affected by many factors.

Table 1. RF classification accuracy evaluation of random sample sets in the Longkou area.

Random Sample Set	Original Image	MNF Transform	ICA Transform	PCA Transform
Train 25	80.14%	95.66%	88.12%	90.27%
Train 50	84.68%	95.65%	90.98%	90.56%
Train 100	85.95%	96.08%	91.82%	91.27%
Train 150	89.27%	96.49%	92.24%	93.06%
Train 200	90.46%	96.43%	93.41%	93.43%
Train 250	89.42%	96.68%	93.11%	93.53%
Train 300	90.75%	97.18%	94.17%	94.62%

Table 2. RF classification accuracy evaluation of random sample sets in the Honghu area.

Random Sample Set	Original Image	MNF Transform	ICA Transform	PCA Transform
Train 25	63.31%	71.91%	54.48%	66.08%
Train 50	59.06%	73.20%	59.93%	62.38%
Train 100	66.88%	76.45%	65.72%	65.67%
Train 150	68.58%	77.70%	68.27%	69.06%
Train 200	69.92%	78.83%	68.77%	68.86%
Train 250	70.32%	79.34%	69.79%	70.48%
Train 300	71.62%	80.43%	70.54%	70.98%

In this study, the number of categories and the clarity of category boundaries had impacts on classification accuracy. The Longkou area had fewer categories and clearer boundaries between categories, and the highest classification accuracy of the random sample set was 97.18%, which was 6.12% higher than the best overall accuracy of 91.06% for the same RF classification algorithm with only four plantings [18]. On the other hand, the Honghu area had many categories and complex feature information, and the highest classification accuracy of the random sample set was 80.43%.

The characteristic of the MNF transform is that two PCA transforms are carried out considering the influence of noise, which has significant advantages when the category situation is more complex. The classification accuracy of different random sample sets in the two regions showed that the MNF–RF combination had higher classification accuracy than the other classification combinations. For random sample sets with the same numbers of sample points, the MNF–RF classification accuracy (95.66%) in the Longkou area was higher than that of the original image (80.14%), with a maximum improvement of 15.52%, and the MNF–RF classification accuracy (71.91%) in the Honghu area was higher than that of other classification combinations (54.48%), with a maximum improvement of 17.42%.

The image classification effect of PCA transform and ICA transform is greatly affected by the category information of the study area. For the Longkou area, where the category information was relatively simple, the RF classification accuracy of the image after feature transform was higher than that of the original image. Comparing the classification accuracy of different random sample sets in Table 1, the classification accuracy of the PCA transform was more ideal than that of the ICA transform. The classification information in the Honghu area was relatively complex. In Table 2, the classification accuracy of the image after MNF transform was higher than that of the original image, but the accuracy of the images after PCA and ICA transforms was lower than that of the original image. Furthermore, because the sample points of the random sample set were selected randomly, the advantages of the ICA and PCA transforms were not obvious. To sum up, the best classification combination for the random sample sets in the two regions was determined as the MNF–RF combination, and the best classification effect of the two regions is shown in Figure 8.

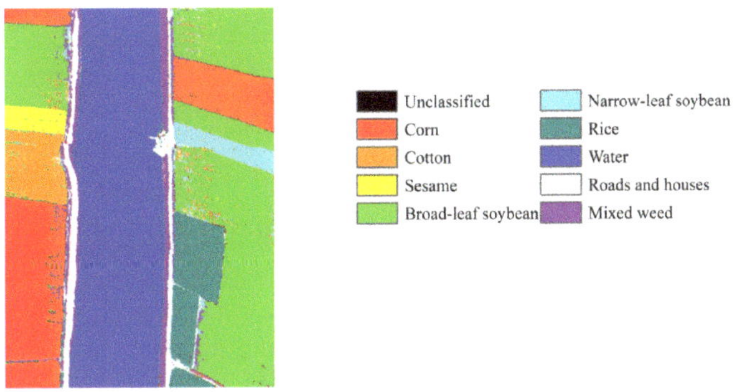

(a)

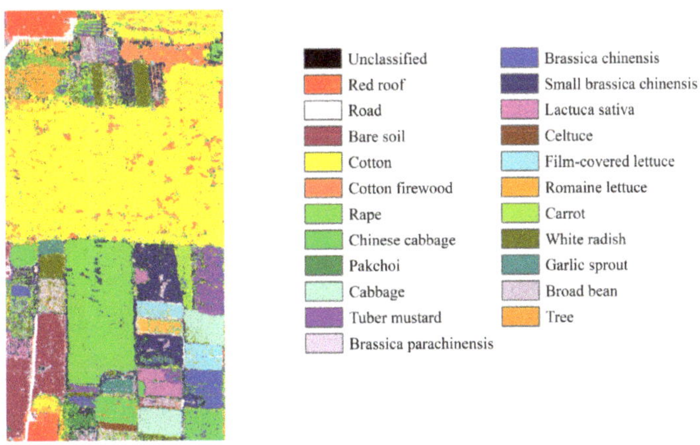

(b)

Figure 8. (**a**) Classification results of the random sample set in the Longkou area; (**b**) classification results of the random sample set in the Honghu area.

4.3. Comparison of Stability

As shown in Figure 9, the curves of the two regions reflected the following characteristics. The overall classification accuracy was positively correlated with the fluctuation of the number of random sample points. The overall classification accuracy of RF classification after feature transformation showed a gentler trend with the number of sample points than the original image and was less affected by the number of sample points. MNF–RF curves in both figures were the smoothest, indicating that MNF–RF was least affected by the number of random sample points, followed by PCA–RF and ICA–RF. The overall classification accuracy fluctuations of MNF–RF in the two regions were less than 0.50% and 3.25%, respectively.

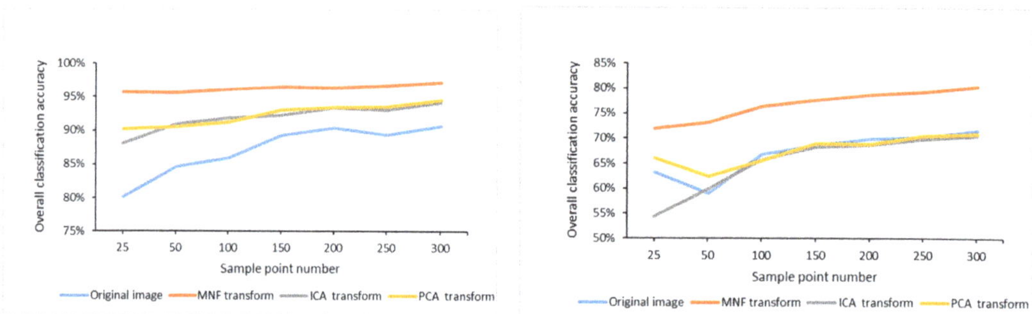

(a) (b)

Figure 9. (**a**) The change of overall classification accuracy with the number of random sample points in the Longkou area; (**b**) the change of overall classification accuracy with the number of random sample points in the Honghu area.

5. Conclusions

Accurate crop classification mapping is an important basis for agricultural production management, agricultural policymaking, and food safety and can provide necessary reference information for agricultural decision-making. Realizing efficient and accurate crop classification is a key step for structural optimization and management plan formulation. It is hoped that the research methods and ideas of this study can provide a reference for the development and structural adjustment of precision agriculture in other countries.

In this study, the MNF–RF combination, which was superior to the ICA–RF and PCA–RF combinations, was the best classification combination in practical application. The best classification accuracies of the MNF–RF random sample set in the two regions were 97.18% and 80.43%, respectively, which were 6.43% and 8.81% higher than the RF classification accuracy of the original image.

The MNF transform takes into account the influence of noise in UAV hyperspectral images to a large extent and has high stability in the face of a different number of random sample points. Therefore, while the curve of RF overall classification accuracy changed with the number of random sample points in the two regions, the MNF transform curve was the gentlest. The overall classification accuracies of MNF–RF in the Longkou and Honghu areas varied by less than 0.50% and 3.25%, respectively, with the fluctuation of sample points.

In conclusion, it is feasible to use the MNF–RF classification combination to perform fine crop classification using UAV hyperspectral image data. In actual use, the MNF–RF combination classification can be used to select fewer random sample points to complete the fine classification of crops with high precision, accuracy, and stability. This reduces the sampling workload, greatly improves efficiency, and saves much manpower and material resources.

In future research work, the impact of different classification combinations and random sample numbers on the individual classification accuracy of each crop can be clarified. In follow-up work, we can further compare the performance of RF classification with that of neural networks and support vector machines using the same transformations as employed with the RF classifier.

Author Contributions: Conceptualization: Zhihua Wang, Zhan Zhao; Methodology: Zhihua Wang, Zhan Zhao; Formal analysis and investigation: Zhihua Wang, Zhan Zhao, Chenglong Yin; Writing—original draft preparation: Zhihua Wang; Writing—review and editing: Zhihua Wang, Zhan Zhao, Chenglong Yin; Resources: Zhan Zhao; Supervision: Zhan Zhao. All authors have read and agreed to the published version of the manuscript.

Funding: This work was funded by the National Science and Technology Fundamental Resources Survey Project (2019FY202502).

Institutional Review Board Statement: Not applicable.

Informed Consent Statement: Not applicable.

Data Availability Statement: Not applicable.

Acknowledgments: Thanks to the RSIDEA team of the State Key Laboratory of Surveying, Mapping, and Remote Sensing Information Engineering, Wuhan University, which constructed a set of publicly shared hyperspectral and high-spatial resolution remote sensing imagery and ground object fine classification data set: WHU-HI. We thank Professor Kunshan Chen of Guilin University of Technology and Associate Professor Ying Yang of Nanjing University of Science and Technology for the cultivation of the first author's logical thinking which has promoted the further perfection of the paper. We would like to express our gratitude and respect to the editors and anonymous reviewers for their valuable comments and suggestions, which helped improve the manuscript greatly.

Conflicts of Interest: The authors declare no conflict of interest.

References

1. Weiguang, G. Practice and Application of Information Technology in Precision Agriculture—Review of Low-Altitude Remote Sensing Technology and Its Application in Precision Agriculture. *Chin. J. Agric. Resour. Reg. Plan.* **2021**, *42*, 144–152.
2. Cai, Y.; Guan, K.; Peng, J.; Wang, S.; Seifert, C.; Wardlow, B.; Li, Z. A high-performance and in-season classification system of field-level crop types using time-series Landsat data and a machine learning approach. *Remote Sens. Environ.* **2018**, *210*, 35–47. [CrossRef]
3. Wei, L.; Yu, M.; Zhong, Y.; Zhao, J.; Liang, Y.; Hu, X. Spatial–Spectral Fusion Based on Conditional Random Fields for the Fine Classification of Crops in UAV-Borne Hyperspectral Remote Sensing Imagery. *Remote Sens.* **2019**, *11*, 780. [CrossRef]
4. Qingzhan, Z.; Ping, J.; Xuewen, W.; Lihong, Z.; Jianxin, Z. Classification of Protection Forest Tree Species Based on UAV Hyperspectral Data. *Trans. Chin. Soc. Agric. Mach.* **2021**, *52*, 190–199.
5. Hui, L.; Jing, H.; Junjie, L. Monitoring of Corn Canopy Blight Disease Based on UAV Hyperspectral Method. *Spectrosc. Spectr. Anal.* **2020**, *40*, 1965–1972.
6. Pascucci, S.; Pignatti, S.; Casa, R.; Darvishzadeh, R.; Huang, W. Special Issue "Hyperspectral Remote Sensing of Agriculture and Vegetation". *Remote Sens.* **2020**, *12*, 3665. [CrossRef]
7. Fei, S.; Yu, X.; Lan, M.; Li, L.; Xia, X.; He, Z.; Xiao, Y. Research on Winter Wheat Yield Estimation Based on Hyperspectral Remote Sensing and Ensemble Learning Method. *Sci. Agric. Sin.* **2021**, *54*, 3417–3427.
8. Liu, Y.; Zhang, G.; Wang, X.; Zhou, C.; Yang, Z.; Wu, X.; Zhang, J. Classification study of Mikania micrantha kunth from UAV hyperspectral image band selection. *Bull. Surv. Mapp.* **2020**, *4*, 34–39. [CrossRef]
9. Li, P.; Zhang, X.; Wang, W.; Zheng, H.; Yao, X.; Zhu, Y.; Cao, W.; Cheng, T. Assessment of Terrestrial Laser Scanning and Hyperspectral Remote Sensing for the Estimation of Rice Grain Yield. *Sci. Agric. Sin.* **2021**, *54*, 2965–2976.
10. Lan, Y.; Zhu, Z.; Deng, X.; Lian, B.; Huang, J.; Huang, Z.; Hu, J. Monitoring and classification of citrus Huanglongbing based on UAV hyperspectral remote sensing. *Trans. Chin. Soc. Agric. Eng.* **2019**, *35*, 92–100.
11. Okwuashi, O.; Ndehedehe, C.E. Deep support vector machine for hyperspectral image classification. *Pattern Recognit.* **2020**, *103*, 107298. [CrossRef]
12. Wang, C.; Zhao, Q.; Ma, Y.; Ren, Y. Crop Identification of Drone Remote Sensing Based on Convolutional Neural Network. *Trans. Chin. Soc. Agric. Mach.* **2019**, *50*, 161–168.
13. Du, P.; Xia, J.; Xue, Z.; Tan, K.; Su, H.; Bao, R. Review of hyperspectral remote sensing image classification. *J. Remote Sens.* **2016**, *20*, 236–256. [CrossRef]

14. Fang, X.; Wen, Z.; Chen, J.; Wu, S.; Huang, Y.; Ma, M. Remote sensing estimation of suspended sediment concentration based on Random Forest Regression Model. *Natl. Remote Sens. Bull.* **2019**, *23*, 756–772.
15. Yang, X.; Su, H.; Li, W.E.; Huang, L.; Wang, X.; Yan, X. Seasonal-spatial variations in satellite-derived global subsurface temperature anomalies. *Natl. Remote Sens. Bull.* **2019**, *23*, 997–1010.
16. Wang, L.; Kong, Y.; Yang, X.; Xu, Y.; Liang, L.; Wang, S. Classification of land use in farming areas based on feature optimization random forest algorithm. *Trans. Chin. Soc. Agric. Eng.* **2020**, *36*, 244–250.
17. Li, H.; Wang, L.; Xiao, S. Random forest classification of land use in hilly and mountaineous areas of southern China using multi-source remote sensing data. *Trans. Chin. Soc. Agric. Eng.* **2021**, *37*, 244–251.
18. Yang, H.; Du, J.; Ruan, P.; Zhu, X.; Liu, H.; Wang, Y. Vegetation Classification of Desert Steppe Based on Unmanned Aerial Vehicle Remote Sensing and Random Forest. *Trans. Chin. Soc. Agric. Mach.* **2021**, *52*, 186–194.
19. Hughes, G. On the mean accuracy of statistical pattern recognizers. *IEEE Trans. Inf. Theory* **1968**, *14*, 55–63. [CrossRef]
20. Richter, R.; Reu, B.; Wirth, C.; Doktor, D.; Vohland, M. The use of airborne hyperspectral data for tree species classification in a species-rich Central European forest area. *Int. J. Appl. Earth Obs. Geoinf.* **2016**, *52*, 464–474. [CrossRef]
21. Dabiri, Z.; Lang, S. Comparison of Independent Component Analysis, Principal Component Analysis, and Minimum Noise Fraction Transformation for Tree Species Classification Using APEX Hyperspectral Imagery. *ISPRS Int. J. Geoinf.* **2018**, *7*, 488. [CrossRef]
22. Sijing, Y.; Changqing, S.; Feng, C.; Leina, Z.; Changxiu, C.; Chao, Z.; Jianyu, Y.; Dehai, Z. Cultivated land health-productivity comprehensive evaluation and its pilot evaluation in China. *Trans. Chin. Soc. Agric. Eng.* **2019**, *35*, 66–78.
23. Zhong, Y.; Hu, X.; Luo, C.; Wang, X.; Zhao, J.; Zhang, L. WHU-Hi: UAV-borne hyperspectral with high spatial resolution (H2) benchmark datasets and classifier for precise crop identification based on deep convolutional neural network with CRF. *Remote Sens. Environ.* **2020**, *250*, 112012. [CrossRef]
24. Zhong, Y.; Wang, X.; Xu, Y.; Wang, S.; Jia, T.; Hu, X.; Zhao, J.; Wei, L.; Zhang, L. Mini-UAV-Borne Hyperspectral Remote Sensing: From Observation and Processing to Applications. *IEEE Geosci. Remote Sens. Mag.* **2018**, *6*, 46–62. [CrossRef]
25. Bajwa, S.; Kulkarni, S. Hyperspectral Data Mining. In *Hyperspectral Remote Sensing of Vegetation*; CRC Press: London, UK, 2011.
26. Jolliffe, I. Principal Component Analysis. In *International Encyclopedia of Statistical Science*; Lovric, M., Ed.; Springer: Berlin/Heidelberg, Germany, 2011; pp. 1094–1096.
27. Green, A.; Berman, M.; Switzer, P.; Craig, M. A transformation for ordering multispectral data in terms of image quality with implications for noise removal. *IEEE Trans. Geosci. Remote Sens.* **1988**, *26*, 65–74. [CrossRef]
28. Harsanyi, J.C.; Chang, C.-I. Hyperspectral image classification and dimensionality reduction: An orthogonal subspace projection approach. *IEEE Trans. Geosci. Remote Sens.* **1994**, *32*, 779–785. [CrossRef]
29. Underwood, E. Mapping nonnative plants using hyperspectral imagery. *Remote Sens. Environ.* **2003**, *86*, 150–161. [CrossRef]
30. Nascimento, J.; Bioucas-Dias, J. Vertex component analysis: A fast algorithm to unmix hyperspectral data. *IEEE Trans. Geosci. Remote Sens.* **2005**, *43*, 898–910. [CrossRef]
31. Hyvarinen, A. Fast and robust fixed-point algorithms for independent component analysis. *IEEE Trans. Neural Netw.* **1999**, *10*, 626–634. [CrossRef]
32. Breiman, L. Random forests. *Mach. Learn.* **2001**, *45*, 5–32. [CrossRef]
33. Rodriguez-Galiano, V.F.; Ghimire, B.; Rogan, J.; Chica-Olmo, M.; Rigol-Sanchez, J.P. An assessment of the effectiveness of a random forest classifier for land-cover classification. *ISPRS J. Photogramm. Remote Sens.* **2012**, *67*, 93–104. [CrossRef]
34. Zhang, Z.; Zhang, X.; Xin, Q.; Yang, X. Combining the Pixel-based and Object-based Methods for Building Change Detection Using High-resolution Remote Sensing Images. *Acta Geod. Cartogr. Sin.* **2018**, *47*, 102–112.
35. Liu, S.; Jiang, Q.; Yue, M.A.; Xiao, Y.; Yuanhua, L.I.; Cui, C. Object-oriented Wetland Classification Based on Hybrid Feature Selection Method Combining with Relief F, Multi-objective Genetic Algorithm and Random Forest. *Trans. Chin. Soc. Agric. Mach.* **2017**, *48*, 119–127.
36. Yun, H.E.; Huang, C.; He, L.I.; Liu, Q.; Liu, G.; Zhou, Z.; Zhang, C. Land-cover classification of random forest based on Sentinel-2A image feature optimization. *Resour. Sci.* **2019**, *41*, 992–1001.

International Journal of *Geo-Information*

MDPI

Article

Improving LST Downscaling Quality on Regional and Field-Scale by Parameterizing the DisTrad Method

Taha I. M. Ibrahim [1,2], Sadiq Al-Maliki [1,*], Omar Salameh [1], István Waltner [1] and Zoltán Vekerdy [1,3]

1 Institute of Environmental Sciences, Hungarian University of Agriculture and Life Sciences, Pater Karoly, u. 1., 2100 Godollo, Hungary; taha.ibrahim.mohammed.ahmed.ibrahim@phd.uni-mate.hu (T.I.M.I.); omar.salameh@phd.uni-mate.hu (O.S.); waltner.istvan@uni-mate.hu (I.W.); vekerdy.zoltan@uni-mate.hu (Z.V.)

2 Agricultural Engineering Research Center, Agricultural Research Corporation (ARC), Ministry of Agricultural, Wad Madni P.O. Box 126, Sudan

3 Faculty of Geo-Information Science and Earth Observation, University of Twente, 7514 AE Enschede, The Netherlands

* Correspondence: sadiq.al-maliki@phd.uni-mate.hu

Abstract: Many scientists have been investigating Land Surface Temperature (LST) because of its relevance in water management science due to its direct influence on the hydrological water cycle. This effect stems from being one of the most significant variables influencing evapotranspiration. One of the most important reasons for the evapotranspiration retrieved from MODIS data's limited suitability for scheduling and planning irrigation schemes is the lack of spatial resolution. As a result, high-resolution LST is required for estimating evapotranspiration. The goal of this study is to improve the resolution of the available LST data, to improve evapotranspiration (ET_a) estimation using statistical downscaling with Normalized Difference Vegetation Index (NDVI) as a predictor. The DisTrad (Disaggregation of Radiometric Surface Temperature) method was used for the LST downscaling procedure, which is based on aggregating the NDVI map to the LST map resolution and then calculating the coefficient of variation of the native NDVI map within the aggregated pixel and classifying the aggregated map into three classes: NDVI < 0.2 for the bare soil, 0.2 ≤ NDVI ≤ 0.5 for the partial vegetation, and NDVI > 0.5 for the full vegetation. DisTrad uses 25% of the pixels with the lowest coefficient of variation from each class to calculate the regression coefficients. In this work, adjustments to the DisTrad method were implemented to enhance downscaling LST and to examine the impacts of that alteration on the evapotranspiration estimation. The linear regression model was tested as an alternative to the original second-order polynomial. In using 10% of the pixels instead of the originally proposed 25% with the lowest coefficient of variation values, it is assumed that a group of pixels with a lower coefficient of variation represents a more homogeneous area, thus it gives more accurate values. The downscaled LST map retrieval was validated using Landsat 8 thermal maps (100 m). Applying the modified DisTrad approach to disaggregate Landsat LST to 30 m (NDVI resolution) yielded an R^2 of 0.72 for the 10%, 0.74 for the 25% and 0.61 for the second-order polynomial lowest coefficient of variation compared to native LST Landsat, which means that 10% can be used as an alternative. Applying the downscaled LST map to estimate ET_a yielded R^2 0.84 in both cases, compared to ET_a yielded from the native Landsat LST. These results prove that using the robust linear regression provided better results than using polynomial regression. With the downscaled Land Surface Temperature data, it was possible to create detailed ET_a maps of the small agricultural fields in the test area.

Keywords: land surface temperature; downscaling; DisTrad; evapotranspiration; Landsat 8; MODIS; linear regression; thermal sensors; temporal resolution; spatial resolution

Citation: Ibrahim, T.I.M., Al-Maliki, S.; Salameh, O.; Waltner, I.; Vekerdy, Z. Improving LST Downscaling Quality on Regional and Field-Scale by Parameterizing the DisTrad Method. *ISPRS Int. J. Geo-Inf.* **2022**, *11*, 327. https://doi.org/10.3390/ijgi11060327

Academic Editors: Dimitris Triantakonstantis, Panagiotis Tziachris and Wolfgang Kainz

Received: 17 March 2022
Accepted: 27 May 2022
Published: 30 May 2022

1. Introduction

Evapotranspiration (ET) is a challenging parameter to estimate, but Land Surface Temperature (LST) and soil moisture are crucial parameters to assess it [1–4]. This effect is one of the most prominent factors affecting evapotranspiration [5–8]. Deriving Land Surface Temperature from thermal remote sensing is one of the most promising means of achieving the mission, whether at the regional or global scale [1,9,10]. The emergence of thermal sensors with high spatial resolution has improved and raised the quality of calculating LST [11,12], as these sensors have a spatial resolution of 60–100 m. However, with this high spatial resolution, poor temporal resolution due to 16 days revisit time for these sensors remains a challenge [13,14]. One of these sensors is the Landsat 8 OLI TIRS, which has two thermal bands with 100 m resolution and Landsat 7 thermal band with 60 m resolution [15,16]. On the other hand, several sensors with low spatial resolution produce high temporal resolution images with a repetition time varying from less than an hour to three days [2,13,17,18]. The poor spatial resolution of the evapotranspiration retrieved from MODIS data is one of the most important reasons for its limited suitability for scheduling and planning irrigation [18,19]. In developing countries, the agricultural lands are partitioned into small fields of less than 1 square kilometer.

Applying coarse-resolution thermal data to the small fields, one faces a problem called the thermal mixing effect, resulting from the differences in the thermal properties of the land cover classes within one pixel. Therefore, this problem needs to be addressed, e.g., by finding a relationship between the temporal and spatial resolutions of the thermal images [20].

The downscaling process is defined as increasing spatial resolution by finding values of the smaller pixels as a function of the original measurement with the coarse pixel size and some additional information, or, in a simplified way, and it can be described as decreasing the pixel size [21,22]. Many different approaches have been followed for the downscaling of LST [20], but the most popular approach is the disaggregation of LST based on a co-variable. The approach improves the resolution of LST, whether spatially or temporally, and makes it suitable for many applications, e.g., all applications that include the surface energy balance (SEB) [7,23]. One of the famous disaggregation methods is DisTrad (Disaggregation Procedure of Radiometric Surface Temperature [24]). Its principle is to find a mathematical relationship between the Radiometric Surface Temperature, and the Normalized Difference Vegetation Index NDVI through the inverse relationship between the LST and the NDVI [2,24–27]. A further developed approach is TsHARP (Temperature Sharpening), which is a modulation of the DisTrad, based on a linear relationship between the vegetation fractional cover (FC) and LST [28].

To parameterize the regression equation between the vegetation index and the Surface Temperature, Kustas et al. [24] have developed a process of aggregating the NDVI map and then calculating the coefficient of variation (CV) of the native NDVI values within each pixel in the aggregated map, using 25% of the aggregated pixels with the lowest coefficients of variation for defining the coefficients of the regression equation [24]. This approach was successfully tested on homogenous areas, but when used on heterogeneous areas, the problem of representativity occurs. Accordingly, the approach needs to be improved for heterogeneous regions [20,29]. This work aims to improve the DisTrad approach for downscaling LST for heterogeneous areas and improve the temporal and spatial distribution of the calculated high-resolution evapotranspiration maps.

The study's specific goals are as follows: (a) Compare linear vs. polynomial based statistical downscaling, (b) compare 25% and 10% aggregated pixels with the lowest coefficients of variation, (c) improve ETa spatial and temporal resolution for the agricultural sector.

2. Materials and Methods

2.1. Study Area

The construction of the Aswan High Dam caused the inundation of the old town of Wadi Halfa by Lake Nasser. The New Halfa Agricultural Scheme was initiated in response as Sudan's largest resettlement project. The New Halfa Agricultural Scheme is a 185,000-ha agricultural settlement scheme on the western side of Kassala State, roughly 400 km east of Khartoum [30]. The project is located on the Butana plain, along the Atbara River. At the time of its construction, the New Halfa Scheme was Sudan's second-largest irrigation project after the Gezira Scheme, which is still the world's largest irrigation scheme. Sudan's irrigation agency manages the water through the Khasm el Girba dam on the Atbara River [31]. The irrigation system is gravity-fed, with the main canal transporting water to the project area via a network of subsidiary canals and motorized pumps in the small scheme areas. The irrigation system includes main canals, branch canals, minor canals, quaternary canals, and tertiary farm ditches. Field irrigation is done using the traditional basin (Angaya) approach, dividing the field into small sections. There are significant water losses in the system, reducing the available freshwater supplies, like evaporation, conveyance losses due to infiltration, etc. [31]. The dam was initially intended to store 1.3 billion cubic meters of water. However, by 1976, the reservoir's storage capacity had been decreased to 0.8 billion cubic meters due to significant siltation originating from the upstream catchment of the river Atbara in Ethiopia's highlands [30]. The reservoir's capacity is now about 0.6 billion m^3. During the growing season, the water in the smaller canals typically flows permanently. Farmers, however, have complained that some regions receive more water than others. Since the reservoir's capacity is dwindling, the irrigated area is shrinking too. Each agricultural settler was given a 15-feddan hawasha (6.3 ha) (tenancy) to cultivate cotton, wheat, sorghum, and groundnuts. Mostly cotton was chosen since it is the most important cash crop for the government to provide hard currency and profit for the tenants. Groundnuts are the scheme's second most significant cash crop, and wheat and sorghum were grown with any surplus sold [31]. Figure 1 shows the location map of the study area.

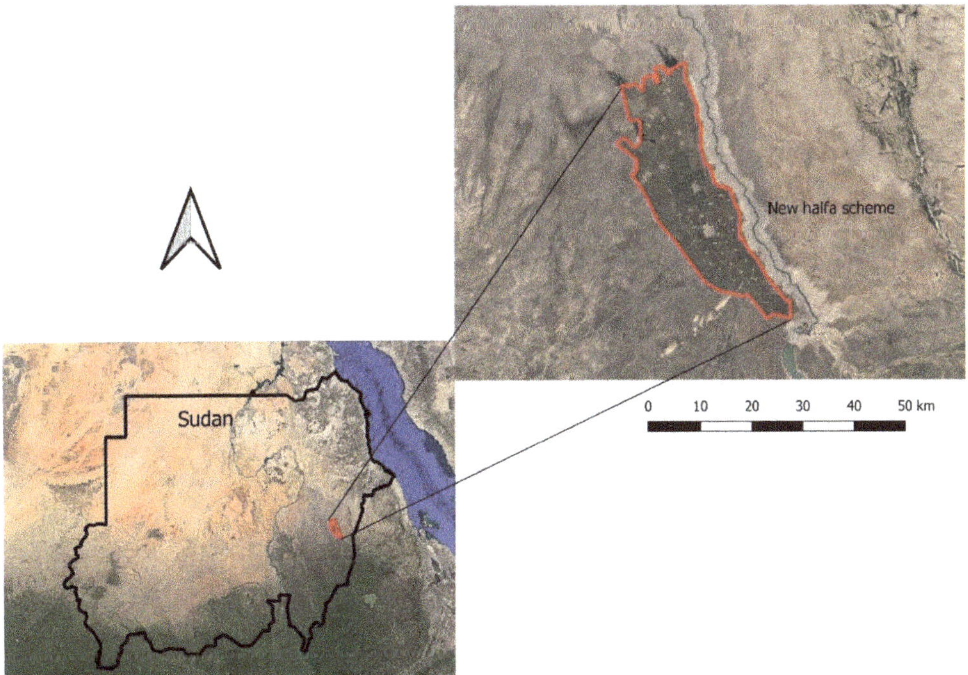

Figure 1. Location map describes the study area.

2.2. The DisTrad Downscaling Procedure for Radiometric Surface Temperature

For finding a mathematical relationship between the Radiometric Surface Temperature and the Normalized Difference Vegetation Index (NDVI) [24] suggested aggregating the fine-resolution NDVI map to the same coarse-resolution as the LST map and then defining the regression coefficients of Equation (1) with a least square fitting using a second-order polynomial.

$$LST^{*}_{CR} = a + b\ NDVI_{CR} + c\ NDVI^{2}_{CR} \tag{1}$$

where:

LST^{*}_{CR}: Land surface temperature at the coarser resolution (°C).

$NDVI_{CR}$: Aggregated normalized difference vegetation index at the coarser resolution.

The LST is influenced by the vegetation cover and other factors, such as soil moisture [32]. When these factors are overlooked during the downscaling process, the results are affected by the spatial variation of the factor values. To overcome this problem, Kustas et al. [24] used the difference between the aggregated downscaled LST and the original one (Equation (2)) for each pixel to estimate the error that represented the influence of other factors, such as soil moisture, and then used it for correction by Equation (3).

$$\Delta\hat{T}_{CR} = LST_{CR} - L\hat{S}T_{CR} \tag{2}$$

where:

$\Delta\hat{T}_{CR}$: Residual of the land surface temperature (°C).

LST_{CR}: Land surface temperature is defined from the satellite measurement (°C).

$L\hat{S}T_{CR}$: Aggregated downscaled land surface temperature (°C).

$$LST_{FR} = LST^{*}_{FR} + \Delta\hat{T}_{CR} \tag{3}$$

where:

LST_{FR}: Corrected, the downscaled land surface temperature at the fine resolution (°C).

LST^{*}_{FR}: Land surface temperature calculated by Equation (1) from the original fine-resolution NDVI (°C).

In practice, for defining the NDVI-LST relationship, the coarse-resolution NDVI map was divided into three groups. The coefficient of variation of the fine-resolution NDVI values within each coarse-resolution pixel was calculated. To avoid the influence of heterogeneity, this map was divided into three classes: NDVI < 0.2 for the bare soil, 0.2 < NDVI < 0.5 for the partial vegetation, and NDVI > 0.5 for the full vegetation. Finally, 25% of the pixels with the lowest coefficient of variation were selected from each group to obtain the correlation [24].

2.2.1. DisTrad Modification

The study area is described as a mixed landscape with various distinct land cover types (dense vegetation, medium dense vegetation, low-density vegetation, bare soil, urban areas, and water). The DisTrad approach is based on the correlation between the LST and NDVI. The original approach uses a second-order polynomial regression, assuming a non-linear relationship between the two variables. However, in certain cases, outliers at the edges of the value range may strongly affect the second-order polynomial. To circumvent this issue, we tested whether a linear regression would improve the robustness of the regression equation.

Furthermore, the original DisTrad technique recommends using 25% of the aggregated pixels with the lowest variability of variation for parameterizing the regression equation. Nonetheless, in the case of a heterogonous area (e.g., due to small agricultural fields relative to the coarse pixel size), where the CV of the original NDVI values within most of the coarse-resolution pixels is relatively high, we tested the effect of the use of only 10% of the aggregated pixels with the lowest coefficients of variation in defining the parameters of the regression equation.

The revised DisTrad sharpening procedure showed a higher correlation between the observed temperature and the MODIS downscaled temperature. Ref. [33] used the correlation between the impervious percentage and the temperature estimation residuals generated from the two MODIS/Terra 5-min land surface temperature products. Ref. [34] also investigated the use of different percentages of aggregated pixels to determine the optimal index pixel fraction to be used in the downscaling application by evaluating the use of 100%, 75%, 50%, or 25% of the pixels with the lowest CV values within the simulated 960 m image. The higher correlation was obtained when using 100% of the pixels $R^2 = 0.65$ compared to 75%, 50%, and 25% where $R^2 = 0.63, 40, 34$ respectively.

Modification Summary

1. Use linear regression instead of polynomial regression by assuming that polynomial is more sensitive for outliers.
2. Use 10% of the aggregated pixels instead of using 25% of the aggregated pixels assuming that based on the heterogeneity of the study area, the 10% of the aggregated pixels will give a stronger correlation between the NDVI and LST in the upper and lower tail in the distribution of the pixels.

The Validation

To validate this modification, we used the NDVI as shown in Figure 2 and LST calculated from the Landsat 8 image on the date 18 January 2018, and the steps are as follows:

1. LST from the Landsat 8 was aggregated to a coarser resolution (1000 m).
2. NDVI from Landsat 8 was aggregated to a coarse resolution (1000 m).
3. The modification was applied to LST1000m and NDVI1000M to downscale LST to fine resolution.
4. LSTnative was used to validate LSTdown.

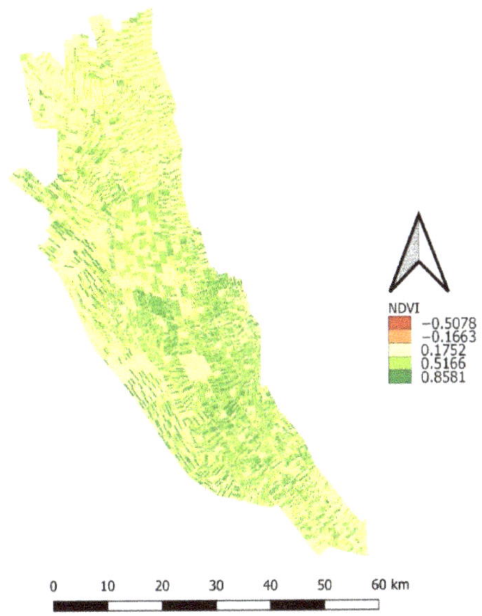

Figure 2. NDVI derived from Landsat 8 18 January 2018.

The same modification was applied to MODIS (MOD11A1 V6. Product) for 37 images to create the time series.

2.3. Evapotranspiration Estimation

Evapotranspiration is the most important parameter in water balance in arid and semi-arid regions. Nonetheless, it is regarded as an element with poor temporal resolution despite its significance. As a result, its temporal resolution must be improved, especially if it is to be utilized for irrigation scheduling and planning. Therefore, it is necessary to downscale the land surface temperature since it is one of the most crucial factors in estimating evapotranspiration, which is the most dynamic parameter. In order to assess the ability of the LST downscaling to improve the temporal resolution of evapotranspiration, the Surface Energy Balance System model (SEBS) was used to estimate evapotranspiration. The original land surface temperature from Landsat 8, downscaled Landsat 8 land surface temperature, resampled MODIS land surface, and downscaled MODIS Land Surface Temperature were all considered to estimate evapotranspiration.

2.3.1. The Surface Energy Balance System

The Surface Energy Balance System (SEBS) model, designed by [35] to estimate evapotranspiration needs two types of data: The first group describes the physical properties of the land surface, including albedo, emissivity, temperature, fractional vegetation coverage, leaf area index, and the height of the vegetation (or roughness height). These data can be retrieved from satellite images. The second group describes the state of the atmosphere at a reference height, comprising air pressure, temperature, humidity, wind speed, downward solar radiation, and downward longwave radiation. These parameters can be determined from meteorological data measured directly or using a model. The data are also utilized to calculate roughness length for heat transfer and evaporative fraction.

SEBS is a physically based energy balance model that does not require prior knowledge of turbulent heat fluxes. To compute instantaneous relative evaporation, the model employs energy balance at wet and dry limiting conditions. The SEBS model is based on the energy balance equation (Equation (4)).

$$R_n = \lambda E + G_0 + H \tag{4}$$

where: R_n is the net radiation, λE is the turbulent latent heat flux (λ is the latent heat of vaporization and E is the actual evapotranspiration), G_0 is the soil heat flux, and H is the turbulent sensible heat flux (All units are $W{\cdot}m^{-2}$ or $J{\cdot}s^{-1}{\cdot}m^{-2}$).

The calculation of net radiation flux on the land surface, R_n ($W{\cdot}m^{-2}$), is given by Equation (5).

$$R_n = (1-\alpha)R_{swd} + \varepsilon\ R_{lwd} - \varepsilon\ \sigma\ {T_0}^4 \tag{5}$$

where: α is the albedo (-), R_{swd} is the downward solar radiation ($W{\cdot}m^{-2}$), ε is the emissivity of the surface (-), R_{lwd} is the downward longwave radiation ($W{\cdot}m^{-2}$), σ is the Stefan–Boltzmann constant ($W{\cdot}m^{-2}{\cdot}K^{-4}$), and T_0 is the surface temperature (K).

The soil heat flux depends on land surface characteristics, soil water content, and other factors. Equation (6) gives the calculation of soil heat flux by the SEBS model:

$$G_0 = R_n(\Gamma_c + (1-f_c)\cdot(\Gamma_s - \Gamma_c)) \tag{6}$$

in which it is assumed that the ratio of soil heat flux to net radiation $\Gamma_c = 0.05$ for full vegetation canopy and $\Gamma_s = 0.315$ for bare soil. An interpolation is then performed between these limiting cases using the fractional canopy cover, f_c·(-).

The sensible heat flux is calculated using Equation (7).

$$H = ((R_n - G_0) - \frac{\rho C_p}{r_{ew}} \cdot \frac{(e_s - e)}{\gamma})/1 + \frac{\Delta}{\gamma}. \tag{7}$$

R_n net radiation, G_0 = soil heat flux, ρ = air density at constant pressure, C_p = specific heat of the air, $(e_s - e)$ = vapor pressure deficit of the air, γ = psychrometric constant, Δ = slope of the saturation vapor pressure.

We will use SEBS to estimate evaporation fraction by making energy balance at limiting cases at the dry limit and the wet limit as shown in Equation (8):

$$\Lambda_r = 1 - \frac{H - H_{wet}}{H_{dry} - H_{wet}} \tag{8}$$

where: the H_{wet} is sensible heat flux at the wet limit and H_{dry} sensible heat flux at the dry limit.

Since the ET ratio (evaporative fraction) Λ can be assumed constant during a day, the daily ET_{24} (mm) can be estimated using the following equations:

$$\Lambda = \frac{\lambda E}{R_n - G_0} = \frac{\Lambda_r - \lambda\, E_{wet}}{R_n - G}. \tag{9}$$

$$E_{daily} = \Lambda_0^{24} \cdot 8.64 \cdot 10^7 \cdot \frac{R_N - G_0}{\lambda \rho w} \tag{10}$$

where: Λ_r is relative evaporation, Λ_0^{24} daily evaporative fraction, ρw density of water measured in kilograms per cubic meter, and λ is the latent heat of vaporization.

2.3.2. Preparation of the Input Data for SEBS

To estimate actual evapotranspiration in SEBS the following data need to be prepared.

Normalized Different Vegetation Index (NDVI)

The NDVI was derived from Landsat 8 using Equation (11).

$$NDVI = \frac{\rho_5 - \rho_4}{\rho_5 + \rho_4} \tag{11}$$

where: ρ_5 is the reflectance in band 5, and ρ_4 is the reflectance in band 4.

Fraction of Vegetation Cover (FVC)

FVC can be derived from NDVI maps (Equation (12)). In this study, FVC was estimated as proposed by [36] for fully vegetated cover and bare soil, as presented in Equation (12).

$$FVC = \frac{NDVI - NDVI_s}{NDVI_v - NDVI_s} \tag{12}$$

where $NDVI_s$ represents the NDVI of bare soil, NDVI is the value of the actual pixel, and $NDVI_s$. corresponds to the NDVI value of the full vegetation canopy coverage.

Emissivity

The broad band land surface emissivity (ε) is calculated based on the FCV using Equation (13), which was introduced by [37].

$$\varepsilon = 0.004 \times FVC + 0.986 \tag{13}$$

Albedo

Albedo is the reflectance of a surface over a wide range of wavelengths. It indicates the reflected fraction of incoming radiation as a function of absorbed radiation. The broadband albedo was calculated using the Landsat 8 (OLI sensor) algorithm at the visible and NIR bands 2 to 7 in this study using Equation (14) [38].

$$\alpha_{OLI} = 0.362\rho_2 + 0.13\rho_4 + 0.373\rho_5 + 0.085\rho_6 + 0.072\rho_7 - 0.0018 \tag{14}$$

where: α_{OLI} is the shortwave albedo for Landsat, and ρ_i is the reflectance of bands i = 2, 4, 5, 6, and 7.

Metrological Data

ERA5 provides data on planetary boundary layer height, incoming shortwave radiation, specific humidity, and pressure, while the ERAinterm data set provides data on sunlight hours.

The air temperature and wind speed inputs will be obtained from the New Halfa meteorological station. These data will be used as point measurements that will be considered representative of the study area since it is a flat area with a gentle slope to the north.

2.3.3. Retrieval of Actual Evapotranspiration in SEBS

The evapotranspiration was calculated using the SEBS model extension, which is included in the ILLWIS program. Daily evapotranspiration maps for 40 days were produced using the downscaled LST.

2.3.4. Data and Processing

The datasets were downloaded from open data sources, as shown in Table 1.

Table 1. Data sources.

Data	Source	Spatial Resolution	Temporal Resolution
Landsat 8	https://espa.cr.usgs.gov/ordering/new/ (23 March 2020)	30 m	16 days
MODIS MOD11A1 V6	https://earthexplorer.usgs.gov/ (23 March 2020)	1 km	daily
NDVI	https://espa.cr.usgs.gov/ordering/new/ (23 March 2020)	30 m	16 days
Sunshine duration	https://apps.ecmwf.int/datasets/data/interim-full-daily/levtype=sfc/ (23 March 2020)	80 km	Daily
SRTM DEM	https://earthexplorer.usgs.gov/ (23 March 2020)	30 m	-
Other climatic data	https://www.ecmwf.int/en/forecasts/datasets/reanalysis-datasets/era5 (23 March 2020)	9 km	Daily

2.3.5. SEBS Validation

Since actual ET data were not available in the irrigated fields, the ETa estimated by SEBS was validated by comparing it to potential evapotranspiration estimated using crop coefficient and reference ET. This was based on the premise that under irrigated field conditions and adequate conditions for the ET, the crop's potential ET is approximately equal to the crop's actual evapotranspiration. Potential evapotranspiration (ETp) was estimated in the New halfa scheme as a product of kc and the reference evapotranspiration (ETo) values determined from the net radiation, relative humidity, air temperature, and wind speed using the Penman–Monteith method. The advantage of using remote sensing is that water used by the soil–water–vegetation system can be calculated directly without the need to quantify additional complicated hydrological processes.

2.3.6. Statistical Justification

The coefficient of determination, root mean square error (RMSE) Equation (15) and mean absolute error (MAE) Equation (16) are statistical measures of accuracy used to evaluate the performance of the downscaling modification and the evapotranspiration estimation.

$$RMSE=\sqrt{\frac{\sum_{i=1}^{n}(LST_{dow}-LST_{native})^2}{2}} \quad (15)$$

$$MAE = \frac{\sum_{i=1}^{n} [abs(LST_{dow} - LST_{native})]}{n} \quad (16)$$

where: LST_{dow} is the modeled variable, LST_{native} is the observed variable, and n is the number of observations.

3. Results and Discussion

3.1. LST and NDVI Regression

The research was carried out during the growing season of the crops. (December–March). LST is inversely linked to NDVI in Figure 3, as was shown by, among others, [39,40].

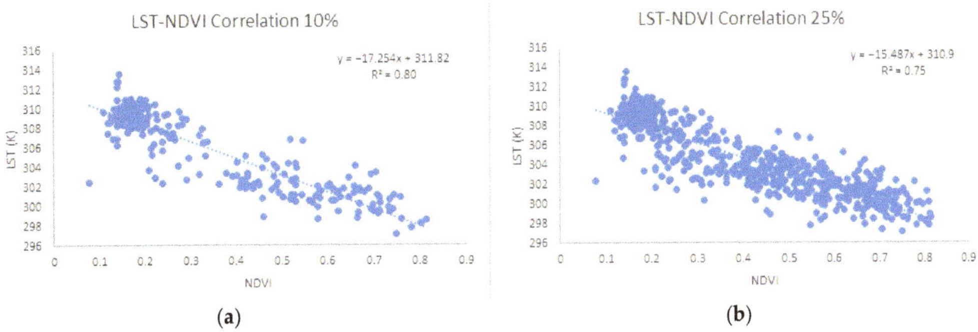

Figure 3. (**a**,**b**) Correlation between the NDVI and LST daily average for 10% and 25% methods, respectively.

Using a coefficient of determination (R^2) indicates a stronger correlation between LST and NDVI, with values ranging from 0.86 to 0.84, all statistically significant. As shown in Figure 4, the results show an agreement when using 10% or 25% of the aggregated pixels for the downscaling process. Using 10% of the pixels gives a good result where R^2 values of both methods, 10% and 25% were (0.84, 0.86). However, it was higher than R^2 between LST and NDVI for the native LST R^2 (0.69). We attribute that to LST from the native Landsat image. It has a spatial resolution of 100 m and is resampled to 30 m resolution. As a result, the fields in the downscaled LST maps are more detailed.

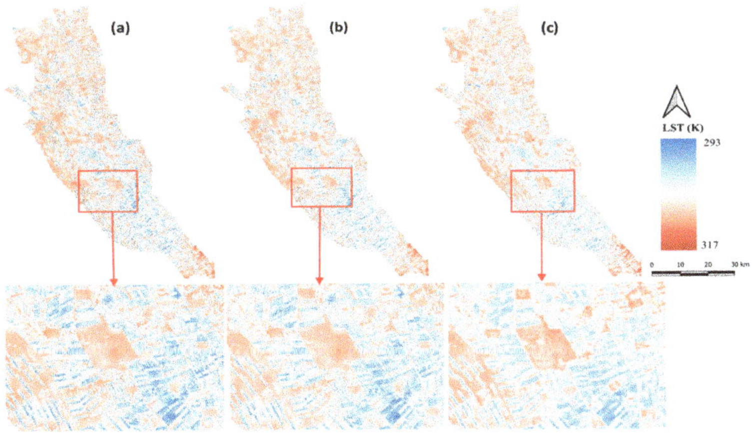

Figure 4. (**a**) LST Landsat10%, (**b**) LST Landsat 25%, and (**c**) LST Landsat native.

3.2. Effects of LST Downscaling on Landsat 8 Image

Figures 3 and 4 show native LST (LSTnative) and downscaled LST (LSTdown) images of a subset of the study area with a target resolution of 30 m. Lower temperatures are associated with dense vegetation (and water surfaces), whereas higher temperatures are associated with urban and bare soil regions distant from cultivated land and dry sandy riverbanks. The mixed land cover zones have a moderate temperature. These downscaled LST patterns are identical to the original in contrast, tone, and saturation Figures 3 and 5.

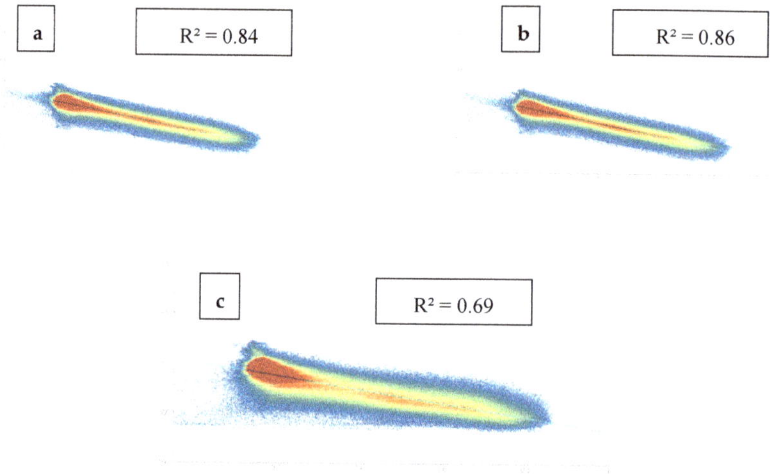

Figure 5. (**a**–**c**) Scatter plot for the relation between NDVI and LST 10%, LST 25%, and LST native, respectively.

However, the simulated LST is unable to determine the temperature of the water bodies, which is why LSTdow is greater than LSTnative. Because the water surface has no relationship with NDVI, the error surface from visual interpretation exhibits greater error corresponding to a water body. These results agree with the results found by [20]. Similarly, bare soil shows higher error rates, followed by urban landscapes. Similarly, bare soil exhibits the greatest error, followed by urban settings. A minimum error was found in the cultivated area. According to [28,40], homogenous vegetated areas have a lower value of downscaling error due to the accurate fitting of regression models over such regions. Table 2 shows the minimum, maximum, (Root Mean Square Error) RMSE, and (Mean Error) ME calculated from the difference between LSTdown and LSTnative. The minimum and maximum temperature values of LSTdown surfaces also agree with the LSTnative.

Table 2. Statistical analysis for the relation between native LST and LST25% and LST10%.

Method	Max ME	Min ME	Mean Error	RME
LST 25%	9.37	−5.12	−0.011	0.89
LST 10%	10.16	−5.63	−0.012	0.98

The minimum and maximum temperatures are very close to LSTnative at 10% and 25% of the data used for correlation, and the coefficient of determination (R^2) when using 10% is 0.72. It is R^2 = 0.74 for the 25%. Ref. [13] found that the RMSE increased with the increase of the target resolution, which is an indicator of decreasing accuracy in the finer resolution as shown in Figure 6. The same result was observed by [28]. This is because of the increasing degree of subpixel variability. Based on this statistical analysis, this modification yields superior results since the other scientists were attempting to achieve coarser resolution than the goal of this research.

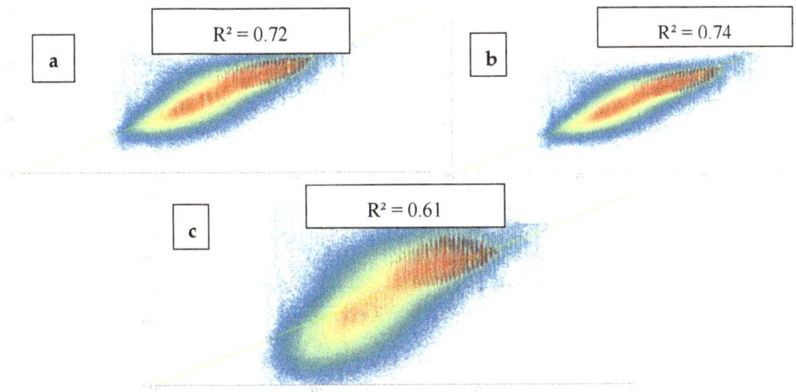

Figure 6. (**a–c**) Scatter plot between native LST with LST10%, LST 25%, and LST25% polynomial, respectively.

The DisTrad method recommended the use of 25% of aggregated pixels with the lowest coefficient of variation for defining the regression equation. Still, in the case of the heterogonous area, like the small fields in the study region, the coefficient of determination can be low due to the high number of mixed pixels. To overcome these problems, 10% of the aggregated pixels with the lowest coefficient of variation were used to fit the correlation.

The result shows that using 10% of the data with the lowest coefficient of variation gave a higher correlation than using 25% of the data with the lowest coefficient of variation, which R^2 was 0.75 and 0.80 for LST25% and 10%, respectively, as shown in Figure 6.

3.3. Effects of Downscaling LST on ET_a Estimation

Concerning the effect of downscaling the land surface temperature on evapotranspiration estimation, we found that the downscaling using only 10% of the pixels results in a good correlation due to the high degree of convergence between the results from these two methods and the native Land Surface Temperature, where the coefficient of determination for both methods 10% and 25% were (84.5 and 84.1) and yielded an average RMSE (0.3 and 0.28 mm/day, respectively). The evapotranspiration maps produced using downscaled land surface temperature had a higher spatial resolution than those produced using native land surface temperature, as the details and boundaries of small fields are more accurate on these maps than on the maps produced using native land surface temperature, as shown in Figures 7 and 8 [19] produced RMSE in the similar range of 0.16 and TsHARP = 0.55 mm/day using the non-linear disaggregation approach (NL-DisTrad).

3.4. Application of Downscaling Model on MODIS Data

After applying the downscaling model to both methods on the Landsat image, the downscaled surface temperature gives reasonable results at an accuracy of 30 m. MODIS data at 1000 m spatial resolution is available with a temporal resolution of one day (17 January 2018). The optical data of 250 m and 500 m are also available with temporal resolution every day, but the obstacle to using optical data from MODIS is that the results obtained from these sensors are not suitable for use in the water management process for small fields due to their poor spatial resolution. Therefore, it is preferable to use optical data from Landsat because they have a high spatial resolution of 30 m. We find that the NDVI from a single Landsat image is suitable for the downscaling model of more than one MODIS image, where it is assumed that the NDVI does not change significantly during this period. Therefore, the NDVI produced by Landsat was used to model the land surface temperature with MODIS products. By using the visual interpretation of the maps resulting from the downscaling process, we find that the downscaling process gives the same effect that appeared when the model was used on the image of Landsat, where the downscaling led to the clarification of the parameters of small fields and the indication of their limits. We also find that the land surface temperature pattern follows the NDVI pattern in terms of density. The low temperatures are in the fields with high NDVI density, while the high temperatures appear in the fields with low NDVI density or in the bare soil and urban areas. This method was followed due to the lack of land surface temperature data for the surface on that day to be taken as a reference. As several scientists indicated, we needed data taken on the same day to be considered a reference and used to validate these results [41,42]. Some scientists pointed out that the difference in the sensors could also affect the results [20]. On the other hand, when comparing the modifications applied to the Kustas method, the parameterization improves the correlation between the vegetation cover and LST, as shown in Figure 9 (a) NDVI and LST25% correlation, (b) NDVI and LST10% correlation and Figure 10 ETa (a) derived from MODIS LST resampled and (b) downscaled, respectively. MODIS LST (c) resampled and (d) downscaled. However, when we tested the MODIS land surface temperature using Landsat 8 data on 18 January 2018, the findings revealed an average RMSE of 1.3 K.

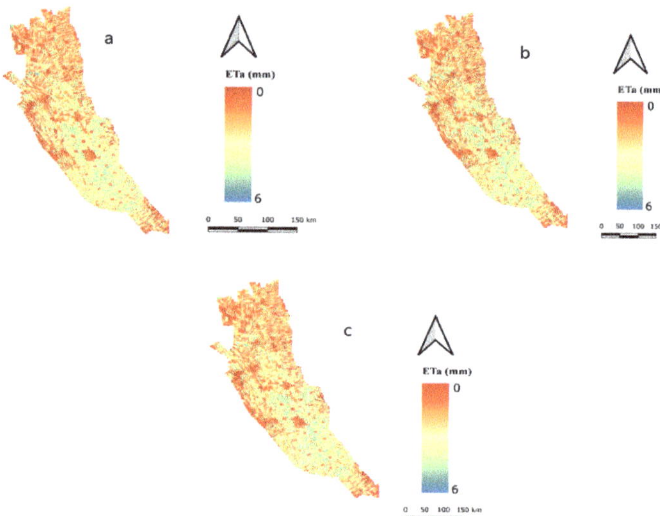

Figure 7. (**a**–**c**) Eta for native LST, LST 10%, and LST 25%, respectively.

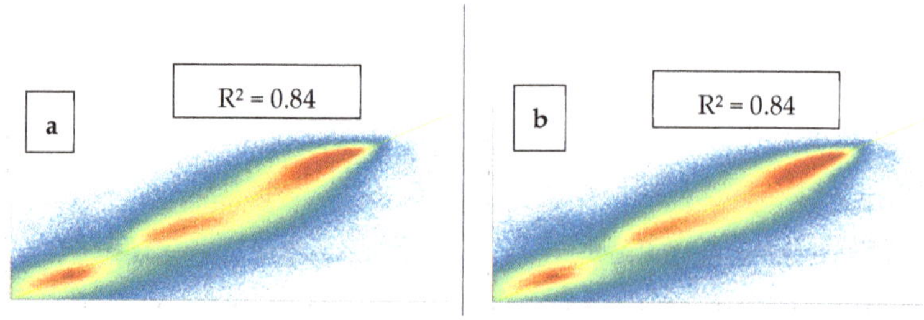

Figure 8. Correlation scatter plot of (**a**) ETa (LST native) against ETa (LST10%) and (**b**) ETa (LST native) against ETa (LST25%).

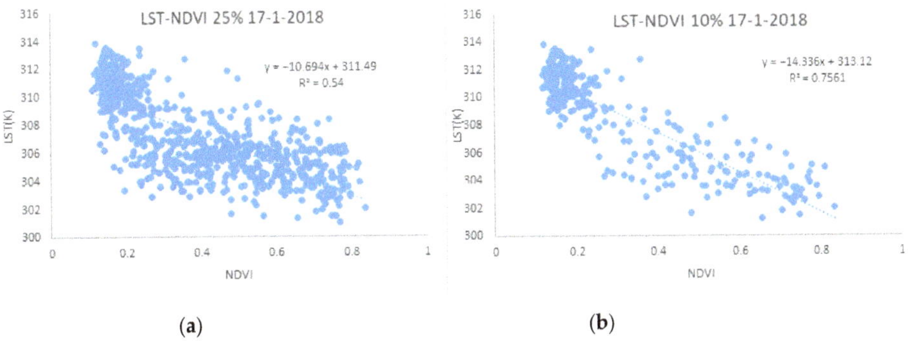

Figure 9. (**a**) NDVI and LST25% correlation, (**b**) NDVI and LST10% correlation.

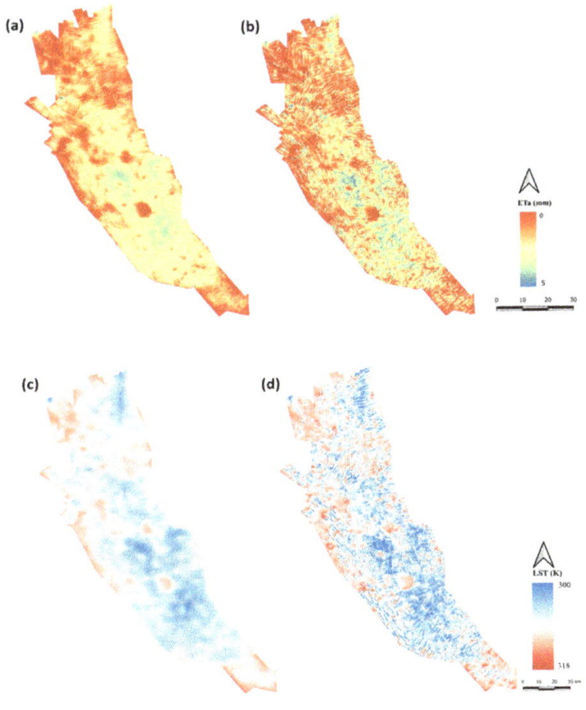

Figure 10. ETa (**a**) derived from MODIS LST resampled and (**b**) downscaled, respectively. MODIS LST (**c**) resampled and (**d**) downscaled.

3.5. Model Validation

To demonstrate that downscaling improves the temporal resolution of evapotranspiration, 40 downscaled land surface temperature maps were used to estimate actual evapotranspiration for wheat crops from 18 December 2017 to 24 February 2018. Eight pixels from eight different fields were chosen, and the mean calculated for these pixels was compared to potential evapotranspiration. Figures 11 and 12 shows the daily evapotranspiration (mm·day^{-1}) calculated using the SEBS approach, ranging from 1.5 mm·d^{-1} at the start of the season to 5.2 mm·d^{-1} in the mid-season.

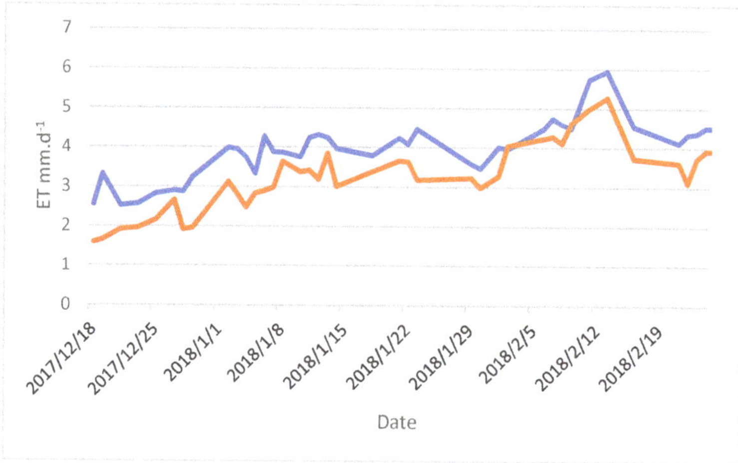

Figure 11. Comparison between daily ETp (blue line) VS. ETa (red line) for 18 August 2017 to 24 February 2018.

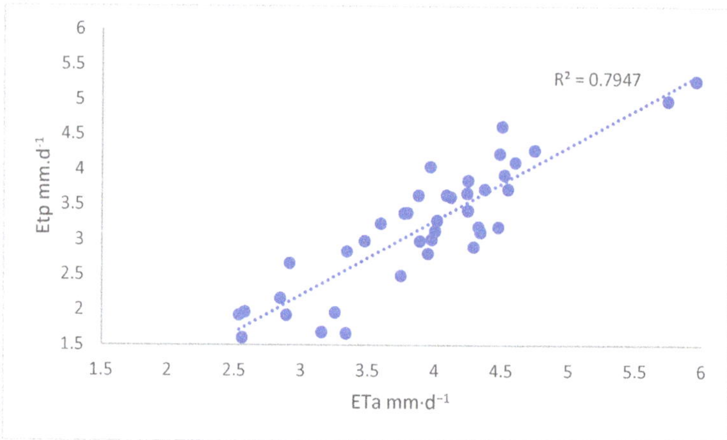

Figure 12. Correlation between ETp and ETa (mm·d^{-1}).

The result compared to potential evapotranspiration ETp was estimated from reference evapotranspiration based on the metrological parameters and crop coefficient. As illustrated in Figure 11, SEBS underestimated evapotranspiration. However, as shown in Figure 11, there was a strong correlation between the two products, with R^2 = 79. Figure 12, on the other hand, depicts realistic trends in the accuracy of ETa generated from SEBS. Figure 12 likewise depicts steady trends in ETa over the course of the wheat growing season. From the above-mentioned results, we can indicate that the LST downscaling improves the temporal evapotranspiration resolution.

4. Conclusions

This article demonstrates how to reparametrize the DisTrad technique for downscaling land surface temperature over small fields in arid and semi-arid environments by addressing the relationship between land surface temperature and vegetation cover at high resolution. The study area is described as a mixed land cover class. It has a variety of distinct land cover types (dense vegetation, medium vegetation, low vegetation, bare soil, urban areas, and water). Due to the complexity of the vegetation cover in this region, where all the fields have a small area, it is difficult to find pixels with a 1 km resolution with dense vegetation cover except in a few pixels. As a result, it was suggested that we use a smaller number of pixels for the downscaling process. It was also suggested to use a linear regression instead of a second-order polynomial since the linear regression is more robust at the edges of the curve and avoids the extreme values resulting from the second-order polynomial. Good results were achieved, which can improve the downscaling process in regions with characteristics like the study area. Compared to the polynomial regression, the LST linear regression resulted in an R^2 = 0.74, 0.72 for 25% and 10%, respectively.

The downscaled land surface temperature evolved into the Surface Energy Balance system module to estimate the actual evapotranspiration at higher temporal resolutions (30 m). This parametrization was applied to the DisTrad method to improve the temporal and spatial resolution of LST and ET, giving RMSE 0.3 and 0.28 mm/day, respectively. Previous studies by [19] used the non-linear method (NL-DisTrad) to aggregate land surface temperature and compared it with the TsHARP method, and integrated it into the model to estimate ET_a. Both methods yielded RMSE = 0.16 for (NL-DisTrad) and 0.55 mm/day for TsHARP another research done by [43] used g the Jarvis model to simulate surface resistance for downscaling evapotranspiration in Zhangye Oasis the result obtained in this work generated RMSE = 0.50 mm/day. Finally, we may extrapolate from the aforementioned findings that this change can enhance and provide dependable Eta maps for irrigation scheduling and agricultural planning. Based on the above-mentioned results, we can conclude that the DisTrad modification has improved the evapotranspiration spatial and temporal resolution.

Author Contributions: Conceptualization, Taha I. M. Ibrahim and Sadiq Al-Maliki; methodology, Taha I. M. Ibrahim; software, Omar Salameh; validation, Zoltán Vekerdy, István Waltner; formal analysis, Zoltán Vekerdy; in-vestigation, Taha I. M. Ibrahim; data curation, Taha I. M. Ibrahim and Sadiq Al-Maliki; writing—original draft preparation, Taha I. M. Ibrahim and Sadiq Al-Maliki; writing—review and editing, Zoltán Vekerdy, István Waltner; supervision Zoltán Vekerdy, István Waltner. All authors have read and agreed to the published version of the manuscript.

Funding: This research was funded by Tempus Public Foundation (TPF)/Stipendium Hungaricum Scholarships Program.

Institutional Review Board Statement: Not applicable.

Informed Consent Statement: Not applicable.

Data Availability Statement: The datasets used in this study are available from the corresponding author on reasonable request. All data and materials are available for publication.

Acknowledgments: We acknowledge a Tempus Public Foundation (TPF)/Stipendium Hungaricum Scholarships Program, Hungarian University of Agriculture and Life Sciences/Doctoral School of Environmental Sciences.

Conflicts of Interest: The authors declare no conflict of interest. The funders had no role in the design of the study; in the collection, analyses, or interpretation of data; in the writing of the manuscript, or in the decision to publish the results.

References

1. Bah, A.R.; Norouzi, H.; Prakash, S.; Blake, R.; Khanbilvardi, R.; Rosenzweig, C. Spatial Downscaling of GOES-R Land Surface Temperature over Urban Regions: A Case Study for New York City. *Atmosphere* **2022**, *13*, 332. [CrossRef]
2. Wang, S.; Luo, Y.; Li, X.; Yang, K.; Liu, Q.; Luo, X.; Li, X. Downscaling Land Surface Temperature Based on Non Linear Geographically Weighted Regressive Model over Urban Areas. *Remote Sens.* **2021**, *13*, 1580. [CrossRef]
3. Song, C.; Hu, G.; Wang, Y.; Qu, X. Downscaling ESA CCI Soil Moisture Based on Soil and Vegetation Component Temperatures Derived From MODIS Data. *IEEE J. Sel. Top. Appl. Earth Obs. Remote Sens.* **2022**, *15*, 2175–2184. [CrossRef]
4. Yang, Y. A Scale-Separating Framework for Fusing Satellite Land Surface Temperature Products. *Remote Sens.* **2022**, *14*, 983. [CrossRef]
5. Pan, X.; Zhu, X.; Yang, Y.; Cao, C.; Zhang, X.; Shan, L. Applicability of Downscaling Land Surface Temperature by Using Normalized Difference Sand Index. *Sci. Rep.* **2018**, *8*, 9530. [CrossRef]

6. Zakšek, K.; Oštir, K. Downscaling land surface temperature for urban heat island diurnal cycle analysis. *Remote Sens. Environ.* **2012**, *117*, 114–124. [CrossRef]
7. Zhan, W.; Chen, Y.; Zhou, J.; Wang, J.; Liu, W.; Voogt, J.; Zhu, X.; Quan, J.; Li, J. Disaggregation of remotely sensed land surface temperature: Literature survey, taxonomy, issues, and caveats. *Remote Sens. Environ.* **2013**, *131*, 119–139. [CrossRef]
8. Li, W.; Wu, H.; Duan, S.B.; Li, Z.L.; Liu, Q. Selection of Predictor Variables in Downscaling Land Surface Temperature using Random Forest Algorithm. In Proceedings of the 2019 IEEE International Geoscience and Remote Sensing Symposium, Yokohama, Japan, 28 July–2 August 2019; pp. 1817–1820.
9. Zhang, Z.; He, G.; Wang, M.; Long, T.; Wang, G.; Zhang, X. Validation of the generalized single-channel algorithm using Landsat 8 imagery and SURFRAD ground measurements. *Remote Sens. Lett.* **2016**, *7*, 810–816. [CrossRef]
10. Yu, P.; Zhao, T.; Shi, J.; Ran, Y.; Jia, L.; Ji, D. Data Descri tor Global spatiotemporally continuous MODIS land surface temperature dataset. *Sci. Data* **2022**, *9*, 143. [CrossRef]
11. Bartkowiak, P.; Castelli, M.; Notarnicola, C. Downscaling land surface temperature from MODIS dataset with random forest approach over alpine vegetated areas. *Remote Sens.* **2019**, *11*, 1319. [CrossRef]
12. Nugraha, A.S.A.; Gunawan, T.; Kamal, M. Downscaling land surface temperature on multi-scale image for drought monitoring. In Proceedings of the 6th Geoinformation Science Symposium 2019, Yogyakarta, Indonesia, 26–27 August 2019; p. 6.
13. Hutengs, C.; Vohland, M. Downscaling land surface temperatures at regional scales with random forest regression. *Remote Sens. Environ.* **2016**, *178*, 127–141. [CrossRef]
14. Peng, W.; Yuan, X.; Gao, W.; Wang, R.; Chen, W. Urban Climate Assessment of urban cooling effect based on downscaled land surface temperature: A case study for Fukuoka, Japan. *Urban Clim.* **2021**, *36*, 100790. [CrossRef]
15. U.S. Geological Survey. *Landsat 8 Data Users Handbook*; NASA: Washington, DC, USA, 2016; Volume 8, p. 97.
16. Guo, F.; Hu, D.; Schlink, U. Remote Sensing of Environment A new nonlinear method for downscaling land surface temperature by integrating guided and Gaussian filtering. *Remote Sens. Environ.* **2022**, *271*, 112915. [CrossRef]
17. Yang, Y.; Li, X.; Pan, X.; Zhang, Y.; Cao, C. Downscaling land surface temperature in complex regions by using multiple scale factors with adaptive thresholds. *Sensors* **2017**, *17*, 744. [CrossRef] [PubMed]
18. Pu, R. Assessing scaling effect in downscaling land surface temperature in a heterogenous urban environment. *Int. J. Appl. Earth Obs. Geoinf.* **2020**, *96*, 102256. [CrossRef]
19. Bindhu, V.M.; Narasimhan, B.; Sudheer, K.P. Development and verification of a non-linear disaggregation method (NL-DisTrad) to downscale MODIS land surface temperature to the spatial scale of Landsat thermal data to estimate evapotranspiration. *Remote Sens. Environ.* **2013**, *135*, 118–129. [CrossRef]
20. Mukherjee, S.; Joshi, P.; Garg, R. A comparison of different regression models for downscaling Landsat and MODIS land surface temperature images over heterogeneous landscape. *Adv. Space Res.* **2014**, *54*, 655–669. [CrossRef]
21. Sattari, F.; Hashim, M.; Pour, A.B. Thermal sharpening of land surface temperature maps based on the impervious surface index with the TsHARP method to ASTER satellite data: A case study from the metropolitan Kuala Lumpur, Malaysia. *Measurement* **2018**, *125*, 262–278. [CrossRef]
22. Atkinson, P.M. Downscaling in remote sensing. *Int. J. Appl. Earth Obs. Geoinf.* **2013**, *22*, 106–114.
23. Mao, Q.; Peng, J.; Wang, Y. Resolution Enhancement of Remotely Sensed Land Surface Temperature: Current Status and Perspectives. *Remote Sens.* **2021**, *13*, 1306. [CrossRef]
24. Kustas, W.P.; Norman, J.M.; Anderson, M.C.; French, A.N. Estimating subpixel surface temperatures and energy fluxes from the vegetation index-radiometric temperature relationship. *Remote Sens. Environ.* **2003**, *85*, 429–440. [CrossRef]
25. Yang, C.; Zhan, Q.; Lv, Y.; Liu, H. Downscaling Land Surface Temperature Using Multiscale Geographically Weighted Regression over Heterogeneous Landscapes in Wuhan, China. *IEEE J. Sel. Top. Appl. Earth Obs. Remote Sens.* **2019**, *12*, 5213–5222. [CrossRef]
26. Li, H.; Wu, G.; Xu, F.; Li, S. Landsat-8 and Gaofen-1 image-based inversion method for the downscaled land surface temperature of rare earth mining areas. *Infrared Phys. Technol.* **2021**, *113*, 103658. [CrossRef]
27. Xu, S.; Zhao, Q.; Yin, K.; He, G.; Zhang, Z.; Wang, G.; Wen, M.; Zhang, N. Spatial downscaling of land surface temperature based on a multi-factor geographically weighted machine learning model. *Remote Sens.* **2021**, *13*, 1186. [CrossRef]
28. Agam, N.; Kustas, W.P.; Anderson, M.; Li, F.; Neale, C.M. A vegetation index based technique for spatial sharpening of thermal imagery. *Remote Sens. Environ.* **2007**, *107*, 545–558. [CrossRef]
29. Ebrahimy, H.; Azadbakht, M. Downscaling MODIS land surface temperature over a heterogeneous area: An investigation of machine learning techniques, feature selection, and impacts of mixed pixels. *Comput. Geosci.* **2019**, *124*, 93–102. [CrossRef]
30. Laxén, J. *Is Prosopis a Curse or a Blessing? An Ecological-Economic Analysis of an Invasive Alien Tree Species in Sudan*; University of Helsinki, Viikki Tropical Resources Institute (VITRI): Helsinki, Finland, 2007.
31. Wallin, M. *Resettled for Development—The Case of New Halfa Agricultural Scheme, Sudan*; The Nordic Africa Institute: Uppsala, Sweden, 2014; Volume 59.
32. Srivastava, P.K.; Han, D.; Ramirez, M.R.; Islam, T. Machine Learning Techniques for Downscaling SMOS Satellite Soil Moisture Using MODIS Land Surface Temperature for Hydrological Application. *Water Resour. Manag.* **2013**, *27*, 3127–3144. [CrossRef]
33. Essa, W.; Verbeiren, B.; van der Kwast, J.; Batelaan, O. Improved DisTrad for downscaling thermal MODIS imagery over urban areas. *Remote Sens.* **2017**, *9*, 1243. [CrossRef]
34. Essa, W.; van der Kwast, J.; Verbeiren, B.; Batelaan, O. Downscaling of thermal images over urban areas using the land surface temperature-impervious percentage relationship. *Int. J. Appl. Earth Obs. Geoinf.* **2013**, *23*, 95–108. [CrossRef]

35. Su, Z. The Surface Energy Balance System (SEBS) for estimation of turbulent heat fluxes. *Hydrol. Earth Syst. Sci.* **2002**, *6*, 85–100. [CrossRef]
36. Jiménez-Muñoz, J.C.; Sobrino, J.A.; Plaza, A.; Guanter, L.; Moreno, J.; Martínez, P. Comparison between fractional vegetation cover retrievals from vegetation indices and spectral mixture analysis: Case study of PROBA/CHRIS data over an agricultural area. *Sensors* **2009**, *9*, 768–793. [CrossRef] [PubMed]
37. Sobrino, J.A.; Jiménez-Muñoz, J.C.; Paolini, L. Land surface temperature retrieval from LANDSAT TM 5. *Remote Sens. Environ.* **2004**, *90*, 434–440. [CrossRef]
38. Liang, S.; Shuey, C.J.; Russ, A.L.; Fang, H.; Chen, M.; Walthall, C.L.; Daughtry, C.S.T.; Hunt, R., Jr. Narrowband to broadband conversions of land surface albedo: II. Validation. *Remote Sens. Environ.* **2003**, *84*, 25–41. [CrossRef]
39. Karnieli, A.; Bayasgalan, M.; Bayarjargal, Y.; Agam, N.; Khudulmur, S.; Tucker, C.J. Comments on the use of the Vegetation Health Index over Mongolia. *Int. J. Remote Sens.* **2006**, *27*, 2017–2024. [CrossRef]
40. Jeganathan, C.; Hamm, N.; Mukherjee, S.; Atkinson, P.; Raju, P.; Dadhwal, V. Evaluating a thermal image sharpening model over a mixed agricultural landscape in India. *Int. J. Appl. Earth Obs. Geoinf.* **2011**, *13*, 178–191. [CrossRef]
41. Njuki, S.M. Assessment of Irrigation Performance by Remote Sensing in the Naivasha Basin, Kenya. Master's Thesis, University of Twente, Enschede, The Netherlands, February 2016; p. 65.
42. Kyalo, D.K. Sentinel-2 and MODIS Land Surface Temperature-Based Evapotranspiration for Irrigation Efficiency Calculations. Master's Thesis, University of Twente, Enschede, The Netherlands, February 2017; p. 50.
43. Tan, S.; Wu, B.; Yan, N. A method for downscaling daily evapotranspiration based on 30-m surface resistance. *J. Hydrol.* **2019**, *577*, 123882. [CrossRef]

International Journal of Geo-Information

MDPI

Article

Prediction and Uncertainty Capabilities of Quantile Regression Forests in Estimating Spatial Distribution of Soil Organic Matter

Melpomeni Nikou [1,2] and Panagiotis Tziachris [2,*]

[1] Department of Meteorology-Climatology, School of Geology, Aristotle University of Thessaloniki, 541 24 Thessaloniki, Greece; nikomelp@geo.auth.gr

[2] Soil and Water Resources Institute, Hellenic Agricultural Organization—Demeter, 570 01 Thessaloniki, Greece

* Correspondence: p.tziachris@swri.gr

Abstract: One of the core tasks in digital soil mapping (DSM) studies is the estimation of the spatial distribution of different soil variables. In addition, however, assessing the uncertainty of these estimations is equally important, something that a lot of current DSM studies lack. Machine learning (ML) methods are increasingly used in this scientific field, the majority of which do not have intrinsic uncertainty estimation capabilities. A solution to this is the use of specific ML methods that provide advanced prediction capabilities, along with innate uncertainty estimation metrics, like Quantile Regression Forests (QRF). In the current paper, the prediction and the uncertainty capabilities of QRF, Random Forests (RF) and geostatistical methods were assessed. It was confirmed that QRF exhibited outstanding results at predicting soil organic matter (OM) in the study area. In particular, R^2 was much higher than the geostatistical methods, signifying that more variation is explained by the specific model. Moreover, its uncertainty capabilities as presented in the uncertainty maps, shows that it can also provide a good estimation of the uncertainty with distinct representation of the local variation in specific parts of the area, something that is considered a significant advantage, especially for decision support purposes.

Keywords: quantile regression forests; random forests; geostatistics; machine learning; soil organic matter; prediction uncertainty

Citation: Nikou, M.; Tziachris, P. Prediction and Uncertainty Capabilities of Quantile Regression Forests in Estimating Spatial Distribution of Soil Organic Matter. *ISPRS Int. J. Geo-Inf.* **2022**, *11*, 130. https://doi.org/10.3390/ijgi11020130

Academic Editor: Wolfgang Kainz

Received: 7 December 2021
Accepted: 8 February 2022
Published: 11 February 2022

1. Introduction

Digital soil mapping (DSM), also known as predictive soil mapping or pedometric mapping, refers to the creation of digital maps that include spatial soil information, such as soil type or soil properties. These maps are created from the combination of multiple parameters (soil, climate, relief etc.) and usually depict the spatial distribution of soil phenomena along with relative information (e.g., estimation uncertainty).

DSM makes extensive use of geographic information systems (GIS), global positioning systems (GPS), remotely sensed spectral data, topographic data derived from digital elevation models (DEMs), predictive or inference models, and software for data analysis. To cope with the large amount of data used in DSM, semi-automated techniques and technologies are used to acquire, process, and visualize these data. Machine learning (ML) and artificial intelligence (AI) are some innovative state of the art technologies that are increasingly used in soil mapping and their uptake is transforming the way soil scientists produce their maps [1]. ML that emerged in the 1990s as a tool for DSM [2] is defined as the computer-assisted practice of using data-driven (and mostly non-linear) statistical models which resorts to a large amount of input data to learn a pattern and make a prediction [1].

According to Leo Breiman [3] two statistical modeling paradigms were distinguished: a data model and an algorithmic model. A data model is an abstract model that organizes elements of data and standardizes how they relate to one another and to the properties of real-world entities, whereas an algorithmic model is a model that uses mathematical algorithms based on the elements of data and estimates the parameters. One broadly used algorithm for that kind of model is Random Forests (RF). The RF is an ensemble learning method for classification, regression and other tasks that operates by constructing a multitude of decision trees at training time and is extensively used in DSM [4]. For example, it provided the best results in estimating soil OM [5], shortened the training time during the soil OM modeling process and improved the model's accuracy and its predictive ability [6]. Finally, according to John et al. [7], RF was the best performing model among other ML algorithms such as artificial neural network (ANN), support vector machine (SVM), and cubist regression.

Most of the time, DSM products represent estimates of spatially distributed soil properties. These estimations comprise an element of uncertainty that is not evenly distributed over the area covered by DSM. If we quantify the uncertainty spatially explicitly, this information can be used to improve the quality of DSM by optimizing the sampling design [8]. Wadoux et al. [1] stated that while the (spatial) cross-validation results might show strong agreement between predicted and measured soil property or class and therefore validate a ML model with very high predictive abilities, an uncertainty quantification would show unrealistic predictions characterized by a large uncertainty. However, most ML methods RF included do not provide uncertainty estimates by default and only 30% of the recent soil studies in their literature review quantified the uncertainty associated with the prediction.

One ML method that intrinsically addresses the lack of uncertainty estimates is Quantile Regression Forests (QRF). The QRF is an extension of RF developed by Nicolai Meinshausen [9] that provides non-parametric estimates of the median predicted value as well as prediction quantiles. It therefore allows spatially explicit non-parametric estimates of model uncertainty by providing information for the full conditional distribution of the response variable, and not only about the conditional mean [10]. As a result, QRF can potentially combine the high accuracy of RF with the built-in uncertainty estimates. However, QRF is not broadly used in soil studies, despite its advantages.

Vaysse and Lagacherie [11], for example, conducted an experiment in which they employed QRF in a temperate Mediterranean area with a comparable soil organic carbon (SOC) dataset in terms of areal extent, observation density, and distribution homogeneity. They claim that QRF outperforms RK when it comes to interpreting uncertainty patterns and is better suited than other modeling methods when spatial sampling is sparse. In Dharumarajan's study [12] the QRF model was used for the estimation of several important soil qualities of Northern Karnataka according to GlobalSoilMap criteria. The QRF model caught maximum variability for most of the soil parameters, and the predicted soil values were dependable with minimum errors. The QRF was also used for the production of global maps of soil properties explicitly highlighting the importance of quantitative and qualitative evaluation and uncertainty communication [13]. Finally, in Veronesi's study in 2019 [14], RF and QRF generated the most reliable confidence intervals for predicting SOC. Even though this is potentially important for practical uses, the confidence intervals were also very wide, so they suggest that these intervals should be handled carefully.

In the current study, the prediction capability along with the uncertainty assessment capacity of the QRF is examined. The popular geostatistical methods of Ordinary Kriging (OK) and Kriging with External Drift (KED) were compared with the ML methods of RF and QRF, in the case of soil OM. Prediction maps of soil OM along with maps of uncertainties were also produced and presented. The study area that was chosen is in northern Greece, at the regional unit of Kastoria and next to the shore of Lake Orestiada. A total number of 414 samples of soil were collected in randomly sampled unique locations in autumn during a six-year period. GPS receivers were used to identify the sampling positions. A high-resolution Digital Elevation Model (DEM) was used to derive topographic products such as aspect, slope, altitude etc. along with Sentinel-2 imagery, for each year from our study period, to produce Normalized Difference Vegetation Index (NDVI) and Normalized Difference Water Index (NDWI) that were used as input data. Finally, the effect of the above-mentioned covariates to the prediction of soil OM was assessed based on the importance score of the applied machine learning methods.

2. Materials and Methods

2.1. Area of Study and Soil Sampling

The study area was chosen in northern Greece, near the shore of Lake Orestiada, in the regional unit of Kastoria (Figure 1). Its coordinates in the World Geodetic System of 1984 (WGS84) include the area between 40°28′42.41″ N and 40°32′35.61″ N latitudes and longitudes of 21°19′4.01″ E and 21°23′8.18″ E longitudes.

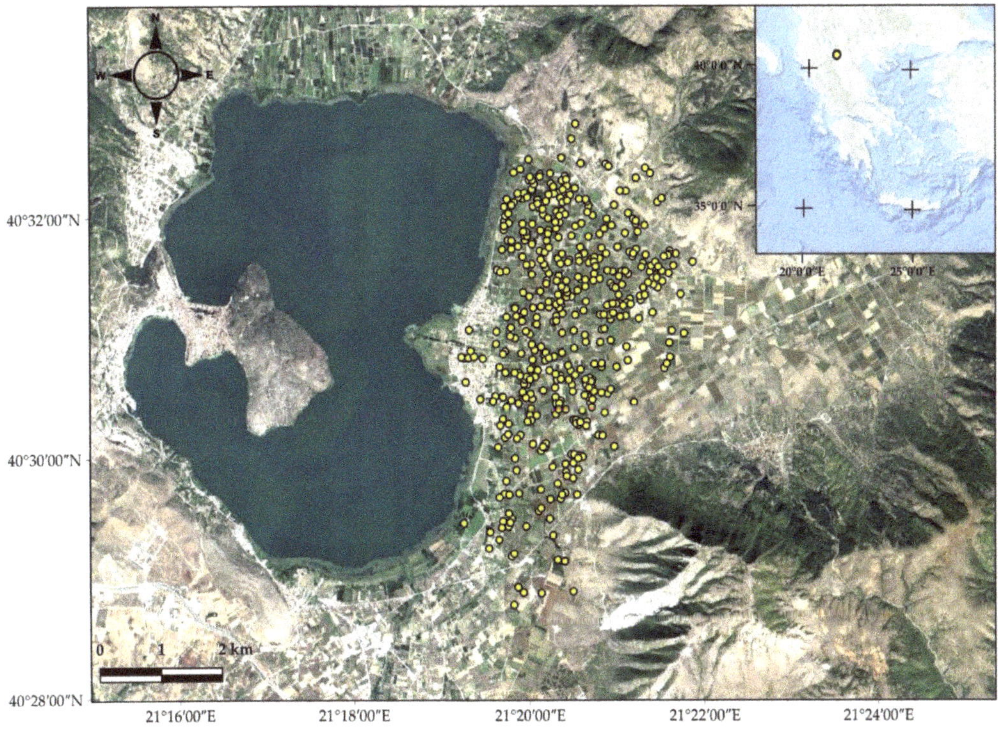

Figure 1. The study area on the shore of Lake Orestiada, in the regional unit of Kastoria, Greece.

While the region of interest is flat, the mean altitude is around 640 m above sea level, ranging from 620 m near the lake to 700 m further north. The climate is temperate and often warm, with harsh winters that frequently keep the temperature below zero throughout the day. The average annual temperature is 11.5 °C, with 636 mm of precipitation. Summers are hot and dry, with a relative humidity of 50% to 55%. Apple trees and beans are the primary agricultural crops.

During a six-year period, a total of 414 soil samples were collected in randomly sampled distinct places around the study area (2012 to 2019). A total of 30 cm of top soil was gathered during late fall season (around end of November). The sampling positions were determined using Global Positioning System (GPS) devices. The minimum distance between two sampling places varies between 60 and 480 m, with an average of 90 m.

2.2. Soil, Environmental and Satellite Covariates

In this study, soil variables, environmental variables and satellite images were selected (Table 1) as potential inputs in the models. Regarding soil covariates, the 414 soil samples that were collected from the area were analyzed for Clay (C) with soil hydrometer

(Bouyoucos) method [15], Magnesium (Mg) with ammonium acetate method and Zinc (Zn) with DTPA method [16]. Moreover, an organic matter (OM) analysis (wet oxidation method) from the same locations was conducted, to calibrate the models and to assess the prediction results. In more detail, during the soil sampling procedure, a composite soil sample consisting of several sub-samples up to a depth of 30 cm was obtained from each field parcel and the soil samples were dried and analyzed in the laboratory of the Soil and Water Resources Institute in Thessaloniki, Greece.

Table 1. Variables used in the study.

	Variables	Category
1	Clay (C)	soil
2	Organic Matter (OM)	soil
3	Magnesium (Mg)	soil
4	Zinc (Zn)	soil
5	Elevation (altitude)	environmental
6	Slope	environmental
7	Aspect	environmental
8	SAGA wetness index (TWI)	environmental
9	Negative Topographic Openness (openn)	environmental
10	Positive Topographic Openness (openp)	environmental
11	Deviation from Mean Value (devmean)	environmental
12	Multiresolution index of Valley Bottom Flatness (vbf)	environmental
13	NDVI 2016	satellite
14	NDWI 2016	satellite
15	NDVI 2017	satellite
16	NDWI 2017	satellite
17	NDVI 2018	satellite
18	NDWI 2018	satellite
19	NDVI 2019	satellite
20	NDWI 2019	satellite

The environmental covariates were derived from the second version of the Advanced Spaceborne Thermal Emission Radiometer-Global Digital Elevation Model version 2 (ASTER GDEM2). The release of the ASTER GDEM2 has enriched the availability of free-of-charge DEM sources, which are especially useful for developing countries, and prompted users to assess its quality and accuracy [17]. The ASTER GDEM2 is consisted of $1^\circ \times 1^\circ$ tiles (30 m resolution) in the World Geodetic System 1984 (WGS84), that was reprojected to Greek Geodetic Reference System 1987 (GGRS87) for this study [18].

Moreover, satellite indices were derived from Sentinel-2 imagery. More specifically Normalized Difference Vegetation Index (NDVI) and Normalized Difference Water Index (NDWI) from 2016 to 2019 were collected at approximately the same time period that the soil data were collected (end of November). The well known and widely used NDVI is a simple, but effective index for quantifying green vegetation. Red light is actively absorbed by healthy plants, while near infrared is reflected. To determine the state of a plant's health, we must compare the values of red and infrared light absorption and reflection [7,19]. The NDWI is a vegetation index sensitive to the water content of vegetation and is complementary to the NDVI. High NDWI values indicate a high plant water content and a high plant fraction coating. Low vegetation content and cover with low vegetation correspond to low NDWI values. The NDWI rate will drop during times of water stress [20].

2.3. Data Preparation and Assessment

Initially, the topographic data and satellite indices along with the soil analysis data were combined and spatially overlayed on the sampling locations. The overall dataset was assessed for outliers and missing values. From the initial 414 points, 403 points remained at the end that were used as inputs for the models in the study.

From the full set of variables, only a subset was used in the study. The variables were eliminated using Akaike Information Criteria (stepAIC) technique and Principal Component Analysis (PCA) and were also assessed for multicollinearity. The remaining variables were C, OM, ZN, MG, Vdepth, Altitude, NDVI_2016, NDVI_2017, and NDWI_2019 (Table 2).

Table 2. Descriptive statistics of the auxiliary variables from the 403 locations in study area.

	C (%)	OM (%)	ZN (%)	MG (%)	Vdepth (m)	Altitude (m)	NDVI_2016	NDVI_2017	NDWI_2019
mean	17.43	2.13	1.91	305.88	31.64	635.33	0.49	0.52	0.63
sd	7.07	0.76	1.43	148.04	7.06	7.39	0.13	0.14	0.10
median	16.00	2.00	1.51	275.00	31.46	634.00	0.50	0.55	0.66
trimmed	16.89	2.06	1.69	290.45	31.64	634.42	0.50	0.53	0.65
mad	5.93	0.68	1.01	127.50	7.14	5.93	0.11	0.13	0.07
min	2.00	0.64	0.12	44.00	4.67	624.00	0.10	0.12	0.24
max	48.00	5.24	10.75	905.00	46.58	680.00	0.75	0.90	0.79
skew	0.89	0.98	2.12	1.05	−0.17	2.49	−0.80	−0.57	−1.44
kurtosis	1.49	1.11	6.86	1.16	0.15	10.30	0.75	−0.05	1.96

The maps of the spatial distributions of the environmental covariates that were derived from ASTER GDEM2 and were used in the study (Vdepth and Altitude), are presented next (Figure 2).

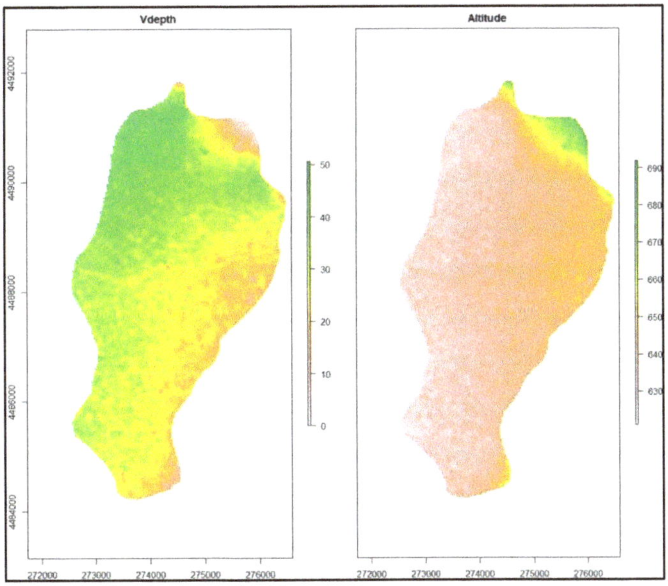

Figure 2. Topographic covariates of the study area.

The maps of the satellite covariates that were used in the study (NDVI_2016, NDVI_2017, and NDWI_2019) are the following (Figure 3).

Finally, the soil covariates (C, MG, ZN) were interpolated from the known point locations of the full dataset with the use of OK and their spatial distribution for the overall study area was estimated (Figure 4).

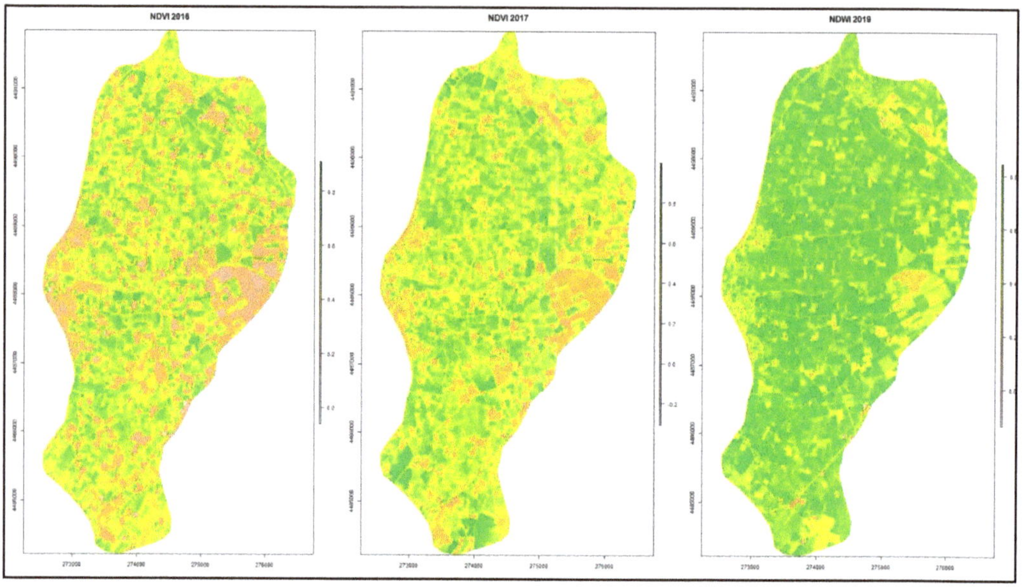

Figure 3. Satellite covariates of the study area.

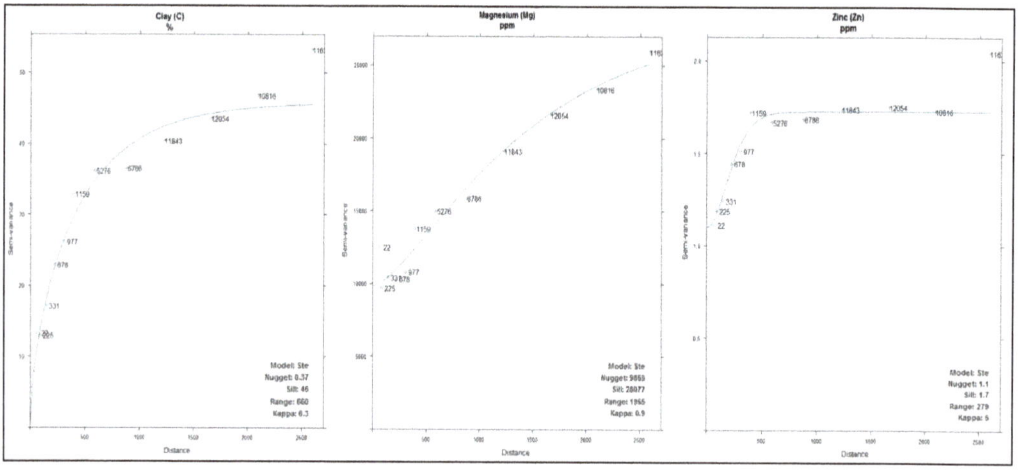

Figure 4. Semivariograms and fitted model of soil covariates of the study area.

For all the soil parameters, the Matern semivariogram model with M. Stein's parameterization (Ste) was applied as the fitted model using gstat's default parameters. Regarding C, its range was 660 m and exhibited a strong spatial dependence with a nugget to sill ratio of 0.8% [21]. The Mg had a range of 1955 m with strong spatial dependence (nugget to sill 3.5%) whereas Zn had a range of 279 m with moderate spatial dependence with a nugget to sill close to 65%.

The produced kriging maps from the soil covariates were used for the estimation of soil OM in the area by the models of the current study (KED, RF, QRF) (Figure 5).

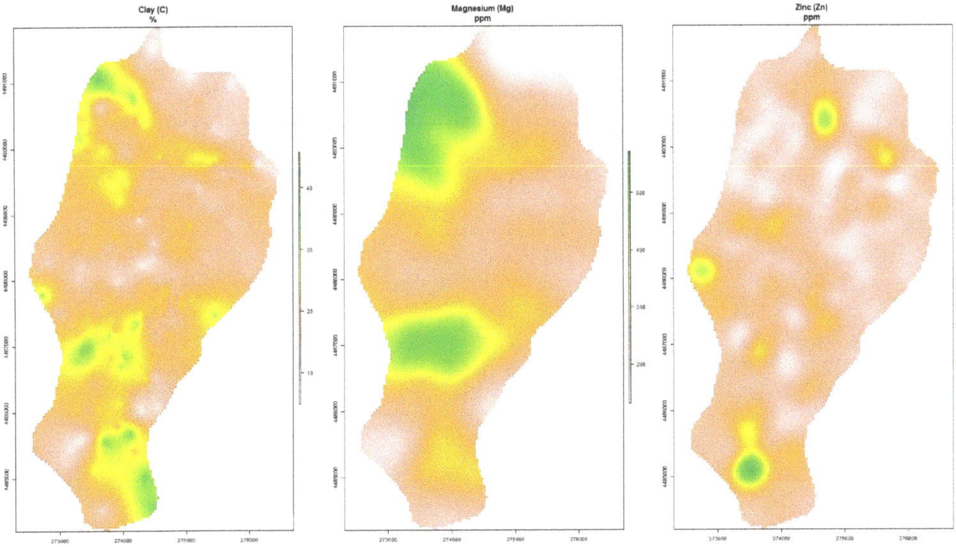

Figure 5. Spatial distribution of soil covariates of the study area.

2.4. Ordinary Kriging (OK) and Kriging with External Drift (KED)

Ordinary Kriging is a type of kriging in which the values' weights sum to one. It is linear because its estimates are a linear combination of the available data. It is also unbiased because it attempts to keep the mean residual to be zero and it tries to minimize the residual variance [22]. The OK implicitly evaluates the mean in a moving neighborhood with local second-order stationarity and its variance is equal to the sum of the simple kriging variance (assuming a known mean) plus the variance due to uncertainty about the true mean value [23].

Universal kriging (UK), Kriging with External Drift and Regression-Kriging (RK) belong to the group of the so-called 'hybrid' [24], i.e., non-stationary geo-statistical methods [23]. In classical geostatistics, spatial prediction for non-stationary processes is accomplished by taking into account a spatial trend (also known as "drift") that is either modeled solely as a function of the coordinates (in UK) or defined "externally" through some auxiliary variables (in KED) [25].

KED solves kriging weights by extending the covariance matrix with auxiliary variables so that the universality conditions are integrated into the kriging system; here, the difficulty is obtaining satisfactory residual variogram in the presence of drift [26].

The implementations of both OK and KED in the current study was done with the gstat package in R.

2.5. Random Forests (RF) and Quantile Regression Forests (QRF)

The idea of developing the RF method is based on the combination of the Bagging and Random Subspace Method, utilizing their advantages and compensating their disadvantages, with impressive results [3,27].

According to Breiman (2001) [3], in the case of classification, "A random forest is a classifier consisting of a collection of tree-structured classifiers {h(x, Θk), k = 1, ... } where the {Θk} are independent identically distributed random vectors and each tree casts a unit vote for the most popular class at input x". In case of regression Breiman states that "...random forests for regression are formed by growing trees depending on a random vector Θ such that the tree predictor h(x, Θ) takes on numerical values as opposed to class labels. The output values are numerical, and we assume that the training set is independently drawn from the distribution of the random vector Y, X".

The RF for regression is extensively used in DSM (e.g., [28–33]) with very positive results in the prediction of different soil parameters. More importantly though, it works equally well with skewed and normally distributed variables, without the need of statistical assumptions or restrictions that other methods demand. Therefore, it is easier and more straightforward to use. It just requires special attention in optimizing the hyperparameters to get the best results. One major drawback of some of the well-known ML methods (RF, ANN etc.) is their lack of intrinsic uncertainty estimation capabilities. So, apart from prediction maps, no prediction error variance can be estimated, contrary to classical geostatistics methods. The main reason for this is that most ML methods, RF included, provide only mean value predictions.

A possible solution to this deficiency came from Nicolai Meinshausen [9], who generalized the standard RF to provide information for the full conditional distribution of the response variable, and not only about the conditional mean. This ML algorithm is called Quantile Regression Forests (QRF) and gives a non-parametric and accurate way of estimating conditional quantiles for high-dimensional predictor variables. The key difference between QRF and RF is as follows: for each node in each tree, RF keeps only the mean of the observations that fall into this node and neglect all other information. In contrast, QRF keeps the values of all observations in this node, not just their mean, and assesses the conditional distribution based on this information.

In the current study, the ranger package in R language was used to implement the ML models. Ranger is a rapid implementation of RF or recursive partitioning that is especially well suited for high-dimensional data.

The assessment of the optimal hyperparameters of a ML model is a crucial step for the estimation of the best ML models for each specific use case. The ideal hyperparameter settings have a direct impact on the model's performance. Although there are various automatic optimization methods, their strengths and drawbacks change when applied to different types of situations [34]. In the current study, the random search method was performed (a 10 k-fold with 3 repeats), in which random combinations of parameters were employed from a range of values and used as hyperparameters. The ML model with the set of parameters that gave the highest accuracy was considered to be the best and used for prediction. The overall dataset (403 samples) was split in two distinct datasets: the training dataset (70% of the data) that was used for estimating the models hyperparameters and the testing dataset (30% of the data) that was used for assessing the different models. The specific hyperparameters for RF that were optimized are presented in Table 3.

Table 3. RF and QRF hyperparameters.

Hyperparameters	Packages	Description
mtry	ranger	The number of random features used in each tree.
num.trees	ranger	The number of grown trees.
min.node.size	ranger	Minimal node size.
splitrule	ranger	A switch for linear output units.

2.6. Uncertainty

The DSM products represent estimates of spatially distributed soil properties. These estimations comprise an element of uncertainty that is not evenly distributed over the area covered by DSM [8]. These flaws are being addressed by combining soil data at sites with spatially exhaustive environmental factors using quantitative models (e.g., [35–37]) DSM products are repeatable and enable continuous data display due to their quantitative nature. Models can also be updated for multiple reasons, and uncertainty can be measured [38,39].

Measurements, digitisation, data input, interpretation, categorization, generalisation, and interpolation are all common sources of mistake [40]. Modeling bias, parameterization, or even measurement mistakes connected with the input data can all cause uncertainty in digital soil maps [41]. Nelson et al. [42] recommends doing an error budget to assess the contribution of each error using a combination of geostatistical and Monte-Carlo simulations to gain a better understanding of uncertainty. The distinction between model error and spatially explicit uncertainty must also be considered [43]. The average squared difference between the estimated value and the actual value is known as model error, which is frequently assessed as the mean square error (MSE) [44,45]. However, spatially explicit uncertainty, often known as "local error", refers to the quantification of model output prediction intervals (e.g., [11,44,46]).

The prediction is linked to an explicit measure of the uncertainty. In many circumstances, such as in a decision-making process, it is just as important to quantify prediction uncertainty as it is to make the prediction itself, thus uncertainty maps are necessary (e.g., [47,48]). In DSM, uncertainty analysis is crucial in deciding whether the predicted soil map is dependable enough to be applied in agricultural production systems or decision-making. Uncertainty analysis also involves acknowledging the model's limitations, which is a step toward model interpretability [1]. As Heuvelink [49] states, we are very interested in prediction intervals in soil mapping, i.e., the range that is likely to contain the value yet to be measured. However, a very small amount of DSM studies estimate uncertainty. According to Wadoux [1], only around 30% of the studies presented in their paper quantified the prediction's uncertainty.

In the current study the uncertainty was estimated for only three out of four methods: OK, KED, QRF. The RF does not provide uncertainty estimation capabilities per se. The geostatistical methods of OK and KED provide by default the variance that it was used for the uncertainty assessment. Mainly, the standard deviation was calculated and its range was depicted in the maps. For the QRF the range was defined as one standard deviation above and below the median value. This range was used to create the uncertainty map of the study area

2.7. Error Assessment

Different metrics (Table 4) were employed to estimate model performance based on the difference between the observations and the predictions at the testing data set.

Table 4. Measurements to assess model performance.

Metrics	Equation	
Mean absolute error (*MAE*)	$MAE = \frac{\sum_{i=1}^{n} \lvert y_i - x_i \rvert}{n}$	(1)
Root mean square error (*RMSE*)	$RMSE = \sqrt{\frac{\sum_{i=1}^{n} [y_i - x_i]^2}{n}}$	(2)
Coefficient of determination (R^2)	$R^2 = 1 - \frac{SSE}{SSTO}$	(3)
Mean bias error (*MBE*)	$MBE = \frac{\sum_{i=1}^{n} (y_i - x_i)}{n}$	(4)

The root mean square error (RMSE) and the mean absolute error (MAE) were estimated, based on the measured value y_i and its prediction x_i in y_i locations of the samples (Equations (1) and (2)). The MAE is the average of the absolute values of the differences between the forecast and the corresponding observation over the verification sample. Since the MAE is a linear score, all individual differences are weighted equally in the average. The RMSE is a quadratic scoring rule that calculates the average magnitude of the error. Because errors are squared before being averaged, the RMSE gives large errors a relatively high weight. As a result, the RMSE is most useful when large errors are especially undesirable. The MAE and RMSE both have a range of 0 to ∞. They're negative scores, thus the lower the number, the better. The coefficient of determination (R^2) (Equation (3)) represents a model's ability to predict or explain an outcome. The R^2 indicates the percentage of variance in the predicted variable and the measured variable where SSE is the sum of squares of errors and SSTO the total sum of squares. The coefficient of determination ranges from 0 to 1, where in 0 (zero) no variation is explained by the model and in 1 (one) all variation is explained by the model. A high R^2 value, in general, implies that the model is a good fit for the data, though fit interpretations vary depending on the context of analysis. Finally, the mean bias error (MBE) was used as a measurement of the bias estimation of the models (Equation (4)).

2.8. Software

For the statistical analysis of the current study, the R (version 4.0.3) statistical software and the caret package were used [50]. Also, the ranger package [51] was utilized for RF and QRF. The geostatistics were implemented with the gstat package [52]. Finally, the Saga-GIS software (https://saga-gis.sourceforge.io/en/index.html (accessed on 18 November 2021)) was used for the environmental indices.

3. Results

3.1. Semivariograms and Fitting Parameters of OK and KED

Initially, OK and KED were implemented using the training dataset for the prediction of the soil OM in the study area. In the case of OK, based on the empirical semivariogram, the Matern semivariogram model with M. Stein's parameterization (Ste) was fitted (Figure 6) using the weighted least square fit of gstat package. The range was 207 m with a nugget at 0.35 and sill at 0.49. There was a moderate spatial dependence based on the nugget to sill ratio (71%).

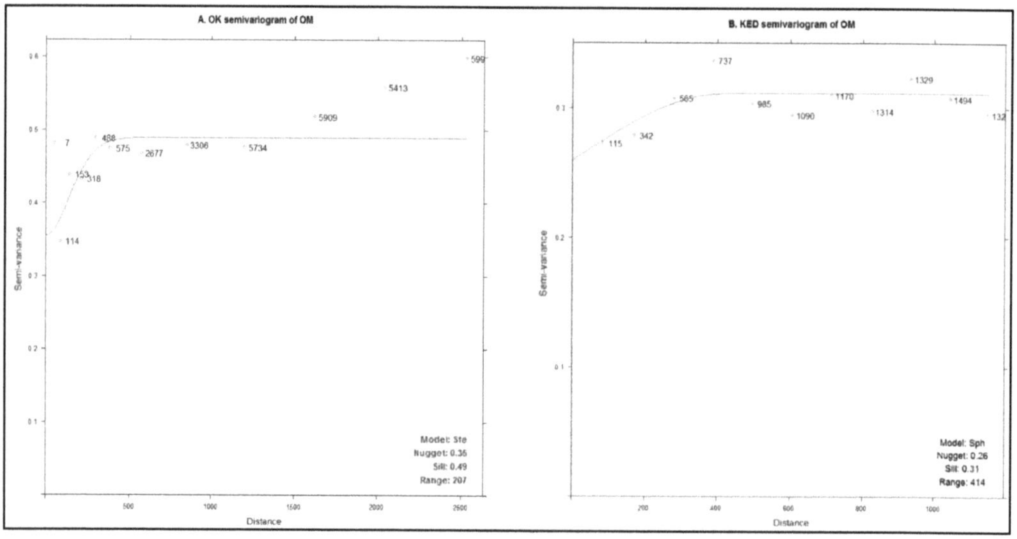

Figure 6. Empirical semivariogram and fitted model of OK and KED.

Regarding KED, the spherical model was used for fitting with the default method of gstat based on weighted least squares fit. The range was at 414 m, double than the OK range. In this case there was a weak spatial dependence with a nugget to total sill ratio of 83%.

3.2. RF and QRF Hyperparameters' Optimization Results

As it is already stated (Section 2.5), ML models need the assessment of their optimal hyperparameters in order to provide the best prediction results. In the case of RF and QRF as defined by the ranger library, four hyperparameters need to be estimated (Table 3). An iterative process (trial and error) was used with the random search optimization method, where different random values of these parameters were introduced from a range of values. The R^2 of the ML models was assessed using a 10-fold cross-validation method that was repeated 3 times in the training data set (Figure 7). The hyperparameters that returned the highest R^2 were finally chosen (Table 5).

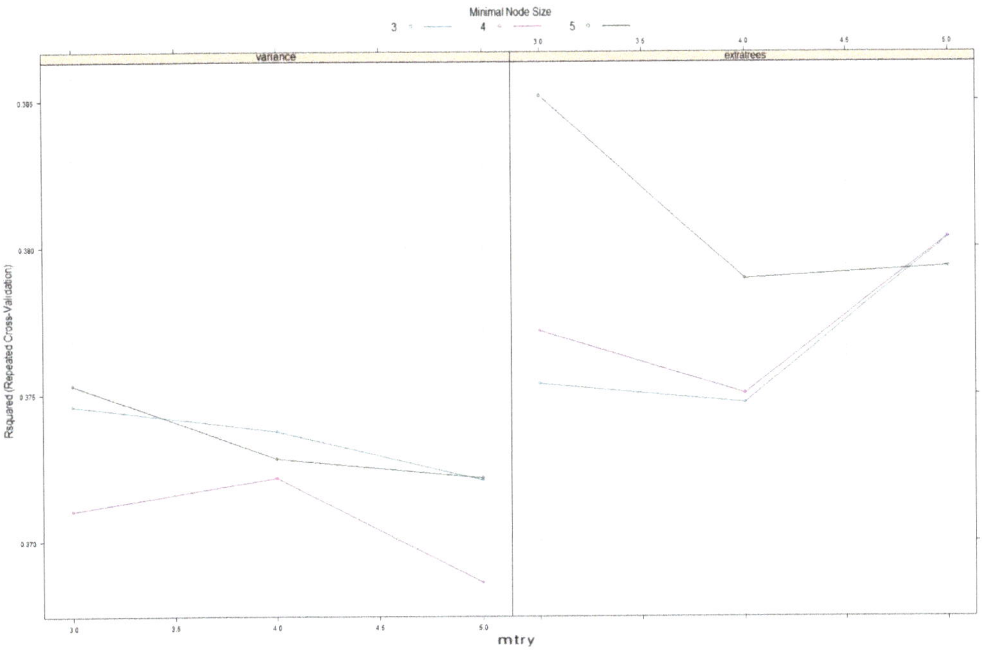

Figure 7. Hyperparameters assessment results based on R^2 in the training dataset.

For the splitrule hyperparameter, only "variance"and "extratrees" methods were used due to unrecoverable errors from "maxstat" and "beta" values.

Table 5. Hyperparameters' range and values used in the current study for RF and QRF.

Parameter	Range	Value Used
mtry	3–5	3
num.trees	500, 1000	500
min.node.size	3–5	5
splitrule	"variance", "extratrees", "maxstat" *, "beta" *	extratrees

* "Maxstat" and "beta" resulted unrecoverable errors and they were not used.

The specific optimal hyperparameters were introduced in the RF and QRF models and used to estimate their prediction capabilities on the testing dataset.

3.3. Feature Importance of the ML Models

The feature importance of the RF and QRF was estimated (Figure 8) with the permutation technique [3], that is defined as the decrease in the model score when a single feature value is randomly shuffled. A feature is "important" if shuffling its values increases the model error (strong effect on the prediction) and "unimportant" if shuffling its values leaves the model error unchanged (low or no effect on the prediction).

Regarding the importance scores, both RF and QRF concede that the soil covariates exhibited the highest importance, something that was expected due to comparable findings of a previous study [28] in a nearby area. In the current study specifically, Zn had the highest score with C second and Mg third. The Altitude from the topographic indices was next, along with the NDVI of 2016 and Vdepth. The last positions were occupied by NDVI of 2017 and NDWI of 2019.

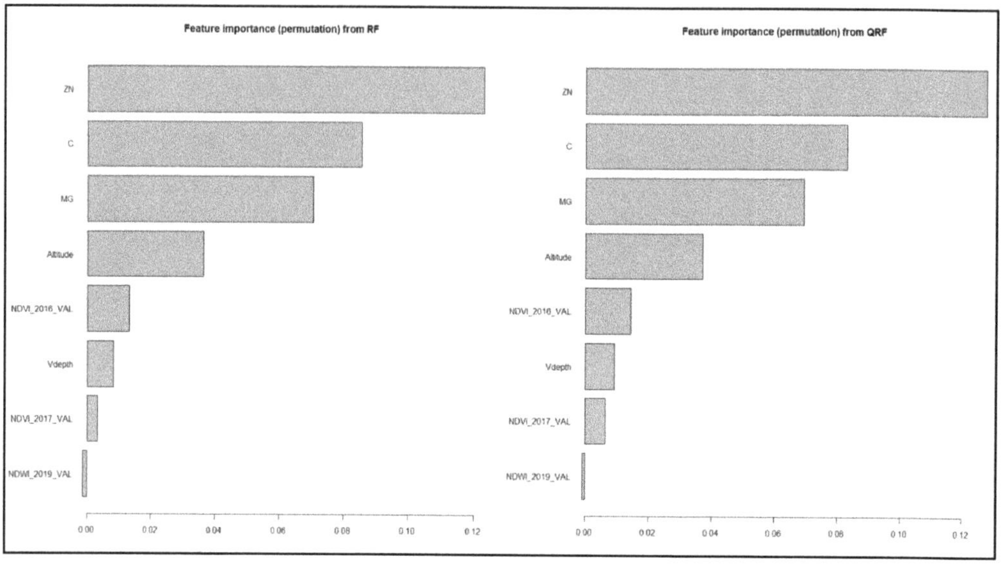

Figure 8. Hyperparameters assessment results based on R^2 in the training dataset.

3.4. Prediction Results

The dataset was partitioned into two random but spatially balanced sets of 70% for training the models and 30% for testing. The difference between the observations of the soil OM and their predictions in the testing data set was used to assess the prediction accuracy of the different models and they are presented in Table 6 and Figure 9.

Table 6. Prediction results for soil OM.

Model	RMSE	R^2	MAE	MBE
Ordinary Kriging (OK)	0.783	0.127	0.586	−0.002
Kriging with external drift (KED)	0.618	0.452	0.455	−0.022
Random Forests (RF)	0.615	0.538	0.453	−0.020
Quantile Regression Forests (QRF)	0.635	0.532	0.459	−0.046

As it is presented in the results (Table 6), OK was the least accurate model with very low R^2 (0.127) and high RMSE and MAE, something that was expected due to its lack of capacity to incorporate auxiliary information. The OK prediction capability is based only on the variable's (OM) spatial autocorrelation, hence the poor current results. It presented the smaller bias though based on the MBE (−0.002).

The KED combines the predictive capabilities of the trend that is based on the auxiliary variables, along with the kriging interpolation. Therefore, the results are decent, with low RMSE (0.618) and MAE (0.455) that are very close to RF and even slightly better than QRF. However, the coefficient of determination (0.452) is much worse than the ML methods. The bias was also small close to zero (−0.022), however higher than the OK.

The ML models exhibited higher prediction capabilities than geostatistical models. More specifically, the best results were achieved by the RF. Especially its R^2 was the highest (0.538) among the models with an improvement of around 20% from KED. Regarding RMSE and MAE, RF's results were best with the lowest values overall. The model bias was low (−0.020) close to KED's value.

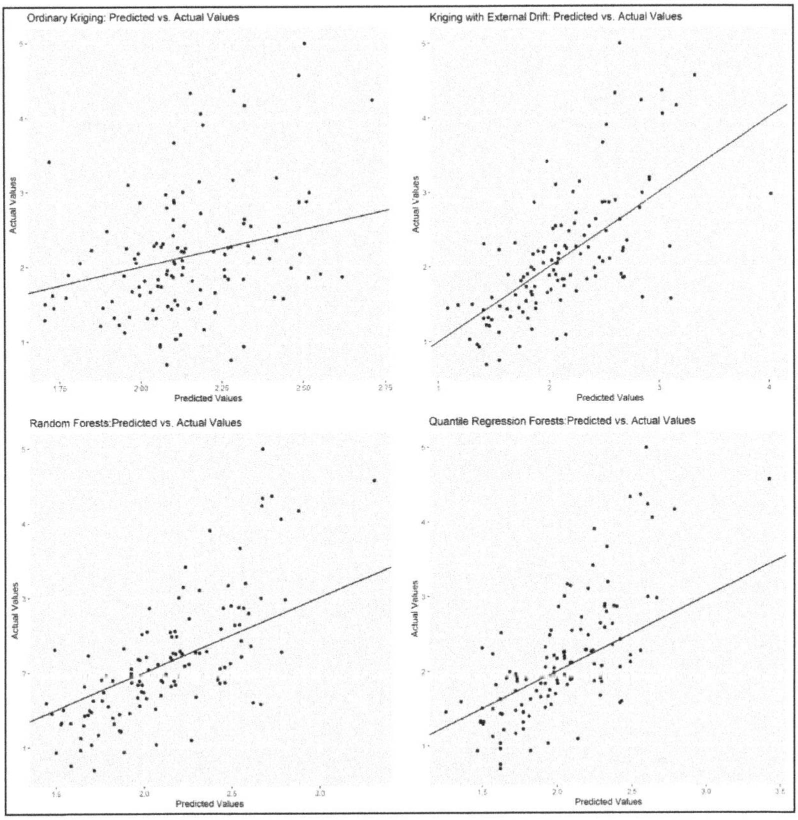

Figure 9. Predicted vs observed scatter plots.

The QRF model also exhibited very good prediction capabilities with high R^2 (0.532) very close to RF and quite low RMSE and MAE, close to RF and KED. The MBE was higher than the other models (−0.046), however still low and close to zero. Thus, RF and QRF can both be used interchangeably for predicting soil OM with similar results in the current study.

3.5. Maps of Prediction and Uncertainty

Next, two sets of maps were produced from the different models of the current study. The first set consists of four prediction maps that present the spatial distribution of soil OM in the area, one for each method: OK, KED, RF and QRF (Figure 10). The second set of maps consist of three maps with OK, KED and QRF methods that depict the spatial distribution of prediction's uncertainty in the area (Figure 11). In this case the RF was not used due to its lack of uncertainty capabilities.

The prediction map of OK exhibited interpolation results with prediction patterns that are relatively uniform all over the study area. The main reason for this is that OK is based solely on the spatial autocorrelation of OM using global model parameters that smooths the results. The KED model produces a map that changes more abruptly due to its covariates' effect, leading to multiple areas with higher and lower local values than the OK. Regarding ML methods (RF, QRF) their maps had even more contrast than the OK and KED, due to their capability of producing patterns that match data as much as possible by better fitting to the dataset. Among them, the QRF seem to present slightly more abrupt patterns than RF with areas with slightly lower and higher values.

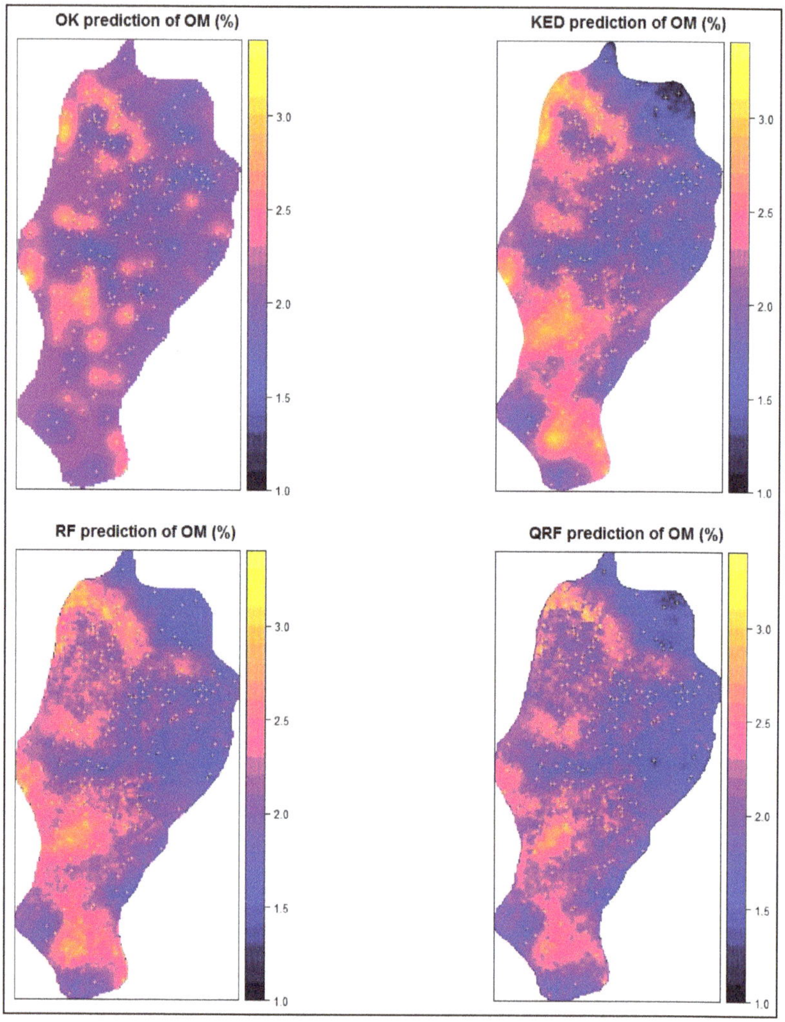

Figure 10. Prediction maps of soil OM.

As it was already mentioned, apart from the prediction results, prediction's uncertainty is a crucial parameter that needs to be estimated in the different locations of the study area. In the current study the uncertainty maps were calculated for only the 3 out of 4 methods. The RF does not support uncertainty estimation. The OK and KED intrinsically provide the error variance by which the standard deviation was calculated and its range was presented in the maps. For QRF the range was defined as one standard deviation above and below the median and it was used to create the uncertainty map of the study area (Figure 11).

Based on the uncertainty maps, it is obvious that OK has a smooth and equally distributed uncertainty range in the area with a mean value approximately at 0.8%. So, in each location the real OM value is approximately ±0.4% above or below the predicted value.

The KED uncertainty map has an overall lower uncertainty range than the OK (around 0.6%), that is almost equally distributed in the overall study area, similarly to OK. There are some slightly increased range values in the northern-west area along with some small patches in between (lighter blue areas) due to the covariates minor effect on the uncertainty results.

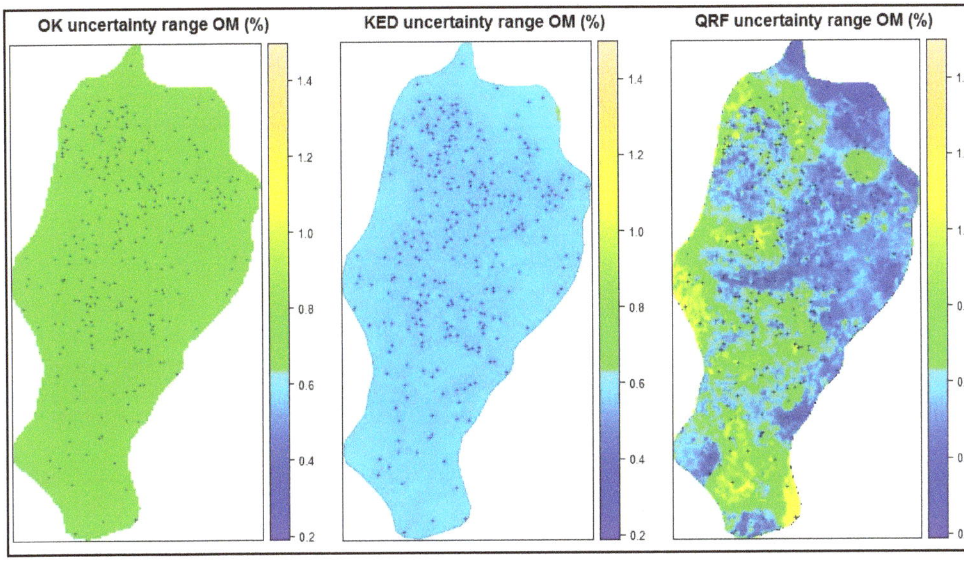

Figure 11. Uncertainty maps of soil OM.

In the case of QRF, the uncertainty map is more diversified than the previous ones. There are distinct regions of very low uncertainty like the one in the north or in the center of the area (with dark blue color) and regions of higher uncertainty like the ones in the south or close to the lake (with yellow color). This clear depiction of the uncertainty in a local scale and the straightforward definition of possible uncertainty zones, is a major advantage over the geostatistical methods especially for decision support purposes.

4. Discussion

One of the core tasks in the DSM studies is the estimation and presentation of the spatial distribution of different soil variables in the study area using different interpolation methods. Apart from that though, estimating and presenting the uncertainty of these interpolation methods are equally important in order to assess the overall work, something that is lacking in some of the recent DSM studies, especially the ones that are based on ML.

The ML methods are increasingly used in DSM, based on their outstanding prediction capabilities that outperforms the classic geostatistical methods, without the drawbacks of statistical assumptions and the restrictions of other methods. However, most of them do not have intrinsic uncertainty estimation capabilities. The RF is a very promising ML method used in multiple DSM studies that nevertheless lacks built-in uncertainty estimation capacity. An interesting alternative is QRF that seems to provide advanced prediction capabilities similar to RF along with innate uncertainty estimation metrics.

In the current paper, it was confirmed that QRF exhibited outstanding results at predicting soil OM in the study area, very close to RF method. Especially R^2 was much higher than the geostatistical methods, something that signifies that more variation is explained by the specific model. Moreover, its uncertainty capabilities as presented in the uncertainty maps, shows that it can also provide a very efficient estimation of the uncertainty in the study area. Uncertainty map with QRF exhibit stronger contrast compared to uncertainty maps of OK, and KED, with distinct representation of the local variation of the uncertainty such as small regions with higher or lower uncertainty. Based on this map it is very easy for a user to define clusters of uncertainty zones and categorize its effect locally. This is a real significant advantage, especially for decision support purposes in which users are interested not only in the prediction accuracy but also in the variation of the error range in different parts of the area.

Author Contributions: Conceptualization, Panagiotis Tziachris; Methodology, Panagiotis Tziachris; Software, Panagiotis Tziachris and Melpomeni Nikou; Data Curation, Melpomeni Nikou; Writing—Original Draft Preparation, Panagiotis Tziachris and Melpomeni Nikou; Writing—Review and Editing, Panagiotis Tziachris and Melpomeni Nikou. All authors have read and agreed to the published version of the manuscript.

Funding: This research received no external funding.

Data Availability Statement: Not applicable.

Acknowledgments: The authors are grateful to the staff of the Soil and Water Resources Institute of Hellenic Agricultural Organization—Demeter in Thessaloniki, Greece for their support in this study (I. Metaxa, A. Bountla, K. Oikonomou, G. Strikos, C. Tsotras and I. Palavatsios).

Conflicts of Interest: The authors declare no conflict of interest.

References

1. Wadoux, A.M.J.C.; Minasny, B.; McBratney, A.B. Machine learning for digital soil mapping: Applications, challenges and suggested solutions. *Earth-Sci. Rev.* **2020**, *210*, 103359. [CrossRef]
2. Lagacherie, P. *Digital Soil Mapping: A State of the Art*; Digital Soil Mapping with Limited Data; Springer: Berlin, Germany, 2008; pp. 3–14. [CrossRef]
3. Breiman, L. Random Forests. *Mach. Learn.* **2001**, *45*, 5–32. [CrossRef]
4. Kam, T.H. Random Decision Forests Tin Kam Ho Perceptron training. *Proc. 3rd Int. Conf. Doc. Anal. Recognit.* **1995**, *1*, 278–282.
5. Wiesmeier, M.; Barthold, F.; Blank, B.; Kögel-Knabner, I. Digital mapping of soil organic matter stocks using Random Forest modeling in a semi-arid steppe ecosystem. *Plant Soil* **2011**, *340*, 7–24. [CrossRef]
6. Liu, J.; Dong, Z.; Xia, J.; Wang, H.; Meng, T.; Zhang, R.; Han, J.; Wang, N.; Xie, J. Estimation of soil organic matter content based on CARS algorithm coupled with random forest. *Spectrochim. Acta-Part A Mol. Biomol. Spectrosc.* **2021**, *258*, 119823. [CrossRef]
7. John, K.; Isong, I.A.; Kebonye, N.M.; Ayito, E.O.; Agyeman, P.C.; Afu, S.M. Using machine learning algorithms to estimate soil organic carbon variability with environmental variables and soil nutrient indicators in an alluvial soil. *Land* **2020**, *9*, 487. [CrossRef]
8. Stumpf, F.; Schmidt, K.; Goebes, P.; Behrens, T.; Schönbrodt-Stitt, S.; Wadoux, A.; Xiang, W.; Scholten, T. Uncertainty-guided sampling to improve digital soil maps. *Catena* **2017**, *153*, 30–38. [CrossRef]
9. Meinshausen, N.; Ridgeway, G. Quantile regression forests. *J. Mach. Learn. Res.* **2006**, *7*, 983–999.
10. Freeman, E.A.; Moisen, G.G. An application of quantile random forests for predictive mapping of forest attributes. *For. Invent. Anal. Symp.* **2015**, *931*, 362.
11. Vaysse, K.; Lagacherie, P. Using quantile regression forest to estimate uncertainty of digital soil mapping products. *Geoderma* **2017**, *291*, 55–64. [CrossRef]
12. Dharumarajan, S.; Vasundhara, R.; Suputhra, A.; Lalitha, M.; Hegde, R. Prediction of Soil Depth in Karnataka Using Digital Soil Mapping Approach. *J. Indian Soc. Remote Sens.* **2020**, *48*, 1593–1600. [CrossRef]
13. Poggio, L.; De Sousa, L.M.; Batjes, N.H.; Heuvelink, G.B.M.; Kempen, B.; Ribeiro, E.; Rossiter, D. SoilGrids 2.0: Producing soil information for the globe with quantified spatial uncertainty. *Soil* **2021**, *7*, 217–240. [CrossRef]
14. Veronesi, F.; Schillaci, C. Comparison between geostatistical and machine learning models as predictors of topsoil organic carbon with a focus on local uncertainty estimation. *Ecol. Indic.* **2019**, *101*, 1032–1044. [CrossRef]
15. Bouyoucos, G.J. Directions for Making Mechanical Analyses of Soils by the Hydrometer Method. *Soil Sci.* **1936**, *42*, 225–230. [CrossRef]
16. Lindsay, W.L.; Norvell, W.A. Development of a DTPA Soil Test for Zinc, Iron, Manganese, and Copper. *Soil Sci. Soc. Am. J.* **1978**, *42*, 421–428. [CrossRef]
17. Suwandana, E.; Kawamura, K.; Sakuno, Y.; Kustiyanto, E.; Raharjo, B. Evaluation of aster GDEM2 in comparison with GDEM1, SRTM DEM and topographic-map-derived DEM using inundation area analysis and RTK-DGPS data. *Remote Sens.* **2012**, *4*, 2419–2431. [CrossRef]
18. del Rosario Gonzalez-Moradas, M.; Viveen, W. Evaluation of ASTER GDEM2, SRTMv3.0, ALOS AW3D30 and TanDEM-X DEMs for the Peruvian Andes against highly accurate GNSS ground control points and geomorphological-hydrological metrics. *Remote Sens. Environ.* **2020**, *237*, 111509. [CrossRef]
19. Roy, D.P.; Li, Z.; Zhang, H.K. Adjustment of sentinel-2 multi-spectral instrument (MSI) red-edge band reflectance to nadir BRDF adjusted reflectance (NBAR) and quantification of red-edge band BRDF effects. *Remote Sens.* **2017**, *9*, 1325. [CrossRef]
20. Hill, M.J. Vegetation index suites as indicators of vegetation state in grassland and savanna: An analysis with simulated SENTINEL 2 data for a North American transect. *Remote Sens. Environ.* **2013**, *137*, 94–111. [CrossRef]

21. Cambardella, C.A.; Moorman, T.B.; Novak, J.M.; Parkin, T.B.; Karlen, D.L.; Turco, R.F.; Konopka, A.E. Field-Scale Variability of Soil Properties in Central Iowa Soils. *Soil Sci. Soc. Am. J.* **1994**, *58*, 1501–1511. [CrossRef]
22. Peng, X.; Wang, K.; Li, Q. A new power mapping method based on ordinary kriging and determination of optimal detector location strategy. *Ann. Nucl. Energy* **2014**, *68*, 118–123. [CrossRef]
23. Wackernagel, H. *Multivariate Geostatistics*; Springer: Berlin, Germany, 1998; ISBN 9783662035528.
24. McBratney, A.B.; Odeh, I.O.A.; Bishop, T.F.A.; Dunbar, M.S.; Shatar, T.M. An Overview of Pedometric Techniques for Use in Soil Survey. *Geoderma* **2000**, *97*, 293–327. [CrossRef]
25. Ignaccolo, R.; Mateu, J.; Giraldo, R. Kriging with external drift for functional data for air quality monitoring. *Stoch. Environ. Res. Risk Assess.* **2014**, *28*, 1171–1186. [CrossRef]
26. Webster, R.; Oliver, M.A. *Geostatistics for Environmental Scientists*; Wiley: Chichester, UK, 2001; ISBN 0471965537.
27. Dietterichl, T.G. Ensemble Learning. In *The Handbook of Brain Theory and Neural Networks*; Arbib, M., Ed.; MIT Press: Cambridge, MA, USA, 2002; pp. 405–408.
28. Tziachris, P.; Aschonitis, V.; Chatzistathis, T.; Papadopoulou, M. Assessment of spatial hybrid methods for predicting soil organic matter using DEM derivatives and soil parameters. *Catena* **2019**, *174*, 206–216. [CrossRef]
29. Dharumarajan, S.; Hegde, R.; Singh, S.K. Spatial prediction of major soil properties using Random Forest techniques-A case study in semi-arid tropics of South India. *Geoderma Reg.* **2017**, *10*, 154–162. [CrossRef]
30. Wang, D.; Zhu, A.X. Soil mapping based on the integration of the similarity-based approach and random forests. *Land* **2020**, *9*, 174. [CrossRef]
31. Stum, A.K.; Boettinger, J.L.; White, M.A.; Ramsey, R.D. Random Forests Applied as a Soil Spatial Predictive Model in Arid Utah. *Digit. Soil Mapp.* **2010**, 179–190. [CrossRef]
32. Fernandes, D.; Machado, T.; Silva, H.G.; Curi, N.; Duarte De Menezes, M. Soil type spatial prediction from Random Forest: Different training datasets. *Sci. Agric.* **2019**, *76*, 243–254.
33. Shukla, G.; Garg, R.D.; Srivastava, H.S.; Garg, P.K. An effective implementation and assessment of a random forest classifier as a soil spatial predictive model. *Int. J. Remote Sens.* **2018**, *39*, 2637–2669. [CrossRef]
34. Yang, L.; Shami, A. On hyperparameter optimization of machine learning algorithms: Theory and practice. *Neurocomputing* **2020**, *415*, 295–316. [CrossRef]
35. McBratney, A.B.; Mendonça Santos, M.L.; Minasny, B. On digital soil mapping. *Geoderma* **2003**, *117*, 3–52. [CrossRef]
36. MacMillan, R.A. Experiences with Applied DSM: Protocol, Availability, Quality and Capacity Building BT-Digital Soil Mapping with Limited Data. In *Digital Soil Mapping with Limited Data*; Hartemink, A.E., McBratney, A., Mendonça-Santos, M., Eds.; Springer: Dordrecht, The Netherlands, 2008; pp. 113–135; ISBN 978-1-4020-8592-5.
37. Scull, P.; Franklin, J.; Chadwick, O.A.; McArthur, D. Predictive soil mapping: A review. *Prog. Phys. Geogr.* **2003**, *27*, 171–197. [CrossRef]
38. Heuvelink, G.B.M. Identification of field attribute error under different models of spatial variation. *Int. J. Geogr. Inf. Syst.* **1996**, *10*, 921–935. [CrossRef]
39. Kempen, B.; Brus, D.J.; Stoorvogel, J.J.; Heuvelink, G.B.M.; de Vries, F. Efficiency Comparison of Conventional and Digital Soil Mapping for Updating Soil Maps. *Soil Sci. Soc. Am. J.* **2012**, *76*, 2097–2115. [CrossRef]
40. Arrouays, D.; McKenzie, N.; Hempel, J.; de Forges, A.R.; McBratney, A.B. *GlobalSoilMap: Basis of the Global Spatial Soil Information System*; CRC Press: Boca Raton, FL, USA, 2014.
41. Minasny, B.; McBratney, A.B. Uncertainty analysis for pedotransfer functions. *Eur. J. Soil Sci.* **2002**, *53*, 417–429. [CrossRef]
42. Nelson, M.A.; Bishop, T.F.A.; Triantafilis, J.; Odeh, I.O.A. An error budget for different sources of error in digital soil mapping. *Eur. J. Soil Sci.* **2011**, *62*, 417–430. [CrossRef]
43. Kasraei, B.; Heung, B.; Saurette, D.D.; Schmidt, M.G.; Bulmer, C.E.; Bethel, W. Quantile regression as a generic approach for estimating uncertainty of digital soil maps produced from machine-learning. *Environ. Model. Softw.* **2021**, *144*, 105139. [CrossRef]
44. Malone, B.P.; McBratney, A.B.; Minasny, B. Empirical estimates of uncertainty for mapping continuous depth functions of soil attributes. *Geoderma* **2011**, *160*, 614–626. [CrossRef]
45. Wang, Z.; Bovik, A.C. Mean squared error: Love it or leave it? *IEEE Signal Process. Mag.* **2009**, *26*, 98–117. [CrossRef]
46. Feizizadeh, B.; Jankowski, P.; Blaschke, T. A GIS based spatially-explicit sensitivity and uncertainty analysis approach for multi-criteria decision analysis. *Comput. Geosci.* **2014**, *64*, 81–95. [CrossRef]
47. Hengl, T.; Toomanian, N. Maps are not what they seem: Representing uncertainty in soilproperty maps. In Proceedings of the Accuracy 2006: 7th International Symposium on Spatial Accuracy Assessment in Natural Resources and Environmental Sciences, Lisbon, Portugal, 5–7 July 2006; pp. 805–813.
48. Goovaerts, P. Geostatistical modelling of uncertainty in soil science. *Geoderma* **2001**, *103*, 3–26. [CrossRef]
49. Heuvelink, G.B.M. Uncertainty quantification of globalsoilmap products. Basis of the global spatial soil information, system. In Proceedings of the First GlobalSoilMap Conference, Orleans, France, 7–9 January 2014; pp. 335–340. [CrossRef]
50. Kuhn, M. Building predictive models in R using the caret package. *J. Stat. Softw.* **2008**, *28*, 1–26. [CrossRef]
51. Wright, M.N.; Ziegler, A. Ranger: A fast implementation of random forests for high dimensional data in C++ and R. *J. Stat. Softw.* **2017**, *77*, 1–17. [CrossRef]
52. Pebesma, E.J. Multivariable geostatistics in S: The gstat package. *Comput. Geosci.* **2004**, *30*, 683–691. [CrossRef]

International Journal of
Geo-Information

MDPI

Article

Random Forests with Bagging and Genetic Algorithms Coupled with Least Trimmed Squares Regression for Soil Moisture Deficit Using SMOS Satellite Soil Moisture

Prashant K. Srivastava [1,*], George P. Petropoulos [2], Rajendra Prasad [3] and Dimitris Triantakonstantis [4]

1 Remote Sensing Laboratory, Institute of Environment and Sustainable Development, Banaras Hindu University, Varanasi 221005, India
2 Department of Geography, Harokopio University of Athens, El. Venizelou 70, Kallithea, 17671 Athens, Greece; gpetropoulos@hua.gr
3 Department of Physics, Indian Institute of Technology, Banaras Hindu University, Varanasi 221005, India; rprasad.app@iitbhu.ac.in
4 Department of Soil Science of Athens, Institute of Soil and Water Resources, Hellenic Agricultural Organization "DIMITRA", 1 Sofokli Venizelou, Lykovrisi, 14123 Attiki, Greece; trdimitrios@gmail.com
* Correspondence: prashant.iesd@bhu.ac.in

Abstract: Soil Moisture Deficit (SMD) is a key indicator of soil water content changes and is valuable to a variety of applications, such as weather and climate, natural disasters, agricultural water management, etc. Soil Moisture and Ocean Salinity (SMOS) is a dedicated mission focused on soil moisture retrieval and can be utilized for SMD estimation. In this study, the use of soil moisture derived from SMOS has been provided for the estimation of SMD at a catchment scale. Several approaches for the estimation of SMD are implemented herein, using algorithms such as Random Forests (RF) and Genetic Algorithms coupled with Least Trimmed Squares (GALTS) regression. The results show that for SMD estimation, the RF algorithm performed best as compared to the GALTS, with Root Mean Square Errors (RMSEs) of 0.021 and 0.024, respectively. All in all, our study findings can provide important assistance towards developing the accuracy and applicability of remote sensing-based products for operational use.

Keywords: SMOS; soil moisture deficit; rainfall-runoff model; random forest; Genetic Algorithm

Citation: Srivastava, P.K.; Petropoulos, G.P.; Prasad, R.; Triantakonstantis, D. Random Forests with Bagging and Genetic Algorithms Coupled with Least Trimmed Squares Regression for Soil Moisture Deficit Using SMOS Satellite Soil Moisture. *ISPRS Int. J. Geo-Inf.* **2021**, *10*, 507. https://doi.org/10.3390/ijgi10080507

Academic Editor: Wolfgang Kainz

Received: 28 April 2021
Accepted: 19 July 2021
Published: 27 July 2021

1. Introduction

Soil moisture is of key importance in numerous fields from weather, climate, agricultural to hydrological sciences [1]. It is also considered as an Essential Climatic Variable (ECV) in 2010 [2–5]. To cater the needs of hydro-meteorologist and remote sensing community one after the other, two dedicated global space satellites have been launched to offer measurements of the soil moisture in the globe: (a) the Soil Moisture and Ocean Salinity (SMOS) mission, launched by the European Space Agency in November 2009, and (b) the Soil Moisture Active Passive (SMAP), launched in 2016 by the National Aeronautics and Space Administration (NASA) [6]. In agricultural applications, the estimation of soil moisture from the Earth's surface is important for a successful management of soil water content and irrigation scheduling. At first, soil moisture can be assessed by utilizing the in-situ probes, an approach which is adequate with the local or point based applications. However, retrieving large scale soil moisture under vegetation conditions is a challenging task because of large spatial variability and heterogeneity in the crop types [7,8]. Different crop types are characterized by different moisture contents, which make many practical applications difficult [9]. Therefore, alternate approaches are needed for an effective agricultural water management [10].

In hydrology, a common soil moisture indicator is the soil moisture deficit (SMD) or depletion, which is directly related to the ratio between actual and potential evapotranspiration (PE) [11–13]. It shows the water required to lift the soil-water content of the crop root zone to field capacity [14,15]. SMD is a vital indicator of water availability, representing the constraint variable in crop yields. The moisture deficit impact on agricultural yield has been examined at multiple scales regarding different soil management practices [16–18]. It is important for crops monitoring because it has a strong impact in sowing, harvesting and finally in crop yield. Timely information of SMD is useful in crop insurance practices and to protect equity against natural disasters such as drought, floods and other socio-economic factors [19]. When a drought occurs, a high value of SMD is generally evident for a prolonged time, while during a flood, SMD is very low [20,21]. There have been a number of studies on the Probability Distribution Model (PDM) [22,23] and their results could be used as a suitable option for SMD generation using the minimal datasets, such as rainfall, evapotranspiration and discharge information. Recently, a few studies by Srivastava et al., 2013, 2014 [3,10] indicated that the soil moisture is strongly linked with the SMD and could be used for estimating this variable.

Some recent research has also indicated significant non-linear relationships between the SMD and soil moisture. In this regard, genetic programming (GP) [24] and RFsregression [25] can be used as non-linear techniques for modeling SMD by using soil moisture estimated from SMOS satellite. Among machine learning, the GP and RFs are evolutionary algorithms that can be used for the estimation of SMD. It is inspired by the biological evolution to find computer programs that perform a user-defined task, generally based on chromosome crossing over for getting a better offspring [26]. The RFs algorithm is a non-parametric technique that is capable of synthesizing regression functions based on discrete or continuous datasets in the field of remote sensing [27–29]. RFs is a strong model in regard to the outliers and can be executed efficiently on large datasets for the regression problems [30]. The RFRmodel's robustness over Support Vector Regression (SVR) and Artificial Neural Network Regression (ANNR) has been demonstrated wheat crop biomass estimation [31]. Soil mapping was performed by utilizing an RFs model in Africa with relatively accurate results. External parameter orthogonalization, coupled with RFs, SVM, partial least squares regression and ANN models, was applied on a larger soil database and satisfactory results were obtained. With the advancement in machine learning models, the Genetic Algorithm coupled with Least Trimmed Squares (GALTS) [32] has been developed, which uses a small number of trials to achieve the objective functions; as a result, it reduces the variance and biases in the datasets.

To our knowledge, a very small number of studies have focused on comparing the use of RFs and GP for SMD estimation. In viewof the above, in this study, a first-time comprehensive evaluation of the use of the RFs and GALTS algorithms are explored in retrieving the SMD using SMOS data for soil moisture information. To validate the model, the benchmark SMD obtained using the PDM model has been used. Before using the output, i.e., SMD, a rigorous sensitivity and uncertainty analysis of PDM has been performed using the state-of-the-art Generalized Likelihood Uncertainty Estimation [33]. The knowledge gained from this study can potentially assist in evaluating the relation between the rainfall runoff model and satellite soil moisture. Furthermore, exploring this relationship is an essential step to help further developing the accuracy and applicability of such products for operational uses exploiting EO technology.

2. Study area and Datasets

2.1. Study Area

The Brue catchment (total area 135.5 km^2) located in the south-west of England is chosen as the study area (Figure 1). The lowland wet grassland of the catchment forms part of the unique landscape of the Somerset Levels and Moors and the region is internationally and nationally designated for its conservation and landscape value. The other minor cover types found in the catchment are forest and urban areas. The area is characterized by a non-complex topography area for the most part, with most of the catchment area covered by agricultural land (95.22%), with a small number of forest patches (3.12%) and urban areas (1.66%). In the Brue catchment, the distribution of soil indicates that most of the area is composed of a clayey type (49%) of soil followed by coarse loam (29%) and silt (21%). The selected study area has been used previously in many studies and equipped with maintained meteorological and flow station by British Atmospheric Data Centre and Environment Agency, respectively [34–37]. The whole terrain comes under topography, with no steep slopes and with an average altitude of 105 m AMSL.

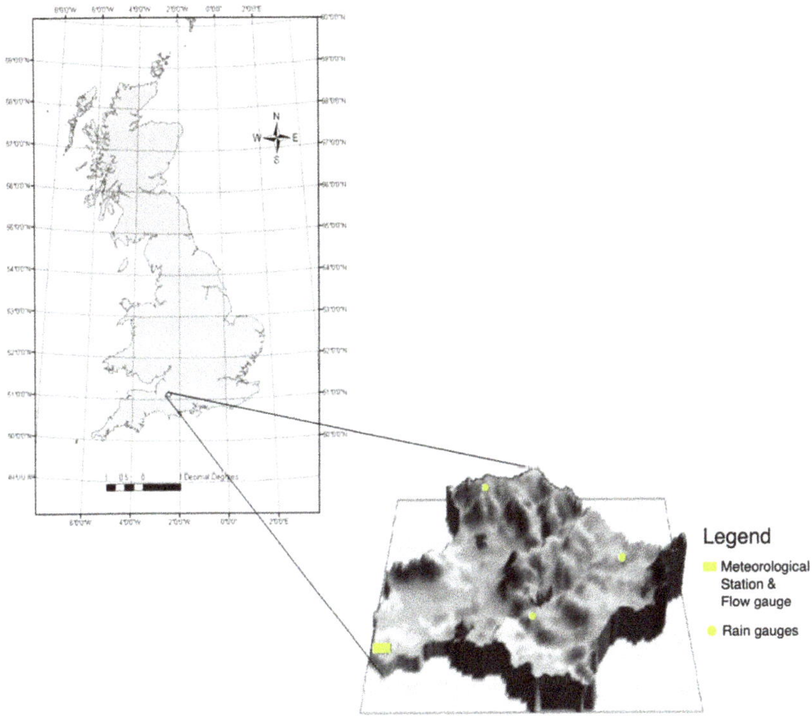

Figure 1. Geographical location of the study area.

2.2. Satellite Data

The SMOS satellite contains the MIRAS instrument—a dual polarized 2D interferometer that operates at a frequency of 1.4 GHz (L-band) [38], launched jointly by the European Space Agency (ESA), the National Centre for Space Studies (CNES-Centre National d'Etudes Spatiales) and the Industrial Technological Development Centre (CDTI–Centro para el Desarrollo Technológico Industrial). The radiometric resolution of the instrument is ~40 km with the soil moisture retrieval unit in $m^3\ m^{-3}$ (i.e., volumetric). In the current study, Level 2 SMOS soil moisture products are used, generated by the SMOS level 2 processor. Acquisition of all the datasets started from February 2011 to January 2012. Soil moisture or surface salinity consist of Level 2 products swath-based maps computed from Level 1c products. The conversion from Level 1c brightness temperatures to Level 2 maps includes a first step to mitigate the impact of Faraday rotation, Sun/Moon/galactic glint, atmospheric attenuation, etc. SMOS acquires data in Icosahedral Snyder Equal Area projection (ISEA 4H9 grid) [39]. Each point (or node) of this grid is known as a DGG that has fixed coordinates and is assigned with an identificator, "DGG Id." For the development of the model, the SMOS pixel with its centroid over the catchment is extracted and considered for the subsequent analysis. The Beam (v 4.9) (developed under ESA Envisat project by

Brockmann Consult GmBh, Germany) open-source software package with SMOS 2.1.3 plugin was used for the extraction. The ascending SMOS data products are selected in the catchment area to minimize the variables influencing soil moisture retrieval, such as vertical soil-vegetation temperature gradients.

3. Probability Distributed Model (PDM)

The PDM model from CEH Wallingford [40] is used as a rainfall runoff simulation model in this study for SMD estimation, as it can calculate soil moisture in the system with suitable time step and data inputs required for use in a water resources assessment. In this study, the lumped rainfall-runoff models called as PDM (Probability Distributed Model) has been used for benchmarking SMD estimation using ground-based datasets. Daily datasets from February 2011 to January 2012 has been used for the model calibration and validation. The PDM model is a fairly general conceptual rainfall-runoff model which transforms rainfall and evaporation data to flow at the catchment outlet and is well tested [41]. It has evolved as a toolkit of model functions that together constitute a lumped rainfall-runoff model capable of representing a variety of catchment-scale hydrological behaviors. The model formulations are adjusted for automatic parameter assessment. For real-time flow forecasting applications, the PDM model is complemented by updating methods based on error prediction and state-correction approaches [36]. The PDM model require three main inputs—evapotranspiration, rainfall and river flow—and the output products are flow and SMD [42]. The SMD is produced to describe the effect of drying on the catchment area on the actual evapotranspiration (ET). The SMD routine in PDM is based on [40]:

$$\frac{E_i'}{E_i} = 1 - \left\{ \frac{(S_{\max} - S(t))}{S_{\max}} \right\}^{b_e} \tag{1}$$

where $\frac{E_i'}{E_i}$ is the ratio of actual ET to potential ET; $(S_{\max} - S(t))$ is the Soil Moisture Deficit; b_e is an exponent in the actual evaporation function; $S_{\max}$ is the total available storage; and $S(t)$ is storage at a particular time t. The Sensitivity analysis (SA) and uncertainty analysis (UA) of the PDM model over the Brue catchment are given in detail by [33]. From the study, it can be concluded that PDM can be used for SMD estimation with less uncertainty, and therefore, it can be used as reference data.

4. Random Forests and Genetic Algorithm Coupled with Least Trimmed Squares (GALTS)

R programming language is used for all the algorithms implementation, which is open-source software. The techniques involve RFsand Genetic Algorithm coupled with Least Trimmed Squares (GALTS) regression for SMD estimation.

4.1. Genetic Algorithm Using Least Trimmed Square (GALTS)

GALTS draws random candidate solutions (or chromosomes) for which search methods are appropriate for use in nonlinear or non-differentiable optimization problems [32]. Genetic Algorithms (GAs) perform a parallel search to cope with the local optima problem. Offspring are selected based on their survival in subsequent generations. However, the minimization of the objective function is a complicated problem in GAs. The authors of one paper [43] developed a method for regression parameter assessment using a Least Trimmed Square (LTS) calculation in a broad dataset. It is an appropriate algorithm because it has been successfully demonstrated in several test beds in the literature, including studies similar to ours, and references have been provided. The purpose of this study is not to perform a comparison between algorithms, which could be a very interesting direction to pursue in a follow-up paper. Some researchers also used least median of squares (LMS) regression for outlier detection, but its convergence rate is slow [43]. On the contrary, LTS has the same breakdown point as LMS, but the convergencerate is quicker than LMS. The objective function of the LTS regression is defined as:

$$\min_{\beta}\sum_{i=1}^{h} r_i^2 \tag{2}$$

where r_i^2 is the ith ordered squared residual and h is a custom integer, which is approximately $n/2$.

4.2. Random Forests

RFs, which was proposed by [30], is a machine learning algorithm which adds an additional layer of randomness to bagging and fits regression trees to random subsets of the input data [44]. The main advantage with RFs is that it is designed to produce accurate predictions that do not overfit the data [30], and differs from bootstrap in constructing the multiple trees. In this study, bagging is used in tandem with random feature selection. RFs for regression are produced by growing trees depending on a random vector Θ, such that the tree predictor $h(x,\Theta)$ takes on numerical values as opposed to class labels with assumption that that the training set is independently drawn from the distribution of the random vector Y, X. The mean-squared generalization error for any numerical predictor $h(x)$ is given by Equation (3):

$$E_{X,Y}(Y - h(X))^2 \tag{3}$$

The RFpredictor is formed by taking the average over k of the trees $\{h(\mathbf{x},\Theta k)\}$ as:

$$E_{X,Y}(Y - av_k h(X,\Theta_k))^2 \to E_{X,Y}(Y - E_\Theta h(X,\Theta))^2 \tag{4}$$

The forest trees can be simplified by using the term $PE^*(tree)$, where it can be defined as Equation (5):

$$PE^*(tree) = E_\Theta E_{X,Y}(Y - h(X,\Theta))^2 \tag{5}$$

It can be used for estimating $PE^*(forest)$ (Equation (6)):

$$PE^*(forest) \le \bar{\rho} PE^*(tree) \tag{6}$$

$PE^*(forest)$ can be summarized by Equation (7):

$$PE^*(forest) = E_{X,Y}[E_\Theta(Y - h(X,\Theta)]^2 = E_\Theta E_{\Theta'} E_{X,Y}(Y - h(X,\Theta))(Y - h(X,\Theta')) \tag{7}$$

The term on the right in (Equation (7)) is a covariance and can be written as:

$$E_\Theta E_{\Theta'}(\rho(\Theta,\Theta')sd(\Theta)sd(\Theta'))$$

where $sd(\Theta) = \sqrt{E_{X,Y}(Y - h(X,\Theta))^2}$; the weighted correlation can be defined as (Equation (8)):

$$\bar{\rho} = E_\Theta E_{\Theta'}(\rho(\Theta,\Theta')sd(\Theta)sd(\Theta'))/E_\Theta sd(\Theta))^2 \tag{8}$$

Then, this condition will follow (Equation (9)):

$$PE^*(forest) \le \bar{\rho}(E_\Theta sd(\Theta))^2 \le \bar{\rho} PE^*(tree) \tag{9}$$

The requirements for accurate RFs are a low correlation between residuals and low error trees. The RFs decreases the average error of the trees employed by the factor $\bar{\rho}$.

5. Performance Evaluation

The estimated SMD of the RFs and GALTS models are compared against the locally estimated SMD through PDM. This is performed in terms of the correlation (r), the Root

Mean Square Error (*RMSE*) and the absolute bias (*Bias*). The bias measures the over- and underestimation of the model output (Equation (10)). The *r* can be calculated by using Equation (11), while the *RMSE* can be estimated by using Equation (12):

$$Bias = [(\overline{y} - \overline{x})] \tag{10}$$

$$r = \frac{n\sum xy - (\sum x)(\sum y)}{\sqrt{n(\sum x^2) - (\sum x)^2}\sqrt{n(\sum y^2) - (\sum y)^2}} \tag{11}$$

$$RMSE = \sqrt{\left(\frac{1}{n}\sum_{i=1}^{n}[y_i - x_i]^2\right)} \tag{12}$$

where x_i i is the observed flow, y_i is is the simulated flow and $\overline{x}$ is the mean.

6. Results and Discussion

6.1. SMD and Soil Moisture Temporal Variations

The temporal pattern between SMOS soil moisture, PDM SMD, RFs and GALTS simulated SMD during calibration and validation is represented through Figure 2. The plots show a high level of temporal variability over the entire monitoring cycle, with a daily step beginning from the first day of January. The distinctive dry periods can be demonstrated, when a distinctive rise in the SMD occurred in the plot. During the period, the trend and pattern in the datasets are very close to the one estimated by the PDM. In general, soil moisture retrieved by SMOS is highly responsive with significant fluctuations over the whole period even to the small variations in SMD. During December and January, a very high value of soil moisture is reported. Over April–May to August–September, rising temperatures and high evaporation cause the soil to dry out, resulting in a rise in SMD values. Increasing temperature caused a substantial SMD development prominent from April to the beginning of August (normally, the driest and warmest months of the year). Drying out follows an exponential decay and an inverse relation can be seen with SMOS soil moisture. According to soil moisture analysis, the wettest months are November and December, while the driest months are March and May. Soil moisture is poor during the winter and near field capacity by mid-April.

6.2. Optimisation of RFs and GALTSAlgorithms

For reliable results, the RFs and GALTS techniques must be optimized, which necessitates a preliminary review of parameters before using them for the final prediction (Figure 3). For RFs, initially, we started with a very small number of trees (around 50) and measured the performance of the technique with respect to the measured SMD. This process was repeated up to a total of 1250 number of forest trees. The mean squared sum of residual is used as an objective function for optimizing the number of trees. For any artificial intelligence technique, this form of trial-and-error approach is widely used to select the right parameters [42]. A similar approach is used for optimizing the number of nodes in RFs. After all the analysis, the optimum values of parameters are obtained as 500 with respect to number of trees and 6 for the number of nodes in the RFs algorithm.

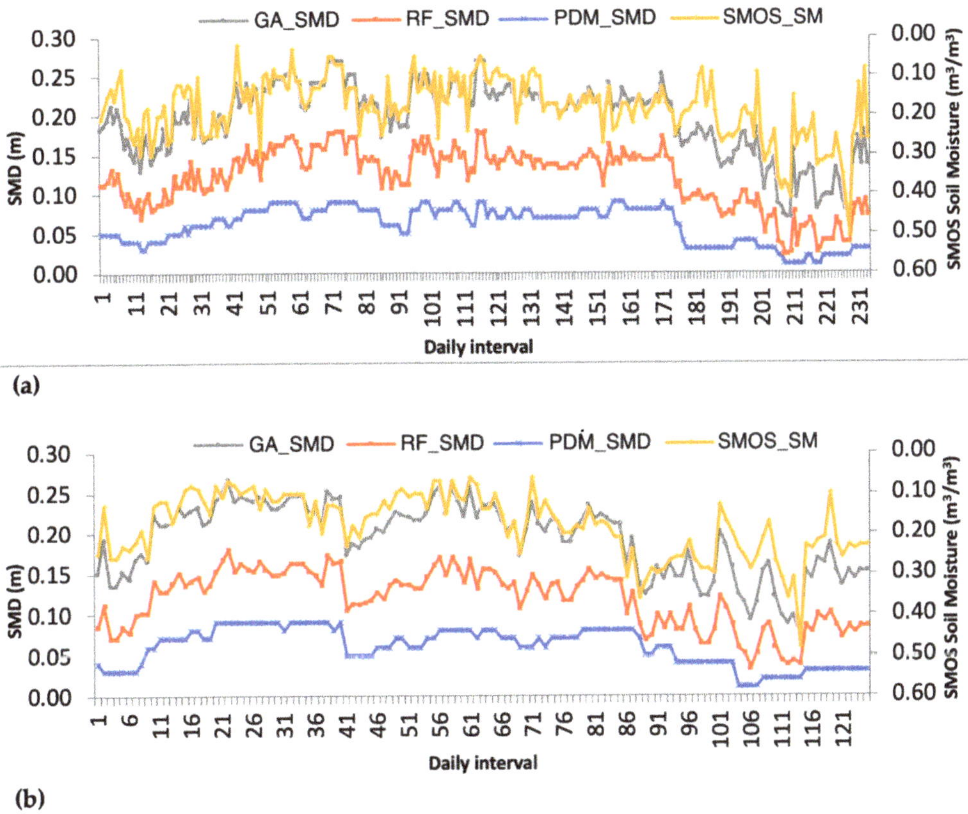

Figure 2. Temporal trend in SMOS soil moisture and SMD from different approaches (**a**) calibration and (**b**) validation.

For optimization of GALTS, two important parameters, i.e., Csteps and population size, are taken into account. As recommended by [43], the sum of squared residuals is taken as an objective function for the optimization of parameters. In case of Csteps, initially, we began with 1 and gradually increase it up to total 40 numbers of Csteps. After a comparison with the SMD, a total of 10 Csteps were found to be appropriate for GALTS. After repeating the procedure with the population sizes from 5–100, a population size of 5 is found optimum for SMD prediction.

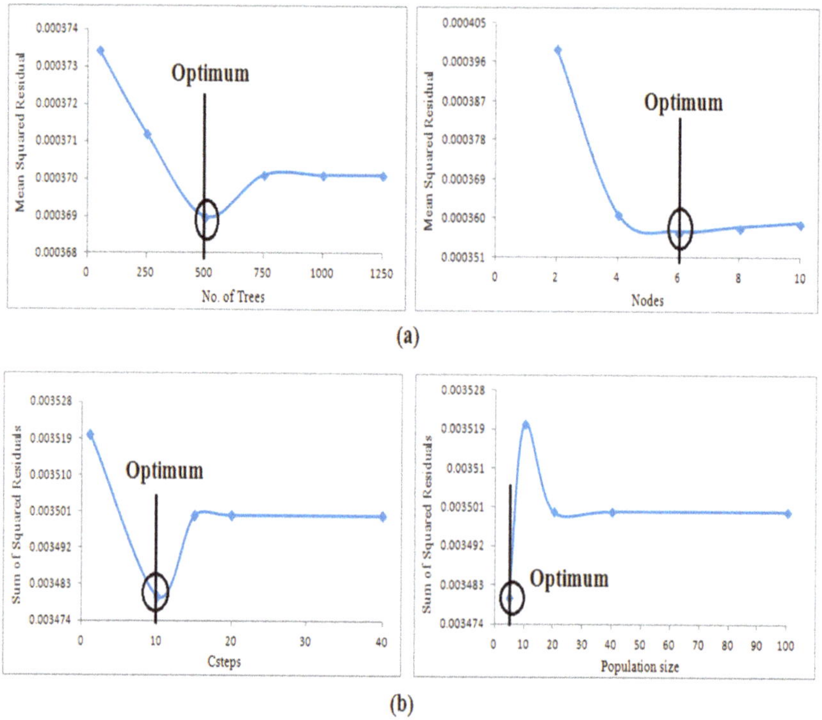

Figure 3. Optimization of (**a**) number of trees and nodes in RFs (**b**) Csteps and population size in GALTS.

6.3. Performances of RFs and GALTS for SMD Prediction

For SMD prediction using the SMOS soil moisture, RFs and GALTS methods are used, while scatter plots depicting the output of various approaches in terms of correlations are given in Figure 4. For developing an appropriate algorithm, the calibration and validation datasets are split into two groups first. Two-thirds of the data from each month is used for calibration, while the other third is used for validation, ensuring that all calibration and validation data are indicative of both seasons. After deriving the relationships, the validation results are used to test the algorithms. The notable things observed in the validation correlation plot are some lower correlations of RFs-simulated SMD, while the correlation of GA-simulated SMD indicate a higher value with the PDM SMD. The best correlation statistics are obtained GALTS with a value of 0.64, followed by a value of 0.49 with RFs-estimated SMD.

The scatter plots produced from all the approaches at the calibration and validation phases are depicted through Figure 5. Looking at the values of the *Bias*, the RFs performed well with the least overestimation and low RMSE, although the value of r is lesser than GALTS. Higher bias and RMSE values obtained by GALTS indicated that the simulated results are overestimated as compared with the RFs. The analysis indicates that the outcome of the RFs can be used for estimation of SMD from SMOS soil moisture. Some binning can be seen in the plot, which could be due to saturation of the SMD values predicted during the particular time period. Furthermore, as PDM is a conceptual model based on water balance, when there are small changes in soil moisture in the region, this will lead to a very negligible variation in the SMD, which cannot be clearly demarcated in the scatter plots.

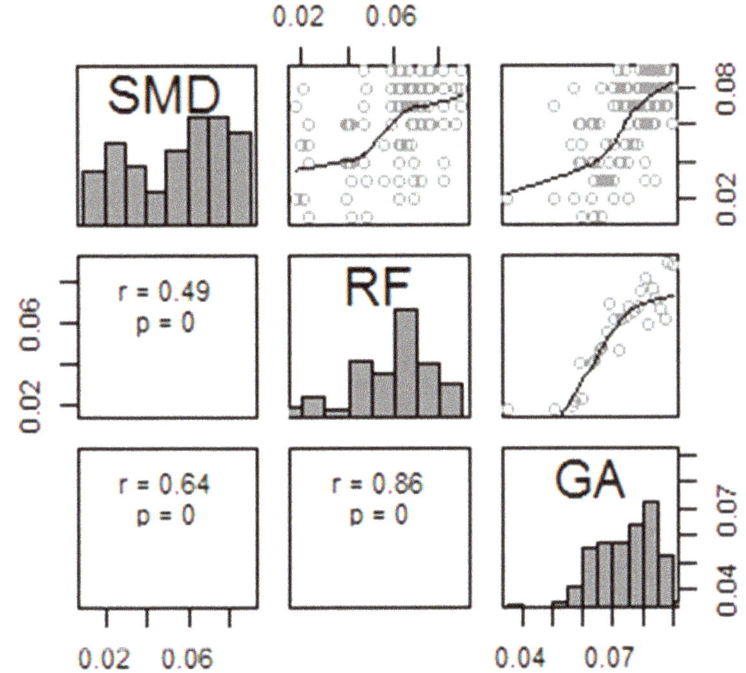

Figure 4. Correlation matrix plot between RFs, GALTS and PDM SMD during validation (the black line indicates the fitting line and the histogram shows the variation of the particular parameter).

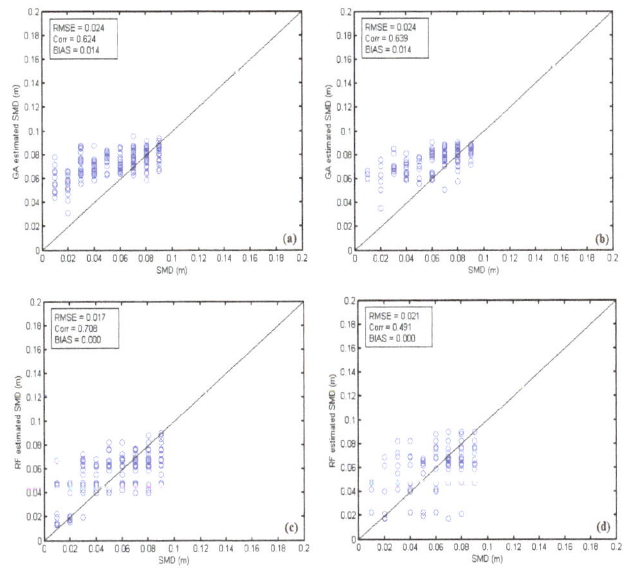

Figure 5. Performances of RFs and the Genetic algorithm during calibration (**a–c**) and validation (**b–d**).

6.4. Other Relevant Studies

In another study, to estimate the SMD [3], soil moisture was estimated from certain algorithms, such as Single channel calculation (SCA). It was found that the inclusion of a physics-based equation for soil moisture retrieval and the incorporation of linear and non-linear models improves the SMD prediction. Moreover, better accuracy is produced with locally calibrated roughness parameters rather than default global values, because they represent the conditions more precisely.

In other research [10], the SMD was estimated by using soil moisture by SMOS and WRF-NOAH LSM as well. Four data fusion models were evaluated: Linear Weighted Algorithm (LWA), Multiple Linear Regression (MLR), Kalman Filter (KF) and Artificial Neural Network (ANN). The LWA method utilizes combinations between soil moisture products. The MLR method uses as a dependent variable the SMD produced by the Probability Distributed Model and as predictors the SMOS and WRF-NOAH LSM. The validation of the different models against SMD produced by the Probability Distributed Model was achieved using ground-based observations. All the data fusion algorithms enhance the information quality and, subsequently, they are appropriate for SMD calculation for hydrological applications. More specifically, the analysis indicated a higher suitability of ANN and KF for data fusion than the LWA or MLR technique. These methods can be used for the data fusion, giving profitable information on the appropriateness of SMOS and WRF-NOAH LSM for SMD calculation.

To assess the suitability of WRF for SMD calculation, ref. [45] used three domains with spatial resolutions of 81 km, 27 km and 9 km. The best domain (Innermost WRF domain of 9 km) was used for estimating the soil moisture deficit (SMD). Two approaches were used in this study. The first method is based on a continuous time series analytical relationship between WRF soil moisture and the SMD, while the second is centered on the effect of vegetation cover on SMD retrieval, represented in terms of growing and non-growing seasons. The results indicated that both approaches could be useful for soil moisture and SMD calculation at the catchment level.

Concluding the relevant studies of SMD estimation, ref. [18] used a process-based Soil and Water Assessment Tool (SWAT) model to simulate the multi-annual fluctuations of soil moisture anomalies (deficits or excesses). For two potential horizons, i.e.,2021–2050 and 2071–2100, the analysis used an ensemble of nine bias-corrected EURO-CORDEX forecasts under two Representative Concentration Pathways: RCP4.5 and 8.5. SWAT, which collects significant SMD and soil moisture excess episodes for various crops, according to the findings. The severity of soil moisture deficits increased for spring cereals, potato and maize. Furthermore, soil moisture excesses were more dependent on RCP and potential horizon selection for potato and maize.

The proposed method could be very useful for ungauged catchments, with as thelimitations that it is developed in an environment that is influenced by a temperate oceanic climate, and hence, it may not be suitable for tropical climates. New optimization parameters and coefficients will be required if the method is implemented in tropical climates or other types of climates. In this case, catchment is very small, so even three gauges are sufficient to develop an appropriate model; however, for larger catchments, a dense gauge networks may be needed for developing some suitable models for SMD prediction.

7. Conclusions

SMD is an integral component for most of the hydrological models and needed for designing appropriate routing schemes. The results indicate that the SMOS soil moisture has proven its importance in hydrological sciences and can be used for the prediction of SMD. The findings of this research show that satellite soil moisture, such as from SMOS, can be used for SMD estimation using sophisticated techniques, such as RFs and GALTS. Performance indices indicated that the RFs performed well with a low bias and small RMSE, although the value of the correlation is lowerthan GALTS. Higher bias and RMSE values obtained by GALTS indicated that the simulated results are overestimated as compared to the RFs. The overall performance of RFs showed that using SMOS soil moisture data, the RFs can be used to estimate SMD. However, more work is needed in the future to reduce the biases in the predictions, as some over estimation is evident in the SMD estimation. Thus, in the future, attempts will be made on the bias correction schemes, so that an operational forecast can be provided for SMD. Moreover, in the future, SMD estimation will be done in other geographical areas as well, so that valuable information and insights can be gathered using other promising satellites such as the Soil Moisture Active Passive satellite (SMAP).

Author Contributions: Conceptualization, Prashant K. Srivastava; Data curation, Prashant K. Srivastava; Formal analysis, Prashant K. Srivastava; Investigation, Prashant K. Srivastava and George P. Petropoulos; Method-ology, Prashant K. Srivastava and George P. Petropoulos; Resources, George P. Petropoulos and Dimitris Triantakonstantis; Visualization, Prashant K. Srivastava, George P. Petropoulos, Rajendra Prasad; Writing—original draft, Prashant K. Srivastava, Dimitris Triantakonstantis; Writing—review & editing, Prashant K. Srivastava, George P. Petropoulos, Dimitris Triantakonstantis and Rajendra Prasad All authors have read and agreed to the published version of the manuscript.

Funding: This research received no external funding.

Institutional Review Board Statement: Not Applicable.

Informed Consent Statement: Not Applicable.

Data Availability Statement: SMOS data used in this study can be downloaded from ESA, while ground based Brue catchment datasets can be obtained from British Atmospheric Data Centre on request.

Acknowledgments: The Commonwealth Scholarship Commission, British Council, United Kingdom, and Ministry of Human Resource Development, Government of India provided the required assistance and funding for this study, which the authors gratefully acknowledge. The authors are grateful to the European Space Agency, who provided the SMOS data. Moreover, the authors would like to thank the British Atmospheric Data Centre for its assistance with the ground observation information. Lastly, special thanks to Yann Kerr, Director, CESBIO, France, for providing the training on the SMOS datasets to the first author.

Conflicts of Interest: The authors declare no conflict of interest.

References

1. Tsatsaris, A.; Kalogeropoulos, K.; Stathopoulos, N.; Louka, P.; Tsanakas, K.; Tsesmelis, D.E.; Krassanakis, V.; Petropoulos, G.P.; Pappas, V.; Chalkias, C. Geoinformation Technologies in Support of Environmental Hazards Monitoring under Climate Change: An Extensive Review. *ISPRS Int. J. Geo-Inf.* **2021**, *10*, 94. [CrossRef]
2. Smith, J.M. *Mathematical Modelling and Digital Simulation for Engineers and Scientists*; John Wiley & Sons, Inc.: Hoboken, NJ, USA, 1986.
3. Srivastava, P.K.; Han, D.; Rico-Ramirez, M.A.; O'Neill, P.; Islam, T.; Gupta, M. Assessment of SMOS soil moisture retrieval parameters using tau–omega algorithms for soil moisture deficit estimation. *J. Hydrol.* **2014**, *519*, 574–587. [CrossRef]
4. Petropoulos, G.P.; McCalmont, J.P. An operational in-situ soil moisture and soil temperature monitoring network for west Wales, UK: The WSMN network. *Sensors* **2017**, *17*, 1481–1491. [CrossRef]
5. Maurya, S.; Srivastava, P.K.; Yaduvanshi, A.; Petropoulos, G.P.; Zhuo, L.; Mall, R.K. Future projections of soil erosion using Coupled Model Intercomparison Project and Earth Observation datasets. *J. Hydrol.* **2021**, *594*. [CrossRef]
6. Petropoulos, G.P.; Sandric, I.; Hristopulos, D.; Carlson, T.N. Evaporative fluxes and Surface Soil Moisture Retrievals in a Mediterranean setting from Sentinel-3 and the "simplified triangle". *Remote. Sens.* **2020**, *12*, 3192. [CrossRef]
7. Petropoulos, G.P.; Ireland, G.; Barrett, B. Surface Soil Moisture Retrievals from Remote Sensing: Current Status, Products & Future Trends. *Phys. Chem. Earth* **2015**, *83–84*, 36–56.
8. Srivastava, P.K.; Pandey, P.C.; Petropoulos, G.P.; Kourgialas, N.K.; Pandley, S.; Singh, U. GIS and remote sensing aided information for soil moisture estimation: A comparative study of interpolation technique. *Resources* **2019**, *8*, 70. [CrossRef]
9. Silva-Fuzzo, D.; Carlson, T.N.; Kourgialas, N.; Petropoulos, G.P. Coupling Remote Sensing with a water balance model for soybean yield predictions over large areas. *Earth Sci. Inform.* **2020**, *13*, 345–359. [CrossRef]
10. Srivastava, P.K.; Han, D.; Rico-Ramirez, M.A.; Al-Shrafany, D.; Islam, T. Data fusion techniques for improving soil moisture deficit using SMOS satellite and WRF-NOAH land surface model. *Water Resour. Manag.* **2013**, *27*, 5069–5087. [CrossRef]
11. Moran, M.; Clarke, T.; Inoue, Y.; Vidal, A. Estimating crop water deficit using the relation between surface-air temperature and spectral vegetation index. *Remote Sens. Environ.* **1994**, *49*, 246–263. [CrossRef]

12. Narasimhan, B.; Srinivasan, R. Development and evaluation of Soil Moisture Deficit Index (SMDI) and Evapotranspiration Deficit Index (ETDI) for agricultural drought monitoring. *Agric. For. Meteorol.* **2005**, *133*, 69–88. [CrossRef]
13. North, M.R.; Petropoulos, G.P.; Rentall, D.V.; Ireland, G.I.; McCalmont, J.P. Quantifying the prediction accuracy of a 1-D SVAT model at a range of ecosystems in the USA and Australia: Evidence towards its use as a tool to study Earth's system interactions. *Earth Surf. Dyn. Discuss.* **2015**, *6*, 217–265.
14. Calder, I.; Harding, R.; Rosier, P. An objective assessment of soil-moisture deficit models. *J. Hydrol.* **1983**, *60*, 329–355. [CrossRef]
15. Rushton, K.; Eilers, V.; Carter, R. Improved soil moisture balance methodology for recharge estimation. *J. Hydrol.* **2006**, *318*, 379–399. [CrossRef]
16. Taylor, S.A. Use of mean soil moisture tension to evaluate the effect of soil moisture on crop yields. *Soil Sci.* **1952**, *74*, 217–226. [CrossRef]
17. Feddes, R.A.; Kowalik, P.J.; Zaradny, H. *Simulation of Field Water Use and Crop Yield*; Centre for Agricultural Publishing and Documentation: Wageningen, The Netherlands, 1978.
18. Piniewski, M.; Marcinkowski, P.; O'Keeffe, J.; Szcześniak, M.; Nieróbca, A.; Kozyra, J.; Kundzewicz, Z.W.; Okruszko, T. Model-based reconstruction and projections of soil moisture anomalies and crop losses in Poland. *Theor. Appl. Climatol.* **2020**, *140*, 691–708. [CrossRef]
19. Codjoe, S.N.A.; Atidoh, L.K.; Burkett, V. Gender and occupational perspectives on adaptation to climate extremes in the Afram Plains of Ghana. *Clim. Chang.* **2012**, *110*, 431–454. [CrossRef]
20. Halcrow, H.G. Actuarial structures for crop insurance. *J. Farm Econ.* **1949**, *31*, 418–443. [CrossRef]
21. Selirio, I.; Brown, D. Soil moisture-based simulation of forage yield. *Agric. Meteorol.* **1979**, *20*, 99–114. [CrossRef]
22. Cabus, P. River flow prediction through rainfall–runoff modelling with a probability-distributed model (PDM) in Flanders, Belgium. *Agric. Water Manag.* **2008**, *95*, 859–868. [CrossRef]
23. Tripp, D.R.; Niemann, J.D. Evaluating the parameter identifiability and structural validity of a probability-distributed model for soil moisture. *J. Hydrol.* **2008**, *353*, 93–108. [CrossRef]
24. Koza, J.R. *Genetic Programming II*; MIT Press: Cambridge, MA, USA, 1994; Volume 17.
25. Liaw, A.; Wiener, M. Classification and regression by random Forest. *R News* **2002**, *2*, 18–22.
26. Islam, T.; Rico-Ramirez, M.A.; Han, D. Tree-based genetic programming approach to infer microphysical parameters of the DSDs from the polarization diversity measurements. *Comput. Geosci.* **2012**, *48*, 20–30. [CrossRef]
27. Svetnik, V.; Liaw, A.; Tong, C.; Culberson, J.C.; Sheridan, R.P.; Feuston, B.P. Random forest: A classification and regression tool for compound classification and QSAR modeling. *J. Chem. Inf. Comput. Sci.* **2003**, *43*, 1947–1958. [CrossRef] [PubMed]
28. Rodriguez-Galiano, V.F.; Ghimire, B.; Rogan, J.; Chica-Olmo, M.; Rigol-Sanchez, J.P. An assessment of the effectiveness of a random forest classifier for land-cover classification. *ISPRS J. Photogramm. Remote Sens.* **2012**, *67*, 93–104. [CrossRef]
29. Islam, T.; Rico-Ramirez, M.A.; Srivastava, P.K.; Dai, Q. Non-parametric rain/no rain screening method for satellite-borne passive microwave radiometers at 19–85 GHz channels with the Random Forests algorithm. *Int. J. Remote Sens.* **2014**, *35*, 3254–3267. [CrossRef]
30. Breiman, L. Random forests. *Mach. Learn.* **2001**, *45*, 5–32. [CrossRef]
31. Kumar, P.; Prasad, R.; Gupta, D.K.; Mishra, V.N.; Vishwakarma, A.K.; Yadav, V.P.; Bala, R.; Choudhary, A.; Avtar, R. Estimation of winter wheat crop growth parameters using time series Sentinel-1A SAR data. *Geocarto Int.* **2018**, *33*, 942–956. [CrossRef]
32. Satman, M.H. A genetic algorithm based modification on the LTS algorithm for large data sets. *Commun. Stat. Simul. Comput.* **2012**, *41*, 644–652. [CrossRef]
33. Srivastava, P.K.; Han, D.; Rico-Ramirez, M.A.; Islam, T. Sensitivity and uncertainty analysis of mesoscale model downscaled hydro-meteorological variables for discharge prediction. *Hydrol. Process.* **2014**, *28*, 4419–4432. [CrossRef]
34. Borga, M. Accuracy of radar rainfall estimates for streamflow simulation. *J. Hydrol.* **2002**, *267*, 26–39. [CrossRef]
35. Mellor, D.; Sheffield, J.; O'Connell, P.E.; Metcalfe, A.V. A stochastic space-time rainfall forecasting system for real time flow forecasting II: Application of SHETRAN and ARNO rainfall runoff models to the Brue catchment. *Hydrol. Earth Syst. Sci.* **2000**, *4*, 617–626. [CrossRef]
36. Bell, V.A.; Moore, R.J. The sensitivity of catchment runoff models to rainfall data at different spatial scales. *Hydrol. Earth Syst. Sci.* **2000**, *4*, 653–667. [CrossRef]
37. Moore, R.J.; Jones, D.A.; Cox, D.R.; Isham, V.S. Design of the HYREX raingauge network. *Hydrol. Earth Syst. Sci.* **2000**, *4*, 521–530. [CrossRef]
38. Kerr, Y.H.; Waldteufel, P.; Wigneron, J.-P.; Martinuzzi, J.-M.; Font, J.; Berger, M. Soil moisture retrieval from space: The Soil Moisture and Ocean Salinity (SMOS) mission. *Geosci. Remote. Sens. IEEE Trans.* **2001**, *39*, 1729–1735. [CrossRef]
39. Pinori, S.; Crapolicchio, R.; Mecklenburg, S. Preparing the ESA-SMOS (soil moisture and ocean salinity) mission-overview of the user data products and data distribution strategy. In Proceedings of the 2008 Microwave Radiometry and Remote Sensing of the Environment, Florence, Italy, 11–14 March 2008; pp. 1–4.
40. Moore, R.J. The PDM rainfall-runoff model. *Hydrol. Earth Syst. Sci.* **2007**, *11*, 483–499. [CrossRef]
41. Liu, J.; Han, D. Indices for calibration data selection of the rainfall-runoff model. *Water Resour. Res.* **2010**, *46*, W04512. [CrossRef]
42. Srivastava, P.K.; Han, D.; Ramirez, M.R.; Islam, T. Machine Learning Techniques for Downscaling SMOS Satellite Soil Moisture Using MODIS Land Surface Temperature for Hydrological Application. *Water Resour. Manag.* **2013**, *27*, 3127–3144. [CrossRef]

43. Rousseeuw, P.J.; Van Driessen, K. Computing LTS regression for large data sets. *Data Min. Knowl. Discov.* **2006**, *12*, 29–45. [CrossRef]
44. Gislason, P.O.; Benediktsson, J.A.; Sveinsson, J.R. Random forests for land cover classification. *Pattern Recognit. Lett.* **2006**, *27*, 294–300. [CrossRef]
45. Srivastava, P.K.; Han, D.; Rico-Ramirez, M.A.; O'Neill, P.; Islam, T.; Gupta, M.; Dai, Q. Performance evaluation of WRF-Noah Land surface model estimated soil moisture for hydrological application: Synergistic evaluation using SMOS retrieved soil moisture. *J. Hydrol.* **2015**, *529*, 200–212. [CrossRef]

International Journal of
Geo-Information

MDPI

Article

Spectral Index for Mapping Topsoil Organic Matter Content Based on ZY1-02D Satellite Hyperspectral Data in Jiangsu Province, China

Yayu Yang [1,2], Kun Shang [1,*], Chenchao Xiao [1], Changkun Wang [3,4] and Hongzhao Tang [1]

1 Land Satellite Remote Sensing Application Center, Ministry of Natural Resources of P.R. China, Beijing 100048, China; yayuyang@cug.edu.cn (Y.Y.); xiaochenchao@lasac.cn (C.X.); tanghz@pku.edu.cn (H.T.)
2 College of Resource, China University of Geosciences, Wuhan 430074, China
3 State Key Laboratory of Soil and Sustainable Agriculture, Institute of Soil Science, Chinese Academy of Sciences, Nanjing 210008, China; ckwang@issas.ac.cn
4 University of Chinese Academy of Sciences, Beijing 100049, China
* Correspondence: shangkun@lasac.cn; Tel.: +86-186-0054-7036

Citation: Yang, Y.; Shang, K.; Xiao, C.; Wang, C.; Tang, H. Spectral Index for Mapping Topsoil Organic Matter Content Based on ZY1-02D Satellite Hyperspectral Data in Jiangsu Province, China. *ISPRS Int. J. Geo-Inf.* **2022**, *11*, 111. https://doi.org/10.3390/ijgi11020111

Academic Editors: Dimitris Triantakonstantis, Panagiotis Tziachris and Wolfgang Kainz

Received: 7 December 2021
Accepted: 2 February 2022
Published: 4 February 2022

Abstract: Estimation of soil organic matter content (SOMC) is essential for soil quality evaluation. Compared with traditional multispectral remote sensing for SOMC mapping, the distribution of SOMC in a certain area can be obtained quickly by using hyperspectral remote sensing data. The Advanced Hyper-Spectral Imager (AHSI) onboard the ZY1-02D satellite can simultaneously obtain spectral information in 166 bands from visible (400 nm) to shortwave infrared (2500 nm), providing an important data source for SOMC mapping. In this study, SOMC-related spectral indices (SIs) suitable for this satellite were analyzed and evaluated in Shuyang County, Jiangsu Province. A series of SIs were constructed for the bare soil and vegetation-covered (mainly rice crops and tree seedlings) areas by combining spectral transformations (such as reciprocal and square root) and dual-band index formulas (such as ratio and difference), respectively. The optimal SIs were determined based on Pearson's correlation coefficient (ρ) and satellite data quality, and applied to SOMC level mapping and estimation. The results show that: (1) The SI with the highest ρ in the bare soil area is the ratio index of original reflectance at 654 and 679 nm ($OR\text{-}RI_{(654,679)}$), whereas the SI in the vegetation area is the square root of the difference between the reciprocal reflectance at 551 and 1998 nm ($V\text{-}RR\text{-}DSI_{(551,1998)}$); (2) the spatial distribution trend of regional SOMC results obtained by linear regression models of $OR\text{-}RI_{(654,679)}$ and $V\text{-}RR\text{-}DSI_{(551,1998)}$ is consistent with the samples; (3) based on the optimal SIs, support vector machine and tree ensembles were used to predict the SOMC of bare soil and vegetation-covered areas of Shuyang County, respectively. The determination coefficient of the soil–vegetation combined prediction results is 0.775, the root mean square error is 3.72 g/kg, and the residual prediction deviation is 2.12. The results show that the proposed SIs for ZY1-02D satellite hyperspectral data are of great potential for SOMC mapping.

Keywords: hyperspectral satellite data; soil organic matter; spectral transformation; spectral analysis; spectral index

1. Introduction

Soil organic matter (SOM) is essential for soil nutrients and a key indicator for evaluating soil environment [1] and food security [2]. It provides nutrients for crop growth [3–5] and plays a vital role in soil formation, conservation, environmental protection, and sustainable development of agriculture and forestry [2]. Changes in the abundance and properties of SOM profoundly impact many processes in the soil ecosystem. Therefore, SOM monitoring is necessary to keep track of soil environmental conditions. Traditional estimation of soil organic matter content (SOMC) mainly relies on collecting a large number of soil samples and laboratory chemical analysis. This method is relatively accurate, but the alteration of soil's natural status and mass loss is inevitable during the sampling and measuring process [6]. It is also time consuming and laborious, and it is difficult to achieve large-scale investigation in a short time [7]. Soil proximal sensing methods can achieve non-destructive measurements, yet only soil properties at the sampling point can be obtained [8]. Therefore, it is challenging to obtain the spatial distribution of SOM at a regional scale [9].

Satellite remote sensing has the advantages of fast monitoring, wide coverage and low cost [9]. In particular, hyperspectral satellite data have rich spectral information, making it possible to predict the spatial distribution of SOMC on a large scale and with high accuracy [10]. The ZY1-02D satellite, launched in September 2019, carries an Advanced Hyperspectral Imager (AHSI) that can simultaneously acquire spectral information in 166 bands from the visible (400 nm) to the shortwave infrared (2500 nm) with a swath width of 60 km [11]. AHSI can obtain spectral details of the absorption features of SOM from hundreds of continuous narrow bands [10,12], making it an important data source for SOM mapping.

In recent years, many researchers have studied the monitoring of SOMC by satellite remote sensing. For example, Tiwari et al. [13] used artificial neural network (ANN) algorithms to model and map soil organic carbon (SOC) in New Delhi, India, based on EO1-Hyperion images. Emadi et al. [14] drew SOC in northern Iran using ANN, random forest (RF) and support vector machine (SVM) algorithms based on Landsat 8 and MODIS data. The Food and Agriculture Organization of the United Nations (FAO) offered [15] step-by-step guidance for developing 1 km grids for SOC stocks in the Soil Organic Carbon Mapping cookbook. Venter et al. [16] used satellite-driven high-resolution maps to map SOC stocks and trends over South Africa and the depth reached 30 cm. European Union's Copernicus programme used the Sentinel satellite to estimate SOC in European croplands [17]. Endsley et al. [18] estimated a daily global carbon budget including surface (0–5cm depth) SOC with the SMAP Level 4 Carbon (L4C) product. However, satellite remote sensing can only be used to estimate the topsoil organic matter content, and the accurate assessment of the actual SOC quantity stored in other layers is still a huge challenge.

The absorption characteristics of soil spectral reflectance are mainly due to the overtones and combinations of fundamental vibrations caused by the stretching and bending of N-H, O-H and C-H groups [19]. The significant negative correlation between SOM and soil spectral reflectance is the foundation for predicting the spatial distribution of SOM [20]. In the bare soil area, researchers have proposed many different forms of soil spectral indices (SIs) to estimate SOMC. For example, Krishnan et al. [21] used the first derivative absorption rate index (FDAI), absorption ratio index I (ARI I) and absorption ratio index II (ARI II) to estimate the SOMC in Illinois, USA. Based on the absorption characteristics of soil biochemical components, Bartholomeus et al. [22] constructed multiple SIs that have strong correlations with SOC. Jin et al. [23] analyzed the correlations between the difference index (DI), ratio vegetation index (RI), normalized difference vegetation index (NDI) and SOMC. However, most of these indices were constructed based on laboratory soil spectra. Due to the difference between laboratory soil samples and field soils and the influence of the observation scale, it is challenging to obtain satisfying results when these indices are directly applied to satellite data.

In the vegetation-covered area, SOM plays a vital role in crop growth. For example, the humic acid in SOM is an essential source of SOC in the vegetation-covered area. Humic acid can stimulate plant growth and absorb colloidal cations [3,24,25]. Cellulose, lignin and other components of vegetation are usually affected by SOM, which leads to changes in the absorption characteristics of the vegetation canopy spectrum [26]. In the vegetation-covered area, according to this indirect interaction mechanism between SOM and vegetation canopy spectrum, researchers tried to relate SOMC with vegetation SIs that can indicate chlorophyll content, moisture and cellulose content. In previous studies, some vegetation SIs have been applied to evaluate SOMC, such as the normalized difference vegetation index (NDVI), enhanced vegetation index (EVI), moisture stress index (MSI) and cellulose absorption index (CAI) [26–29]. However, these indices are mainly designed to indicate the physicochemical parameters of the vegetation and have very limited ability to characterize the SOM.

Therefore, in this study, Shuyang County of Jiangsu Province was selected as the study area and satellite-based SIs were constructed that are sensitive to topsoil organic matter content in bare soil and vegetation-covered areas. In Section 2, the study area, soil sampling data, and satellite data are introduced, and the construction process of SIs and evaluation methods are described in detail. Then, in Section 3, the correlations between the transformed spectra and SOMC are analyzed, the optimal SIs are obtained, and the application effects of the optimal SIs are evaluated. In Section 4, the considerations for screening the SIs, the necessity of constructing SIs separately in bare soil and vegetation-covered areas, and the impact of land-use types on SOMC are discussed.

2. Materials and Methods

2.1. Study Area

The study area, Shuyang County (33°53′–34°25′ N, 118°30′–119°10′ E), is located in the northern part of Jiangsu Province (as shown in Figure 1a). The county spans 60 km from east to west and 55 km from north to south, covering an area of 2298 km^2. The terrain is higher in the west and lower in the east, and the ground elevation is mostly between 4.5 and 7 m. The river network is dense. The whole area lies in a temperate monsoon climate zone with abundant sunshine and average annual precipitation of 937.60 mm [30–32]. The soil texture of the topsoil (0–15 cm) in Shuyang County is mainly silt, and the contents of clay and sand are relatively small. The soil types include cambisols, regosols, luvisols, greyzems, anthrosols, gleysols, fluvisols, nitisols and arenosols (Figure 1b) [33,34], and the crops are mainly rice, wheat, maize and soybean [35]. The land-use types of cultivated land in the study area mainly include paddy fields, dry land and nurseries (Figure 1c).

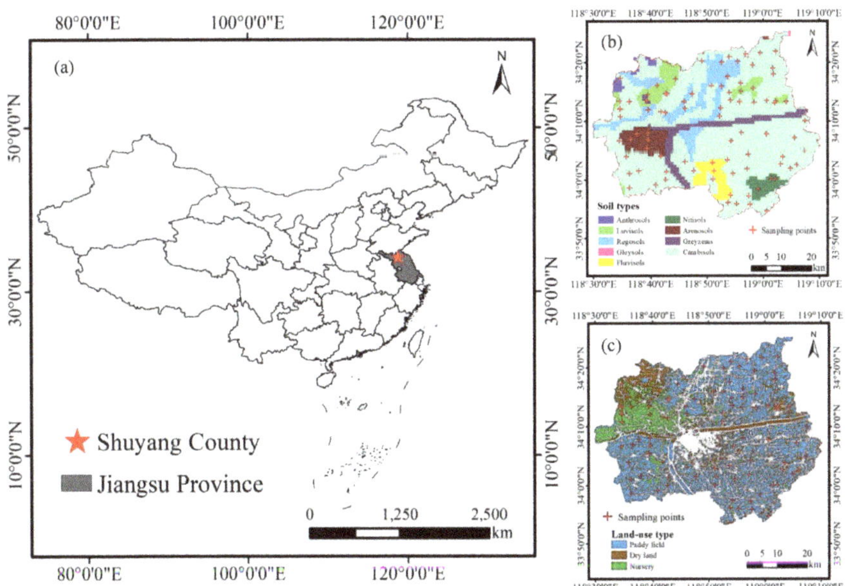

Figure 1. Overview of the study area and sampling points: (**a**) location, (**b**) soil types, and (**c**) land-use types.

2.2. Hyperspectral Satellite Data Acquisition and Preprocessing

The hyperspectral satellite image used in this study (Figure 2a) was acquired by the AHSI of ZY1-02D on October 22, 2020. The detailed parameters of AHSI/ZY1-02D data are shown in Table 1. The spectrum of AHSI ranges from 0.4 to 2.5 μm with 166 bands. The spectral resolution of the 76 bands in the visible and near-infrared (VNIR) region is 10 nm, whereas the spectral resolution of the 90 bands in the shortwave infrared (SWIR) region is 20nm. The swath width of AHSI is 60 km, and the spatial resolution is 30 m.

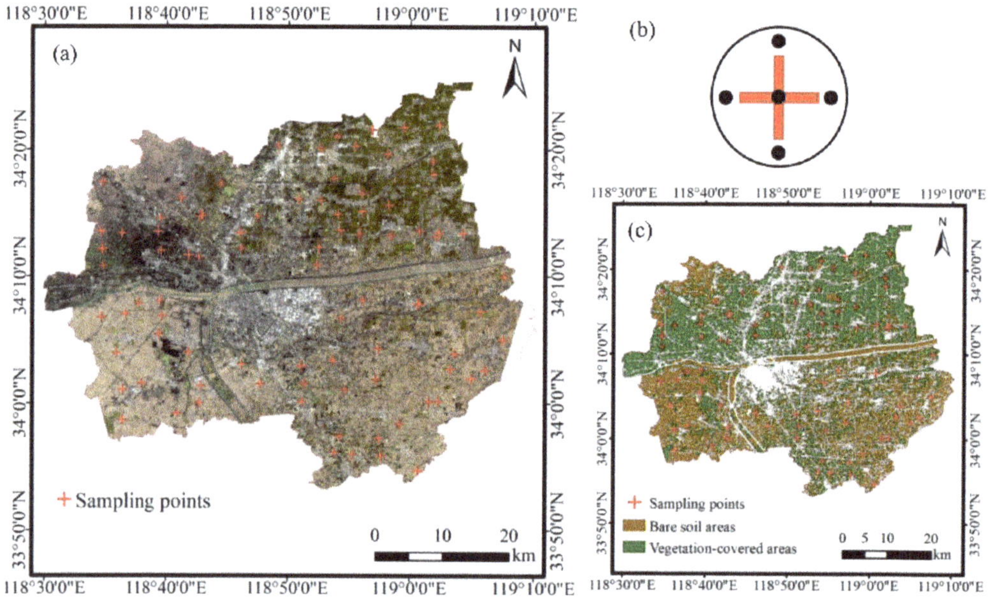

Figure 2. (**a**) The true-color satellite image (R: 662 nm, G: 551 nm, B: 482 nm) of Advanced Hyperspectral Imager (AHSI) data of ZY1-02D satellite obtained on 22 October 2020, in Shuyang County, (**b**) the schemata the for five-point method, (**c**) distribution of bare soil and vegetation-covered areas.

Table 1. The parameters of Advanced Hyperspectral Imager (AHSI) data of ZY1-02D satellite.

Spectral range	400–2500 nm
Number of bands	166
Spatial resolution	30 m
Spectral resolution	VNIR 10 nm
	SWIR 20 nm
Swath width	60 km
Orbital period	55 days

Radiometric calibration and atmospheric correction (using the MODTRAN radiative transfer model) were carried out on the L1A data to obtain surface reflectance. Due to the low spectral response or atmospheric absorption of the sensor, only 130 bands in the spectral ranges of 464–997 nm, 1022–1106 nm, 1173–1324 nm, 1459–1778 nm and 1963–2468 nm were used in the study. A Savitzky–Golay filter [36] (the number of points on the left and right sides of the filter core center point was 3, the order of the derivative was set to 0, and the degree of smoothing polynomial was 1) was used to remove noise and smooth the spectrum, followed by orthorectification. Then the images were seamlessly mosaicked and cropped to obtain the preprocessed reflectance image of the whole Shuyang County. After this, we generated a binary map of cultivated and non-cultivated land of the study area according to the ground survey data. Finally, we masked the reflectance image with the binary map, and only the cultivated land area was kept for subsequent experiments. All the images were co-registered and rectified to the world geodetic system (WGS) 1984 datum.

Shuyang County was in the dry season when the satellite images acquired. The bare soil area in the image is mainly the soil that has just been plowed after the maize harvest, and the vegetation-covered area includes rice crops and tree seedlings. At the imaging time, the vegetation was in the growing season. The spectra of vegetation pixels have the typical vegetation spectral absorption characteristics (such as the red edge and green peak). Firstly, the training samples of bare soil and vegetation-covered areas were selected based on the comparison between pixel spectra and the standard spectral library (such as the USGS spectral library), in addition to the visual interpretation of the high-resolution satellite images (Gaofen-1 and Gaofen-2). Subsequently, the red edge normalized difference vegetation index (RENDVI) [37,38] was calculated based on the reflectance image of the whole study area. The classification accuracy of 96.59% was calculated using 88 randomly selected test samples. It can be seen from Figure 2c that the bare soil area was mainly distributed in the south, whereas the vegetation-covered area was distributed in the north.

2.3. Ground Sampling and Soil Measurements

In July 2020, 92 surface soil samples (0–15 cm) were collected through a random sampling scheme (the minimum distance between the sampling points was set as 2 km). At each sampling point (red crosses in Figures 1 and 2), five soil samples were collected in a circular area with a diameter of 10 m using the five-point method (Figure 2b), and the five samples were then mixed into one composite sample [39]. A one-kilogram composite sample of each spot was taken, dried indoors, and passed through a 2 mm sieve to remove gravel and plant residues. Then, the SOC content of the samples was determined in the fine fraction by the potassium dichromate-volumetric method [40]. Finally, the SOMC was obtained by multiplying a transformation coefficient of 1.724 (Van Bemmelen factor) to the SOC content [41]. According to Figure 2c, 92 sampling points were divided into two categories, with 38 sampling points in the bare soil area and 54 sampling points in the vegetation-covered area.

2.4. Methods

2.4.1. Research Process

The flow chart of this research is shown in Figure 3. The whole process can be divided into three main steps: (1) obtaining SOMC and reflectance spectra of soil samples, (2) constructing the optimal SIs, and (3) evaluating the application effect of the optimal SIs.

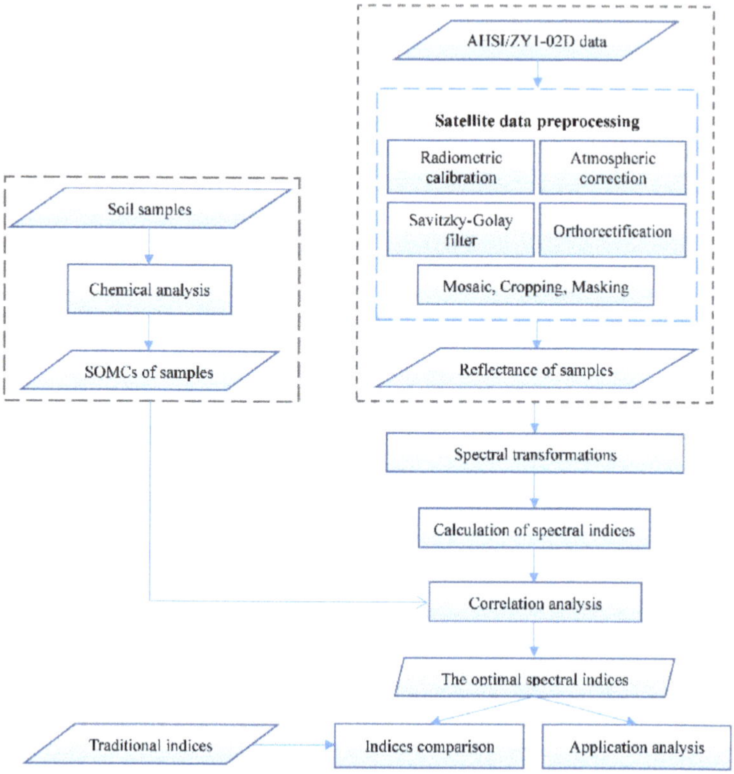

Figure 3. Flowchart of the research.

2.4.2. Construction of the Optimal SIs

Considering that the reflectance of some wavelengths is not sensitive to SOMC, spectral transformations are implemented on the original reflectance (OR). The spectral transformations can enhance the detailed information of the spectrum and help to extract the spectral features of SOM [39,42]. Before the transformation, low-quality bands are removed to avoid more noise and uncertainty. It should be noted that image spectroscopy is a powerful tool for estimating the spatial distribution of SOMC. In addition to obtain spectral information, it can also observe continuous spatial distribution information [43,44]. Therefore, the impact of each operation on the image quality needs to be considered in the construction of SIs. It is necessary to ensure that surface reflectance images and the optimal SIs do not lose the ability to intuitively express surface information. In this study, the images with better quality were those without noticeable mosaic lines, stripe noise and other phenomena that blur the actual feature information in terms of visual effects.

In Table 2, the formulas of different spectral transformations are listed. We calculated the correlations between SOMC and the reflectance spectra of different spectral transformations (including OR). Then, we analyzed the distribution of the correlation between different spectra and SOMC, and checked the quality of the transformed reflectance image. The spectra with higher correlations and better image quality were selected to calculate SIs. Next, further statistical analyses were conducted on all index formulas, and finally, we selected four index formulas, as shown in Table 3. The correlation between each type of spectral index (SI) and SOMC was analyzed, and the indices with higher correlation and better image quality were chosen as the optimal SIs.

Table 2. The spectral transformations involved in this research [1].

Type	Expression
Original reflectance (OR)	R
Reciprocal reflectance (RR)	$1/R$
Square root reflectance (SRR)	$\sqrt{R}$
First-order differential reflectance (FDR)	$R'_{\lambda_i} = \frac{1}{2} \times \left(\frac{R_{\lambda_{i+1}} - R_{\lambda_i}}{\lambda_{i+1} - \lambda_i} + \frac{R_{\lambda_i} - R_{\lambda_{i-1}}}{\lambda_i - \lambda_{i-1}} \right)$
First-order differential of reciprocal reflectance (RFDR)	$(1/R_{\lambda_i})' = \frac{1}{2} \times \left(\frac{(1/R_{\lambda_{i+1}}) - (1/R_{\lambda_i})}{\lambda_{i+1} - \lambda_i} + \frac{(1/R_{\lambda_i}) - (1/R_{\lambda_{i-1}})}{\lambda_i - \lambda_{i-1}} \right)$
First-order differential of square root reflectance (SRFDR)	$(\sqrt{R_{\lambda_i}})' = \frac{1}{2} \times \left(\frac{\sqrt{R_{\lambda_{i+1}}} - \sqrt{R_{\lambda_i}}}{\lambda_{i+1} - \lambda_i} + \frac{\sqrt{R_{\lambda_i}} - \sqrt{R_{\lambda_{i-1}}}}{\lambda_i - \lambda_{i-1}} \right)$
Second-order differential reflectance (SDR)	$R''_{\lambda_i} = \frac{1}{2} \times \left(\frac{R'_{\lambda_{i+1}} - R'_{\lambda_i}}{\lambda_{i+1} - \lambda_i} + \frac{R'_{\lambda_i} - R'_{\lambda_{i-1}}}{\lambda_i - \lambda_{i-1}} \right)$

[1] $i-1$ and $i+1$ denote the former and latter band of band i, λ is the wavelength.

Table 3. The construction formulas of spectral indices (SIs) [1].

SIs	Expression
Difference index (DI)	$p - q$
Ratio index (RI)	$\frac{p}{q}$
Normalized difference index (NDI)	$\frac{p-q}{p+q}$
Square root index of difference (DSI)	$\sqrt{p - q}$

[1] p and q are the transformed spectra values corresponding to any two bands and $p \neq q$.

Pearson's correlation coefficient (ρ) was adopted to determine the correlation between two variables in the correlation analysis. Its formula is as follows:

$$\bullet\ \rho_{X,Y} = \frac{cov(X,Y)}{\sigma_X \sigma_Y} = \frac{E((X - \mu_X)(Y - \mu_Y))}{\sigma_X \sigma_Y}, \tag{1}$$

where σ_X, σ_Y are the standard deviations of the X and Y arrays, respectively; μ_X, μ_Y are the mean values of X, Y, respectively.

2.4.3. Application Assessment of the Optimal SIs

In order to evaluate the ability of the optimal SIs to characterize SOMC, we analyzed the application performance from two aspects. One compares the optimal SIs with the traditional SOM indices. In bare soil and vegetation-covered areas, we selected an optimal SI with the highest correlation. We calculated the correlation between other traditional indices and SOMC for comparison, and the bases for the selection of traditional indices are listed in Table 4. The linear regression (LR) model describing the relationship between each index and SOMC was established individually, and the corresponding determination coefficient (R^2) was calculated to evaluate the performance of each index. In addition, the LR models of the optimal SIs here were also applied to obtain the inversion map of SOMC for recognition of SOMC levels.

The other utilizes the optimal SIs to construct the prediction models of SOMC to evaluate their application performance. All the optimal SIs were selected to construct the corresponding model in the bare soil and vegetation-covered areas. Considering the data redundancy caused by the high correlation between indices, principal component analysis (PCA) was adopted for dimensionality reduction. PCA linearly transforms predictors to remove redundant dimensions, and generates a new set of variables called principal components [45,46].

In terms of the inversion model, SVM and tree ensembles (TE) show good performance in soil attributes' estimation by satellite remote sensing. George et al. [47] used different hyperspectral indices generated by EO-1 Hyperion data and the SVM method to draw various soil salinity severity levels in the Mathura region of the Indo-Gangetic plain of India, and the overall classification accuracy was 78.13%. Based on GF-5 data, Meng et al. [48] used the TE model to predict SOMC in Mingshui County, Heilongjiang Province, China, and the root mean square error of the model was 3.36 g/kg. In this study, six SVM models (linear SVM, quadratic SVM, cubic SVM, fine Gaussian SVM, medium Gaussian SVM, coarse Gaussian SVM) and two TE models (boosted trees, bagged trees) were compared for SOMC estimation in bare soil and vegetation-covered areas. In order to avoid overfitting, the K-fold cross-validation method was used to verify the modeling accuracy. For the bare soil and vegetation-covered areas, the model with the highest accuracy of each was applied to the image of the corresponding area. The inversion results of bare soil and vegetation-covered areas were merged to obtain the SOMC distribution of the whole area.

Furthermore, we adopted the R^2, the root mean square error (RMSE), and the residual prediction deviation (RPD) to evaluate the SOMC characterization ability of the inversion model constructed by the optimal SIs. In general, a well-established model usually has a high R^2 and RPD, and a low RMSE [23,39]. According to the research of Yuan et al. [49], the RPD values are divided into five levels to interpret the model performance, RPD < 1.4 (unacceptable), 1.4 ≤ RPD < 1.8 (fair), 1.8 ≤ RPD < 2.0 (good), 2.0 ≤ RPD < 2.5 (very good), and RPD ≥ 2.5 (excellent). Their calculation formulas are as follows:

$$\bullet\ R^2 = 1 - \frac{\sum_{i=1}^{n}(\hat{y}_i - y_i)^2}{\sum_{i=1}^{n}(y_i - \overline{y})^2}, \tag{2}$$

$$\bullet\ RMSE = \sqrt{\frac{\sum_{i=1}^{n}(y_i - \hat{y}_i)^2}{n}}, \tag{3}$$

$$\bullet\ RPD = \frac{\sqrt{\sum_{i=1}^{n}(y_i - \overline{y})^2/n}}{RMSE}, \tag{4}$$

where y_i is the actual value, the $\overline{y}$ is the mean value of y_i, $\hat{y}_i$ is the fitted value, and n is the number of samples.

Table 4. The formulas of the traditional SIs.

Index Type	Abbreviation	Formula	Properties	References
Soil SI	SOC1	$\frac{1}{\sum_{400nm}^{700nm} R}$	SOMC	[22]
	SOC2	$\frac{1}{(R_{600nm}-R_{400nm})/(600-400)}$	SOMC	[22]
	SOC3	$\frac{1}{(R_{2200nm}-R_{2138nm})/(2200-2138)}$	SOMC	[22]
	NSMI	$\frac{R_{1800nm}-R_{2119nm}}{R_{1800nm}+R_{2119nm}}$	Soil moisture	[50,51]
Vegetation SI	CAI	$\frac{R_{2000nm}+R_{2200nm}}{2} - R_{2100nm}$	Cellulose Absorption	[52–54]
	NDLI	$\frac{\log(1/R_{1754nm})-\log(1/R_{1680nm})}{\log(1/R_{1754nm})+\log(1/R_{1680nm})}$	Lignin concentration	[53,55–57]
	MSI	$\frac{R_{1610nm}}{R_{868nm}}$	Leaf water content	[58,59]
	SATVI	$\frac{R_{1610nm}-R_{665nm}}{R_{1610nm}+R_{665nm}} \times 2 - \frac{R_{2190nm}}{2}$	Total vegetation cover	[58,60]

3. Results

3.1. Descriptive Statistics of Samples

Table 5 illustrates the statistics of SOMC of soil samples in Shuyang County. The SOMC exhibits a range from 10.27 to 47.80 g/kg. The mean value and the standard deviation are 25.17 and 7.88 g/kg, respectively. The moderate coefficient of variation is 31.32%. These samples are further categorized into bare soil and vegetation-covered areas. The minimum SOMC of samples from both regions are similar, and both are moderate variations. However, the maximum SOMC in the vegetation-covered area can reach 47.80g/kg, which is much higher than that in the bare soil area. Moreover, the mean of

SOMC in the vegetation-covered area is about 3.00 g/kg greater than that in the bare soil area.

Table 5. The statistics of soil organic matter content (SOMC) in Shuyang County.

	All Samples	Samples in Bare Soil Areas	Samples in Vegetation-Covered Areas
Number of samples	92	38	54
Range (g/kg)	10.27–47.80	10.27–34.40	10.96–47.80
Mean (g/kg)	25.17	23.45	26.36
Standard deviation(g/kg)	7.88	6.65	8.49
Coefficient of variation (%)	31.32	28.36	32.22

3.2. Spectral Characteristics of the Pixel Reflectance of the Sample Sites

Based on the processed images, the reflectance spectra of the pixels at 92 sampling points were extracted. In the bare soil area, it can be seen from Figure 4a that the reflectance spectrum decreases with the increase in SOMC in the entire 400–2500 nm range. In the range of 500–600 nm, the slope of the curve decreases with the increase in SOMC. There is a spectral absorption in the range of 2150–2250 nm. These are consistent with the results of previous studies. Bartholomeus et al. [22] found that in the 400–700 nm range, there was a flatter spectral profile at higher SOMC, and Jin et al. [23] found that there was a spectral absorption at 2200 nm. On the whole, the extracted reflectance spectral curves can clearly depict the soil spectral variation patterns at different SOMCs.

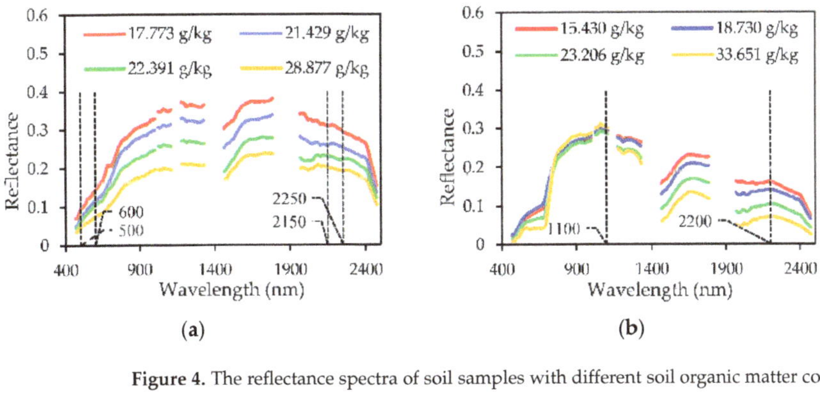

Figure 4. The reflectance spectra of soil samples with different soil organic matter content (SOMC) in (**a**) bare soil and (**b**) vegetation-covered area.

In the vegetation-covered area, as shown in Figure 4b, the reflectance spectrum decreases with the increase in SOMC in the range of 1100–2500 nm. In previous studies, Bartholomeus et al. [53] also found that in the range of 2000–2200 nm, with the increase in SOMC, cellulose also increased, resulting in low reflectance in this spectral region. In general, the extracted reflectance spectral curves can well demonstrate the vegetation spectral variation patterns at different SOMCs.

3.3. Correlation between Transformed Spectra and SOMC

Because complex spectral transformations such as the second-order differential may introduce more noise into the image [39], these bands need to be removed (as shown in Table 6). In OR, RR and SRR, most of the bands removed are due to the low spectral response or atmospheric absorption of the sensor, and the total number of low-quality bands is 36. However, after differential transformation, the number of low-quality bands is significantly increased. Among these, FDR is the least increased. The number of low-quality bands in SDR reaches 147, increasing the most, and its image quality is relatively poor. The correlations between SOMC and the transformed spectra are shown in Table 7.

Table 6. The wavelength range and the number of low-quality bands of each transformed spectrum.

Spectral Transformation Type	Wavelengths of Bands of Low Quality (nm)	The Number of Bands Removed
OR/RR/SRR	395–456; 1006–1014; 1122–1156; 1341–1442; 1795–1947; 2484–2501	36
FDR	395–559; 585–637; 679; 748–1089; 1122–1156; 1207–1308; 1341–1442; 1475–1526; 1660–1762; 1795–1947; 1981; 2014–2082; 2233–2384; 2484–2501	125
RFDR	395–473; 550–662; 748–757; 774; 791–808; 885; 920; 996–1022; 1122–1156; 1341–1442; 1475–1510; 1762–1947; 1981; 2014–2031; 2132–2250; 2334–2401; 2434–2501	86
SRFDR	395–508; 524–559; 576–637; 748–1089; 1122–1156; 1190–1308; 1341–1442; 1475–1526; 1660–1712; 1762; 1795–1947; 1981; 2014–2082; 2216–2401; 2450–2501	127
SDR	395–456; 473–671; 696–1073; 1106; 1122–1156; 1207–1291; 1341–1442; 1492–1745; 1795–1947; 1998; 2031–2216; 2249–2367; 2434–2501	147

Table 7. Correlation with SOMC after different spectral transformations.

TRs	Nb [1]	Bare Soil Region			Vegetation-Covered Area		
		Nsb [2]	Maximum $\lvert\rho\rvert$	WLsb [3] (nm)	Nsb [2]	Maximum $\lvert\rho\rvert$	WLsb [3] (nm)
OR	130	0			52	0.558 ***	533–610, 696–1106 (722)
RR	130	0			62	0.573 ***	524–1106 (722)
SRR	130	0			56	0.563 ***	524–636, 696–1106 (722)
FDR	41	3	0.350 *	662, ***1459***, 1779	6	0.502 ***	687, 696, 1106, 1173, ***1190***, 132445
RFDR	80	2	0.370 *	2048, ***2317***	3	0.381 **	687, 1694, ***1745***
SRFDR	39	5	0.462 **	654, ***662***, 671, 1459, 1779	5	0.477 ***	516, 1106, ***1173***, 1324, 1745
SDR	19	6	0.493 **	1173, 1190, 1308, 1425, ***1762***, 1779	4	0.462 ***	1173, ***1190***, 1762, 1779

[1] Nb stands for the number of available bands. [2] Nsb stands for the number of sensitive bands. [3] WLsb stands for the wavelength of the sensitive band. ***Value***: Wavelength of the maximum correlation coefficient. * Correlation is significant at 0.05 level. ** Correlation is significant at 0.01 level. *** Correlation is significant at 0.001 level.

In the bare soil area, OR, RR and SRR do not have the bare soil spectra that have significant correlations with SOMC, whereas differential transformations can obtain some sensitive bands ($\lvert\rho\rvert > 0.320$, $p < 0.05$). The first-order differential transformation on the basis of OR, RR and SRR can effectively improve the correlation with SOMC. For spectra after differential transformations of different orders, the results show that the higher the order, the higher the correlation with SOMC, indicating the differential transformation is an effective means to improve the correlation with SOMC for soil spectra. As shown by the overall results of the different transformations, it can be found that these sensitive wavelengths are mainly concentrated around 660, 1459 and 1779 nm, which is consistent with the studies of Jin et al. [23], Wang et al. [39] and Guo et al. [61].

In the vegetation-covered area, OR, RR and SRR can improve the correlation with SOMC and increase the number of sensitive bands ($|\rho| > 0.268$, $p < 0.05$). However, the first-order differential transformation based on RR and SRR will decrease the correlation with SOMC and reduce the number of sensitive bands. For spectra after differential transformations of different orders, the results show that the higher the order, the smaller the correlation with SOMC. It can be seen from the overall results of different transformations that these sensitive wavelengths are mainly concentrated around 680 nm and 1745–1779 nm, which are consistent with the sensitive wavelengths obtained in the studies of Bartholomeus et al. [53], Zhang et al. [26] and Fourty et al. [55]. Considering the correlation with SOMC and the image quality, OR, RR, FDR and RFDR were chosen to calculate the SIs.

3.4. Correlation between SIs and SOMC

Sixteen types of SIs were calculated by four transformed spectra (OR, RR, FDR and RFDR) and four index formulas (DI, RI, NDI and DSI). All available bands of the transformed spectra were substituted into the calculation. Then, their correlations with the SOMC of soil samples in bare soil and vegetation-covered areas were analyzed. The correlation coefficient matrix diagrams are shown in Figures 5 and 6. In addition, the maximum correlation coefficient ($|\rho|_{max}$) between each type of SI and SOMC in bare soil and vegetation-covered areas is listed in Tables 8 and 9. Considering the performance of various types of SIs in the bare soil area and vegetation-covered area, $|\rho|$ greater than 0.600 was utilized as the threshold to obtain the optimal SIs.

Table 8. The maximum correlation index of different types of SIs in the bare soil area.

Type		Number of SIs with $|\rho|$>0.6	$|\rho|_{max}$	Index Formula Corresponding to $|\rho|_{max}$
OR	DI	1	0.612 ***	$R_{1459nm} - R_{2014nm}$
	RI	6	0.627 ***	R_{654nm} / R_{679nm}
	NDI	5	0.626 ***	$(R_{654nm} - R_{679nm})/(R_{654nm} + R_{679nm})$
	DSI	0	0.548 ***	$\sqrt{R_{1543nm} - R_{2216nm}}$
RR	DI	3	0.614 ***	$1/R_{1493nm} - 1/R_{2014nm}$
	RI	6	0.623 ***	$(1/R_{679nm})/(1/R_{654nm})$
	NDI	5	0.626 ***	$\frac{1/R_{654nm}-1/R_{679nm}}{1/R_{654nm}+1/R_{679nm}}$
	DSI	0	0.426 **	$\sqrt{1/R_{559nm} - 1/R_{671nm}}$
FDR	DI	0	0.518 ***	$R'_{662nm} - R'_{1779nm}$
	RI	0	0.488 **	$R'_{1190nm} / R'_{2216nm}$
	NDI	1	0.603 ***	$(R'_{662nm} - R'_{2216nm})/(R'_{662nm} + R'_{2216nm})$
	DSI	0	0.555 ***	$\sqrt{R'_{662nm} - R'_{1779nm}}$
RFDR	DI	0	0.488 **	$(1/R_{2048nm})' - (1/R_{2081nm})'$
	RI	0	0.587 ***	$(1/R_{765nm})'/(1/R_{1291nm})'$
	NDI	2	0.625 ***	$\frac{(1/R_{1106nm})'-(1/R_{2048nm})'}{(1/R_{1106nm})'+(1/R_{2048nm})'}$
	DSI	0	0.491 **	$\sqrt{(1/R_{1728nm})' - (1/R_{2301nm})'}$

** Correlation is significant at 0.01 level. *** Correlation is significant at 0.001 level.

Table 9. The maximum correlation index of different types of SIs in the vegetation-covered area.

Type		Number of SIs with $\mid\rho\mid$>0.6	$\mid\rho\mid_{max}$	Index Formula Corresponding to $\mid\rho\mid_{max}$
OR	DI	0	0.576 ***	$R_{722nm} - R_{2468nm}$
	RI	19	0.604 ***	R_{525nm}/R_{2267nm}
	NDI	3	0.601 ***	$(R_{525nm} - R_{2250nm})/(R_{525nm} + R_{2250nm})$
	DSI	0	0.569 ***	$\sqrt{R_{722nm} - R_{2468nm}}$
RR	DI	84	0.634 ***	$1/R_{722nm} - 1/R_{1291nm}$
	RI	0	0.587 ***	$(1/R_{525nm})/(1/R_{1510nm})$
	NDI	3	0.601 ***	$\frac{1/R_{525nm}-1/R_{2250nm}}{1/R_{525nm}+1/R_{2250nm}}$
	DSI	83	0.639 ***	$\sqrt{1/R_{551nm} - 1/R_{1998nm}}$
FDR	DI	0	0.575 ***	$R'_{671nm} - R'_{1173nm}$
	RI	0	0.490 **	R'_{1173nm}/R'_{1627nm}
	NDI	0	0.489 ***	$(R'_{705nm} - R'_{1173nm})/(R'_{705nm} + R'_{1173nm})$
	DSI	0	0.581 ***	$\sqrt{R'_{671nm} - R'_{1173nm}}$
RFDR	DI	0	0.421 **	$(1/R_{1745nm})' - (1/R_{2301nm})'$
	RI	0	0.393 **	$(1/R_{868nm})'/(1/R_{1577nm})'$
	NDI	0	0.468 ***	$\frac{(1/R_{688nm})'-(1/R_{1745nm})'}{(1/R_{688nm})'+(1/R_{1745nm})'}$
	DSI	0	0.348 **	$\sqrt{(1/R_{1241nm})' - (1/R_{1627nm})'}$

** Correlation is significant at 0.01 level. *** Correlation is significant at 0.001 level.

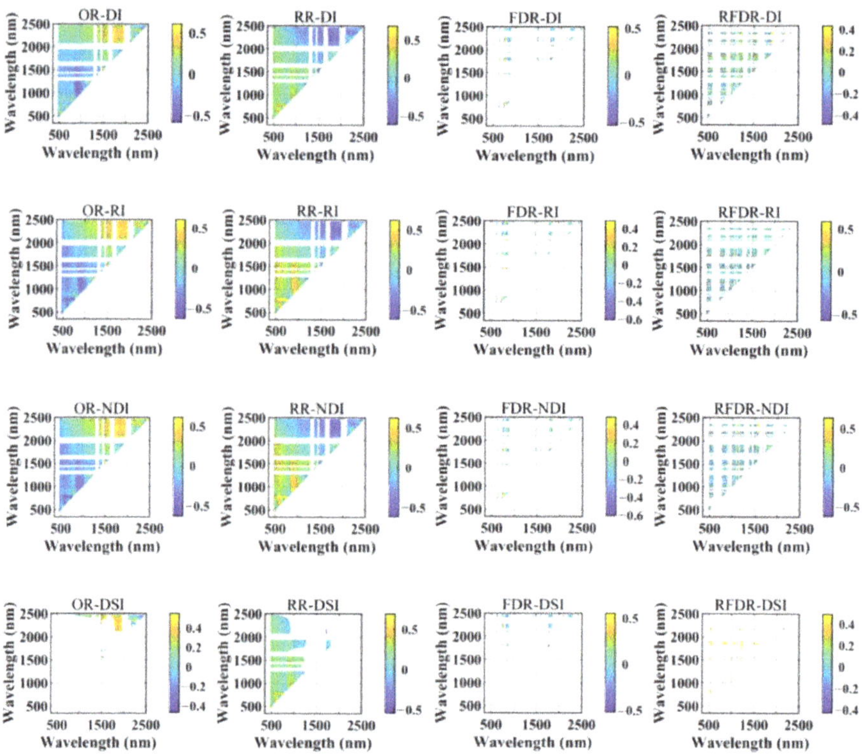

Figure 5. Correlation coefficient matrix of spectral indices (SIs) and SOMC of soil samples in the bare soil area.

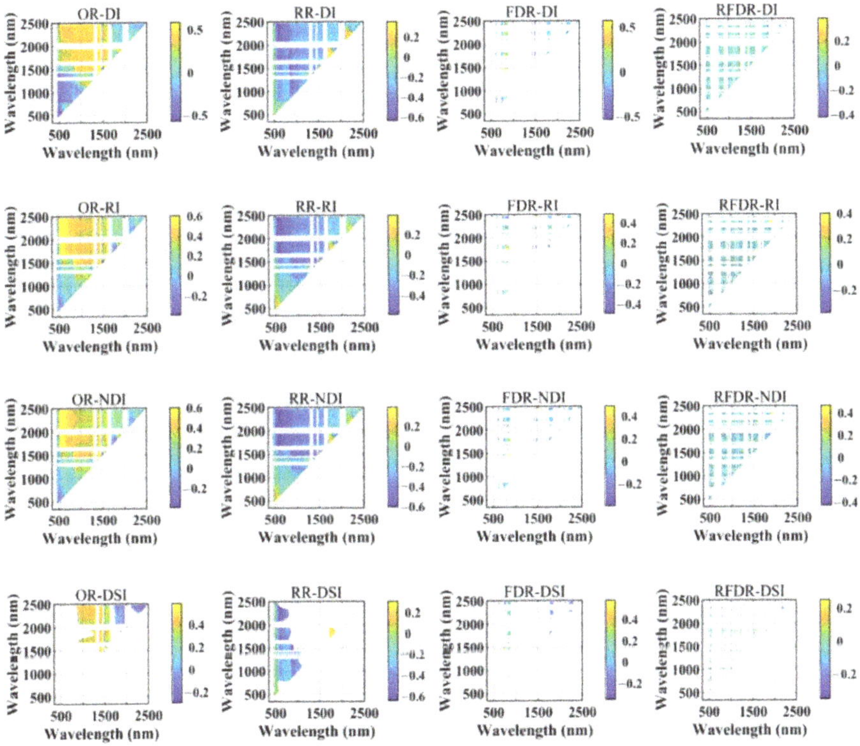

Figure 6. Correlation coefficient matrix of SIs and SOMC of soil samples in the vegetation-covered area.

In the bare soil area, it can be seen from Figure 5 and Table 8 that for the same types of spectral transformation, NDI can maximize the correlation between SOMC and three types of transformed spectra (RR, FDR and RFDR). In addition, RI can also achieve a high correlation between SOMC and three types of transformed spectra (OR, RR and RFDR). In four index formulas, DSI has the lowest correlation with SOMC. Among the four types of SIs constructed based on RR, there are 14 SIs with a correlation greater than 0.6 with SOMC, followed by 12 SIs based on OR. For the same index formula, the indices calculated from OR and RR generally have higher correlations with SOMC than FDR and RFDR. In the bare soil area, OR-RI has the highest correlation with SOMC at the combination of 654 and 679 nm (OR-$RI_{(654,679)}$), with $|\rho|$ as high as 0.627.

In the vegetation-covered area, it can be seen from Figure 6 and Table 9 that for the same types of spectral transformation, DSI can obtain the highest correlation between SOMC and two types of transformed spectra (RR and FDR). In addition, DI also attains a high correlation between SOMC and three types of transformed spectra (RR, FDR and RFDR). Among the four types of SIs constructed based on RR, the number of SIs with a correlation greater than 0.6 with SOMC is as many as 180, far more than other spectra. For the same index formula, the indices calculated from OR and RR have significantly higher correlations with SOMC than FDR and RFDR. In the vegetation-covered area, RR-DSI has the highest correlation with SOMC at the combination of 551 and 1998 nm (RR-$DSI_{(551,1998)}$), with $|\rho|$ as high as 0.639.

Because the values of OR-NDI and RR-NDI are opposite to each other, and the wavelengths used are the same, it is sufficient to choose one of them. When modeling, the optimal SIs obtained by OR-NDI were adopted. Therefore, there are 24 optimal SIs for the bare soil area and 189 for the vegetation-covered area.

3.5. Application of the Optimal SIs

3.5.1. Characterization of SOMC in Soil Samples

According to Tables 8 and 9, OR-RI$_{(654,679)}$ and RR-DSI$_{(551,1998)}$ have the highest correlation with SOMC in bare soil and vegetation-covered areas, respectively (the letter 'V' is added to RR-DSI$_{(551,1998)}$ as a prefix (V-RR-DSI$_{(551,1998)}$), which represents the optimal SI in the vegetation-covered area). These two indices were selected to construct LR models and compare them with other traditional indices (Table 4). The correlations of SOMC and different SIs are shown in Figure 7.

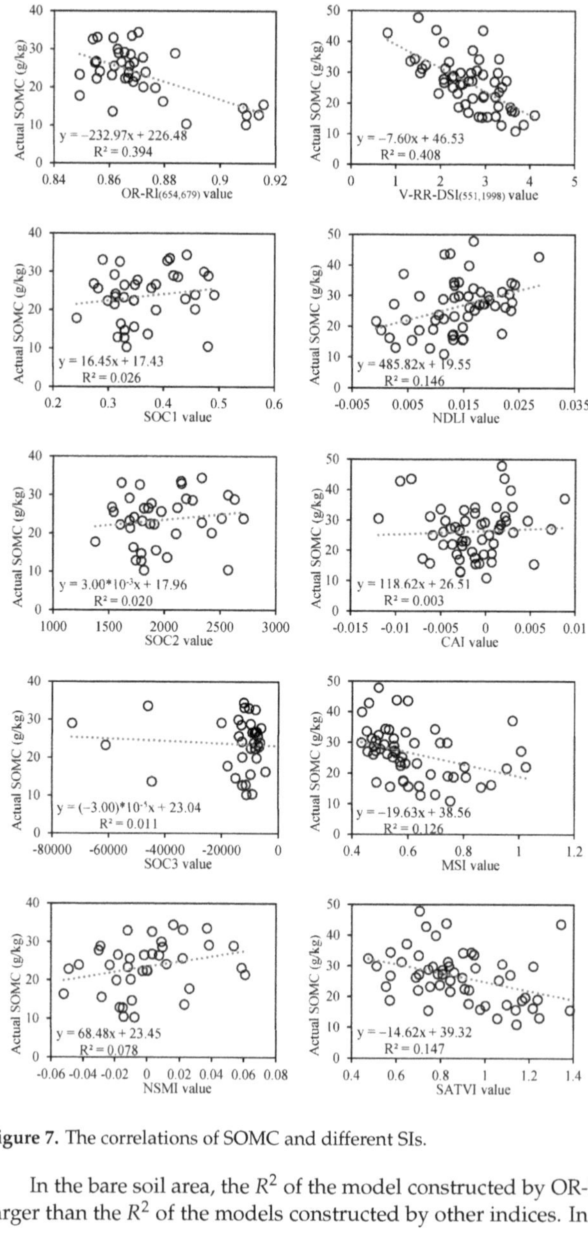

Figure 7. The correlations of SOMC and different SIs.

In the bare soil area, the R^2 of the model constructed by OR-RI$_{(654,679)}$ is 0.398, much larger than the R^2 of the models constructed by other indices. In the vegetation-covered

area, the R^2 of the model constructed by V-RR-DSI$_{(551,1998)}$ is 0.408, whereas the other indices are all less than 0.2. In general, the OR-RI$_{(654,679)}$ and V-RR-DSI$_{(551,1998)}$ proposed in this study well characterize SOMC in bare soil and vegetation-covered areas, respectively.

3.5.2. Recognition of SOMC Levels in Soil Samples

According to the linear relations shown in Figure 7, SOMC in the bare soil area was calculated by OR-RI$_{(654,679)}$, whereas SOMC in the vegetation-covered area was calculated by V-RR-DSI$_{(551,1998)}$. After the SOMCs of bare soil and vegetation-covered areas were merged, SOMC of the whole cultivated land in Shuyang County was obtained. Based on the numerical distribution of SOMC, the calculated and measured SOMC was similarly divided into five levels by the equal interval method, as shown in Figure 8.

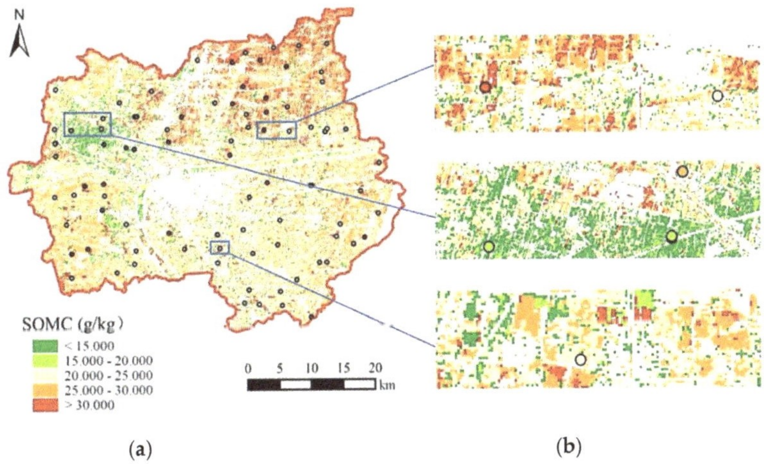

Figure 8. (**a**) Recognition of SOMC levels in soil samples in cultivated land area of Shuyang County and (**b**) details of partial enlargement.

On the whole, the SOMC levels in the cultivated area of Shuyang County are consistent with the trend shown by SOMC of the soil samples (Figure 8a). They all display that the spatial distribution of the SOMC level in Shuyang County is high in the northeast, low in the northwest and intermediate in the south. The areas with the highest SOMC level are concentrated in vegetation-covered areas in the northeast, where most areas exceed 30 g/kg. The lowest SOMC level is found in bare soil areas in the northwest, mostly below 15 g/kg. The detailed image (Figure 8b) shows that the results calculated using OR-RI$_{(654,679)}$ and V-RR-DSI$_{(551,1998)}$ can well recognize soil samples with different SOMC levels.

3.5.3. Estimation of SOMC in Soil Samples

As mentioned in Section 3.4, 24 soil SIs and 189 vegetation SIs were employed in the prediction model. The modeling data of bare soil and vegetation-covered areas are divided into 10 folds for cross-validation, respectively. PCA with a 99.99% explanatory variance rate was used for dimensionality reduction and a new set of variables was generated for modeling. Among the eight inversion models introduced in Section 2.4.3, the linear SVM and least-squares boosting tree algorithms obtained the highest prediction accuracy in bare soil and vegetation-covered areas, respectively.

The prediction accuracy of SOMC based on the soil-vegetation combined prediction is shown in Figure 9. Overall, the predicted results have a good linear relationship with the SOMC of samples. The R^2, RMSE and RPD of the soil-vegetation combined prediction results are 0.775, 3.72 g/kg and 2.12, respectively, suggesting the model is very good.

Specifically, the RMSE of bare soil and vegetation-covered samples is 4.59 and 2.96 g/kg, respectively. Most scatters are very close to the 1:1 line, especially in the vegetation-covered area. About 88.04% of the predicted results are distributed within the expected error (80% precision lines). It can be seen that the optimal SIs proposed in this study have a great potential in predicting SOMC in both bare soil and vegetation-covered areas.

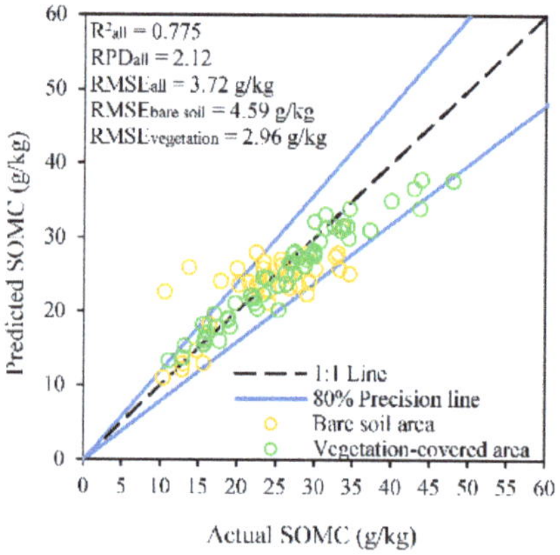

Figure 9. The prediction accuracy of SOMC in the cultivated land of Shuyang County.

4. Discussion

4.1. Image Quality of Transformed Spectra and SIs

Previous studies have shown that spectral transformation and index calculation of the original reflectance spectrum can improve the correlation between the spectrum and SOMC. For example, Wei et al. [7] improved the correlation between soil reflectance and SOMC by integer-order differential spectral transformations. This is also confirmed by the results of this study. As shown in Table 7, the OR of soil has no significantly correlated bands with SOMC, whereas the correlation coefficient between SDR and SOMC is greater than 0.450, which is significant at the 0.01 level.

However, the satellite imaging process is often affected by atmospheric absorption, sensor instability and other factors [62], which inevitably leads to a certain degree of noise contamination in the data (such as Gaussian noise, stripe noise [63], impulse noise, deadline noise and mixed noise) [64]. With the increase in the complexity of spectral transformation and the number of bands involved in the calculation, the uncertainty of image quality also increases. In this study, the image quality after single-band transformation, such as the reciprocal and square root, is better than after first-order differentiation. As the order of the differential transformation increases, the image quality decreases. This is mainly due to the increase in the bands involved in the calculation (as shown in Figure 10). Similar results have been reported in previous studies. For example, Wang et al. [39] found that exploiting the image spectral information by higher-order fractional derivatives was disadvantageous.

Therefore, in this study, in addition to calculating the correlation between reflectance and SOMC, the image quality of transformed reflectance was also taken into account during the construction of SIs. The bands with heavy noise were removed. In order to reduce the influence of noise on prediction results, simple SIs based on single-band spectral transformation should be used first.

In addition, this study was mainly based on the autumn AHSI/ZY1-02D data of Shuyang County. The optimal SIs obtained are suitable for the images of bare soil types of cambisols, regosols and luvisols, and the images of the rice-growing season in the Yangtze River Delta. Therefore, the application performance of the optimal SIs in different regions, vegetation growing seasons and soil types data remains to be further studied.

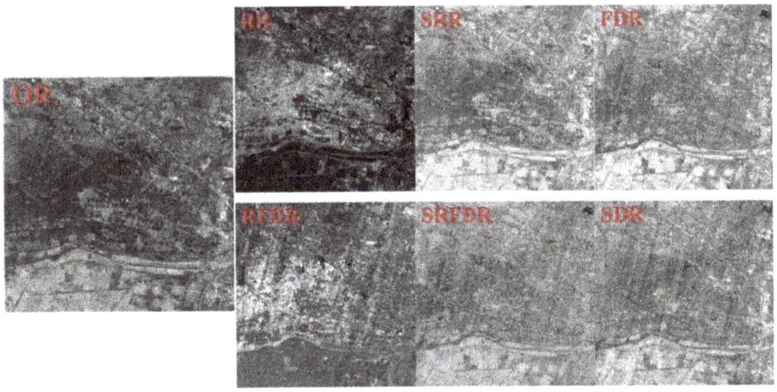

Figure 10. Reflectance images after different spectral transformations at 645 nm (from the AHSI/ZY1-02D data). The meanings of OR, RR, SRR, FDR, RFDR, SRFDR and SDR are consistent with Table 2.

4.2. Advantages of Constructing SIs Separately in Bare Soil Area and Vegetation-Covered Area

Bare soil and vegetation co-exist in most cultivated land. However, the spectral response mechanisms of the two are different. Therefore, the SIs based on bare soil and vegetation-covered areas were constructed separately. In this study, the sensitive spectral wavelengths of SOM in the bare soil area are concentrated at 550–600 nm, 650–700 nm, 1100–1200 nm, 1450–1500 nm, around 2000 nm, and 2200–2300 nm (Figure 5), which are the absorption positions of soil parameters such as soil moisture, iron and SOM. These sensitive wavelengths can also explain the interaction between SOM and soil moisture and iron oxides to a certain extent. In the vegetation-covered area, the distributions of sensitive wavelengths are 500–600 nm, 700–750 nm, 1100–1300 nm, and 1900–2350 nm (Figure 6). The absorption at 1120–1200 nm and 1940–2100 nm is mainly caused by lignin, cellulose, moisture and other parameters of plants. These findings are consistent with the studies of Hong et al. [65] and Lu et al. [66]. The direct mechanism of SOM on the spectrum of bare soil pixels relies on the stretching and vibration of chemical bonds [19]. In the vegetation-covered area, SOM affects the vegetation growth, cellulose, lignin and other physicochemical parameters, and further affects the vegetation canopy spectrum. In bare soil and vegetation-covered areas, the distribution of sensitive wavelengths to SOMC is quite different, indicating the necessity of constructing SIs separately.

The $|\rho|$ of the optimal SI with the highest correlation obtained in the bare soil area is 0.627 (OR-RI$_{(654,679)}$), and that in the vegetation area is 0.639 (V-RR-DSI$_{(551,1998)}$). For comparison, index construction and correlation analysis were also conducted for all sample points together. Figure 11 illustrates the correlation between SOMC and SIs constructed by different combinations of spectral transformations and index formulas. It can be seen that the highest correlation coefficient is 0.501 ($p < 0.001$), which is lower than the correlation coefficient of the indices constructed separately, revealing the advantage of constructing SIs separately.

$\|\rho\|_{max}$	DI	RI	NDI	DSI
OR	0.429	0.486	0.476	0.425
RR	0.501	0.446	0.476	0.492
FDR	0.462	0.385	0.380	0.353
RFDR	0.328	0.366	0.388	0.337

Figure 11. The maximum correlation coefficient between SOMC of SIs constructed by different combinations of spectral transformations and index formulas.

4.3. The Impacts of Soil Types and Cultivated Land-Use Type on SOMC

The soil types in the cultivated land of the study area are shown in Figure 1b, and are mainly cambisols, regosols and luvisols. According to the Harmonized World Soil Database (version 1.1) provided by FAO, there was no significant difference in topsoil organic carbon content among these three types of soil. Based on the results of the distribution of SOMC levels obtained by OR-$RI_{(654,679)}$ and V-RR-$DSI_{(551,1998)}$ (Figure 8a), the statistical results of SOMC of different soil types are shown in Table 10. The average values of the three soil types are 24.39, 24.39 and 24.05 g/kg, respectively, and the difference among them is very small, which is consistent with previous studies [33].

Table 10. The statistics of SOMC in different soil types.

Soil Type	Area Proportion (%)	The Minimum Value of SOMC (g/kg)	The Maximum Value of SOMC (g/kg)	Average SOMC (g/kg)
Cambisols	70.28	$7.02{*}10^{-4}$	49.42	24.39
Regosols	9.38	0.03	45.51	24.39
Luvisols	5.67	$2.61{*}10^{-3}$	45.44	24.05
Other soil types [1]	14.68	$7.16{*}10^{-3}$	48.75	23.41

[1] Other soil types include greyzems, anthrosols, gleysols, fluvisols, nitisols and arenosols.

The land-use types of cultivated land in the study area mainly include paddy fields, dry land and nurseries (Figure 1c). Based on the results of the distribution of SOMC levels obtained by OR-$RI_{(654,679)}$ and V-RR-$DSI_{(551,1998)}$ (Figure 8a), the statistical results of SOMC of different cultivated land-use types are shown in Table 11. Figure 12a–c show SOMC distribution in the paddy field, dry land and nurseries, respectively.

Table 11. The statistics of SOMC in different land-use types.

Land-Use Type	The Minimum Value of SOMC (g/kg)	The Maximum Value of SOMC (g/kg)	Average SOMC (g/kg)
Paddy field	$1.00{*}10^{-3}$	49.42	25.34
Dry land	0.02	48.75	23.23
Nursery	0.11	45.43	21.73
All	$1.00{*}10^{-3}$	49.42	24.23

It can be seen from Figure 12 that the SOMC values of the three land types are all normally distributed. Overall, SOMC is the highest in paddy fields, followed by the dryland, and is lower in nurseries. In the paddy field, the SOMC is most concentrated at 25–26 g/kg. In the dryland, SOMC is most concentrated at 23–24 g/kg, whereas the SOMC in the nursery is most concentrated at 21–22 g/kg. The maximum values of SOMC distribution in the paddy field, dry land, and nursery are 49.42, 48.75, 45.43 g/kg, and the average values are 25.34, 23.23, 21.73 g/kg, respectively. There have been similar reports in previous studies. The studies of Bird et al. [67] and Huang et al. [68] showed that a low mineralization rate and the high humification coefficient of fresh organic matters in flooded soils resulted in SOC content in paddy soil that was 11–56% higher than that in dry land soil. Gao et al. [69] found that transforming paddy fields into nurseries would lead to a decrease in SOC.

West et al. [70] and Wiesmeier et al. [71] mentioned that different land-use types would significantly affect the carbon storage capacity of the soil and the distribution of SOMC, which is also confirmed in this study. The results also show that the SIs proposed in this study can effectively indicate the spatial distribution of SOMC in Shuyang County.

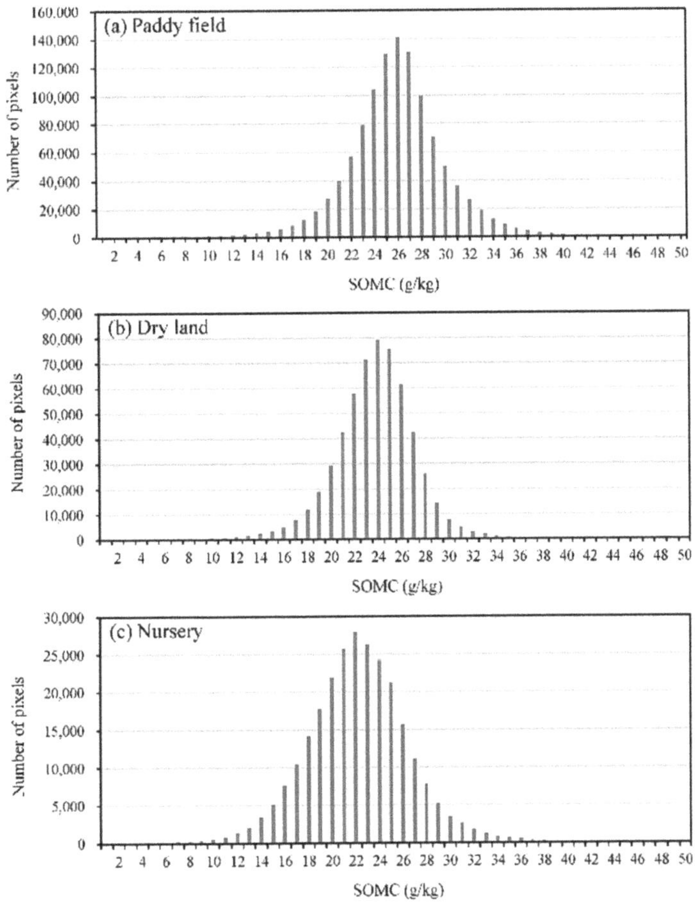

Figure 12. Distribution of SOMC in different land-use types ((**a**) paddy field, (**b**) dry land and (**c**) nursery) in Shuyang County.

5. Conclusions

The AHSI data of the newly launched ZY1-02D satellite provides an important data source for soil quality monitoring. To quickly achieve an SOMC assessment on a large scale, in this study, the SOM SIs suitable for this satellite were constructed. The optimal SIs that are the most sensitive to SOMC in bare soil and vegetation-covered areas were analyzed. Moreover, the application performance of these indices was evaluated. The conclusions obtained are as follows:

1. In the bare soil area, the SIs constructed based on OR and RR have higher correlations with SOMC. For the same transformed spectrum, the SIs calculated by RI and NDI have the highest correlations with SOMC, followed by DI. Among all the constructed SIs, OR-$RI_{(654,679)}$ has the highest correlation with SOMC, and the correlation coefficient is −0.627. In the vegetation-covered area, the correlations between SOMC and the SIs based on RR are higher than those of other transformed spectra. Among the different index formulas, the correlations between the SIs calculated by DSI and DI and SOMC are higher than those of RI and NDI. The correlation coefficient between V-RR-$DSI_{(551,1998)}$ and SOMC is −0.639, which is the highest among all the calculated SIs.

2. The results show that the optimal SIs can be used to present the spatial distribution trend of SOMC and recognize SOMC levels. Based on the optimal SIs, the SOMC predicted by the model has a good linear relationship with the actual SOMC of samples. The R^2, RMSE and RPD of the soil-vegetation combined prediction results are 0.775, 3.72 g/kg and 2.12, respectively.

Overall, the optimal SIs constructed in this study are stable, simple to calculate, and can be used as essential parameters for SOMC estimation, showing great potential in SOMC prediction using AHSI/ZY1-02D data. In addition, the application performances of the optimal SIs in different regions, vegetation growing seasons and soil types remain to be further studied.

Author Contributions: Conceptualization, Kun Shang and Yayu Yang; methodology, Kun Shang; software, Yayu Yang and Hongzhao Tang; validation, Yayu Yang and Kun Shang; formal analysis, Yayu Yang; investigation, Changkun Wang; resources, Kun Shang, Chenchao Xiao, Changkun Wang and Hongzhao Tang; data curation, Kun Shang; writing—original draft preparation, Yayu Yang; writing—review and editing, Kun Shang, Chenchao Xiao and Yayu Yang; visualization, Kun Shang; supervision, Yayu Yang, Kun Shang and Chenchao Xiao; project administration, Kun Shang; funding acquisition, Kun Shang and Hongzhao Tang All authors have read and agreed to the published version of the manuscript.

Funding: This research was funded by the National Natural Science Foundation of China (Grant No. 42001317) and the National Key R&D Program of China (Grant No. 2021YFD1500102, 2018YFB0504800 and 2018YFB0504801).

Data Availability Statement: The data presented in this study are available on request from the corresponding author. The data are not publicly available due to privacy.

Acknowledgments: This work was supported by the National Natural Science Foundation of China (Grant No. 42001317). The authors are grateful to Shihu Zhao and Shuhan Liu for the help with image preprocessing. The authors wish to thank the referees for providing helpful suggestions for the improvement of the manuscript.

Conflicts of Interest: The authors declare no conflict of interest.

References

1. Luo, Z.; Wang, E.; Sun, O.J. Soil carbon change and its responses to agricultural practices in Australian agro-ecosystems: A review and synthesis. *Geoderma* **2010**, *155*, 211–223. [CrossRef]
2. Nocita, M.; Stevens, A.; van Wesemael, B.; Aitkenhead, M.; Bachmann, M.; Barthès, B.; Dor, E.B.; Brown, D.J.; Clairotte, M.; Csorba, A.; et al. Soil Spectroscopy: An Alternative to Wet Chemistry for Soil Monitoring. *Adv. Agron.* **2015**, *132*, 139–159. [CrossRef]
3. Seely, B.; Welham, C.; Blanco, J.A. Towards the application of soil organic matter as an indicator of forest ecosystem productivity: Deriving thresholds, developing monitoring systems, and evaluating practices. *Ecol. Indic.* **2010**, *10*, 999–1008. [CrossRef]
4. Six, J.; Paustian, K. Aggregate-associated soil organic matter as an ecosystem property and a measurement tool. *Soil Biol. Biochem.* **2014**, *68*, A4. [CrossRef]
5. Castaldi, F.; Palombo, A.; Santini, F.; Pascucci, S.; Pignatti, S.; Casa, R. Evaluation of the potential of the current and forthcoming multispectral and hyperspectral imagers to estimate soil texture and organic carbon. *Remote Sens. Environ.* **2016**, *179*, 54–65. [CrossRef]
6. Shi, T.; Liu, H.; Chen, Y.; Wang, J.; Wu, G. Estimation of arsenic in agricultural soils using hyperspectral vegetation indices of rice. *J. Hazard. Mater.* **2016**, *308*, 243–252. [CrossRef] [PubMed]
7. Wei, L.; Yuan, Z.; Wang, Z.; Zhao, L.; Zhang, Y.; Lu, X.; Cao, L. Hyperspectral inversion of soil organic matter content based on a combined spectral index model. *Sensors* **2020**, *20*, 2777. [CrossRef] [PubMed]
8. Rathod, P.H.; Rossiter, D.G.; Noomen, M.F.; Rathod, P.H.; Rossiter, D.G.; Noomen, M.F. Proximal Spectral Sensing to Monitor Phytoremediation of Metal-Contaminated Soils. *Int. J. Phytoremediation* **2013**, *15*, 405–426. [CrossRef] [PubMed]
9. Lagacherie, P.; Mcbratney, A.B. Spatial soil information systems and spatial soil inference systems: Perspectives for digital soil mapping. *Dev. Soil Sci.* **2006**, *31*, 3–22. [CrossRef]
10. Meng, X.; Bao, Y.; Liu, J.; Liu, H.; Zhang, X.; Zhang, Y.; Wang, P.; Tang, H.; Kong, F. Regional soil organic carbon prediction model based on a discrete wavelet analysis of hyperspectral satellite data. *Int. J. Appl. Earth Obs. Geoinf.* **2020**, *89*, 102111. [CrossRef]
11. Shang, K.; Xiao, C.; Gan, F.; Wei, H. Estimation of soil copper content in mining area using ZY1-02D satellite hyperspectral data. *J. Appl. Remote Sens.* **2021**, *15*, 042607. [CrossRef]

12. Wang, J.; He, T.; Lv, C.; Chen, Y.; Jian, W. Mapping soil organic matter based on land degradation spectral response units using Hyperion images. *Int. J. Appl. Earth Obs. Geoinf.* **2010**, *12*, S171–S180. [CrossRef]
13. Tiwari, S.K.; Saha, S.K.; Kumar, S. Prediction Modeling and Mapping of Soil Carbon Content Using Artificial Neural Network, Hyperspectral Satellite Data and Field Spectroscopy. *Adv. Remote Sens.* **2015**, *4*, 63–72. [CrossRef]
14. Emadi, M.; Taghizadeh-Mehrjardi, R.; Cherati, A.; Danesh, M.; Mosavi, A.; Scholten, T. Predicting and mapping of soil organic carbon using machine learning algorithms in Northern Iran. *Remote Sens.* **2020**, *12*, 2234. [CrossRef]
15. *FAO Soil Organic Carbon Mapping Cookbook*; FAO: Rome, Italy, 2017; ISBN 9789251304402.
16. Venter, Z.S.; Hawkins, H.J.; Cramer, M.D.; Mills, A.J. Mapping soil organic carbon stocks and trends with satellite-driven high resolution maps over South Africa. *Sci. Total Environ.* **2021**, *771*, 145384. [CrossRef] [PubMed]
17. Donlon, C.J.; Cullen, R.; Giulicchi, L.; Vuilleumier, P.; Francis, C.R.; Kuschnerus, M.; Simpson, W.; Bouridah, A.; Caleno, M.; Bertoni, R.; et al. The Copernicus Sentinel-6 mission: Enhanced continuity of satellite sea level measurements from space. *Remote Sens. Environ.* **2021**, *258*, 112395. [CrossRef]
18. Arthur Endsley, K.; Kimball, J.S.; Reichle, R.H.; Watts, J.D. Satellite Monitoring of Global Surface Soil Organic Carbon Dynamics Using the SMAP Level 4 Carbon Product. *J. Geophys. Res. Biogeosci.* **2020**, *125*, 1–18. [CrossRef]
19. Clark, R.N.; King, T.V.V.; Klejwa, M.; Swayze, G.A. High Spectral Resolution Reflectance Spectroscopy of Minerals. *J. Geophys. Res. Solid Earth* **1990**, *95*, 12653–12680. [CrossRef]
20. Ben-dor, E. Near-Infrared Analysis as a Rapid Method to Simultaneously Evaluate Several Soil Properties. *Soil Sci. Soc. Am. J.* **1995**, *59*, 364–372. [CrossRef]
21. Krishnan, P.; Alexander, J.D.; Butler, B.J.; Hummel, J.W. Reflectance Technique for Predicting Soil Organic Matter. *Soil Sci. Soc. Am. J.* **1980**, *44*, 1282–1285. [CrossRef]
22. Bartholomeus, H.M.; Schaepman, M.E.; Kooistra, L.; Stevens, A.; Hoogmoed, W.B.; Spaargaren, O.S.P. Spectral reflectance based indices for soil organic carbon quantification. *Geoderma* **2008**, *145*, 28–36. [CrossRef]
23. Jin, X.; Du, J.; Liu, H.; Wang, Z.; Song, K. Remote estimation of soil organic matter content in the Sanjiang Plain, Northest China: The optimal band algorithm versus the GRA-ANN model. *Agric. For. Meteorol.* **2016**, *218*, 250–260. [CrossRef]
24. Prescott, C.E.; Maynard, D.G.; Laiho, R. Humus in northern forests: Friend or foe? *For. Ecol. Manage.* **2000**, *133*, 23–36. [CrossRef]
25. Munson, S.A.; Carey, A.E. Organic matter sources and transport in an agriculturally dominated temperate watershed. *Appl. Geochem.* **2004**, *19*, 1111–1121. [CrossRef]
26. Zhang, Y.; Guo, L.; Chen, Y.; Shi, T.; Luo, M.; Ju, Q.L.; Zhang, H.; Wang, S. Prediction of soil organic carbon based on Landsat 8 monthly NDVI data for the Jianghan Plain in Hubei Province, China. *Remote Sens.* **2019**, *11*, 1683. [CrossRef]
27. Takata, Y. Analysis of spatial and temporal variation of soil organic carbon budget in northern Kazakhstan. *Jpn. Agric. Res. Q.* **2010**, *44*, 335–342. [CrossRef]
28. Zhao, M.S.; Rossiter, D.G.; Li, D.C.; Zhao, Y.G.; Liu, F.; Zhang, G.L. Mapping soil organic matter in low-relief areas based on land surface diurnal temperature difference and a vegetation index. *Ecol. Indic.* **2014**, *39*, 120–133. [CrossRef]
29. Kumar, P.; Pandey, P.C.; Singh, B.K.; Katiyar, S.; Mandal, V.P.; Rani, M.; Tomar, V.; Patairiya, S. Estimation of accumulated soil organic carbon stock in tropical forest using geospatial strategy. *Egypt. J. Remote Sens. Sp. Sci.* **2016**, *19*, 109–123. [CrossRef]
30. Shuyang Meteorological Data China Weather Network. Available online: http://www.weather.com.cn/cityintro/101191302.shtml (accessed on 19 January 2022).
31. Zeng, Y.; Guo, H.; Yao, Y.; Huang, L. The formation of agricultural e-commerce clusters: A case from China. *Growth Chang.* **2019**, *50*, 1356–1374. [CrossRef]
32. Yang, S.C.; Huang, Z.M.; Huang, C.Y.; Tsai, C.C.; Yeh, T.K. A case study on the impact of ensemble data assimilation with GNSS-zenith total delay and radar data on heavy rainfall prediction. *Mon. Weather Rev.* **2020**, *148*, 1075–1098. [CrossRef]
33. Hamonized World Soil Database (Version 1.1). Available online: https://geodata.pku.edu.cn/index.php?c=content&a=show&id=730# (accessed on 19 January 2022).
34. Shi, X.; Yu, D.; Sun, W.; Wang, H.; Zhao, Q.; Gong, Z. Reference benchmarks relating to great groups of genetic soil classification of China with soil taxonomy. *Chin. Sci. Bull.* **2004**, *49*, 1507–1511. [CrossRef]
35. Xin, Z.; Huang, B.; Dong, C.S.; Weixia, S.; Hu, W.Y.; Tian, K. Tempo-spatial variability of soil organic matter and total nitrogen in farmland and its affecting factors in Shuyang county, Jiangsu province. *Soils* **2013**, *45*, 405–411. [CrossRef]
36. Press, W.H.; Teukolsky, S.A. Savitzky-Golay Smoothing Filters. *Comput. Phys.* **1990**, *4*, 669. [CrossRef]
37. Gitelson, A.; Merzlyak, M.N. Spectral reflectance changes associated with autumn senescence of Aesculus hippocastanum L. and Acer platanoides L. leaves. Spectral features and relation to chlorophyll estimation. *J. Plant Physiol.* **1994**, *143*, 286–292. [CrossRef]
38. Sims, D.A.; Gamon, J.A. Relationships between leaf pigment content and spectral reflectance across a wide range of species, leaf structures and developmental stages. *Remote Sens. Environ.* **2002**, *81*, 337–354. [CrossRef]
39. Wang, X.; Zhang, F.; Kung, H.T.; Johnson, V.C. New methods for improving the remote sensing estimation of soil organic matter content (SOMC) in the Ebinur Lake Wetland National Nature Reserve (ELWNNR) in northwest China. *Remote Sens. Environ.* **2018**, *218*, 104–118. [CrossRef]
40. Chen, H.; Pan, T.; Chen, J.; Lu, Q. Waveband selection for NIR spectroscopy analysis of soil organic matter based on SG smoothing and MWPLS methods. *Chemom. Intell. Lab. Syst.* **2011**, *107*, 139–146. [CrossRef]
41. Larney, F.J.; Ellert, B.H.; Olson, A.F. Carbon, ash and organic matter relationships for feedlot manures and composts. *Can. J. Soil Sci.* **2005**, *85*, 261–264. [CrossRef]

42. Lu, Q.; Wang, S.; Bai, X.; Liu, F.; Tian, S.; Wang, M.; Wang, J. Rapid estimation of soil heavy metal nickel content based on optimized screening of near-infrared spectral bands. *Acta Geochim.* **2020**, *39*, 116–126. [CrossRef]
43. Ben-Dor, E.; Inbar, Y.; Chen, Y. The reflectance spectra of organic matter in the visible near-infrared and short wave infrared region (400-2500 nm) during a controlled decomposition process. *Remote Sens. Environ.* **1997**, *61*, 1–15. [CrossRef]
44. Edward, M. Barnes, Kenneth A. Sudduth, John W. Hummel, Scott M. Lesch, D.L.C.; Chenghai Yang, Craig S.T. Daughtry, and W.C.B. Remote- and Ground-Based Sensor Techniques to Map Soil Properties. *Commun. Soil Sci. Plant Anal.* **2015**, *46*, 1668–1676. [CrossRef]
45. Ladoni, M.; Alaviph, S.K.A.; Bahrami, H.A.; Noroozi, A.A. Remote sensing of soil organic carbon in semi-arid region of iran. *Arid L. Res. Manag.* **2010**, *24*, 271–281. [CrossRef]
46. Yadav, J.; Sehra, K. Large Scale Dual Tree Complex Wavelet Transform based robust features in PCA and SVD subspace for digital image watermarking. *Procedia Comput. Sci.* **2018**, *132*, 863–872. [CrossRef]
47. George, J.; Kumar, S. Hyperspectral Remote Sensing in Characterizing Soil Salinity Severity using SVM Technique—A Case Study of Alluvial Plains. *Int. J. Adv. Remote. Sens. GIS* **2015**, *4*, 1344–1360. [CrossRef]
48. Meng, X.; Bao, Y.; Ye, Q.; Liu, H.; Zhang, X.; Tang, H.; Zhang, X. Soil organic matter prediction model with satellite hyperspectral image based on optimized denoising method. *Remote Sens.* **2021**, *13*, 2273. [CrossRef]
49. Yuan, J.; Wang, X.; Yan, C.X.; Wang, S.R.; Ju, X.P.; Li, Y. Soil moisture retrieval model for remote sensing using reflected hyperspectral information. *Remote Sens.* **2019**, *11*, 366. [CrossRef]
50. Haubrock, S.N.; Chabrillat, S.; Lemmnitz, C.; Kaufmann, H. Surface soil moisture quantification models from reflectance data under field conditions. *Int. J. Remote Sens.* **2008**, *29*, 3–29. [CrossRef]
51. Nocita, M.; Stevens, A.; Noon, C.; Van Wesemael, B. Prediction of soil organic carbon for different levels of soil moisture using Vis-NIR spectroscopy. *Geoderma* **2013**, *199*, 37–42. [CrossRef]
52. Daughtry, C.S.T. Agroclimatology: Discriminating crop residues from soil by shortwave infrared reflectance. *Agron. J.* **2001**, *93*, 125–131. [CrossRef]
53. Bartholomeusa, H. Quantitative retrieval of soil organic carbon using laboratory spectroscopy and spectral indices. In Proceedings of the ISPRS Commission VII Symposium 'Remote Sensing: From Pixels to Processes', Enschede, The Netherlands, 8–11 May 2006.
54. Daughtry, C.S.T.; Hunt, E.R.; McMurtrey, J.E. Assessing crop residue cover using shortwave infrared reflectance. *Remote Sens. Environ.* **2004**, *90*, 126–134. [CrossRef]
55. Fourty, T.; Baret, F.; Jacquemoud, S.; Schmuck, G.; Verdebout, J. Leaf optical properties with explicit description of its biochemical composition: Direct and inverse problems. *Remote Sens. Environ.* **1996**, *56*, 104–117. [CrossRef]
56. Melillo, J.M.; Aber, J.D.; Muratore, J.F.; Jun, N. Nitrogen and Lignin Control of Hardwood Leaf Litter Decomposition Dynamics. *Ecology* **2008**, *63*, 621–626. [CrossRef]
57. Serrano, L.; Peñuelas, J.; Ustin, S.L. Remote sensing of nitrogen and lignin in Mediterranean vegetation from AVIRIS data. *Remote Sens. Environ.* **2002**, *81*, 355–364. [CrossRef]
58. Wang, K.; Qi, Y.; Guo, W.; Zhang, J.; Chang, Q. Retrieval and mapping of soil organic carbon using sentinel-2A spectral images from bare cropland in autumn. *Remote Sens.* **2021**, *13*, 1072. [CrossRef]
59. Welikhe, P.; Quansah, J.E.; Fall, S.; McElhenney, W. Estimation of Soil Moisture Percentage Using LANDSAT-based Moisture Stress Index. *J. Remote Sens. GIS* **2017**, *6*, 1–5. [CrossRef]
60. Marsett, R.C.; Qi, J.; Heilman, P.; Biedenbender, S.H.; Watson, M.C.; Amer, S.; Weltz, M.; Goodrich, D.; Marsett, R. Remote sensing for grassland management in the arid Southwest. *Rangel. Ecol. Manag.* **2006**, *59*, 530–540. [CrossRef]
61. Guo, P.; Li, T.; Gao, H.; Chen, X.; Cui, Y. Evaluating Calibration and Spectral Variable Selection Methods for Predicting Three Soil Nutrients Using Vis-NIR Spectroscopy. *Remote Sens.* **2021**, *13*, 4000. [CrossRef]
62. Zhang, Q.; Yuan, Q.; Li, J.; Liu, X.; Shen, H.; Zhang, L. Hybrid noise removal in hyperspectral imagery with a spatial-spectral gradient network. *IEEE Trans. Geosci. Remote Sens.* **2019**, *57*, 7317–7329. [CrossRef]
63. Li, C.; Zhou, C.; Ma, L.; Tang, L.; Wang, X. A stripe noise removal method of interference hyperspectral imagery based on interferogram correction. *Image Signal Process. Remote Sens. XVIII* **2012**, *8537*, 85370A. [CrossRef]
64. Rasti, B.; Scheunders, P.; Ghamisi, P.; Licciardi, G.; Chanussot, J. Noise reduction in hyperspectral imagery: Overview and application. *Remote Sens.* **2018**, *10*, 482. [CrossRef]
65. Hong, Y.; Yu, L.; Chen, Y.; Liu, Y.; Liu, Y.; Liu, Y.; Cheng, H. Prediction of soil organic matter by VIS-NIR spectroscopy using normalized soil moisture index as a proxy of soil moisture. *Remote Sens.* **2018**, *10*, 28. [CrossRef]
66. Lu, Y.; Bai, Y.; Yang, L.; Wang, H. Hyperspectral extraction of soil organic matter content based on principal component regression. *N. Z. J. Agric. Res.* **2007**, *50*, 1169–1175. [CrossRef]
67. Bird, J.A.; Van Kessel, C.; Horwath, W.R. Stabilization of 13C-Carbon and Immobilization of 15N-Nitrogen from Rice Straw in Humic Fractions. *Soil Sci. Soc. Am. J.* **2003**, *67*, 806–816. [CrossRef]
68. Huang, R.; Lan, M.; Liu, J.; Gao, M. Soil aggregate and organic carbon distribution at dry land soil and paddy soil: The role of different straws returning. *Environ. Sci. Pollut. Res.* **2017**, *24*, 27942–27952. [CrossRef] [PubMed]
69. Gao, J.; Pan, G.; Jiang, X.; Pan, J.; Zhuang, D. Land-use induced changes in topsoil organic carbon stock of paddy fields using MODIS and TM / ETM analysis: A case study of Wujiang County, China. *J. Environ. Sci.* **2008**, *20*, 852–858. [CrossRef]

70. West, T.O.; Marland, G. A synthesis of carbon sequestration, carbon emissions, and net carbon flux in agriculture: Comparing tillage practices in the United States. *Agric. Ecosyst. Environ.* **2008**, *91*, 217–232. [CrossRef]
71. Wiesmeier, M.; Von Lützow, M.; Spörlein, P.; Geuß, U.; Hangen, E.; Reischl, A.; Schilling, B.; Kögel-knabner, I. Land use effects on organic carbon storage in soils of Bavaria: The importance of soil types. *Soil Tillage Res.* **2015**, *146*, 296–302. [CrossRef]

International Journal of *Geo-Information*

MDPI

Article

Downscaling of AMSR-E Soil Moisture over North China Using Random Forest Regression

Hongyan Zhang [1,2], Shudong Wang [1], Kai Liu [1,*], Xueke Li [3], Zhengqiang Li [1], Xiaoyuan Zhang [1] and Bingxuan Liu [4]

1 Aerospace Information Research Institute, Chinese Academy of Sciences, Haidian District, Beijing 100094, China; zhanghongyan20@mails.ucas.ac.cn (H.Z.); wangsd@aircas.ac.cn (S.W.); lizq@radi.ac.cn (Z.L.); 3001180141@cugb.edu.cn (X.Z.)
2 University of Chinese Academy of Sciences, Beijing 100049, China
3 Institute at Brown for Environment and Society, Brown University, Providence, RI 02912, USA; xueke_li@brown.edu
4 School of Environmental Science & Engineering, Tianjin University, Tianjin 300350, China; bxliu@stidc.ac.cn
* Correspondence: liuk@aircas.ac.cn

Abstract: Satellite retrieval can offer global soil moisture information, such as Advanced Microwave Scanning Radiometer-Earth Observing System (AMSR-E) data. AMSR-E has been used to provide soil moisture all over the world, with a coarse resolution of 25 km × 25 km. The coarse resolution of the soil moisture dataset often hinders its use in local or regional research. This work proposes a new framework based on the random forest (RF) model while using five auxiliary data to downscale the AMSR-E soil moisture data over North China. The downscaled results with a 1 km spatial resolution are verified against in situ measurements. Compared with AMSR-E data, the correlation coefficient of the downscaled data is increased by 0.17, and the root mean squared error, mean absolute error, and unbiased root mean square error are reduced by 0.02, 0.01, and 0.03 m^3/m^3, respectively. In addition, the comparison results with Multiple Linear Regression and Support Vector Regression downscaled data show that the proposed method significantly outperforms the other two methods. The feasibility of our model is well supported by the importance analysis and leave-one-out analysis. Our study, which combines RF with spatiotemporal search algorithms and efficient auxiliary data, may provide insights into soil moisture downscaling in large areas with various surface characteristics and climatic conditions.

Keywords: downscaling; soil moisture; AMSR-E; random forest; North China

Citation: Zhang, H.; Wang, S.; Liu, K.; Li, X.; Li, Z.; Zhang, X.; Liu, B. Downscaling of AMSR-E Soil Moisture over North China Using Random Forest Regression. *ISPRS Int. J. Geo-Inf.* **2022**, *11*, 101. https://doi.org/10.3390/ijgi11020101

Academic Editors: Dimitris Triantakonstantis, Panagiotis Tziachris and Wolfgang Kainz

Received: 13 December 2021
Accepted: 28 January 2022
Published: 1 February 2022

Publisher's Note: MDPI stays neutral with regard to jurisdictional claims in published maps and institutional affiliations.

1. Introduction

Soil moisture plays a considerable part in the process of global water and energy balance [1], affecting many fields, such as hydrology, meteorology, and agriculture. The characteristics of soil moisture change can reflect many physical and chemical properties of soil, and soil moisture also has a significant impact on terrestrial ecosystems. Nowadays, soil moisture monitoring has been widely used in drought monitoring [2,3], water resources assessment [4,5], crop yield estimation [6], weather forecasting, and hydrological simulation [7,8]. Therefore, obtaining accurate and high-resolution soil moisture data is of great significance for ecosystem balance research and coping with climate change [9,10].

In the past few years, drought disasters in North China have been increasing and have become one of the most serious natural disasters. The rapid growth of industrial, irrigation, and domestic water consumption has led to a sharp increase in water resource consumption, which in turn has led to more serious droughts in northern China [11]. It is significantly urgent to improve our understanding of the spatial and temporal variations of soil moisture and quantify the drought situation. However, research on soil moisture downscaling in northern China is inadequate. Therefore, it is necessary to establish a set of regional soil moisture data with high spatial resolution in North China.

In recent years, the development of remote sensing technology has provided convenience for the acquisition of soil moisture [12]. At the same time, microwave remote sensor is a reliable method to estimate soil moisture (SM) [13], as the microwave is very sensitive to SM changes, less affected by different weather conditions [14], and can realize all-weather and all-coverage Earth observations. At present, a variety of satellite soil moisture products are available, such as Advanced Microwave Scanning Radiometer-Earth Observing System (AMSR-E) [15], Soil Moisture Active Passive (SMAP) [16], and Soil Moisture and Ocean Salinity (SMOS) [17]. The SM data provided by these products is efficient in some large-scale applications [18]. However, for small-scale research, the spatial resolution of these products is too coarse, basically tens of kilometers [19,20], which is not adequate for capturing the spatial heterogeneity of SM. Therefore, these products are very limited in some regional- and local-scale research [21]. Meanwhile, the 1-km spatial resolution has become a common high resolution in many SM downscaling studies [22].

To solve the problem of the rough spatial resolution of satellite remote sensing data, plenty of methods have been presented to improve their resolution. At present, the research focused on downscaling methods and applications can be roughly divided into the following three categories: (1) model-based methods [23], (2) methods based on multi-source satellite remote sensing data fusion [24], and (3) methods using geo-information data [22,25]. Each category can be subdivided into several categories according to specific conditions. Model-based methods are divided into statistical models [26] and land surface models. Based on the land surface models, there are deterministic downscaling [27], statistical downscaling [28], and data assimilation [29]. No matter what kind of model-based method, we need to consider whether the model can be applied to a wider range of research. Furthermore, the model needs a lot of input parameter data, and obtaining this data is difficult, which makes the application have many limitations. The methods based on multi-source satellite data fusion are divided into two methods according to the different fusion data: the fusion method of active and passive microwave data [30] and the fusion method of optical/thermal and microwave data [31]. Earlier scholars have made a lot of attempts and found that the combination of active and passive microwave remote sensing data can obtain more accurate results. Nevertheless, the active microwave data is expensive and needs detailed terrain, vegetation, and other information, so it is a challenge to use and popularize it on a large scale [30]. The downscaling methods using geo-information data are mainly based on the spatial distribution relationship between soil moisture and other land surface variables, such as terrain, soil attributes, and vegetation characteristics [32]. On this basis, the geo-statistical or fractal interpolation model is constructed. In this method, terrain is often used as an auxiliary information source [33]. This method can infer fine-scale changes in soil moisture from the relationship between soil moisture and available auxiliary variables. However, when establishing geo-statistical or analytical interpolation models, a large number of in situ observations need to be used. At the same time, these models have obvious regional characteristics, which obviously limit their applicability [25].

With the improvement of computer performance and the development of artificial intelligence technology, machine learning methods have been introduced into the study of soil moisture downscaling. Machine learning is a kind of data analysis method, which enables computers to learn from data intelligently to find hidden patterns of phenomena [1]. Many studies have shown that compared with other machine learning methods, the random forest (RF) algorithm is more suitable for satellite product downscaling [34,35]. Since the RF model can describe nonlinear correlation and adopt the ensemble method, it becomes more flexible through randomization. At the same time, this method does not require many in situ data to participate in the model. Although the RF model requires some high-resolution auxiliary data, these data are relatively easy to obtain. However, the application of RF in large areas is insufficient, and RF is also very sensitive to surface characteristics and driving parameters, especially in arid and semi-arid areas.

Based on the above issues, this study explores how the random forest method works under large-scale and complex conditions combined with effective parameters. The purpose of our study was to propose 1 efficient method to downscale the soil moisture data of AMSR-E over North China from 25 to 1 km. Our study combines RF with spatiotemporal search algorithms and the effective auxiliary data, which is easy to obtain and relate to soil moisture. We evaluated the downscaled results with in situ observations. The proposed downscaling method was furthermore compared with other commonly used downscaling methods and several methods of earlier studies. Section 2 details the methods and data used in this paper. The results of the study are described in Section 3. The discussion part is given in Section 4. The discussion section provides the importance analysis of explanatory factors, leave-one-out analysis, comparison of different methods, and limitations of the study. The conclusion is given at the end of the article (Section 5).

2. Materials and Methods

2.1. Study Area

Figure 1 shows the elevation and dry humidity classification map of the study area. The location of the selected ground stations in this study is also shown in the figure. The study area is North China, between longitude 32° N and 42° N and latitude 93° E and 123° E, with an area of about 14,900 km^2. It includes Shandong, Hebei, Ningxia, Gansu, Shanxi, Shaanxi, Beijing, and Tianjin. The climate zone in this region has a large span, including temperate monsoon climate, subtropical monsoon climate, plateau mountain climate, temperate continental climate, warm temperate semi-humid and semi-arid monsoon climate, etc. The land use types are complex, with about 18 land use types. The altitude of this area ranges from −2 to 5516 m.

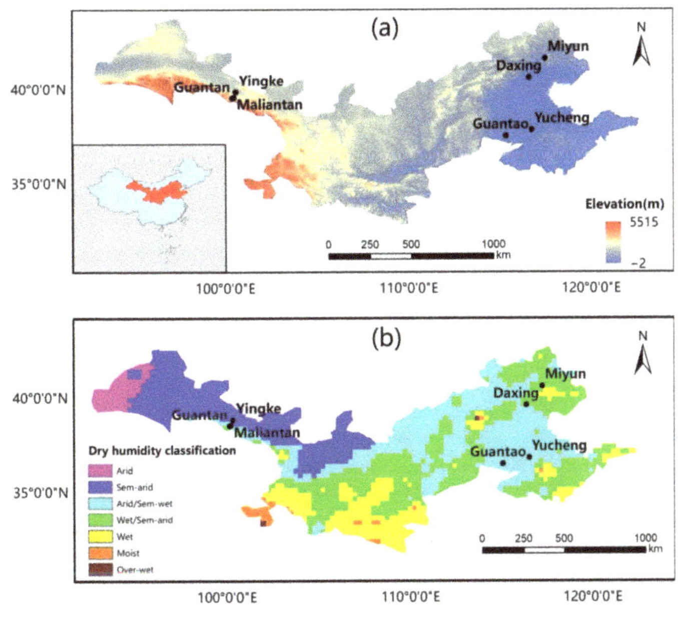

Figure 1. (**a**) Elevation and (**b**) dry humidity classification map of the study area. The black spots are the locations of the selected ground stations in this study.

2.2. Study Data

The spatial downscaling process of soil moisture data based on passive microwave satellite remote sensing data must be supported by the image information of relevant surface features. The image information is provided by the data with higher spatial resolution. In this study, the Moderate Resolution Imaging Spectroradiometer (MODIS), DEM, and precipitation data with higher spatial resolution were used to support the spatial downscaling based on AMSR-E data. At the same time, the ground station data was selected to verify the results. The data used in this study, its specific description, and spatial resolution are shown in Table 1.

Table 1. List of remote sensing soil moisture data and auxiliary data used in this study.

Datasets	Description	Spatiotemporal Resolution
LPRM-AMSR_E	Soil moisture (SM)	25 km/daily
MOD11A1	Land surface temperature (LST)	1 km/daily
MOD13A2	Normalized difference vegetation index (NDVI)	1 km/16 d
MCD43C3	Surface albedo (ALB)	0.05°/daily
SRTM DEM	Elevation	90 m/-
prec_ITPCAS-CMFD	Precipitation	0.1°/daily

2.2.1. AMSR-E Soil Moisture Data

AMSR-E was developed by the National Aeronautics and Space Administration (NASA) and the Japan Aerospace Exploration Agency (JAXA). AMSR-E was carried on the Aqua polar orbit of NASA Earth Observation System (EOS) and launched in May 2002. AMSR-E measures the brightness temperature of vertical and horizontal polarization at 6 frequencies of 6.9, 10.7, 18.7, 23.8, 36.5, and 89.0 GHz.

In this study, level 3 soil moisture data of AMSR-E in 2009 were selected and developed based on the Land Parameter Retrieval Model (LPRM), with a spatial resolution of 25 km [36]. The product was available free of charge from NASA's official website (http://search.earthdata.nasa.uov/search, 19 November 2020).

2.2.2. MODIS Data

Vegetation index, Albedo, and land surface temperature have been extensively used in the downscaling of remote sensing soil moisture data in many studies [37,38]. These parameters are available from Moderate Resolution Imaging Spectroradiometer (MODIS). MODIS on Terra and Aqua satellites is a large space remote sensing instrument developed by NASA. At present, it has been widely used for monitoring various environments, including ocean, land surface, and so on. In this study, three MODIS products were used, including the normalized difference vegetation index (NDVI), land surface temperature (LST), and Albedo.

LST was obtained from MODIS daily product MOD11A1, and NDVI was acquired from 16 day vegetation index product MOD13A2. The spatial resolution of both is 1 km. The Albedo was obtained from the daily Albedo product MCD43C3 with a resolution of 0.05°. These data were acquired from the Land Process Distributed Active Archive Center (LP DAAC) (https://lpdaac.usgs.gov/, 30 April 2021).

2.2.3. DEM

In many studies of soil moisture downscaling, elevation is considered to be one of the most effective topographical features [39,40]. Therefore, we used this variable to downscale the AMSR-E SM data in our proposed approach.

Digital elevation model (DEM) data was acquired from NASA's Space Shuttle Radar Topography Mission (SRTM). The data has a spatial resolution of 90 m and is one of the most used DEM data.

2.2.4. Precipitation Data

The most direct way for the atmosphere to affect land soil moisture is through precipitation [41]. Many regional- and global-scale studies have confirmed the tele-connection between precipitation and soil moisture variability [42].

The precipitation data is the China meteorological forcing dataset (CMFD) from the national Qinghai Tibet Plateau science data center, which is a set of near land surface meteorological reanalysis data. CMFD is made with the Princeton reanalysis data, GEWEX-SRB radiation data, GLDAS and TRMM precipitation data as the background field, and the conventional meteorological observation data [43,44]. The data set was provided by the National Tibetan Plateau Data Center (http://data.tpdc.ac.cn, 20 April 2021).

2.2.5. In Situ Soil Moisture Observations

To evaluate the accuracy of downscaled soil moisture, field measurement data from 7 locations in 2009 were collected. Field measurements at seven sites were obtained from the Chinese Ecosystem Research Network (CERN) and the Watershed Allied Telemetry Experimental Research (WATER) project. The data set was provided by the National Tibetan Plateau Data Center (http://data.tpdc.ac.cn, 26 April 2021). Each site was equipped with a set of quadratic net radiometers (CNR1 and CNR4) that collected atmospheric down-welling and up-welling longwave radiation. In total, 6 layers of soil moisture sensors were arranged at depths of 0.05, 0.1, 0.20, 0.40, 0.80, and 1.20 m, respectively. Detailed information of the 7 sites is exhibited in Table 2.

Table 2. Information of the field sites.

ID	Site	Land-Cover	Elevation	Longitude	Latitude
1	Guantan	Forest	2835 m	100.25 E	38.53 N
2	Maliantan	Grassland	2817 m	100.30 E	38.55 N
3	Yingke	Cropland	1519 m	100.42 E	38.85 N
4	Daxing	Cropland	20 m	116.42 E	39.62 N
5	Guantao	Cropland	30 m	115.12 E	36.51 N
6	Miyun	Orchard	350 m	117.32 E	40.63 N
7	Yucheng	Cropland	23 m	116.57 E	36.83 N

2.3. *Methods*

2.3.1. Random Forest Model

The RF model [45] is an improved decision tree model, which makes decisions by averaging multiple regression trees [46]. The model is nonlinear and insensitive to outliers. Compared with other machine learning methods, the RF model is more widely used in the downscaling of SM data [47]. The major idea of the RF model is to build the nonlinear function relationship between various input parameters and output soil moisture:

$$\mathrm{SM} = f_{RF}(\mathrm{C}) + \varepsilon \tag{1}$$

where the left SM is soil moisture data, and C is an input vector representing various input parameters; f_{RF} is the nonlinear function that establishes the relationship between input parameters and output SM.

RF establishes several decision trees in the training stage, and then calculates the average prediction value of these trees as the output of the method. RF models can divide the input feature space into a lot of regression trees, called a forest, in which each tree is generated by guided samples. One boot sample contains approximately two-thirds of the training samples, and one-third of the remaining data is used to validate each tree. The result of RF is the average prediction result of each decision tree, as follows:

$$f(SM|C) = \frac{1}{n}\sum_{i=1}^{n} f_i(SM|C) \tag{2}$$

where n is the number of regression trees, $f(SM|C)$ is the ensemble decision tree, and $f_i(SM|C)$ is the sub-decision tree of the original soil moisture (SM) given the bootstrapped data from the training input variables (C).

2.3.2. Soil Moisture Downscaling Framework

In order to downscale AMSR-E soil moisture data, we used known high-resolution data sets closely related to soil moisture, including LST, NDVI, Albedo, DEM, and precipitation. Atmospheric auxiliary variables (LST, Albedo, and precipitation) can maintain the temporal variations of the downscaled SM. The geophysical auxiliary variable DEM is used to capture the spatial variations and patterns of the downscaled SM. The other auxiliary variable, such as NDVI, is also included to illustrate the impact of vegetation variations on the spatiotemporal pattern of the downscaled SM.

In this study, we used the RF models to develop a downscaling framework. The foundation of this method is explained in detail in the Section 2.3.1. The major contribution of this study is to propose a new method that incorporates the above five effective auxiliary data into the RF models to improve the estimation of soil moisture at a finer resolution.

A reliable spatial and temporal window size is the basis of the success of the presented method, as too large of a window may lead to over-fitting problems in the regression model. In this study, we used an adaptive spatiotemporal window size determination method. Two key parameters are used to decide the window size: the temporal day number and the spatial window size. The initial value of the spatial window size needs to be given first. When the corresponding conditions are met, the model will automatically search until termination. More details regarding this strategy can be found in our previous studies [48].

Figure 2 shows the process of the proposed method. The following three steps show the construction of the soil moisture downscaling method proposed in this paper:

Step 1: Firstly, we prepared the required data, including AMSR-E, MODIS (LST, NDVI, Albedo), DEM, precipitation data, and in situ measurements. To establish the downscaling model, the upscaling data of LST, NDVI, Albedo, DEM, and precipitation were consistent with AMSR-E, 25 km.

Step 2: The RF models were established by using the upscaled data and AMSR-E data. In the RF model mentioned in Section 2.3.1, C was LST, NDVI, Albedo, DEM, and precipitation. After modeling, the model was evaluated and analyzed. The SM relationship model built by the proposed method was applied to 1-km resolution variables to obtain high-resolution SM data at 1-km resolution:

$$C = (LST, Albedo, NDVI, precipitation, DEM) \tag{3}$$

Step 3: The downscaled results were verified with in situ measurements and compared with other downscaling methods, namely Multiple Linear Regression (MLR) and Support Vector Regression (SVR).

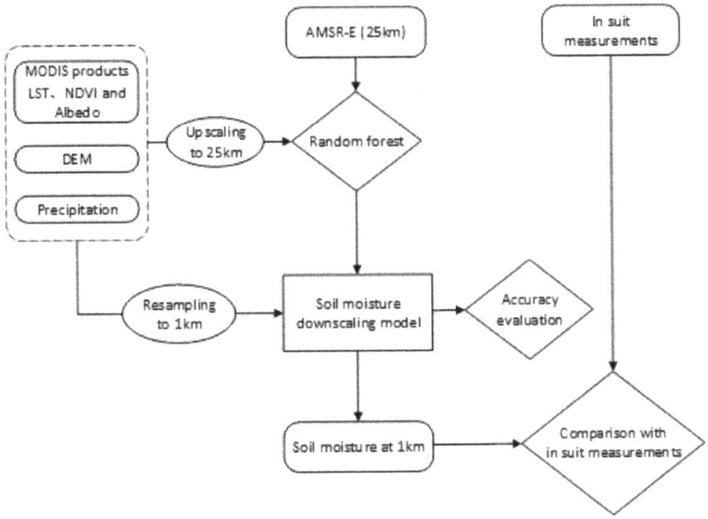

Figure 2. Flowchart of SM downscaling in this study.

2.4. Evaluation

Statistical indexes were used to conduct the quantitative analysis of the original SM and the estimated SM, and to compare the different downscaling methods. This study used the correlation coefficient (R) [49], root mean squared error (RMSE) [50], mean absolute error (MAE) [51], and unbiased root mean square error (ubRMSE) in the paper. The calculation formulas are as follows:

$$R = \frac{\sum(X_i - \overline{X})(Y_i - \overline{Y})}{\sqrt{\sum(X_i - \overline{X})^2 (Y_i - \overline{Y})^2}} \tag{4}$$

$$RMSE = \sqrt{\frac{\sum_{i=1}^{n}(X_i - Y_i)^2}{n}} \tag{5}$$

$$MAE = \frac{1}{n}\sum_{i=1}^{n}|X_i - Y_i| \tag{6}$$

$$\text{ubRMSE} = \sqrt{RMSE^2 - BIAS^2} \tag{7}$$

where n is the number of observations, and X_i and Y_i are the original and estimated SM, respectively.

In addition to the above commonly used indicators for soil moisture downscaling, this study also used the fractions skill score (FSS) [52] to evaluate the spatial agreement between the original data and downscaled data. The FSS was given by the following equation:

$$\text{FSS} = 1 - \frac{MSE_{(n)}}{MSE_{(n)ref}} \tag{8}$$

where n is the window size and MSE is mean square error.

The value of FSS ranges from 0 (total mismatch) to 1 (perfect prediction). In most cases, an FSS value greater than 0.5 is a good predictor [53].

3. Results

3.1. Correlation Analysis between Soil Moisture and Variables

As mentioned earlier, Albedo, NDVI, LST, and precipitation data were used as input variables in the RF-based SM relationship model. According to previous downscaling studies [39,54], these variables are closely related to SM.

This study analyzed the correlation between the used variables and SM according to the linear regression relationship, as shown in Figure 3a. In the whole study area, the correlation between SM and available variables is relatively high. Among them, the correlation between NDVI and SM is the highest, about 0.53. This is because NDVI can describe vegetation growth [55] and soil background conditions. The correlation between precipitation and SM is also high, about 0.40, relating to the fact that precipitation is the most obvious way for the atmosphere to affect SM [41]. Albedo and LST are also highly related to SM, which are −0.46 and −0.23, respectively. Albedo and LST are important variables in monitoring and SM downscaling due to their controlling role in surface energy exchange [56]. Overall, the results show the practicability of using these parameters in soil moisture downscaling.

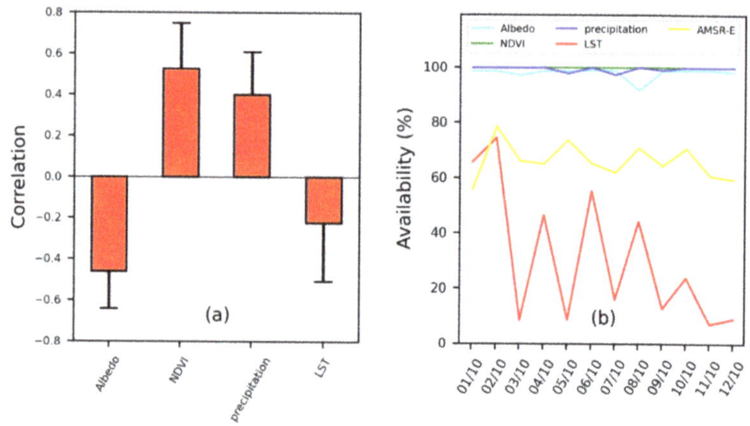

Figure 3. (**a**) The correlation between soil moisture and each parameter factor is the correlation coefficient (R) calculated according to the linear regression between the used variables and AMSR-E. The red bar is the average value, and the black line is the standard deviation. (**b**) The integrity of each parameter.

The data was collected in 2009. When observing AMSR-E data, we found that the integrity of the data on the 10th was relatively high. Therefore, we selected the 10th of each month in 2009. Notice that the dataset used in this study is not very complete. The integrity of each scene data is counted here, as shown in Figure 3b. It can be observed that the integrity of precipitation, NDVI, and Albedo is relatively high; part of AMSR-E is missing; and LST data is the most missing. The correlation between LST and AMSR-E is not particularly high, which may also be related to its high degree of deletion.

3.2. Spatial Distribution of Soil Moisture

3.2.1. Spatial Distribution of AMSR-E and Downscaled Soil Moisture

Figures 4 and 5 show the spatial distribution of AMSR-E soil moisture and the downscaled soil moisture, respectively. It can be found that, spatially, the soil moisture in the south is higher and that in the north is lower. In terms of time, the soil moisture is higher in November, December, and January, and lower in June, July, and August. Because of strong evapotranspiration in summer, the soil moisture is lower in summer and higher in winter. The downscaled results retain similar spatial patterns in the AMSR-E observations in most areas. At the same time, the spatial details of AMSR-E observations are also improved. The spatial distribution characteristics of the soil moisture after downscaling are generally consistent with the AMSR-E.

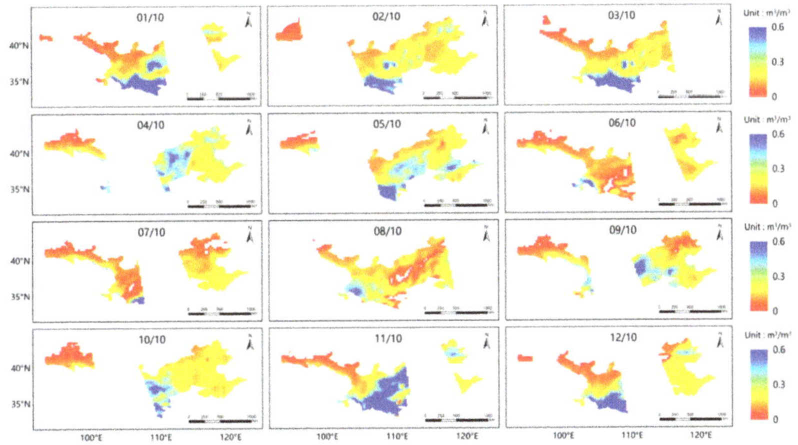

Figure 4. Spatial distribution map of AMSR-E soil moisture on the 10th of each month.

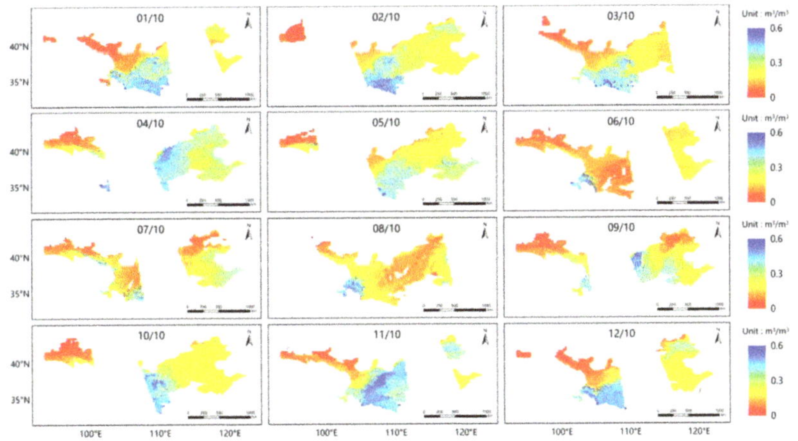

Figure 5. Spatial distribution map of the RF downscaled soil moisture on the 10th of each month.

Comparing the images before and after downscaling, it can be seen visually that the downscaled image is much clearer and more detailed than the AMSR-E image. Overall, the performance of the downscaled data is relatively good, but it is overestimated or underestimated in some places. Specifically, the downscaled image is underestimated in the south with high soil moisture and overestimated in the north with low soil moisture. This is because machine learning methods tend to reduce the range of soil moisture data. In general, the visual consistency between the RF downscaled SM and the original AMSR-E SM is great.

3.2.2. Difference between Downscaled Data and Soil Moisture of AMSR-E

To analyze the difference between AMSR-E and downscaled data, the box charts of 2 kinds of data in 12 months and the box charts classified by soil dry humidity were checked. Figure 6a shows the box diagram of AMSR-E and RF downscaled data for 12 months. It is observed that the differences between the 2 datasets are small in most months, and the values of the 2 data are basically kept within 0.1–0.6. Soil moisture is lower in June, July, and August, and higher in November, December, and January. As mentioned earlier, due to strong evapotranspiration in summer, soil moisture is lower in summer and higher in winter. At the same time, the RF results are generally high and overestimated. Earlier studies have shown similar results [57,58], probably relating to the fact that some high-resolution auxiliary data are added to better capture surface information and describe detailed information after higher resolution.

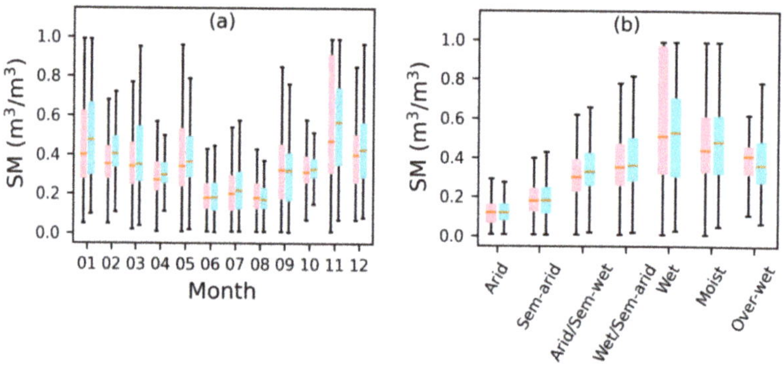

Figure 6. (**a**) Box chart of AMSR-E and the RF downscaled data for 12 months; (**b**) box diagram of AMSR-E and the RF downscaled data after classification by dry humidity (AMSR-E in red and downscaling data in blue).

Figure 6b compares the two datasets after classified statistics according to dry humidity classification. In the seven dry moisture soil types, there is little difference between the two data. Meanwhile, it can be found that the RF results are generally high and overestimated; however, underestimation occurs in the over-wet region.

To further analyze the difference before and after downscaling, we subtracted the AMSR-E data from downscaled data to obtain the difference diagram. Figure 7 shows the difference between downscaled data and AMSR-E data, with FSS values at the bottom left of the graph. It is observed that the differences in most regions are relatively small. The places with underestimated phenomena are generally located in the south area with higher soil moisture. The overestimated phenomenon is located in the central and northern areas. This is consistent with earlier studies [1,46,59], which illustrate that machine learning methods tend to reduce the range of the target data. Therefore, the method will reduce the value of the high soil moisture area in the original AMSR-E data and increase the value of the low soil moisture area. The FSS values for all months are greater than 0.5, which shows good consistency of the data before and after downscaling. In general, the differences between the RF downscaled data and the original AMSR-E data are small, implying that the RF downscaled data and AMSR-E data have similar spatial patterns.

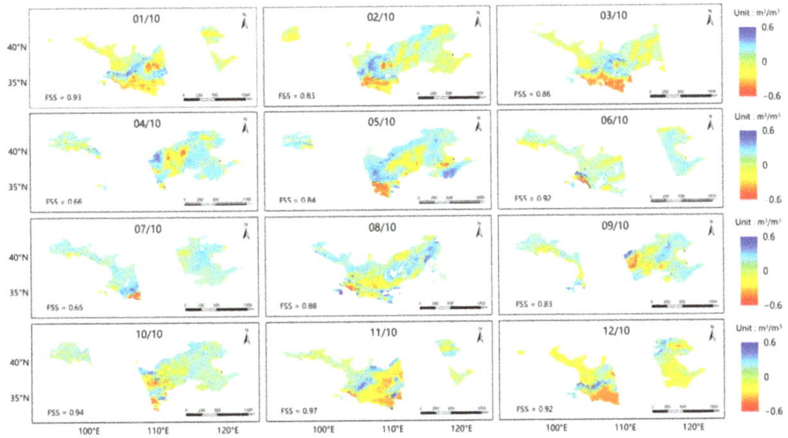

Figure 7. Difference between downscaled data and AMSR-E data (FSS values are shown at the bottom left of the graph).

3.3. Evaluation with In Situ Measurements

In order to prove the effect of RF downscaling, we verified AMSR-E and the RF downscaled data with ground station data, respectively. Figure 8 shows the verification results of the AMSR-E and RF downscaled data with ground station data. It is observed that the performance of R, RMSE, MAE, and ubRMSE after downscaling is better than those of the AMSR-E. Among them, the R of the RF downscaled data is higher. The R of AMSR-E is 0.02, and the R of the RF downscaled result is 0.19. The RMSE, MAE, and ubRMSE of downscaled data are lower. The RMSE, MAE, and ubRMSE of AMSR-E are 0.20, 0.15, and 0.19 m^3/m^3, respectively. The RMSE, MAE, and ubRMSE of the RF downscaled results are 0.18, 0.14, and 0.16 m^3/m^3, respectively. Therefore, compared with AMSR-E data, the verification results of the RF downscaled data with in situ measurements are better and closer to the ground soil moisture. The large differences mainly occur in areas where the in situ soil moisture is large. This may be due to the lesser capacity to capture the evapotranspiration information well, thereby overestimating soil moisture. In general, it shows that the RF downscaling algorithm is reasonable and effective.

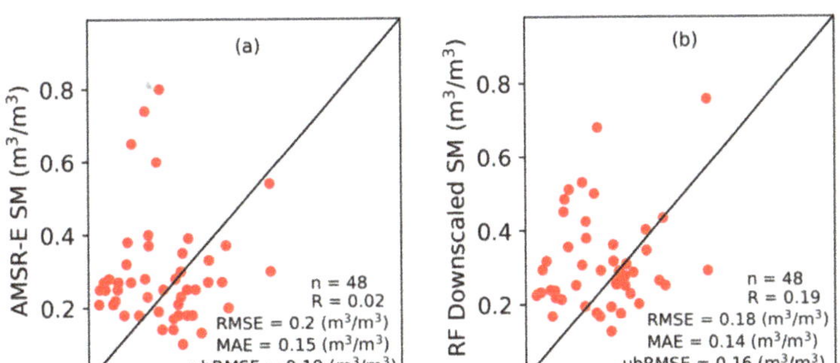

Figure 8. The verification results of (**a**) AMSR-E, (**b**) RF downscaled data with in situ measurements.

4. Discussion

4.1. Importance Analysis of Explanatory Factors

Analyzing the importance of the five input explanatory factors (Albedo, DEM, NDVI, LST, and precipitation) to the downscaled SM results is critical to understand the impact of participating variables on the performance of the proposed RF method. The importance of these variables is described by the mean and standard deviation. We used the importance score, a key index provided by the RF algorithm [57], to analyze the contribution of each variable in the proposed method.

As shown in Figure 9, DEM is identified as the most important variable with an importance value of 0.29, because it can capture the spatial patterns and variations of the downscaled SM [1]. Albedo and NDVI have significant effects on RF modeling, and the importance values are 0.24 and 0.21, respectively, since they can reflect the surface energy flux and vegetation state [60]. The LST has importance with a value of 0.13 relating to the control effect of SM on land surface energy exchange [56]. Precipitation has the same results as LST, because precipitation is a vital parameter directly leading to variations in SM [46]. In general, the results show that it is reasonable to use these auxiliary data for soil moisture downscaling, demonstrating the certain impact of explanatory factors on RF modeling.

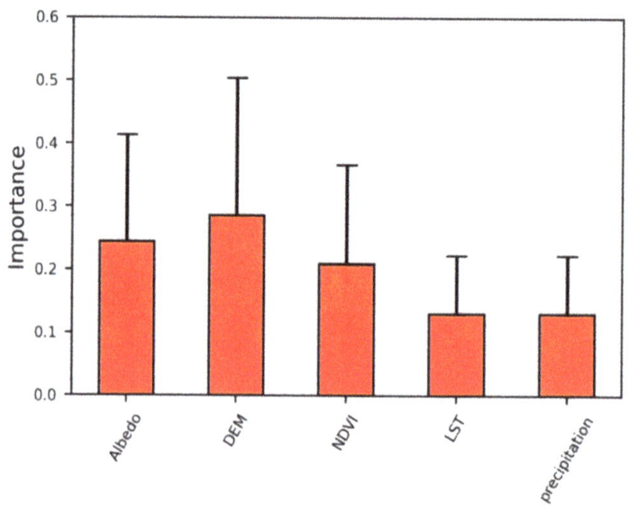

Figure 9. The importance of explanatory factors determined by the model. The red bar is the mean value and the black line is the standard deviation.

4.2. Leave-One-Out Analysis

Leave-one-out analysis was further conducted to verify the rationality and superiority of our model. Specifically, with this approach, the downscaling model was implemented after removing one input variable. Then, the downscaled results were verified with in situ observations. Figure 10 shows the verification results diagram of the leave-one-out approach. The first is the verification results of all parameters participating in random forest downscaling, followed by the verification results of the downscaled results with in situ data after removing these variables (Albedo, DEM, NDVI, LST, and precipitation, respectively). The numbers on the histogram are processed to better show their differences. The results of the random forest with all parameters involved are the best, and RMSE and MAE of RF are the lowest. Without precipitation and NDVI data, RMSE and MAE are increased significantly. Therefore, precipitation and NDVI have the greatest impact on the model results, followed by the Albedo, DEM, and LST data. Compared with the RF results, the RMSE and MAE of no-precipitation increased by 0.026 and 0.012 m^3/m^3, respectively. No-LST had the least impact on the results, and its RMSE and MAE increased by 0.018 and

0.007 m^3/m^3, respectively. Therefore, the addition of these five parameters can improve the performance of our proposed method and obtain better downscaling results.

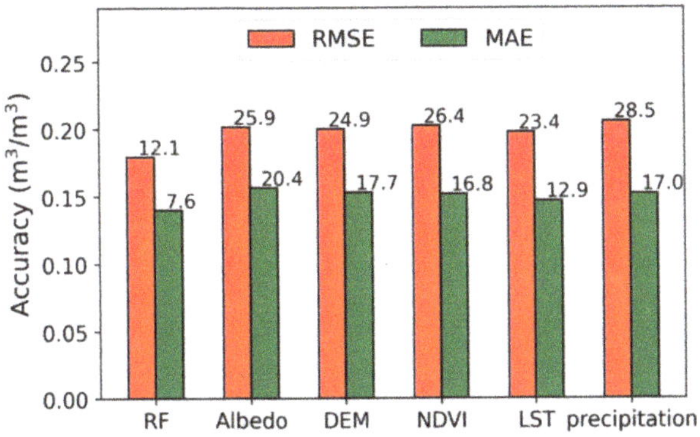

Figure 10. Verification results of the leave-one-out approach (the first is the verification results that all parameters participate in random forest downscaling, followed by the verification results after removing the corresponding parameters).

4.3. Comparison of Different Methods

To further verify the accuracy of the proposed downscaling approach, our study selected several methods that are often used in downscaling research to compare with the proposed RF method. Here, we compare the proposed RF method with Multiple Linear Regression (MLR) [61] and Support Vector Regression (SVR) [62]. The verification results of the downscaled data of these three methods with in situ observations were obtained. It is found from Figure 11 that the proposed RF method has the best results and significantly outperforms the other two methods. The RMSE and MAE of RF are lower than those of the other two methods, with an RMSE and MAE of 0.181 and 0.134 m^3/m^3, respectively. The R of RF is higher with a value of 0.261. At the same time, the performance of MLR is better than SVR, with lower RMSE and higher R.

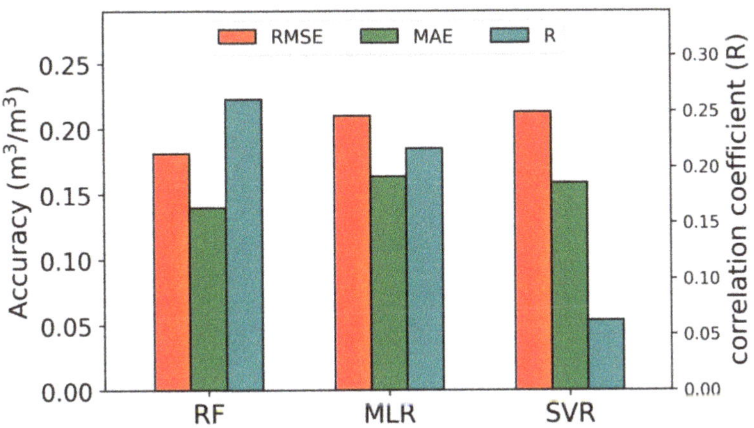

Figure 11. Comparison of three methods (RF is the proposed method, MLR is Multiple Linear Regression, and SVR is Support Vector Regression).

The reason for the above results may be that the MLR model is suitable for simple linear regression. The MLR is prone to over-fitting [63] and is not suitable for large-scale research.

The kernel function of the SVR model plays a vital role in the modeling process [64], but the type of kernel function and other parameters need to be selected artificially in the operation of the model. Hence, this problem may affect the final accuracy and prediction ability of the model and makes the accuracy poor. The RF model has randomness, which makes the model more stable and more generalized in the process of learning and application. The randomness of RF also reduces the occurrence of over-fitting to a great extent. Moreover, there are fewer artificially determined parameters in the operation process of the model, which avoids the interference of human factors.

The proposed method is compared with methods from several earlier studies. Four commonly used machine learning approaches [47] are first used for comparison. The polynomial fitting method [55], which is a popularly used method, is added for the comparative analysis. In addition, the UCLA downscaling method that was proposed by Kim and Hogue [65] is also used in the comparison. The study area, data used, and the corresponding accuracy of each method are shown in Table 3. This study confirms that the proposed approach can be compared with other methods, or even better. The proposed method selected five kinds of effective auxiliary data and used an adaptive spatiotemporal window determination strategy. Compared with other methods, it is more suitable for application to complex large areas.

Table 3. Comparison of different methods for downscaled SM against in situ soil moisture measurements (CART is classification and regression trees, KNN is K-nearest neighbors, BAYE is Bayesian regression, and RF is random forest).

Methods	Study Area	Data Used	R/R^2	RMSE (m^3/m^3)	Reference Studies
CART KNN BAYE RF	Heilongjiang, Jilin, and Liaoning Provinces, China	ESA CCI, MODIS	0.135 0.13 0.081 0.191	0.076 0.074 0.075 0.073	[47]
UCLA method	Southern Arizona, USA	AMSR-E, MODIS	0.27	0.051	[65]
Polynomial fitting method	AACES field, Australia	SMOS, MODIS	0.14–0.21	0.09–0.17	[55]
Proposes method	North China	AMSR-E, MODIS	0.19–0.26	0.18	-

4.4. Limitations of the Study

This study successfully downscaled the AMSR-E data with 25 km resolution to 1 km; however, there are still some shortcomings. First, the study area focused on North China, where there are not many ground stations and verification data. Therefore, the verification results obtained have certain limitations. In future work, more field data can be collected for verification. Second, AMSR-E was selected for the original soil moisture data in this study. The quality of the original data may also affect the results of downscaling. We can use other high-quality data in the future. This study used some common statistical metrics to validate the results. More validation metrics, such as Empirical Orthogonal Functions (EOF) and SPAtial EFficiency metric (SAPAEF) [53,66,67], can be used in future work. In this study, downscaling soil moisture data were rarely applied. Future work can apply downscaling soil moisture data more widely [68–70] and combine it with other directions [71,72]. Finally, the proposed method in this study is mainly based on the random forest. Considering that the use of one method may cause errors, multiple methods can be assembled for downscaling in the future.

5. Conclusions

This study proposed an efficient framework based on the RF model to downscale AMSR-E soil moisture over North China from its original resolution of 25 km to a better resolution of 1 km. This study used an adaptive spatiotemporal window determination strategy and five auxiliary data (LST, NDVI, Albedo, precipitation, and DEM). The participating variables are closely related to soil moisture and were acquired from a satellite remote sensing dataset and ground-based observations. The downscaled image was better able to describe the details of soil moisture, and showed a consistent spatial pattern with the original AMSR-E SM. The main conclusions are given as follows.

(1) The overall R, RMSE, MAE, and ubRMSE between the downscaled SM produced by the proposed model and in situ measurements are 0.19, 0.18, 0.14, and 0.16 m^3/m^3, respectively. These results are better than those of the original AMSR-E SM, with R of 0.02, RMSE of 0.20 m^3/m^3, MAE of 0.15 m^3/m^3, and ubRMSE of 0.19 m^3/m^3, respectively. This means that the proposed downscaling algorithm is reasonable and effective.

(2) DEM, Albedo, precipitation, and LST show high importance for SM downscaling, indicating their efficiency as key variables closely related to SM changes. These results also show that the inclusion of all input variables leads to the best downscaling accuracy. At the same time, these variables combined with machine learning and spatiotemporal search algorithms can be applied to complex large areas.

(3) The proposed RF method significantly outperforms the MLR and SVR. The downscaled data acquired from the proposed downscaling method were also compared with MLR and SVR. The verification results with in situ measurements showed that the proposed approach is better than the other two methods. The R, RMSE, and MAE between the RF downscaled SM and in situ measurements are 0.261, 0.181, and 0.134 m^3/m^3, respectively. They are better than the results of the MLR and SVR. The R, RMSE, and MAE of the MLR downscaled SM are 0.217, 0.201, and 0.164 m^3/m^3, respectively. The R, RMSE, and MAE of the SVR downscaled SM are 0.063, 0.213, and 0.159 m^3/m^3, respectively.

In summary, this study successfully generated 1 km $\times$ 1 km SM data over North China, meeting the need for drought monitoring, hydrological modeling, and some related research. The merit of the proposed method is that there is no limit to land surface conditions, such as semiarid or humid areas. In the meantime, this method can be applied to large areas with various climate and surface characteristics. Future work will focus on collecting other SM data and more in situ observations, and using ensemble machine learning algorithms.

Author Contributions: Conceptualization, Hongyan Zhang, Kai Liu and Shudong Wang; methodology, Hongyan Zhang and Kai Liu; software, Hongyan Zhang, Kai Liu and Xiaoyuan Zhang; validation, Hongyan Zhang and Kai Liu; formal analysis, Hongyan Zhang, Kai Liu and Xueke Li; investigation, Hongyan Zhang, Kai Liu and Bingxuan Liu; resources, Hongyan Zhang and Kai Liu; data curation, Hongyan Zhang and Xiaoyuan Zhang; writing—original draft preparation, Hongyan Zhang; writing—review and editing, Hongyan Zhang, Kai Liu, Xueke Li and Shudong Wang; visualization, Hongyan Zhang and Kai Liu; supervision, Kai Liu, Xueke Li ,Shudong Wang and Zhengqiang Li; project administration, Kai Liu, Xueke Li, Shudong Wang and Zhengqiang Li; funding acquisition, Shudong Wang. All authors have read and agreed to the published version of the manuscript.

Funding: This research was funded by the Inner Mongolia Autonomous Region Science and Technology Achievement Transformation Special Fund Project (No. 2021CG0045) and the National Natural Science Foundation of China (No. 42141007, 41671362).

Data Availability Statement: Data sharing is not applicable.

Acknowledgments: We are grateful to the Atmosphere Archive & Distribution System (LAADS) Distributed Active Archive Center (DAAC) for their MODIS products, National Aeronautics and Space Administration (NASA) for providing AMSR-E and the National Tibetan Plateau Data Center for their precipitation Data.

Conflicts of Interest: The authors declare no conflict of interest.

References

1. Abbaszadeh, P.; Moradkhani, H.; Zhan, X. Downscaling SMAP Radiometer Soil Moisture Over the CONUS Using an Ensemble Learning Method. *Water Resour. Res.* **2019**, *55*, 324–344. [CrossRef]
2. Zhang, T.; Peng, J.; Liang, W.; Yang, Y.T.; Liu, Y.X. Spatial-temporal patterns of water use efficiency and climate controls in China's Loess Plateau during 2000–2010. *Sci. Total Environ.* **2016**, *565*, 105–122. [CrossRef] [PubMed]
3. Bindlish, R.; Crow, W.; Jackson, T.J. Role of Passive Microwave Remote Sensing in Improving Flood Forecasts. *IEEE Geosci. Remote Sens. Lett.* **2009**, *6*, 112–116. [CrossRef]
4. Mao, K.; Shi, J.; Tang, H.; Li, Z.-L.; Wang, X.; Chen, K.-S. A Neural Network Technique for Separating Land Surface Emissivity and Temperature From ASTER Imagery. *IEEE Trans. Geosci. Remote Sens.* **2007**, *46*, 200–208. [CrossRef]
5. Renzullo, L.J.; van Dijk, A.I.J.M.; Perraud, J.M.; Collins, D.; Henderson, B.; Jin, H.; Smith, A.B.; McJannet, D.L. Continental satellite soil moisture data assimilation improves root-zone moisture analysis for water resources assessment. *J. Hydrol.* **2014**, *519*, 2747–2762. [CrossRef]
6. Chen, C.F.; Son, N.T.; Chang, L.Y.; Chen, C.C. Monitoring of soil moisture variability in relation to rice cropping systems in the Vietnamese Mekong Delta using MODIS data. *Appl. Geogr.* **2011**, *31*, 463–475. [CrossRef]
7. Tuttle, S.; Salvucci, G. Empirical evidence of contrasting soil moisture-precipitation feedbacks across the United States. *Science* **2016**, *352*, 825–828. [CrossRef]
8. Liu, Y.Y.; Dorigo, W.A.; Parinussa, R.M.; de Jeu, R.A.; Wagner, W.; McCabe, M.F.; Evans, J.P.; Van Dijk, A.I.J.M. Trend-preserving blending of passive and active microwave soil moisture retrievals. *Remote Sens. Environ.* **2012**, *123*, 280–297. [CrossRef]
9. Krishnan, P.; Black, T.A.; Grant, N.J.; Barr, A.G.; Hogg, E.T.H.; Jassal, R.S.; Morgenstern, K. Impact of changing soil moisture distribution on net ecosystem productivity of a boreal aspen forest during and following drought. *Agric. For. Meteorol.* **2006**, *139*, 208–223. [CrossRef]
10. Dorigo, W.; Wagner, W.; Albergel, C.; Albrecht, F.; Balsamo, G.; Brocca, L.; Chung, D.; Ertl, M.; Forkel, M.; Gruber, A.; et al. ESA CCI Soil Moisture for improved Earth system understanding: State-of-the art and future directions. *Remote Sens. Environ.* **2017**, *203*, 185–215. [CrossRef]
11. Zhao, S.H.; Cong, D.M.; He, K.X.; Yang, H.; Qin, Z.H. Spatial-Temporal Variation of Drought in China from 1982 to 2010 Based on a modified Temperature Vegetation Drought Index (mTVDI). *Sci. Rep.* **2017**, *7*, 17173. [CrossRef]
12. Petropoulos, G.P.; Ireland, G.; Barrett, B. Surface soil moisture retrievals from remote sensing: Current status, products & future trends. *Phys. Chem. Earth* **2015**, *83–84*, 36–56. [CrossRef]
13. Peng, J.; Loew, A. Recent Advances in Soil Moisture Estimation from Remote Sensing. *Water* **2017**, *9*, 530. [CrossRef]
14. Engman, E.T.; Chauhan, N. Status of Microwave Soil-Moisture Measurements with Remote-Sensing. *Remote Sens. Environ.* **1995**, *51*, 189–198. [CrossRef]
15. Njoku, E.; Jackson, T.; Lakshmi, V.; Chan, T.; Nghiem, S. Soil moisture retrieval from AMSR-E. *IEEE Trans. Geosci. Remote Sens.* **2003**, *41*, 215–229. [CrossRef]
16. Entekhabi, D.; Njoku, E.G.; O'Neill, P.E.; Kellogg, K.H.; Crow, W.T.; Edelstein, W.N.; Entin, J.K.; Goodman, S.D.; Jackson, T.J.; Johnson, J.; et al. The Soil Moisture Active Passive (SMAP) Mission. *Proc. IEEE* **2010**, *98*, 704–716. [CrossRef]
17. Berger, M.; Camps, A.; Font, J.; Kerr, Y.; Miller, J.; Johannessen, J.A.; Boutin, J.; Drinkwater, M.R.; Skou, N.; Floury, N.; et al. Measuring ocean salinity with ESA's SMOS mission-Advancing the science. *ESA Bull.* **2002**, *111*, 113–121.
18. Molero, B.; Merlin, O.; Malbeteau, Y.; Al Bitar, A.; Cabot, F.; Stefan, V.; Kerr, Y.; Bacon, S.; Cosh, M.H.; Bindlish, R.; et al. SMOS disaggregated soil moisture product at 1 km resolution: Processor overview and first validation results. *Remote Sens. Environ.* **2016**, *180*, 361–376. [CrossRef]
19. Njoku, E.G.; Entekhabi, D. Passive microwave remote sensing of soil moisture. *J. Hydrol.* **1996**, *184*, 101–129. [CrossRef]
20. Schmugge, T. Applications of passive microwave observations of surface soil moisture. *J. Hydrol.* **1998**, *212*, 188–197. [CrossRef]
21. Jin, Y.; Ge, Y.; Liu, Y.J.; Chen, Y.H.; Zhang, H.T.; Heuvelink, G.B.M. A Machine Learning-Based Geostatistical Downscaling Method for Coarse-Resolution Soil Moisture Products. *IEEE J. Sel. Top. Appl. Earth Obs. Remote Sens.* **2021**, *14*, 1025–1037. [CrossRef]
22. Peng, J.; Loew, A.; Merlin, O.; Verhoest, N.E.C. A review of spatial downscaling of satellite remotely sensed soil moisture. *Rev. Geophys.* **2017**, *55*, 341–366. [CrossRef]
23. de Jeu, R.A.M.; Wagner, W.; Holmes, T.R.H.; Dolman, A.J.; van de Giesen, N.C.; Friesen, J. Global Soil Moisture Patterns Observed by Space Borne Microwave Radiometers and Scatterometers. *Surv. Geophys.* **2008**, *29*, 399–420. [CrossRef]
24. Merlin, O.; Walker, J.P.; Chehbouni, A.; Kerr, Y. Towards deterministic downscaling of SMOS soil moisture using MODIS derived soil evaporative efficiency. *Remote Sens. Environ.* **2008**, *112*, 3935–3946. [CrossRef]
25. Busch, F.A.; Niemann, J.D.; Coleman, M. Evaluation of an empirical orthogonal function-based method to downscale soil moisture patterns based on topographical attributes. *Hydrol. Process.* **2012**, *26*, 2696–2709. [CrossRef]
26. Peng, J.; Borsche, M.; Liu, Y.; Loew, A. How representative are instantaneous evaporative fraction measurements of daytime fluxes? *Hydrol. Earth Syst. Sci.* **2013**, *17*, 3913–3919. [CrossRef]
27. Ines, A.V.M.; Mohanty, B.P.; Shin, Y. An unmixing algorithm for remotely sensed soil moisture. *Water Resour. Res.* **2013**, *49*, 408–425. [CrossRef]
28. Loew, A.; Mauser, W. On the disaggregation of passive microwave soil moisture data using a priori knowledge of temporally persistent soil moisture fields. *IEEE Trans. Geosci. Remote Sens.* **2008**, *46*, 819–834. [CrossRef]

29. Lievens, H.; Tomer, S.K.; Al Bitar, A.; De Lannoy, G.J.M.; Drusch, M.; Dumedah, G.; Franssen, H.J.H.; Kerr, Y.H.; Martens, B.; Pan, M.; et al. SMOS soil moisture assimilation for improved hydrologic simulation in the Murray Darling Basin, Australia. *Remote Sens. Environ.* **2015**, *168*, 146–162. [CrossRef]
30. Njoku, E.G.; Wilson, W.J.; Yueh, S.H.; Dinardo, S.J.; Li, F.K.; Jackson, T.J.; Lakshmi, V.; Bolten, J. Observations of soil moisture using a passive and active low-frequency microwave airborne sensor during SGP99. *IEEE Trans. Geosci. Remote Sens.* **2002**, *40*, 2659–2673. [CrossRef]
31. Merlin, O.; Rudiger, C.; Al Bitar, A.; Richaume, P.; Walker, J.P.; Kerr, Y.H. Disaggregation of SMOS Soil Moisture in Southeastern Australia. *IEEE Trans. Geosci. Remote Sens.* **2012**, *50*, 1556–1571. [CrossRef]
32. Werbylo, K.L.; Niemann, J.D. Evaluation of sampling techniques to characterize topographically-dependent variability for soil moisture downscaling. *J. Hydrol.* **2014**, *516*, 304–316. [CrossRef]
33. Ranney, K.J.; Niemann, J.D.; Lehman, B.M.; Green, T.R.; Jones, A.S. A method to downscale soil moisture to fine resolutions using topographic, vegetation, and soil data. *Adv. Water Resour.* **2015**, *76*, 81–96. [CrossRef]
34. Pelletier, C.; Valero, S.; Inglada, J.; Champion, N.; Dedieu, G. Assessing the robustness of Random Forests to map land cover with high resolution satellite image time series over large areas. *Remote Sens. Environ.* **2016**, *187*, 156–168. [CrossRef]
35. Park, S.; Park, S.; Im, J.; Rhee, J.; Shin, J.; Park, J.D. Downscaling GLDAS Soil Moisture Data in East Asia through Fusion of Multi-Sensors by Optimizing Modified Regression Trees. *Water* **2017**, *9*, 332. [CrossRef]
36. Imaoka, K.; Kachi, M.; Fujii, H.; Murakami, H.; Hori, M.; Ono, A.; Igarashi, T.; Nakagawa, K.; Oki, T.; Honda, Y.; et al. Global Change Observation Mission (GCOM) for Monitoring Carbon, Water Cycles, and Climate Change. *Proc. IEEE* **2010**, *98*, 717–734. [CrossRef]
37. Merlin, O.; Malbeteau, Y.; Notfi, Y.; Bacon, S.; Er-Raki, S.; Khabba, S.; Jarlan, L. Performance Metrics for Soil Moisture Downscaling Methods: Application to DISPATCH Data in Central Morocco. *Remote Sens.* **2015**, *7*, 3783–3807. [CrossRef]
38. Fang, B.; Lakshmi, V. Soil moisture at watershed scale: Remote sensing techniques. *J. Hydrol.* **2014**, *516*, 258–272. [CrossRef]
39. Wilson, D.J.; Western, A.W.; Grayson, R.B. A terrain and data-based method for generating the spatial distribution of soil moisture. *Adv. Water Resour.* **2005**, *28*, 43–54. [CrossRef]
40. Mascaro, G.; Vivoni, E.R.; Deidda, R. Soil moisture downscaling across climate regions and its emergent properties. *J. Geophys. Res. Earth Surf.* **2011**, *116*, D22114. [CrossRef]
41. Crow, W.T.; Berg, A.A.; Cosh, M.H.; Loew, A.; Mohanty, B.P.; Panciera, R.; de Rosnay, P.; Ryu, D.; Walker, J.P. Upscaling Sparse Ground-Based Soil Moisture Observations for the Validation of Coarse-Resolution Satellite Soil Moisture Products. *Rev. Geophys.* **2012**, *50*, RG2002. [CrossRef]
42. Jones, A.R.; Brunsell, N.A. A scaling analysis of soil moisture-precipitation interactions in a regional climate model. *Theor. Appl. Climatol.* **2009**, *98*, 221–235. [CrossRef]
43. Yang, K.; He, J.; Tang, W.J.; Qin, J.; Cheng, C.C.K. On downward shortwave and longwave radiations over high altitude regions: Observation and modeling in the Tibetan Plateau. *Agric. For. Meteorol.* **2010**, *150*, 38–46. [CrossRef]
44. He, J.; Yang, K.; Tang, W.J.; Lu, H.; Qin, J.; Chen, Y.Y.; Li, X. The first high-resolution meteorological forcing dataset for land process studies over China. *Sci. Data* **2020**, *7*, 25. [CrossRef]
45. Breiman, L. Random forests. *Mach. Learn* **2001**, *45*, 5–32. [CrossRef]
46. Long, D.; Bai, L.; Yan, L.; Zhang, C.; Yang, W.; Lei, H.; Quan, J.; Meng, X.; Shi, C. Generation of spatially complete and daily continuous surface soil moisture of high spatial resolution. *Remote Sens. Environ.* **2019**, *233*, 111364. [CrossRef]
47. Liu, Y.X.Y.; Yang, Y.P.; Jing, W.L.; Yue, X.F. Comparison of Different Machine Learning Approaches for Monthly Satellite-Based Soil Moisture Downscaling over Northeast China. *Remote Sens.* **2018**, *10*, 31. [CrossRef]
48. Liu, K.; Su, H.; Li, X.; Chen, S. Development of a 250-m Downscaled Land Surface Temperature Data Set and Its Application to Improving Remotely Sensed Evapotranspiration Over Large Landscapes in Northern China. *IEEE Trans. Geosci. Remote Sens.* **2020**, *10*, 31. [CrossRef]
49. Udovicic, M.; Bazdaric, K.; Bilic-Zulle, L.; Petrovecki, M. What we need to know when calculating the coefficient of correlation? *Biochem. Med.* **2007**, *17*, 10–15. [CrossRef]
50. Tan, S.; Wu, B.F.; Yan, N.N.; Zhu, W.W. An NDVI-Based Statistical ET Downscaling Method. *Water* **2017**, *9*, 995. [CrossRef]
51. Zuo, J.P.; Xu, J.H.; Chen, Y.N.; Wang, C. Downscaling Precipitation in the Data-Scarce Inland River Basin of Northwest China Based on Earth System Data Products. *Atmosphere* **2019**, *10*, 613. [CrossRef]
52. Roberts, N.M.; Lean, H.W.J. Scale-selective verification of rainfall accumulations from high-resolution forecasts of convective events. *Mon. Weather Rev.* **2008**, *136*, 78–97. [CrossRef]
53. Skok, G.; Roberts, N.J. Analysis of fractions skill score properties for random precipitation fields and ECMWF forecasts. *Q. J. R. Meteorol. Soc.* **2016**, *142*, 2599–2610. [CrossRef]
54. Srivastava, P.K.; Han, D.W.; Ramirez, M.R.; Islam, T. Machine Learning Techniques for Downscaling SMOS Satellite Soil Moisture Using MODIS Land Surface Temperature for Hydrological Application. *Water Resour. Manag.* **2013**, *27*, 3127–3144. [CrossRef]
55. Piles, M.; Camps, A.; Vall-Llossera, M.; Corbella, I.; Panciera, R.; Rudiger, C.; Kerr, Y.H.; Walker, J. Downscaling SMOS-Derived Soil Moisture Using MODIS Visible/Infrared Data. *IEEE Trans. Geosci. Remote Sens.* **2011**, *49*, 3156–3166. [CrossRef]
56. Wang, J.R.; Choudhury, B.J. Remote-Sensing of Soil-Moisture Content over Bare Field at 1.4 Ghz Frequency. *J. Geophys. Res. Oceans* **1981**, *86*, 5277–5282. [CrossRef]

57. Im, J.; Park, S.; Rhee, J.; Baik, J.; Choi, M. Downscaling of AMSR-E soil moisture with MODIS products using machine learning approaches. *Environ. Earth Sci.* **2016**, *75*, 1120. [CrossRef]
58. Zhao, W.; Li, A.N. A Downscaling Method for Improving the Spatial Resolution of AMSR-E Derived Soil Moisture Product Based on MSG-SEVIRI Data. *Remote Sens.* **2013**, *5*, 6790–6811. [CrossRef]
59. Liu, Y.X.Y.; Xia, X.L.; Yao, L.; Jing, W.L.; Zhou, C.H.; Huang, W.M.; Li, Y.; Yang, J. Downscaling Satellite Retrieved Soil Moisture Using Regression Tree-Based Machine Learning Algorithms Over Southwest France. *Earth Space Sci.* **2020**, *7*, e2020EA001267. [CrossRef]
60. Park, S.; Im, J.; Park, S.; Rhee, J. Drought monitoring using high resolution soil moisture through multi-sensor satellite data fusion over the Korean peninsula. *Agric. For. Meteorol.* **2017**, *237*, 257–269. [CrossRef]
61. Wilks, D.S. *Statistical Methods in the Atmospheric Sciences*; Academic Press: Cambridge, MA, USA, 2011; Volume 100.
62. Tripathi, S.; Srinivas, V.; Nanjundiah, R.S. Downscaling of precipitation for climate change scenarios: A support vector machine approach. *J. Hydrol.* **2006**, *330*, 621–640. [CrossRef]
63. Xie, W.H.; Yi, S.Z.; Leng, C. A Study to Compare Three Different Spatial Downscaling Algorithms of Annual TRMM 3B43 Precipitation. In Proceedings of the 26th International Conference on Geoinformatics, Kunming, China, 28–30 June 2018. [CrossRef]
64. Ebrahimy, H.; Azadbakht, M. Downscaling MODIS land surface temperature over a heterogeneous area: An investigation of machine learning techniques, feature selection, and impacts of mixed pixels. *Comput. Geosci.* **2019**, *124*, 93–102. [CrossRef]
65. Kim, J.; Hogue, T.S. Improving Spatial Soil Moisture Representation Through Integration of AMSR-E and MODIS Products. *IEEE Trans. Geosci. Remote Sens.* **2012**, *50*, 446–460. [CrossRef]
66. Koch, J.; Demirel, M.C.; Stisen, S.J. The SPAtial EFficiency metric (SPAEF): Multiple-component evaluation of spatial patterns for optimization of hydrological models. *Geosci. Model Dev.* **2018**, *11*, 1873–1886. [CrossRef]
67. Ahmed, K.; Sachindra, D.A.; Shahid, S.; Demirel, M.C.; Chung, E.-S.J.H.; Sciences, E.S. Selection of multi-model ensemble of general circulation models for the simulation of precipitation and maximum and minimum temperature based on spatial assessment metrics. *Hydrol. Earth Syst. Sci.* **2019**, *23*, 4803–4824. [CrossRef]
68. Liu, K.; Wang, S.; Li, X.; Wu, T.J. Spatially disaggregating satellite land surface temperature with a nonlinear model across agricultural areas. *J. Geophys. Res. Biogeosci.* **2019**, *124*, 3232–3251. [CrossRef]
69. Liu, K.; Wang, S.; Li, X.; Li, Y.; Zhang, B.; Zhai, R. The assessment of different vegetation indices for spatial disaggregating of thermal imagery over the humid agricultural region. *Int. J. Remote Sens.* **2020**, *41*, 1907–1926. [CrossRef]
70. Liu, K.; Su, H.; Tian, J.; Li, X.; Wang, W.; Yang, L.; Liang, H. Assessing a scheme of spatial-temporal thermal remote-sensing sharpening for estimating regional evapotranspiration. *Int. J. Remote Sens.* **2018**, *39*, 3111–3137. [CrossRef]
71. Liu, K.; Su, H.; Li, X.; Chen, S.; Zhang, R.; Wang, W.; Yang, L.; Liang, H.; Yang, Y. A thermal disaggregation model based on trapezoid interpolation. *IEEE J. Sel. Top. Appl. Earth Obs. Remote Sens.* **2018**, *11*, 808–820. [CrossRef]
72. Liu, K.; Li, X.; Long, X. Trends in groundwater changes driven by precipitation and anthropogenic activities on the southeast side of the Hu Line. *Environ. Res. Lett.* **2021**, *16*, 094032. [CrossRef]

International Journal of *Geo-Information*

MDPI

Article

Integration of Geostatistical and Sentinal-2AMultispectral Satellite Image Analysis for Predicting Soil Fertility Condition in Drylands

Mohamed S. Shokr [1,*], Yasser S. A. Mazrou [2,3], Mostafa A. Abdellatif [4], Ahmed A. El Baroudy [1], Esawy K. Mahmoud [1], Ahmed M. Saleh [4], Abdelaziz A. Belal [4] and Zheli Ding [5]

1 Soil and Water Department, Faculty of Agriculture, Tanta University, Tanta 31527, Egypt; drbaroudy@agr.tanta.edu.eg (A.A.E.B.); esawy.rezk@agr.tanta.edu.eg (E.K.M.)
2 Applied College-Muhyle, King Khalid University, Abha 62587, Saudi Arabia; ymazrou@kku.edu
3 Department of Agriculture Economic, Faculty of Agriculture, Tanta University, Tanta 31527, Egypt
4 National Authority for Remote Sensing and Space Science (NARSS), Cairo 11843, Egypt; mostafa.abdou@narss.sci.eg (M.A.A.); a.saleh@narss.sci.eg (A.M.S.); abelal@narss.sci.eg (A.A.B.)
5 Haikou Experimental Station, Chinese Academy of Tropical Agricultural Sciences, Haikou 571101, China; dingzheli@zju.edu.cn
* Correspondence: mohamed_shokr@agr.tanta.edu.eg

Citation: Shokr, M.S.; Mazrou, Y.S.A.; Abdellatif, M.A.; El Baroudy, A.A.; Mahmoud, E.K.; Saleh, A.M.; Belal, A.A.; Ding, Z. Integration of Geostatistical and Sentinal-2AMultispectral Satellite Image Analysis for Predicting Soil Fertility Condition in Drylands. *ISPRS Int. J. Geo-Inf.* **2022**, *11*, 353. https://doi.org/10.3390/ijgi11060353

Academic Editors: Dimitris Triantakonstantis, Panagiotis Tziachris and Wolfgang Kainz

Received: 11 April 2022
Accepted: 14 June 2022
Published: 17 June 2022

Abstract: For modelling and predicting soil indicators to be fully operational and facilitate decision-making at any spatial level, there is a requirement for precise spatially referenced soil information to be available as input data. This paper focuses on showing the capacity of Sentinal-2A(S2A) multispectral imaging to predict soil properties and provide geostatistical analysis (ordinary kriging) for mapping dry land soil fertility conditions (SOCs). Conditioned Latin hypercube sampling was used to select the representative sampling sites within the study area. To achieve the objectives of this work, 48 surface soil samples were collected from the western part of Matrouh Governorate, Egypt, and pH, soil organic matter (SOM), available nitrogen (N), phosphorus (P), and potassium (K) levels were analyzed. Multilinear regression (MLR) was used to model the relationship between image reflectance and laboratory analysis (of pH, SOM, N, P, and K in the soil), followed by mapping the predicted outputs using ordinary kriging. Model fitting was achieved by removing variables according to the confidence level (95%).Around 30% of the samples were randomly selected to verify the validity of the results. The randomly selected samples helped express the variety of the soil characteristics from the investigated area. The predicted values of pH, SOM, N, P, and K performed well, with R^2 values of 0.6, 0.7, 0.55, 0.6, and 0.92 achieved for pH, SOM, N, P, and K, respectively. The results from the ArcGIS model builder indicated a descending fertility order within the study area of: 70% low fertility, 22% moderate fertility, 3% very low fertility, and 5% reference terms. This work evidence that which can be predicted from S2A images and provides a reference for soil fertility monitoring in drylands. Additionally, this model can be easily applied to environmental conditions similar to those of the studied area.

Keywords: soil fertility modelling; GIS; ordinary kriging; cLHS; S2A image; drylands

1. Introduction

Soils are very complicated ecosystems formed by biological and non-biological factors that widely fluctuate from one environment to another. Applying suitable agricultural management practices requires the characterization, modelling, and mapping of soil properties [1–4]. Models of soil properties and indicators require the use of accurate, spatially referenced soil information as an input to help facilitate the decisions made at national and international levels [5]. Estimating different soil properties by traditional methods requires intensive field and laboratory work; this is time-consuming, costly, and the results may also be affected by significant uncertainty [6]. Thus, the combination of remote sensing data and field measurements has been proposed as a way to model and map soil properties [7,8]. Furthermore, increasingly high-quality data, including digital elevation models (DEMs) and satellite images, have allowed for novel methods to be established to select the samples distributed within the feature space of environmental variables. Conditioned Latin hypercube sampling (cLHS) is an atypical example [9]. It has been implemented in many mapping studies in terms of guaranteeing the full coverage of a multivariate feature space [10–23]. Some critical soil properties, such as salinity, texture, minerals, and organic matter, have been estimated and investigated using enhanced thematic mapper plus (ETM+), Landsat-7, and Landsat-8 operational land imagers (OLIs) [24]. Visible, near-infrared, and short-wave infrared hyper spectral methods have been used to estimate surface soil properties such as soil organic carbon (SOC), cation exchange capacity, soil texture, and calcium carbonate [25–29]. One of the major challenges today is finding the best way to develop low-cost methods for modelling and mapping physical and chemical soil properties over large areas. Remote sensing (RS) has commonly been utilized, mainly at a regional scale, for soil mapping because of the low costs, quick data acquisition, and availability of high temporal and spatial information quality [30]. Satellites and aerial platforms that are used for capturing bare soil images can be analyzed by traditional methods, such as band ratio and discriminant analysis [30]. An adequate prediction of soil properties using hyperspectral data is not easy due to the soil reflectance phenomenon. To overcome this problem, regression analysis, including multiple regression analysis (MLR), principal component regression (PCR), and partial least-squares (PLS) regression, have been used, with these making connections between soil properties and reflectance data or spectral indices [30,31]. The use of hyperspectral VNIR/SWIR imagery was confirmed by a previous study as being a very important tool for mapping soil properties, however, it cannot be implemented for large areas or for long time periods due to high cost and low availability [5]. The low temporal resolution of VNIR/SWIR multispectral, coarse spatial, and spectral resolutions within these sensors causes deficiencies in spatial prediction performance under the annual crop system, as the time period of bare soil availability is very short [32,33]. The European Space Agency launched multispectral satellite Sentinal-2A data last November (2015), with its development of large area images considered revolutionary [5]. Sentinal-2A (S2A) data was released last November (2015); it contains high temporal, spatial (10 to 60 m), and spectral resolutions (20 to 180 nm) [5,33]. Today, to identify elements that are deficient in a soil, there is a requirement for quick analysis with low cost and high accuracy; satellite image data can complement analysis in situ [34,35]. Precise measurements and efficient methods should be conducted to improve soil and water quality [36]. Measurable information for soil characteristics and their spatial spreading is very important to achieve agriculture sustainability development and reduce negative environmental effects [37–39]. This is very critical in drylands due to these areas having a lot of limiting factors for soil fertility, such as a low content of nitrogen (N), phosphorus (P), potassium(K), soil organic matter (SOM %), and irrigation water scarcity. The mapping of these factors provides a good indicator of land degradation [40–42]. Recently, the development of low-cost approaches is needed for mapping soil properties within a large area. Additionally, it is critical to consider that the management of agricultural lands require a method of rapid analysis to characterize the elements deficient in soil and in crops as well. Therefore, this research aims to assess the potential of S2A imaging and geostatistical analysis (ordinary kriging) to predict selected soil properties and show the distributions of fertile conditions of study area based on the predicted properties. The study areas focus on western parts of Matrouh Governorate (Northwest coast of Egypt), parts of which were used previously as test areas for mapping using Vis-NIR Spectroscopy data. The proposed approach enabled us to model relationships between the spectral reflectance, which was extracted from bands of S2A satellite imagery, and laboratory analysis, to create maps of soil properties by using OK and finally produce of SOC maps.

2. Materials and Methods

The proposed procedure in this work is displayed in Figure 1.

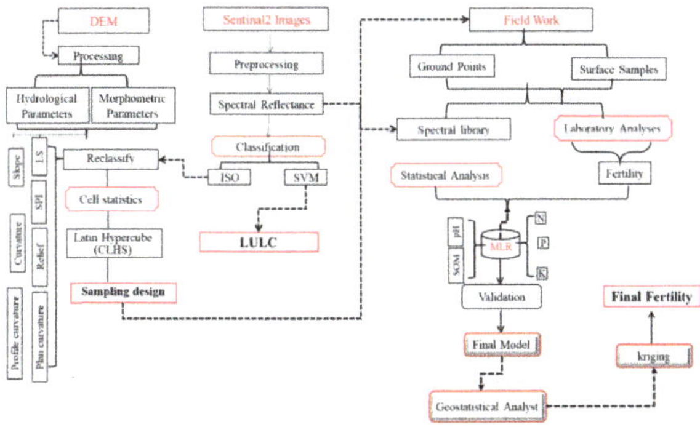

Figure 1. Schematic diagram of current work methodology; note: MLR = multi linear regression.

2.1. Study Area

The experimental area is located in the western part of Matrouh Governorate, Egypt, between longitudes 26°49′58.0″ and 26°58′06″ East, and latitudes 31°13′07″ and 31°26′36″ North with a total area of 213.7 km^2 (Figure 2). The study area has a wide variety of environmental conditions [43,44] and is characterized by a Mediterranean Sea climate, with rainfall fluctuating between 105 and 200 mm per day and the mean temperatures of 8.1 to 18 °C and 20 to 29.2 °C in the winter and summer seasons, respectively [6]. The area suffers from low-density vegetation cover during the summer, with the vegetation cover starting to increase in the winter and spring due to precipitation. The land elevation differs between 11 and 212 m above sea level (asl), as can be seen in Figure 3, with the lowest areas located closest to the Mediterranean Sea and the highest areas appearing in the southern region (Figure 3) [6]. Miocene and Quaternary deposits are the main geological units [44,45]. The main geomorphologic units in the study area are plains (3453.68 ha), wadi (733 ha), terraces (7149.65 ha), basins (399.27 ha), pavement plain, and reference unit (11,064.76 ha) units, according to [46]. Figure 4 shows the three classes of land use and land cover (LULC) highlighted in the investigated area: agriculture, urban, and bare soils. These classes cover 1064.26 (5%), 273.36 (1.3), and 20,210.76 (93.7%) hectares (ha) of the study area, respectively. The soils of the study area are classified into two orders, i.e., Entisols and Aridisols, which are divided into five subgroups: *Typic TorriPsamments, Typic Calcigypsids, TypicHaplogypsids, Typic Haplocalcids* and *Lithic Torriorthents* [44,47].

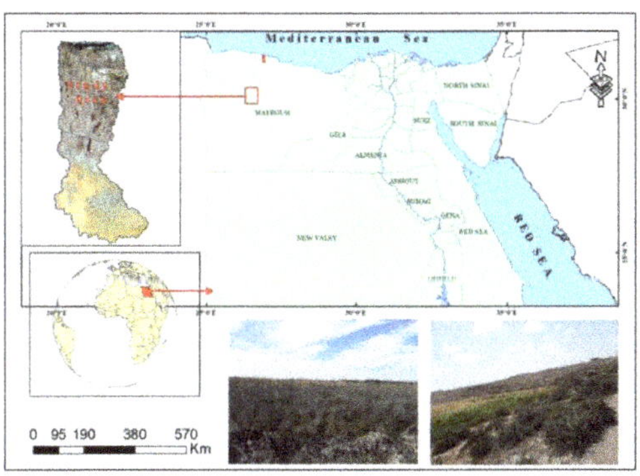

Figure 2. Location of the investigated area.

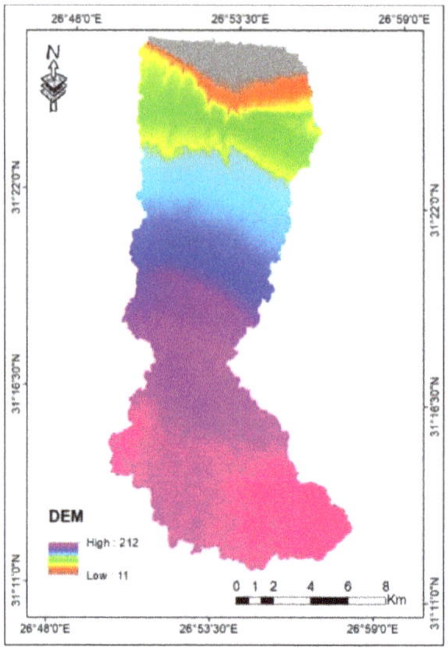

Figure 3. Digital elevation model (DEM) of the investigated area.

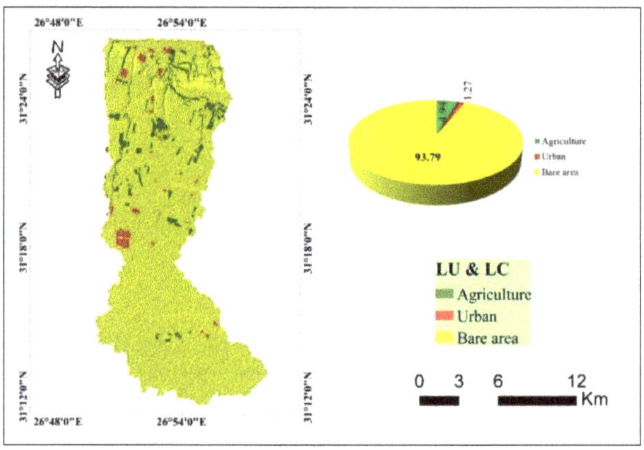

Figure 4. Land use (LU) and land cover (LC) of the study area.

2.2. Soil Sampling and Laboratory Analysis

Conditioned Latin hypercube sampling (cLHS) was utilized to realize the best sampling site distributions based on landscape attributes [9,48]. The sample points number was set at 400, and the iteration number was set at 50,000 as seen in Figures 1 and 5.The selection of the samples (400 points) was based on hydrological and morphometric parameters, i.e., digital elevation model (DEM), slope, curvature, profile curvature, plan curvature, stream power index, slope length, and ISO data classification of the S2A images (Supplementary Materials: Figures S1 and S2) [49], and urban areas were excluded. All soil sample locations were checked in the field; forty-eight locations were chosen to represent different landforms of the study area according to [46] (Figure 6). The soil samples were collected from the study area and transferred to the relevant soil, water, and plant laboratories within the Faculty of Agriculture, Tanta University in accordance with ISO/IEC 17,025:2017 requirements for analysis. The samples were air-dried under normal temperature and then passed through a 2 mm sieve. The soil organic matter proportion (SOM %) was analyzed according to the Walkley and Black method, while soil reaction (pH) was measured using pH meter (V HANNA) [44]. The available nitrogen (N), phosphorus (P) and potassium (K) levels were determined by Kjeldahl spectrophotometry (880 wavelength) and flame photometry [50].

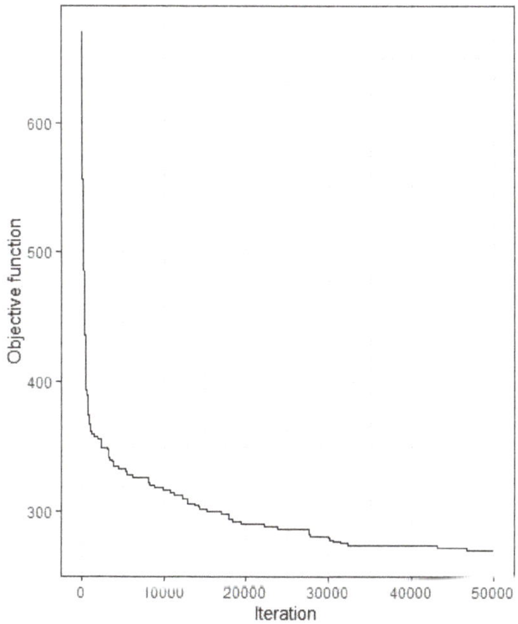

Figure 5. Iteration of selected parameters for CLHs.

Figure 6. Distribution of surface soil samples within the study area.

2.3. Digital Image Processing

A Sentinal-2A image was acquired in the September of 2021, and low-density cloud data was downloaded from the United States Geological Survey (USGS). The multispectral S2A image has 13 spectral bands; four of them have10 m resolution, and six of the bands have a 20 m resolution (Table 1). The image was atmospherically corrected utilizing the ATCOR2 radiative transfer model according to [51]. Universal Transverse Mercator (UTM) coordinates were used for the geo-rectifying of the image. Then, all satellite image analyses and processes were carried out using the standard producers provided by SNAP software [52]. All spectral bands were resampled to 10 m by the S2A resampling function in the SNAP software (Figure 7).

Table 1. Characteristics of the Sentenial2 satellite image.

No.	Band Name	Central Wavelength (nm)	Resolution (m)
1	Coastal aerosol	443.9	60
2	Blue	496.6	10
3	Green	560	10
4	Red	664.5	10
5	Vegetation Red Edge	703.9	20
6	Vegetation Red Edge	740.2	20
7	Vegetation Red Edge	782.5	20
8	NIR	835.1	10
8a	Narrow NIR	864.8	20
9	Water vapour	945	60
10	SWIR–Cirrus	1373.5	60
11	SWIR	1613.7	20
12	SWIR	2202.4	20

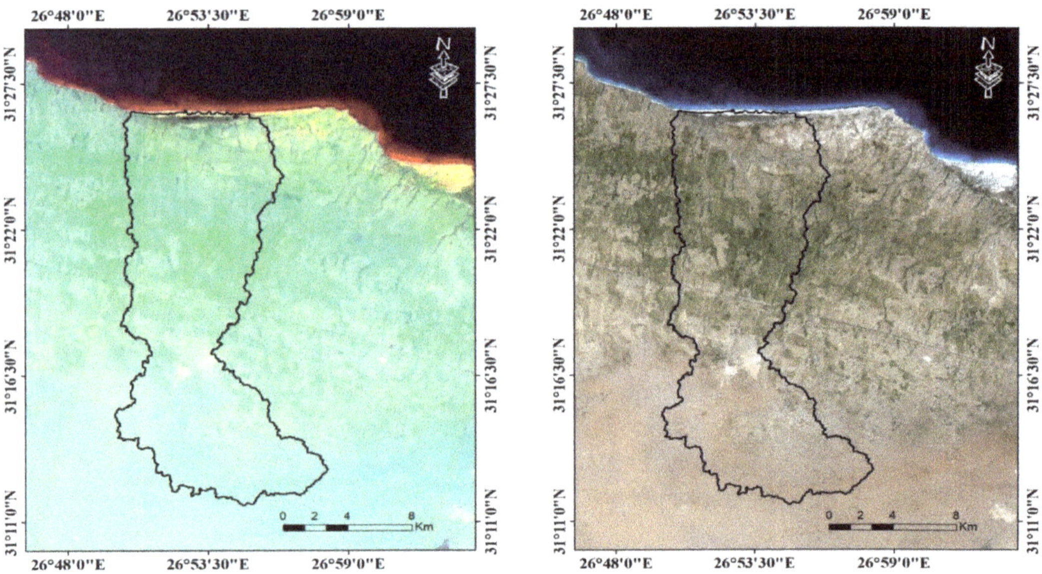

(a) (b)

Figure 7. Image resolutions within the study area: (**a**) 60 m and (**b**) the resampled image with 10 m resolution.

2.4. Procedure of Modelling

Multiple linear regressions (MLRs) were used to predict the dependent variable response, from the many independent variables, as a correlation function between them. The MLR algorithm was implemented using the SPSS V22, with stepwise (backward) analysis, fitting the model via the removal of variables according to the confidence level (95%). Around 70% of samples were used for calibration and 30% for validation. The average of the spectral reflectance from the S2A image was used to determine soil fertility parameters, i.e., pH, SOM, N, P, and K. The output maps were validated by calculating the correlation coefficient (R^2) and the root mean square error (RMSE) (Equations (1) and (2)). In addition, R^2 was utilized to also describe the validation of the model, where x expresses the value of the soil parameter *i*, which was analyzed in the laboratory and used as the background value for the calibration process. *y* represents the predicted value and *n* is the number of soil samples utilized for the calibration [53,54].

$$R^2 = 1 - \frac{\sum_{i=1}^{n}(y_i - \hat{y}_i)^2}{\sum_{i=1}^{n}(y_i - \overline{y_i})^2} \tag{1}$$

where n = the number of samples, y_i = measured value; $\hat{y}_i$ = the predicted value, and $\overline{y_i}$ = the mean ofallvalues.

$$\text{RMSE} = \sqrt{\frac{1}{N}\sum_{I=1}^{N}[Z_1(x_i) - Z_2(x_i)]^2} \tag{2}$$

where $Z_1(x_i)$ = measured values and $Z_2(x_i)$ = predicted values.

2.5. Producing Maps of Soil Properties

The prediction of soil property values was carried out using ordinary kriging (OK). OK is a geostatistical model that predicts selected soil properties in anon sampled location [55,56]. The histogram tool and normal QQ plots were used to check the normal distribution of data for each parameter. In this work, we implemented semi variogram models (Gaussian, spherical, and stable) for each parameter [55,56] according to Equations (3) and (4). The selection of the best model was based on RMSE calculation, the mean standardized error (MSE), and the root mean square standardized error (RMSEE) according to Equations (5) and (6).

The Gaussian function was defined as:

$$(h) = \begin{cases} C_0 + C\left(1 - exp\left(\frac{-h^2}{a^2}\right)\right), h > 0 \\ 0, h = 0 \end{cases} \tag{3}$$

The spherical function was defined as:

$$(h) = \begin{cases} C_0 + C\left(\frac{3h}{2a} - \frac{1}{2} - \left(\frac{h}{a}\right)^3\right), 0 < h \le a \\ C_0 + C, h > a \\ 0, h = 0 \end{cases} \tag{4}$$

where a = the actual ranges for the spherical and circular, h = the spatial lag, C_0 = is the nugget, and C = the partial sill.

$$\text{MSE} = \frac{1}{N}\sum_{i=1}^{N}[Z_1(X_1) - Z_2(X_2)] \tag{5}$$

$$\text{RMSSE} = \sqrt{\frac{1}{N}\sum_{I=1}^{N}[Z_1(x_i) - Z_2(x_i)]^2} \tag{6}$$

2.6. Modelling of SFC in the Study Area

Soil fertility condition (SFC) characterizes the soil nutrient content, i.e., available N, P, K, SOM, and pH (Figure 8), developed using the MLR model. The factors that were selected in this study for calculating fertility status were based on the criteria for crop growth requirements according to [37]. The SFC are used as an indicator for the degree of crop suitability for exact uses [6]. GIS spatial modeling was used for evaluating SFCs based on the following Equation (7).

$$SFC = S * [N * P * K * SOM * pH]^{1/5} \quad (7)$$

where S = score of each parameter.

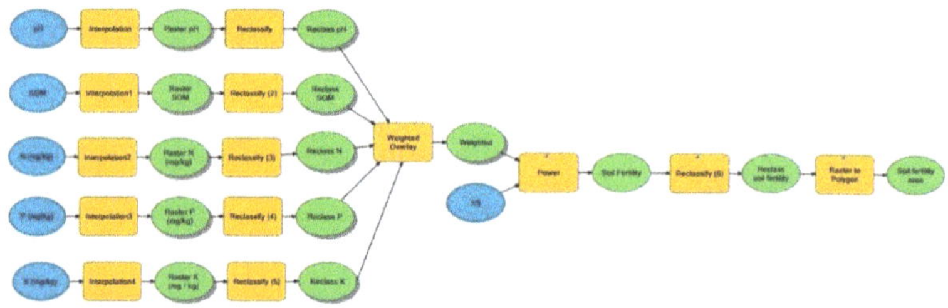

Figure 8. Flow chart devised in this study to model SFC.

The model builder function in Arc GIS 10.7 was used to create the spatial model of SFCs. Selected spatial analysis of the parameters is displayed in a diagram chain [57] (Figure 8). Output from each process is considered as the input for the next process [58]. The subsequent stages were applied in this research to obtain the final SFC map of the investigated area: (a) Interpolation of soil properties to raster layer, with the results from this stage (b) reclassified into five classes (Very low, low, moderate, high, and very high), then (c) scoring each SFC parameter based on Table 2, (d) feeding Equation (7) utilizing the raster calculator tool and finally (f) the outputs from (d) were weighted and overlayed to obtain and display the final SFC map.

Table 2. Score factors of SFC parameters.

Selected Factor	Measuring Unit	1	0.8	0.5	0.2
N	ppm	>80	80–40	40–20	>20
P	ppm	>15	15–10	10–5	<5
K	ppm	>400	400–200	200–100	<100
SOM	%	>2	2–1	1–0.5	<0.5
pH	-	5.5–7	7–7.8	7.9–8.5	>8.5

3. Results and Discussion

3.1. Soil Properties

The results reveal that the values of pH varied from 7.10 to 8.98 within the study area, with an average pH of 7.92. The high value of pH is due to the presence of limestone parent material [6]. These calcareous soils occupy wide areas of Egypt's Western Desert and North Africa [59]. These values of pH indicate a high similarity within the study area (STD = 0.47) [60]. Soil nutrient availability and enzyme and soil organism activity in different kinds of soils are both influenced by values of pH; thus, pH is considered one of the most important factors soil factors [61]. The study area contains a low content of soil organic matter (SOM), ranging from 0.03 to 0.83%. The low values of SOM are logical for arid regions as high temperatures encourage the decomposition of organic matter [60]. For the spatial distribution of SOM %, the highest values were located in the wadi area unit due to the presence of seasonal crops [46]; however, there is a low variation of SOM content with STD ≅ 0.5. Generally, the

Egyptian soils suffer from a deficiency of macronutrients, i.e., N, P, and K have ranges of 3.32 to 87.11, 0.8 to 7.34, and 20.13 to 200.40 ppm, respectively. Concerning SD, the variation in N and K is very wide (STD = 13.04 and 32.27, respectively), with a STD value of1.77 for P. The study area has the climatic and topographic conditions of the northern coast, which is characterized by low values of N, P, and K except for some areas that cultivate seasonal crops. These results are in line with those of [62] (Table 3 and Figure 9).

Table 3. Statistics of measured SFC parameters.

Statistics	M. pH	M. SOM	M. N	M. P	M. K
MAX	8.98	0.83	87.11	7.34	200.40
Mean	7.92	0.38	45.47	4.18	83.37
MIN	7.10	0.03	3.32	0.80	20.13
STD	0.47	0.21	13.04	1.77	32.27

Note: M = measure.

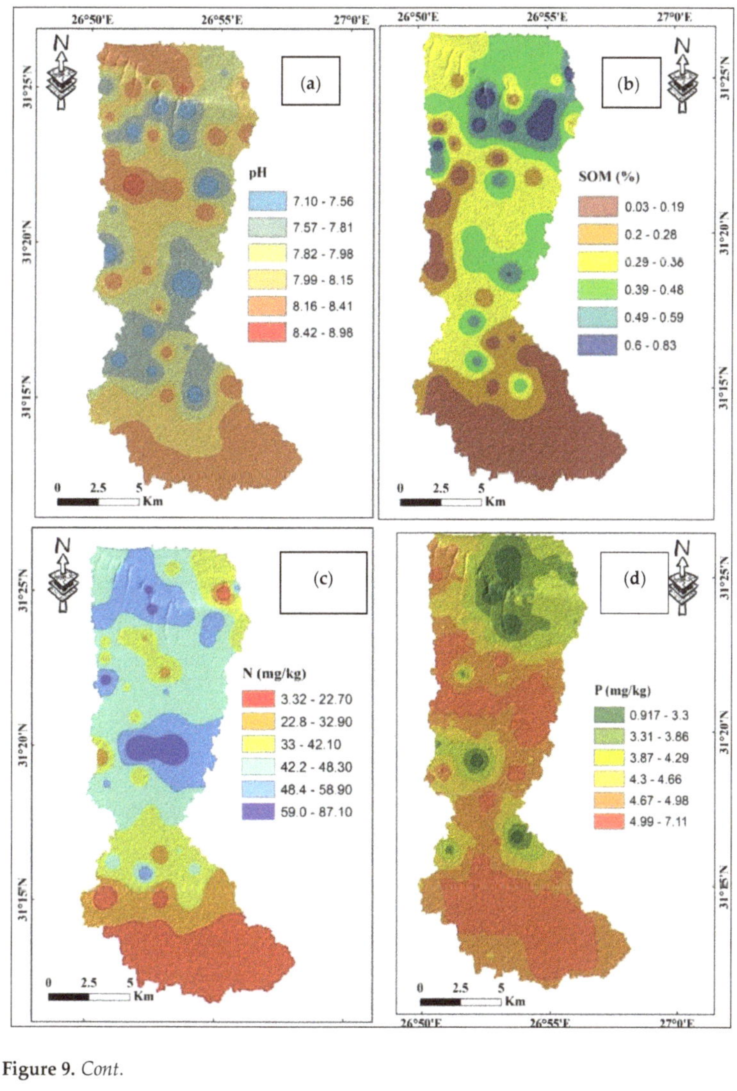

Figure 9. *Cont.*

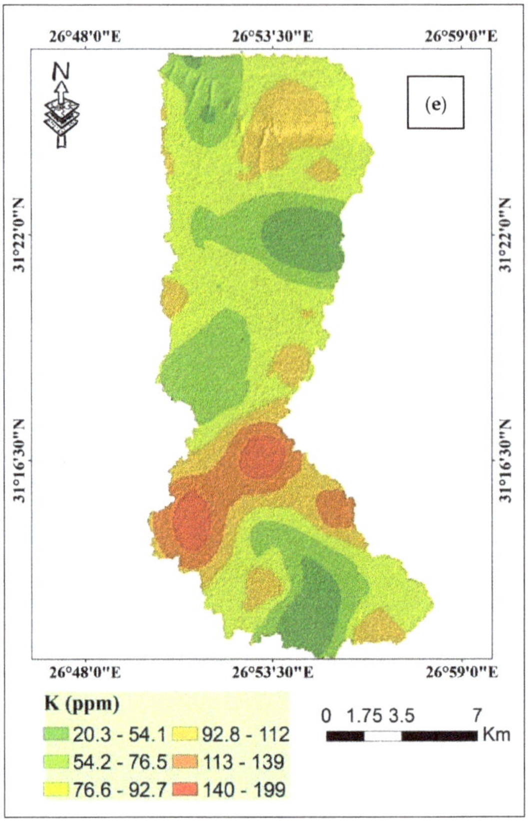

Figure 9. Spatial distribution of measured SOC parameters: (**a**) pH, (**b**) soil organic matter %, (**c**) nitrogen content (ppm), (**d**) phosphorus content (ppm), and (**e**) potassium content (ppm).

3.2. Producing SFC Parameter Using S2A Image

S2A data may help to characterize a given soil gradient, even if only a few soil samples were used during differentiated crop operations (within the process of a precision agriculture) [30]. The following five Equations (8)–(12), were used to obtain the values of pH, SOM %, N, P, and K from the reflectance of the satellite image. OK was then used for the spatial distribution of these selected parameters. An assessment of accuracy was performed by calculating NRMSE (Table 4).

$$\text{pH} = (\text{band } 2 * 3.66) + (\text{band } 3 * 9.744) + (\text{b4} * -0.016) + 4.52 \quad (8)$$

$$\text{SOM }\% = (\text{band } 6 * -0.5567) + (\text{band } 7 * 36.58) + (\text{band } 11 * 0.2359) - 13.43 \quad (9)$$

$$\text{N} = (\text{band } 4 * -78.98) + (\text{band } 5 * 2272.91) - 740.16 \quad (10)$$

$$\text{P} = (\text{band } 6 * -5.76) + (\text{band } 10 * 1089.051) - 1.18 \quad (11)$$

$$\text{K} = (\text{band } 1 * 12{,}930.48) + (\text{band } 2 * -64.75) + (\text{band } 10 * -1107.29) - 468.28 \quad (12)$$

All the studied parameters indicated acceptable results due to R^2 values of 0.75 ± 0.31, 0.82 ± 0.11, 0.74 ± 8.70, 0.50 ± 1.53, and 0.97 ± 7.89 for pH, SOM %, N, P, and K, respectively. NRMSE values were 0.16, 0.14, 0.1, 0.01, and 0.04, respectively (Table 3 and Figure 4). The outputs from our results indicate that MLR is effective for predicting soil properties, with these outputs being consistent with [6,63,64]. The accuracy values (R^2 calibration) were

0.6, 0.7, 0.55, 0.6, and 0.92 for pH, SOM, N, P, and K, respectively. As a whole, spectral reflections are affected by a variety of soil properties and their concentrations [65,66]. Although the determination of respondent parts of a wavelength to dynamic changes in element concentration is very complex, statistical analysis can resolve this issue [6]. The qualitative analysis (visual comparison) and quantitative statistics should be utilized for the assessment of model performance [67], and the results of this research are in line with those of other studies [6,46]. The performance of the model for all selected elements was assessed using determination (adj R). All models showed good fitting with adj R ranging from 0.55 (N) to 0.92 (K).

Table 4. Model validation of soil property values obtained from S2A image.

Selected Parameters	R^2 Calibration	Adjusted R	RMSE	NRMSE	R^2 Validation
pH	0.6	0.54	0.31	0.16	0.75
SOM %	0.7	0.65	0.11	0.14	0.82
N (ppm)	0.55	0.52	8.70	0.1	0.74
P (ppm)	0.6	0.60	1.53	0.01	0.50
K (ppm)	0.92	0.91	7.89	0.04	0.97

3.3. Spatial Distribution of Predicted SOC Parameters Based on OK

The SOC parameter mapping relied on those values which were predicted by applying the above models (Equations (8)–(12)) from the satellite image for each characteristic. OK was utilized for mapping the SOC parameters and the efficacy and ability were measured by calculating RMSE, MSE, and RMSSE, as shown in Table 4. The spatial distribution of pH, and SOM regarding N, P, and K is shown in Figure 9. The predicted pH values ranged from 7.28 to 8.54 in central and northern parts of the study area, whereas the study area was characterized by a low content of SOM with the highest values located in wadi areas due to agricultural activities. The predicated values of N, P, and K ranged from 20.31–66.15, 2.01–6.84, and 24.13–174.59, respectively (Table 5). It can be concluded from these results that the selected soil properties perform as acceptable indicators for soil fertility condition of the study area, with these findings falling in line with those of [6]. The outputs showed that the Gaussian model is fit for SOM, N, and P, whereas the spherical and stable models are suitable for pH and K. It can be concluded from Table 6 that OK has the ability to map selected SOC parameters with high accuracy as RMSEE values are close to 1.00, i.e., 1.01, 0.98, 0.94, 0.95, and 1.00 for pH, SOM, N, P, and K, respectively; these results are consistent with [46,68]. These findings confirm that OK was suitable and reliable for predicting the spatial distribution of selected parameters (Figure 10).

Table 5. Statistics of predicted SFC parameters based on the S2A image.

Statistics	Pred. pH	Pred. SOM	Pred. N	Pred. P	Pred. K
MAX	8.54	0.71	66.15	6.84	174.59
Mean	7.90	0.38	45.47	4.18	83.37
MIN	7.28	0.05	20.31	2.01	24.13
SD	0.35	0.17	9.63	1.40	30.64

Table 6. Geostatistical OK analysis of predicted soil properties (pH, OM, N, P, and K).

Soil Parameters	Transformation	Trend Type	Model Type	Mean	RMSE	MSE	RMSSE	ASE
Pred. pH	Box cox	Constant	Spherical	0	0.34	0	1.01	0.3
Pred. SOM	None	Constant	Gaussian	0	0.16	0.02	0.98	0.2
Pred. N	log	None	Gaussian	5.32	14.1	0.15	0.94	28
Pred. P	log	None	Gaussian	0.01	1.24	0.02	0.95	1.4
Pred. K	Normal score	None	Stable	0.78	26.7	0.03	1.03	25

Note: Pred. = predicted.

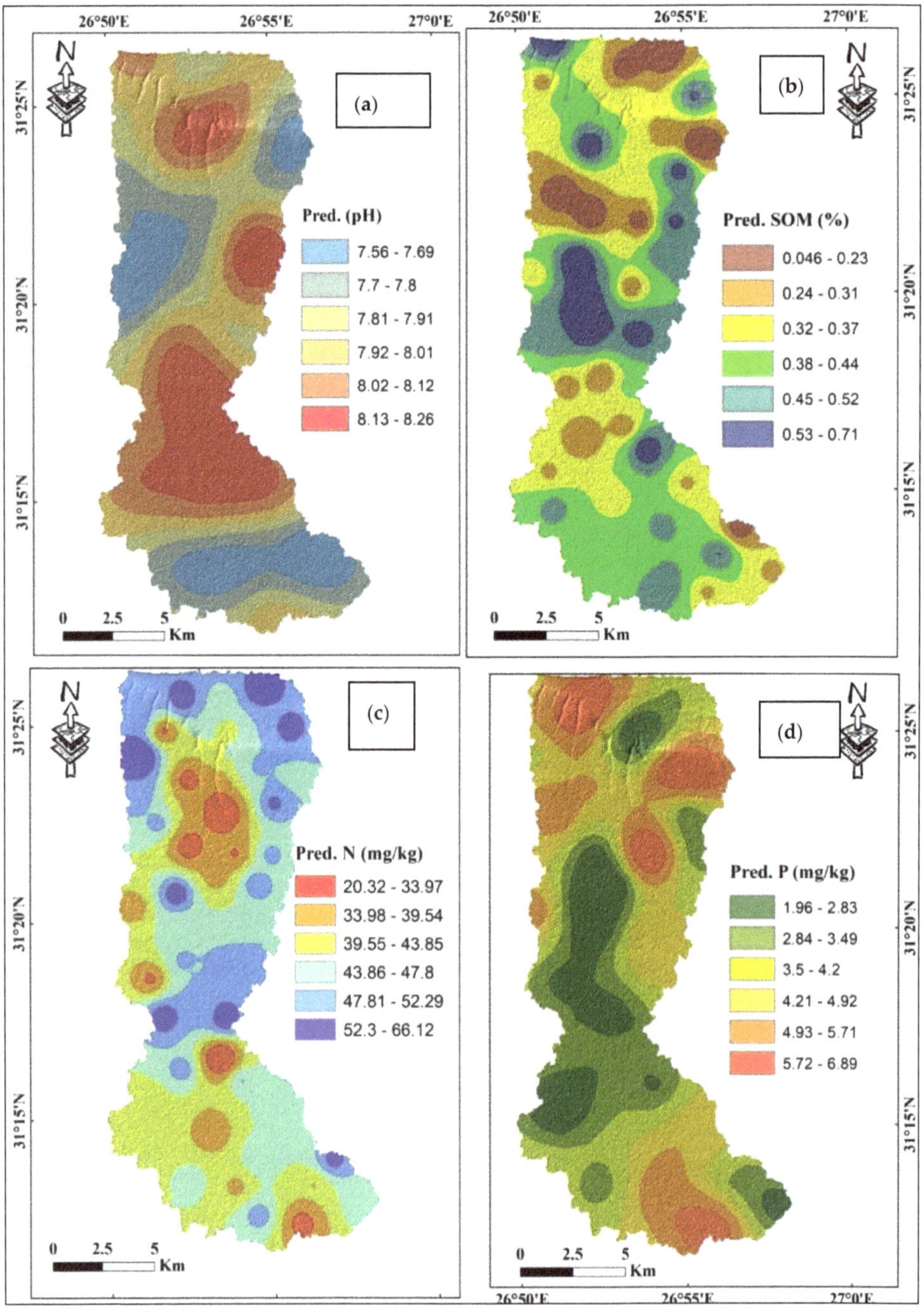

Figure 10. *Cont.*

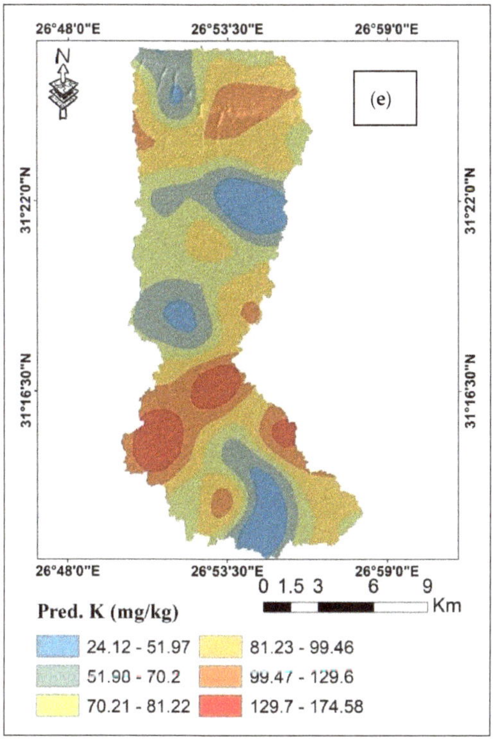

Figure 10. Spatial distribution of predicted SOC parameters: (**a**) pH, (**b**) soil organic matter %, (**c**) nitrogen content (ppm), (**d**) phosphorus content (ppm), and (**e**) potassium content (ppm).

3.4. Multivariate Statistical Analysis (MSA)

A correlation coefficient was used to measure the relationship between the variables and their strength [69,70]. This plays a very important role in dimensionality reduction of sample data and missing value estimations [71]. A Pearson correlation analysis illustrated the correlation between measured and predicted (pred.) soil parameters. There is a significant positive correlation between the pH and pred. pH, where r = 0.75. Furthermore, there is a significant positive correlation between the SOM and pred. SOM (r = 0.8). There is also a significant positive correlation between N and pred. N (r = 0.7), P and pred P (0.5), and K and pred. K (r = 0.92), as seen in Table 7 and Figure 11.

Table 7. Spearman correlations between original and predicated soil properties.

	pH	Pred. pH	OM	Pred. OM	N	Pred. N	P	Pred. P	K	Pred. K
pH	1.00	0.753 **	0.039	−0.124	0.035	0.086	−0.033	−0.051	0.081	0.027
Pred. pH		1.00	0.230	0.111	−0.027	0.081	−0.199	−0.242	0.029	−0.045
OM			1.00	0.791 **	0.311 *	0.177	−0.203	−0.257	−0.033	0.040
Pred. OM				1.00	0.211	0.129	−0.217	−0.084	−0.174	−0.036
N					1.000	0.678 **	−0.057	−0.096	−0.047	0.062
Pred. N						1.000	−0.031	0.006	−0.055	0.070
P							1.000	0.456**	−0.271	−0.165
Pred. P								1.000	−0.223	−0.275
K									1.000	0.925 **
Pred. K										1.00

** *p* values = 0.01 and * = 0.05 level.

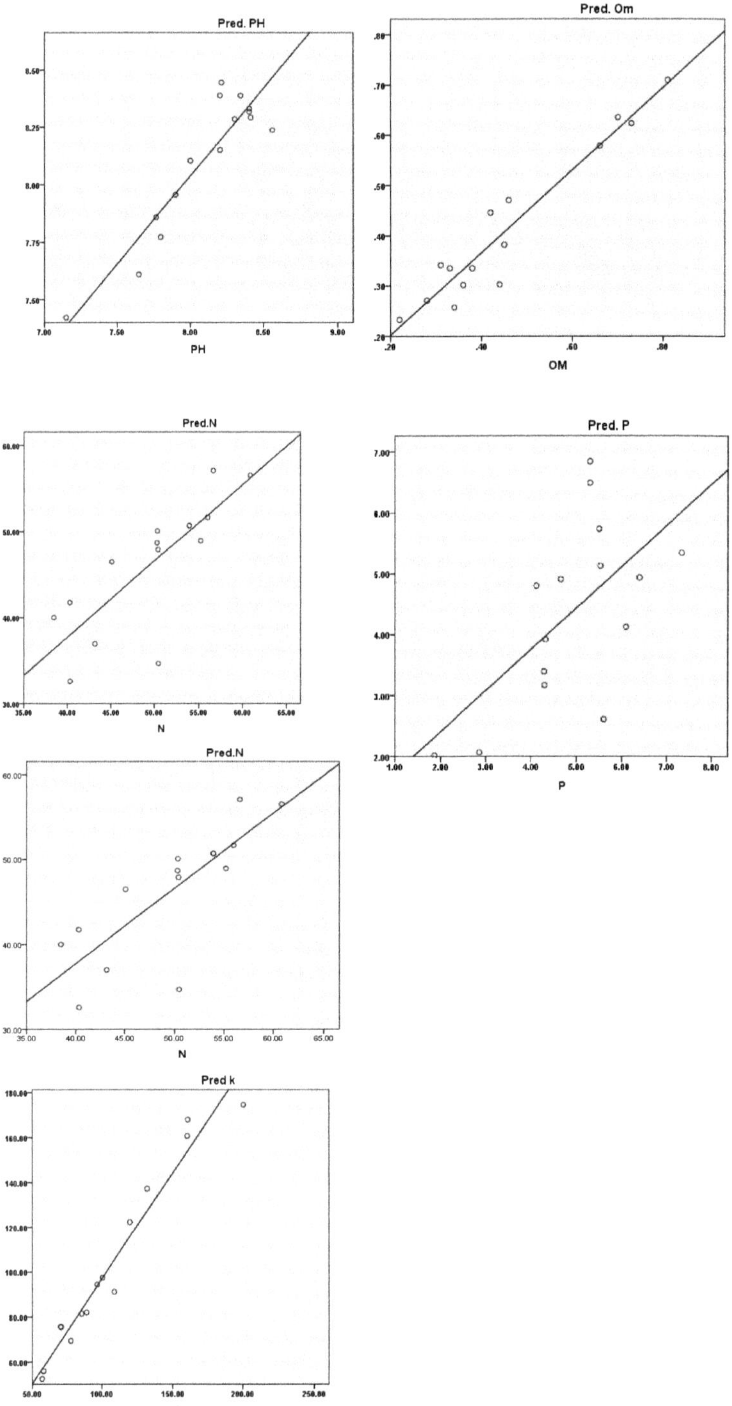

Figure 11. Linear correlation between measured and predicted parameters.

3.5. The SFC of Study Area

The mapping of soil fertility plays an important role in the realization of many applications within relevant research fields, ranging from soil management for sustainability to precision farming concepts [6]. The spatial distribution of SFCs in the study area was produced based on selected pH, SOM, N, P, and K parameters. The study area was classified according to the following three classes: moderate, low, and very low, with areas of 4607.90, 14,900.21, and 705.73 ha, respectively. References terms covered an area of 1155.91 ha (Tables 8 and S1, and Figure 12). The low values of SFCs in the study area are due to the mismanagement of agricultural activities and the location of the study area within an arid zone. Moderate values were noticed in those areas which have agricultural activates and deep profiles with a flat or almost flat surface [71,72]. These results demonstrate that the desert areas of Egypt suffer from a poverty of soil fertility in comparison to the Nile Delta [63,65,73,74] and need suitable management to choose the best crops according to soil quality, water availability, and the climate of an area.

Table 8. Areas of SFC classes in the study area.

SFC Classes	Symbol	Area (Hectare)
Moderate	F3	4607.90
Low	F4	14,900.21
Very low	F5	705.73
References terms	-	1155.91

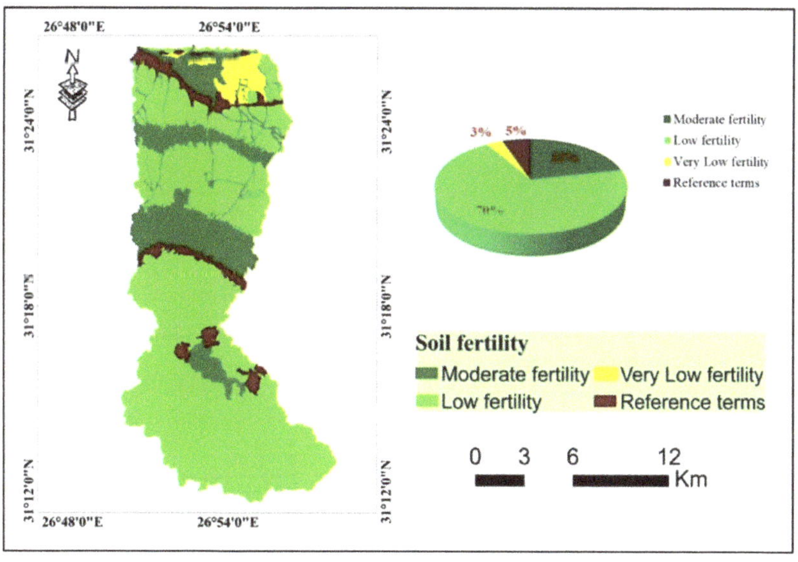

Figure 12. SFC areas of the study area.

4. Conclusions

Reflectance analysis from S2A imagery is a very effective procedure for predicting the pH, SOM, and presence of macro elements (N, P and K) in drylands. The current work aimed to find a relationship between reflectance from S2A imagery and SFC parameter concentrations. MLR was applied as representative of a dryland area to predict soil fertility parameters, with the mapping of those parameters, using ordinary kriging geostatistical analysis, then performed. The results revealed that reflectance within S2A imagery could predict different element concentrations, evidenced by the R2 values of 0.75 ± 0.31, 0.82 ± 0.11, 0.74 ± 8.70, 0.50 ± 1.53, and 0.97 ± 7.89 for pH, SOM %, N, P, and

K, respectively, and NRMSE values of 0.16, 0.14, 0.1, 0.01, and 0.04 also respective to this. The study area was divided into four classes of fertility condition, i.e., low fertility (70%), moderate fertility (22%), very low fertility (3%), and references terms (5%). The procedure herein provides an accurate and low-cost tool for wide implementation in assisting with food security by monitoring soil fertility status in drylands and improving soil conservation. Overcoming dry land ecological and poverty issues is considered one of the most important goals of the global 2030 Agenda for Sustainable Development (SDGs).

Supplementary Materials: The following supporting information can be downloaded at: https://www.mdpi.com/article/10.3390/ijgi11060353/s1, Figure S1: Samples distribution by conditional Latin hypercube (cLHS); Figure S2: Hydrological and metamorphic parameters extracted from R software; Table S1: Weight of final SFC based on GIS model builder.

Author Contributions: Conceptualization Mohamed S. Shokr and Mostafa A. Abdellatif; methodology Mohamed S. Shokr and Mostafa A. Abdellatif; software Mohamed S. Shokr and Mostafa A. Abdellatif; validation, Mohamed S. Shokr and Mostafa A. Abdellatif; formal analysis, Ahmed M. Saleh, Mohamed S. Shokr, Yasser S. A. Mazrouand Mostafa A. Abdellatif; investigation, Mohamed S. Shokr and Mostafa A. Abdellatif; resources Mohamed S. Shokr and Mostafa A. Abdellatif; data curation, Mohamed S. Shokr and Mostafa A. Abdellatif, writing—Mohamed S. Shokr and Ahmed A. El Baroudy, writing—review and editing, Ahmed M. Saleh, Mohamed S. Shokr, Ahmed A. El Baroudy, Esawy K. Mahmoud and Zheli Ding; visualization, Mohamed S. Shokr, Ahmed A. El Baroudyand Zheli Ding; supervision Ahmed M. Saleh, Abdelaziz A. Belal and Esawy K. Mahmoud; project administration Abdelaziz A. Belal; funding acquisition, Yasser S. A. Mazrou and Zheli Ding. All authors have read and agreed to the published version of the manuscript.

Funding: King Khalid University for funding this work, through the Program of Research Groups under grant number (RGP 2/67/43).

Institutional Review Board Statement: Not applicable.

Informed Consent Statement: Not applicable.

Data Availability Statement: Not applicable.

Acknowledgments: The authors extend their appreciation to the Deanship of Scientific Research at King Khalid University for funding this work through the Program of Research Groups under grant number (RGP2/67/43). The authors are grateful to the Senior Foreign Expert Project of China (Grant number G2021034008L).The authors would like to thank the National Authority for Remote Sensing and Space Science (NARSS) for supporting the field survey and remote sensing work.

Conflicts of Interest: The authors declare no conflict of interest.

References

1. Mohamed, E.S.; Abu-Hashim, M.; Belal, A.A.A. Sustainable Indicators in Arid Region: Case Study–Egypt. In *Sustainability of Agricultural Environment in Egypt Soil-Water-Food Nexus, The Handbook of Environmental Chemistry*; Springer Science and Business Media LLC: Berlin/Heidelberg, Germany, 2018; pp. 273–293.
2. AbdelRahman, M.A.; Shalaby, A.; Mohamed, E.S. Comparison of two soil quality indices using two methods based on geographic information system. *Egypt J. Remote Sens. Space Sci.* **2019**, *22*, 127–136. [CrossRef]
3. Abu-hashim, M.; Elsayed, M.; Belal, A.E. Effect of land-use changes and site variables on surface soil organic carbon pool at Mediterranean Region. *J. Afr. Earth Sci.* **2016**, *114*, 78–84. [CrossRef]
4. Hendawy, E.; Belal, A.A.; Mohamed, E.S.; Elfadaly, A.; Murgante, B.; Aldosari, A.A.; Lasaponara, R. The prediction and assessment of the impacts of soil sealing on agricultural land in the North Nile Delta (Egypt) using satellite data and GIS modeling. *Sustainability* **2019**, *11*, 4662. [CrossRef]
5. Vaudour, E.; Gomez, Y.C.; Fouad, P. Lagacherie, Sentinel-2 image capacities to predict common topsoil properties of temperate and Mediterranean agroecosystems. *Remote Sens. Environ.* **2019**, *223*, 21–33, ISSN 0034-4257. [CrossRef]
6. Mohamed, E.S.; Baroudy, A.A.E.; El-beshbeshy, T.; Emam, M.; Belal, A.A.; Elfadaly, A.; Aldosari, A.A.; Ali, A.M.; Lasaponara, R. Vis-NIR Spectroscopy and Satellite Landsat-8 OLI Data to Map Soil Nutrients in Arid Conditions: A Case Study of the Northwest Coast of Egypt. *Remote Sens.* **2020**, *12*, 3716. [CrossRef]
7. Horta, A.; Malone, B.; Stockmann, U.; Minasny, B.; Bishop, T.F.A.; McBratney, A.B.; Pallasser, R.; Pozza, L. Potential of integrated field spectroscopy and spatial analysis for enhanced assessment of soil contamination: A prospective review. *Geoderma* **2015**, *241*, 180–209. [CrossRef]

8. Angelopoulou, T.; Tziolas, N.; Balafoutis, A.; Zalidis, G.; Bochtis, D. Remote sensing techniques for soil organic carbon estimation: A review. *Remote Sens.* **2019**, *11*, 676. [CrossRef]
9. Minasny, B.; McBratney, A.B. A conditioned Latin hypercube method for sampling in the presence of ancillary information. *Comput. Geosci.* **2006**, *32*, 1378–1388. [CrossRef]
10. Brungard, C.W.; Boettinger, J.L. Conditioned Latin hypercube sampling: Optimal sample size for digital soil mapping of arid Rangelands in Utah, USA. In *Digital Soil Mapping*; Boettinger, J.L., Howell, D.W., Moore, A.C., Hartemink, A.E., Kienast-Brown, S., Eds.; Progress in Soil Science; Springer: Dordrecht, The Netherlands, 2010.
11. Brungard, C.W.; Boettinger, J.L.; Duniway, M.C.; Wills, S.A.; Edwards, T.C., Jr. Machine learning for predicting soil classes in three semi-arid landscapes. *Geoderma* **2015**, *239–240*, 68–83. [CrossRef]
12. Worsham, L.; Markewitz, D.; Nibbelink, N.P.; West, L.T. A comparison of three field sampling methods to estimate soil carbon content. *Forest Sci.* **2012**, *58*, 513–522. [CrossRef]
13. Taghizadeh-Mehrjardi, R.; Nabiollahi, K.; Minasny, B.; Triantafilis, J. Comparing data mining classifiers to predict spatial distribution of USDA-family soil groups in Baneh region, Iran. *Geoderma* **2015**, *253–254*, 67–77. [CrossRef]
14. Levi, M.R.; Rasmussen, C. Covariate selection with iterative principal component analysis for predicting physical soil properties. *Geoderma* **2014**, *219*, 46–57. [CrossRef]
15. Taghizadeh-Mehrjardi, R.; Minasny, B.; Sarmadian, F.; Malone, B.P. Digital mappingof soil salinity in Ardakan region, central Iran. *Geoderma* **2014**, *213*, 15–28. [CrossRef]
16. Lacoste, D.; Gaspard, P. Isometric fluctuation relations for equilibrium states with broken symmetry. *Phys. Rev. Lett.* **2014**, *113*, 24. [CrossRef]
17. Scarpone, C.; Schmidt, M.G.; Bulmer, C.E.; Knudby, A. Modelling soil thickness in the critical zone for Southern British Columbia. *Geoderma* **2016**, *282*, 59–69. [CrossRef]
18. Waruru, B.K.; Shepherd, K.D.; Ndegwa, G.M.; Sila, A.M. Estimation of wet aggregation: Indices using soil properties and diffuse reflectance near infrared spectroscopy, an application of classification and regression tree analysis. *Biosyst. Eng.* **2016**, *152*, 148–164. [CrossRef]
19. Adhikari, K.; Hartemink, A.E. Soil organic carbon increases under intensive agriculture in the Central Sands, Wisconsin, USA. *Geoderma Reg.* **2017**, *10*, 115–125. [CrossRef]
20. Domenech, M.B.; Castro-Franco, M.; Costa, J.L.; Amiotti, N.M. Sampling scheme optimization to map soil depth to petrocalcic horizon at field scale. *Geoderma* **2017**, *290*, 75–82. [CrossRef]
21. Jeong, G.; Oeverdieck, H.; Park, S.J.; Huwe, B.; Ließ, M. Spatial soil nutrients prediction using three supervised learning methods for assessment of land potentials in complex terrain. *Catena* **2017**, *154*, 73–84. [CrossRef]
22. Babaei, F.; Zolfaghari, A.A.; Yazdani, M.R.; Sadeghipour, A. Spatial analysis of infiltration in agricultural lands in arid areas of Iran. *CATENA* **2018**, *170*, 25–35. [CrossRef]
23. Nawar, S.; Buddenbaum, H.; Hill, J.; Kozak, J. Modeling and mapping of soil salinity with reflectance spectroscopy and landsat data using two quantitative methods (PLSR and MARS). *Remote Sens.* **2014**, *6*, 10813–10834. [CrossRef]
24. Selige, T.; Bohner, J.; Schmidhalter, U. High resolution topsoil mapping using hyper spectral image and field data in multivariate regression modeling procedures. *Geoderma* **2006**, *136*, 235–244. [CrossRef]
25. Stevens, A.; Udelhoeven, T.; Denis, A.; Tychon, B.; Lioy, R.; Hoffman, L.; Van Wesemael, B. Measuring soil organic carbon in croplands at regional scale using airborne imaging spectroscopy. *Geoderma* **2010**, *158*, 32–45. [CrossRef]
26. Gomaes, C.; Coulouma, G.; Lagacherie, P. Regional Predictions of eight common soil properties and their spatial structures from hyperspectral Vis-NIR data. *Geoderma* **2012**, *189–190*, 176–185.
27. Gomaes, C.; Lagacherie, P.; Bacha, S. Using an VNIR/SWIR hyperspectral image to map topsoil properties over bare soil surfaces in the Cap Bon region (Tunsia). In *Digital Soil Assessment and Beyond*; Minsny, B., Malone, B.P., McBratney, A.B., Eds.; Springer: Berlin/Heidelberg, Germany, 2012; pp. 387–392.
28. Vaudour, E.; Gilliot, J.M.; Bel, L.; Lefevre, J.; Chehdi, K. Regional prediction of soil organic carbon content over temperature croplands using visible near-infrared airborne hyperspectral imagery and synchronous field spectra. *Int. J. Appl. Earth Obs. Geoinf.* **2016**, *49*, 24–38.
29. Baroudy, A.A.E.; Ali, A.M.; Mohamed, E.S.; Moghanm, F.S.; Shokr, M.S.; Savin, I.; Poddubsky, A.; Ding, Z.; Kheir, A.M.S.; Aldosari, A.A.; et al. Modeling Land Suitability for Rice Crop Using Remote Sensing and Soil Quality Indicators: The Case Study of the Nile Delta. *Sustainability* **2020**, *12*, 9653. [CrossRef]
30. Santaga, F.S.; Agnelli, A.; Leccese, A.; Vizzari, M. Using Sentinel-2 for Simplifying Soil Sampling and Mapping: Two Case Studies in Umbria, Italy. *Remote Sens.* **2021**, *13*, 3379. [CrossRef]
31. Ben-Dor, E.; Banin, A. Near-Infrared Analysis as a Rapid Method to Simultaneously Evaluate Several Soil Properties 94. *Soil Sci. Soc. Am. J.* **1995**, *59*, 364–372. [CrossRef]
32. Adeline, K.; Gomez, C.; Gorretta, N.; Roger, J.M. predictive ability of soil properties to spectral degradation from laboratory Vis-NIR spectroscopy data. *Geoderma* **2017**, *288*, 143–153. [CrossRef]
33. Gomez, C.; Adeline, K.; Bacha, S.; Driessn, B.; Gorretta, N.; Lagacherie, P.; Roger, J.M.; Briottet, X. Sensitivity of clay content prediction to spectral configuration of VNIR/SWIR imaging data. *Geoderma* **2015**, *288*, 143–153.
34. Rossel, R.V.; McBratney, A.B. Laboratory evaluation of a proximal sensing technique for simultaneous measurement of soil clay and water content. *Geoderma* **1998**, *85*, 19–39. [CrossRef]

35. Abd-Elmabod, S.K.; Muñoz-Rojas, M.; Jordán, A.; Anaya-Romero, M.; Phillips, J.D.; Jones, L.; Zhang, Z.; Pereira, P.; Fleskens, L.; Van Der Ploeg, M. Climate change impacts on agricultural suitability and yield reduction in a Mediterranean region. *Geoderma* **2020**, *374*, 114453. [CrossRef]
36. Shokr, M.S.; El Baroudy, A.A.E.; Fullen, M.A.; El-Beshbeshy, T.R.; Ali, R.R.; Elhalim, A.; Guerra, A.J.T.; Jorge, M.C.O. Mapping of heavy metal contamination In alluvial soils of the Middle Nile Delta of Egypt. *J. Environ. Eng. Landsc. Manag.* **2016**, *24*, 218–231. [CrossRef]
37. El Baroudy, A.A. Mapping and evaluating land suitability using a GIS-based model. *Catena* **2016**, *140*, 96–104. [CrossRef]
38. Barbosa, A.; Marinho, T.; Martin, N.; Hovakimyan, N. Multi-Stream CNN for spatial resource allocation: A crop management application. In Proceedings of the IEEE/CVF Conference on Computer Vision and Pattern Recognition Workshops, Seattle, WA, USA, 14–19 June 2020; pp. 58–59.
39. Preston, W.; Araújo do Nascimento, C.W.; Agra Bezerra da Silva, Y.J.; Silva, D.J.; Alves Ferreira, H. Soil fertility changes in vineyards of a semiarid region in Brazil. *J. Soil Sci. Plant Nutr.* **2017**, *17*, 672–685. [CrossRef]
40. Shrivastava, P.; Kumar, R. Soil salinity. A serious environmental issue and plant growth promoting bacteria as one of the tools for its alleviation. *Saudi J. Biol. Sci.* **2015**, *22*, 123–131. [CrossRef] [PubMed]
41. Schütz, L.; Gattinger, A.; Meier, M.; Müller, A.; Boller, T.; Mäder, P.; Mathimaran, N. Improving crop yield and nutrient use efficiency via biofertilization—A global meta-analysis. *Front. Plant Sci.* **2018**, *8*, 2204. [CrossRef]
42. Khdery, G.; Gad, A.A.; El-Zeiny, A.M. Spectroscopic Characterization of Plant Cover in El-Fayoum Governorate, Egypt. *Egypt J. Soil Sci.* **2020**, *60*, 397–408. [CrossRef]
43. Abd-Elmabod, S.K.; Jordán, A.; Fleskens, L.; Phillips, J.D.; Muñoz-Rojas, M.; Van der Ploeg, M.; Anaya- Romero, M.; De la Rosa, D. Modelling agricultural suitability along soil transects under current conditions and improved scenario of soil factors. In *Soil Mapping and Process Modeling for Sustainable Land Use Management*; Elsevier: Amsterdam, The Netherland, 2017; pp. 193–219, ISBN 9780128052006.
44. Khalifa, M.E.; Beshay, N.F. Soil classification and potentiality assessment for some rainfed areas at West of Matrouh, Northwestern Coast of Egypt. *Alex. Sci. Exch. J. Int. Q.J. Sci. Agric. Environ.* **2015**, *36*, 325–341.
45. Ministry of Industry and Mineral Resources (MIMR). *The Egyptian Geological Survey and Mining Authority Scale 1:2:000.000*; Ministry of Industry and Mineral Resources: Riyadh, Saudi Arabia, 1981.
46. Abdellatif, M.A.; El Baroudy, A.A.; Arshad, M.; Mahmoud, E.K.; Saleh, A.M.; Moghanm, F.S.; Shaltout, K.H.; Eid, E.M.; Shokr, M.S. A GIS-Based Approach for the Quantitative Assessment of Soil Quality and Sustainable Agriculture. *Sustainability* **2021**, *13*, 13438. [CrossRef]
47. Soil Survey Staff (USDA). Soil Survey Staff (USDA). Soil survey field and laboratory methods manual. In *Soil Survey Investigations Report No. 51*; Version 2.0; Department of Agriculture, Natural Resources Conservation Service: Washington, DC, USA, 2014.
48. Roudier, P. Clhs: A R Package for Conditioned Latin Hypercube Sampling. 2011. Available online: http://cran.rproject.org/web/packages/clhs/index.html (accessed on 14 October 2021).
49. Fidalgo, E.C.C.; Pedreira, B.C.C.G.; Abreu, M.B.; Moura, I.B.; Godoy, M.D.P. *Uso e Cobertura da Terra naBaciaHidrográfica do Rio Guapi-Macacu*; (Documentos 105); Embrapa Solos: Rio de Janeiro, Brazil, 2008.
50. Rukun, L. *Analysis Methods of Soil Agricultural Chemistry*; Agriculutural Science and Technology Press: Beijing, China, 1999.
51. Yang, Y.; Zhu, J.; Tong, X.; Wang, D. The Spatial Pattern Characteristics of Soil Nutrients at the Field Scale. In *Proceedings of the Computer and Computing Technologies in Agriculture II, Volume 1. CCTA 2008. IFIP Advances in Information and Communication Technology*; Li, D., Zhao, C., Eds.; Springer: Boston, MA, USA, 2009; Volume 293. [CrossRef]
52. Bui, Q.-T.; Jamet, C.; Vantrepotte, V.; Mériaux, X.; Cauvin, A.; Mograne, M.A. Evaluation of Sentinel-2/MSI Atmospheric Correction Algorithms over Two Contrasted French Coastal Waters. *Remote Sens.* **2022**, *14*, 1099. [CrossRef]
53. Dangal, S.R.; Sanderman, J.; Wills, S.; Ramirez-Lopez, L. Accurate and Precise Prediction of Soil Properties from a Large Mid-Infrared Spectral Library. *Soil Syst.* **2019**, *3*, 11. [CrossRef]
54. Mohamed, E.; Belal, A.A.; Ali, R.R.; Saleh, A.; Hendawy, E.A. Land degradation. In *The Soils of Egypt*; Springer Cham: Basel, Switzerland, 2019; pp. 159–174.
55. Otto, S.A.; Kadin, M.; Casini, M.; Torres, M.A.; Blenckner, T. A quantitative framework for selecting and validating food web indicators. *Ecol. Indic.* **2018**, *84*, 619–631. [CrossRef]
56. Biswas, A.; Si, B.C. Model averaging for semivariogram model parameters. Adv. *Agrophys. Res.* **2013**, *4*, 81–96.
57. Shokr, M.S.; Abdellatif, M.A.; El Baroudy, A.A.; Elnashar, A.; Ali, E.F.; Belal, A.A.; Attia, W.; Ahmed, M.; Aldosari, A.A.; Szantoi, Z.; et al. Development of A Spatial Model for Soil Quality Assessment under Arid and Semi-Arid Conditions. *Sustainability* **2021**, *13*, 2893. [CrossRef]
58. El Behairy, R.A.; El Baroudy, A.A.; Ibrahim, M.M.; Kheir, A.M.; Shokr, M.S. Modelling and Assessment of Irrigation Water Quality Index Using GIS in Semi-arid Region for Sustainable Agriculture. *Water Air Soil Pollut.* **2021**, *232*, 352. [CrossRef]
59. El-Aziz, S.H.A. Evaluation of land suitability for main irrigated crops in the North-Western Region of Libya. *Eurasian J. Soil Sci.* **2018**, *7*, 73–86. [CrossRef]
60. Ali, R.R.; Moghanm, F.S. Variation of soil properties over the landforms around Idku lake, Egypt. *Egypt J. Remote Sens. Space Sci.* **2013**, *16*, 91–101. [CrossRef]

61. Ostovari, Y.; Ghorbani-Dashtaki, S.; Bahrami, H.A.; Abbasi, M.; Dematte, J.A.M.; Arthur, E.; Panagos, P. Towards prediction of soil erodibility, SOM and $CaCO_3$ using laboratory Vis-NIR spectra: A case study in a semi-arid region of Iran. *Geoderma* **2018**, *314*, 102–112. [CrossRef]
62. Belal, A.A.; Mohamed, E.S.; Abu-hashim, M.S.D. Land Evaluation Based on GIS-Spatial Multi-Criteria Evaluation (SMCE) for Agricultural Development in Dry Wadi, Eastern Desert. *Egypt Int. J. Soil Sci.* **2015**, *10*, 100–116. [CrossRef]
63. Mohamed, E.S. Spatial assessment of desertification in north Sinai using modified MEDLAUS model. *Arab. J. Geosci.* **2013**, *6*, 4647–4659. [CrossRef]
64. Mohamed, E.S.; Ali, A.; El-Shirbeny, M.; Abutaleb, K.; Shaddad, S.M. Mapping soil moisture and their correlation with crop pattern using remotely sensed data in arid region. *Egypt. J. Remote Sens. Space Sci.* **2019**, *23*, 347–353. [CrossRef]
65. Mohamed, E.; Abdellatif, M.; Abd-Elmabod, S.K.; Khalil, M. Estimation of surface runoff using NRCS curve number in some areas in northwest coast, Egypt. In Proceedings of the E3S Web of Conferences, Barcelona, Spain, 10–12 February 2020; EDP Sciences: Les Ulis, France; Volume 167, p. 2002.
66. Zhang, Z.; Yu, D.; Shi, X.; Warner, E.; Ren, H.; Sun, W.; Tan, M.; Wang, H. Application of categorical information in the spatial prediction of soil organic carbon in the red soil area of China. *Soil Sci.* **2010**, *56*, 307–318. [CrossRef]
67. Moriasi, D.N.; Arnold, J.G.; Van Liew, M.W.; Bingner, R.L.; Harmel, R.D.; Veith, T.L. Model evaluation guidelines for systematic quantification of accuracy in watershed simulations. *ASABE* **2007**, *50*, 885–900. [CrossRef]
68. Aldabaa, A.; Yousif, I.A.H. Geostatistical approach for land suitability assessment of some desert soils. *Egypt. J. Soil Sci.* **2020**, *60*, 195–205. [CrossRef]
69. Gao, L.; Zhu, X.; Han, Z.; Wang, L.; Zhao, G.; Jiang, Y. Spectroscopy-Based Soil Organic Matter Estimation in Brown Forest Soil Areas of the Shandong Peninsula, China. *Pedosphere* **2019**, *29*, 810–818. [CrossRef]
70. Fu, C.; Tian, A.; Zhu, D.; Zhao, J.; Xiong, H. Estimation of Salinity Content in Different Saline-Alkali Zones Based on Machine Learning Model Using FOD Pretreatment Method. *Remote Sens.* **2021**, *13*, 5140. [CrossRef]
71. Abdel-Kader, F.H. Digital soil mapping at pilot sites in the northwest coast of Egypt: A multinomial logistic regression approach. *Egypt J. Remote Sens. Space Sci.* **2011**, *14*, 29–40. [CrossRef]
72. Said, M.E.S.; Ali, A.M.; Borin, M.; Abd-Elmabod, S.K.; Aldosari, A.A.; Khalil, M.M.N.; Abdel-Fattah, M.K. On the Use of Multivariate Analysis and Land Evaluation for Potential Agricultural Development of the Northwestern Coast of Egypt. *Agronomy* **2020**, *10*, 1318. [CrossRef]
73. Barseem, M.S.; el Sayed, A.N.; Yousset, A.M. Impact of geologic setting on the groundwater occurrence in wadis El Sanab, Hashem, and Khrega using geoelectrical methods northwestern coast. *Egypt Arab J. Geosci.* **2014**, *7*, 5127–5139. [CrossRef]
74. Abdel-Fattah, M.K.; Abd-Elmabod, S.K.; Aldosari, A.A.; Elrys, A.S.; Mohamed, E.S. Multivariate Analysis for Assessing Irrigation Water Quality. A Case Study of the Bahr Mouise Canal, Eastern Nile Delta. *Water* **2020**, *12*, 2537. [CrossRef]

International Journal of *Geo-Information*

MDPI

Article

Assessment and Mapping Soil Water Erosion Using RUSLE Approach and GIS Tools: Case of Oued el-Hai Watershed, Aurès West, Northeastern of Algeria

Aida Bensekhria * and Rabah Bouhata

State Key Laboratory of Natural Hazards and Regional Planning, Institute of Earth Sciences and Universe, Batna 2 University, Batna 05000, Algeria; r.bouhata@univ-batna2.dz
* Correspondence: a.bensekhria@univ-batna2.dz; Tel.: +213-663902033

Abstract: The problem of soil water erosion is one of the primary causes of agro-pedological heritage degradation. The combined effect of natural factors and inappropriate human actions has weakened the soil, which seriously threatens the region's fertile lands and soils, which can ultimately lead to an irreversible situation of desertification. This study focuses on analysis and mapping of the vulnerability to erosion in Oued el-Hai watershed, Algeria, based on a technical methodology that combines the universal soil loss equation (USLE) with the geographic information system (GIS) tools. The results are organized into three main classes of different rate values, from one area to another, depending on the influence of different factors that control the erosion process. The highest loss rate value is greater than 30 t·ha^{-1}·yr^{-1} and covers 23.2% of the total area, mainly located in the mountainous areas with steep slopes. However, the minimum potential erosion rate value is mainly located on the plain, with an average of 10 t·ha^{-1}·yr^{-1} covering 45.2% of the total area of the watershed. The estimate of potential water erosion has given alarming results. The total area of the watershed could lose a rate of 16.69 t·ha^{-1}·yr^{-1} (on average) each year. The method and results described in this article are valuable for understanding the soil erosion risk and are useful for managing and planning land use that will avoid land degradation. Hence, the results of this study are considered an important document which constitutes a decision support tool in terms of the management and protection of natural resources.

Keywords: Oued el-Hai; Eastern Algeria; soil water erosion; USLE; GIS; watershed; vulnerability

Citation: Bensekhria, A.; Bouhata, R. Assessment and Mapping Soil Water Erosion Using RUSLE Approach and GIS Tools: Case of Oued el-Hai Watershed, Aurès West, Northeastern of Algeria. *ISPRS Int. J. Geo-Inf.* **2022**, *11*, 84. https://doi.org/10.3390/ijgi11020084

Academic Editors: Dimitris Triantakonstantis, Panagiotis Tziachris and Wolfgang Kainz

Received: 10 November 2021
Accepted: 10 January 2022
Published: 24 January 2022

Publisher's Note: MDPI stays neutral with regard to jurisdictional claims in published maps and institutional affiliations.

1. Introduction

Soil is a food resource for more than 7.5 billion inhabitants, a support of human activities, a purification system and a water reserve. It remains largely unknown by those who use it directly and on a daily basis. Soil is increasingly threatened by multiple physical, chemical and biological degradations. The loss of productive soils will amplify food price volatility and potentially push millions of people into poverty. The careful management of soil can increase the food supply, provide support for climate regulation and a pathway to safeguard ecosystem functions [1]. The main threat to soils is erosion by wind and water, as highlighted by the latest United Nations publications, namely the state of soil resources in the world [2], where soil erosion has been identified as one of the major soil threats. Water erosion of soils has become a global problem, where accelerated forms of soil erosion have become a widespread phenomenon and pose a major challenge to achieving the United Nations Sustainable Development Goals [3].

Climate change and ever-increasing erosion in many regions of the world, and especially in arid and semi-arid environments, lead to the degradation of the living conditions of residents and disastrous consequences for natural resources and the population [4]. In Algeria, soil resources are seriously threatened by water erosion [5–12]. Annual water losses due to siltation in dams are estimated at around 20 million m^2 [13], with an average annual specific erosion varying between 2000 and 4000 t/km^2 [14]. However, the intensity of water erosion varies from one region to another: it varied from 47% in the west, 27% in the center and 26% in the eastern part of the country [15]. In 2011, the Ministry of Agriculture and Rural Development and Fisheries (MARDF) estimated that around 14 million hectares of mountain areas in the north of the country were degraded by water erosion. The various bathymetric surveys, carried out over the period 1986–2008 by the National Agency for Dams and Transfers on all of 59 dams in operation, showed that the volume lost by siltation was 898 Mm^3 or 13.4% of total volume holdbacks (Ministry of Water Resources, 2010) [16].

The modeling and forecasting of soil erosion by water has more than seven decades of history. There are many different models; some are simple, and some are more complex, such as the Soil Erosion Model for Mediterranean Regions (SEMMED) [17], the revised model of Morgan, Morgan and Finney (RMMF) [18], the Water Erosion Prediction Project (WEPP) [19], the Soil and Water Assessment Tool (SWAT) [20], and the European Soil Erosion Model EUROSEM [21], as well as the update of existing empirical approaches. Currently, the universal soil loss equation (USLE) [22] and revised USLE (RUSLE) [23] are the most widely applied soils erosion prediction models in the world (have been applied in 109 countries) and under many different conditions. It has been most often applied for over 80 years due to its robustness, simplicity, high degree of flexibility, accuracy and accessibility of data [24–27].

In order to assess the effects of soils water erosion, we used the USLE/RUSLE model, which is one of the most widely used methods in the Mediterranean region: in Algeria [8,11,28]; in Turkey [29,30]; in Morocco [31,32]; in Spain [33]; in Italy [34,35]; in Tunisia [36]; and in Ethiopia [37]. There are many other similar studies on land degradation due to soil erosion [38,39]. The above-mentioned studies show the limitations of the RUSLE model. The estimation of potential soil erosion is improved by the development of the RUSLE model. The RUSLE model has different input factors that can be taken from the literature or from statistical and empirical data by integration with GIS. Results generated by RUSLE are valid for the estimation of soil erosion due to water. Estimation of the quantity of soil erosion is important because the results of this study will be beneficial for researchers and engineers to plan for the control of soil loss in the future. The RUSLE model based on GIS was also used in similar studies [40,41] and provides quantitative data of soil erosion. The method involves applying all factors of the RUSLE equation to georeferenced data. Incorporating remote sensing data and GIS techniques with the USLE/RUSLE model offers researchers the ability to predict the average annual rate of soil erosion at local, regional and national scales. The Oued el-Hai watershed is considered to be a typical area of water erosion, due to many erosion factors, such as climate, geology, topography, plant cover and soil conditions.

The main objective of this study is the quantitative estimation of sensitivity to soil erosion in the Oued el-Hai watershed (NE Algeria). The maps are produced with the integration of remote sensing data and GIS tools using the RUSLE model. The maps will identify the intensity of erosion in different parts and their influencing factors in more concrete standings. To fight against and prevent this phenomenon of soil degradation, it is important to define the risk degree of water erosion and factors that control it. Producing accurate water erosion risk maps helps decision makers adopt best strategies in land rehabilitation and erosion control projects, such as the sustainability of dams, ecosystems, and better agriculture production.

2. Study Site Location

The watershed of Oued el-Hai represents one of the three largest valleys of the Aurès massif in Eastern Algeria, with an impluvium of nearly 1660 km^2. It is part of the large watershed of Chott Melghir. It is located in the southern mountain slope of Aurès, between the longitudes 5°30′ and 6°17′ east and the latitudes 35°5′ and 35°35′ north. It is bounded to the north by the Constantine highlands watershed, to the east by the Oued

Abdi watershed, to the west by the Chott El-Hodna watershed (Figure 1). It is elongated in a NE–SW direction.

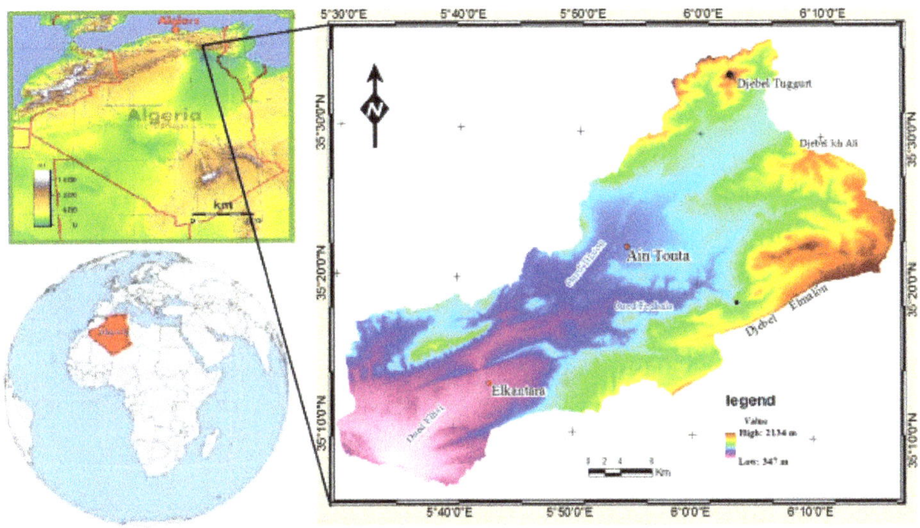

Figure 1. Location map of Oued el-Hai watershed.

The Oued el-Hai watershed is characterized by a series of massifs of a NE–SW orientation, with a decrease in the altitude from north to south, such as the Belezma mountains to the north in which the maximum altitude reaches 2091 m at Dj. Tuggurt. To the west are the Metlili mountains (peak reaches 1496 m). To the south is Dj. Bous at 1789 m, while to the east is Dj. El Malouat2091 m and Dj. El Rherahat1865 m. The prevailing climate is semi-arid to arid, and the precipitation is most often in form of violent and short thunderstorms. The irregularity of rains can occur during the same year or from one year to the next. The natural vegetation of the Oued el-Hai watershed belongs to the arid and semi-arid bioclimatic stage. It is presented in a very degraded form on heights, such as Aleppo pine, Juniper, Holm Oak and Thuja. On the lower slopes, areas not prepared for agriculture use are covered by Alfa and Mugwort. The bottom of the valley is of a gentle slope and occupied by very modest cereal cultivation.

The different geological units extend from Quaternary to Triassic, of which the main formations belong to the Cretaceous [36]. The geological formations are very variable: basically, sedimentary rocks, such as more or less hard limestones, sometimes dolomitic, and marly rocks strongly altered on the surface mainly due to the very harsh climate of the region. The fields occupied by alternating between gray and green marl and clear massive limestone, outcrop at Djebels Tuggurt, Ich Ali and Metlili. The massive limestones occupy the half-block of El Kantara with a thickness of 340 m, Djebels Bouss and El Malou. Hard sandstones and soft marls containing gypsum crystals, outcrop in the extreme north of the basin (Jebel Tichao). Therefore, Oued El-Hai watershed is characterized by components that promote onset of all forms of erosion. Altitudes decrease from upstream to downstream with a vertical drop of 1714 m. The slopes are medium to steep, where the class of slope ranging from 0% to 3 % and 4% to 12% presents 42.96%, whereas the class of slope ranging from 12.5% to 20% presents 48.84% of the total area. Rock formations are of medium to low resistance to erosion. Additionally, soils are skeletal, little evolved with raw minerals and poorly protected by degraded to very degraded vegetation, particularly scrubland. All these factors prove that the study area suffers from the phenomenon of erosion.

3. Materials and Methods

3.1. Sources of Data

The important input parameters used in this study were rainfall, soil and land use (Table 1).

Table 1. Source of data and description.

No	Data Type	Source	Description
1	Rainfall data	National Weather Center	Precipitation data for 30 years (1984–2014) of five precipitation stations
2	Soil data	Aures soil map Laboratory field sample analysis results	The soil map prepared by BNEDER (1984)
3	SRTM–DEM	https://earthexplorer.usgs.gov/ (accessed on 2 January 2022)	N35E005.hgt 23 September 2014 N35E006.hgt 23 September 2014 Resolution 30 m
4	Land use	Satellite image (Landsat) Vegetation cover map	Extracted from satellite image and vegetation cover map

3.2. Method

Several soil erosion models exist with varying degrees of complexity. One of the most widely applied empirical models for assessing the sheet and rill erosion is the universal soil loss equation (USLE), developed by Wischmeier and Smith in 1965.The adopted methodology is based on the estimation of (USLE) parameters. This method of analysis allows us to develop cartographic documents in order to fully understand the areas subjected to water erosion risk. It is an empirical model that combines factors that have a direct or indirect impact on the rate of the non-linear water erosion of soils. Therefore, the RUSLE model was selected and applied in the study area, as it requires land use, a land cover map that can be generated by remote sensing images, management practices, soil types and properties. The other advantage of a selection of RUSLE is that the parameters of this model can be easily integrated with GIS for better analysis. The overall methodology used in the present study is schematically represented in Figure 2.

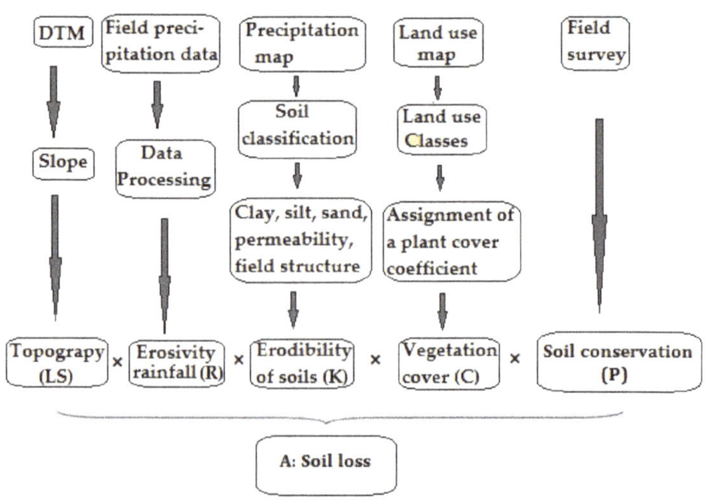

Figure 2. Flow chart of methodology.

When combined with GIS techniques, the revised WISCHMEIER equation [42] allows not only to estimate the rate of gross soil loss, but also to analyze its spatial distribution. According to the USLE model, erosion is a multiplicative function (Equation (1)) taking into account rainfall erosion (factor R) by resistance of the environment (factors C, K, LS, and P). Each factor is a numerical estimate of a specific component that affects the severity of soil erosion at a given location. Five factors are used to calculate soil losses (A): precipitation aggressiveness (R), soil erodibility (K), topography (LS), land cover (C), and anti-erosion practices (P).So the equation is in the form:

$$A = R \cdot K \cdot LS \cdot C \cdot P \tag{1}$$

A = Average annual soil loss (metric $t \cdot ha^{-1} \cdot yr^{-1}$).
R = Rainfall Erosivity Factor (Mega Joules mm/ $t \cdot ha^{-1} \cdot yr^{-1}$).
K = Soil erodibility index to water erosion (metric t/ha/MJ/mm).
LS = Topographic factor depending on the slope and its length (L in meter and S in%).
C = Land cover.
P = Conservation and development factor.

4. Results and Discussion

4.1. Rainfall Erosivity Factor (R)

Rain is the main factor in water erosion. Its erosivity depends mainly on its quantity and intensity or on the kinetic energy which results directly from it. Rain erosivity is often quantified using the R factor of the equation in [43]. In calculating this parameter, we used the maximum intensity of precipitation for 30 min (I30) as well as rain kinetic energy, for all heavy rains exceeding a certain rainfall threshold (E30) (Table 2). As these data are generally not available at conventional weather stations, a simplified method was used to estimate the R factor. According to [37], a good correlation (r2 = 0.95) was found between the average of E30, I30 and the product of annual precipitation multiplied by rains that fell during 1 h and 24 h with a period recurrence of 2 years. In this sense, Wischmeier proposed another formula [38]:

$$R = K \cdot C^{n} \tag{2}$$

$$C = h_1 \cdot h_{24} \cdot H \tag{3}$$

Table 2. Distribution of rainfall in Oued el-Hai watershed.

Station/Parameters	P_j Avg (mm)	Standard Deviation (mm)	H Annual (mm)
Djamourah	27.01	2.754	164.2
Segana	28.24	4.594	183.69
Ain Touta	32.37	3.872	262.37
Tazoult	35.33	5.142	353.0
Chaabet.O.Chlih	40.32	6.14	413.80

R: Rainfall erosivity factor ($MJ \cdot mm \cdot ha^{-1} \cdot h^{-1} \cdot yr^{-1}$).
C: Rainfall factor. Calculated by the multiplication of h1, h24 and H.
h_1: Maximum height of fallen rain in 1 h with a recurrence of 2 years (cm).
h_{24}: Maximum rain height of 24 h with a recurrence of 2 years (cm).
H: Average annual rainfall height (cm).
K, n: climate-related coefficients (in sub-humid to semi-arid climates, K = 0.751, n = 0.80).

In order to determine the fallen precipitation during heavy rains of different duration (t) and desired frequency ($ht\%$), the Montana-type formula [39] is applied.

$$h_{t\%} = h_{j\%} \left(\frac{t}{24}\right)^{b} \tag{4}$$

$$CV = \frac{ecar\ t - type\ P_j}{\overrightarrow{P_j}} \tag{5}$$

$$P_{j\%} = \frac{\overrightarrow{P_j}}{\sqrt{CV^2+1}} \quad (6)$$

t: duration;

h_t: desired frequency;

$h_j\%$: maximum daily frequency rainfall, determined after adjusting series of maximum daily rainfall to Gumbel's theoretical law (see Table 2);

P_j: daily precipitation;

$\overrightarrow{P_j}$: average daily precipitation;

b: climatic exponent determined by adjusting Gumbel law of short-term rains recorded in 20 stations equipped with a pluviograph:

$$b = 1 - \left(\frac{\text{In}I_2 - \text{In}I_1}{\text{In}t_2 - \text{In}t_1}\right) \quad (7)$$

$$I_1 = 25\left(\frac{\text{mm}}{\text{h}}\right), t_1 = 0.5(\text{h}),\ I_2 = \frac{\overrightarrow{P_j}}{24}(\text{mm/h}), t_2 = 2(\text{h}) \quad (8)$$

The values of R for five stations are presented in Table 3 down below.

Table 3. Values of rainfall erosivity factor for five stations.

Parameters/Station	Djamourah	Segana	Ain Touta	Tazoult	Chaabet.O.Chlih
P_j avg (mm)	27.01	28.24	32.37	35.33	40.32
Standard deviation	2.754	4.594	3.872	5.142	6.14
CV	0.102	0.163	0.119	0.146	0.152
$P_j\%$ (mm)	26.87	27.88	32.14	34.97	39.864
b	0.20	0.21	0.25	0.27	0.30
h_1 (cm)	1.428	1.428	1.452	1.482	1.527
h_{24} (cm)	2.687	2.788	3.217	3.497	3.986
H (cm)	16.42	18.369	26.237	35.30	41.380
R ($\text{MJ}\cdot\text{mm}\cdot\text{ha}^{-1}\cdot\text{h}^{-1}\cdot\text{yr}^{-1}$)	20.65	23.27	35.15	48.48	62.60

The application of kriging was carried out by a simple kriging-type interpolation method under Geostatistical Analyst of ArcGis software for the development of R map (Figure 3). In the watershed that is the subject of our study, the R value varies from 21 at the outlet to 62.57 on Jebel Tuggurt with an overall average of 40.3 ($\text{MJ}\cdot\text{mm}\cdot\text{ha}^{-1}\cdot\text{h}^{-1}\cdot\text{yr}^{-1}$).

4.2. Soil Erodibility (K)

The nature of soil is a major parameter of erosion since the removal of particles directly depends on the properties of soil and subsoil. The data available for K factor calculation were the soil map. It was possible to obtain the K factor for a type of soil from the results of soil analyses. It was possible to determine the percentages of clay, silt, sand and organic matter of each major type of soil, by taking samples and then generalizing on the concerned type. The repetition of experiments on different types of soils enabled [22] to develop an equation for calculation of soil erodibility:

$$100\text{K} = 2.1\cdot\text{M}^{1,14}\cdot(10^{-4})\cdot(12 - \text{a}) + 3.25\cdot(\text{b} - 2) + 2.5\cdot(\text{c} - 3) \quad (9)$$

M: calculated by the formula M = (% fine sand + silt). (100 − % clay);

a: percentage of organic matter;

b: permeability code;

c: structure code.

The soils erodibility index of the Oued el-Hai watershed varies from 0.27 to 0.44 in wind input soils and 0.008 to 0.27 in alluvial input soils (Figure 4). We noticed that the forest areas have deferent granulometric characteristics compared to other lands of the same type, in particular, its organic matter components.

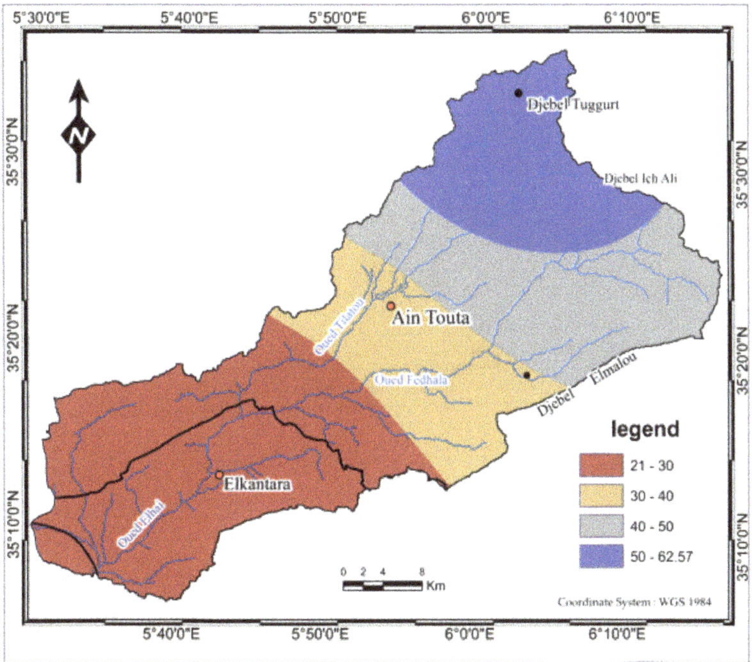

Figure 3. Map of rainfall erosivity factor (R) unit (MJ·mm·ha^{-1}h^{-1}yr^{-1}).

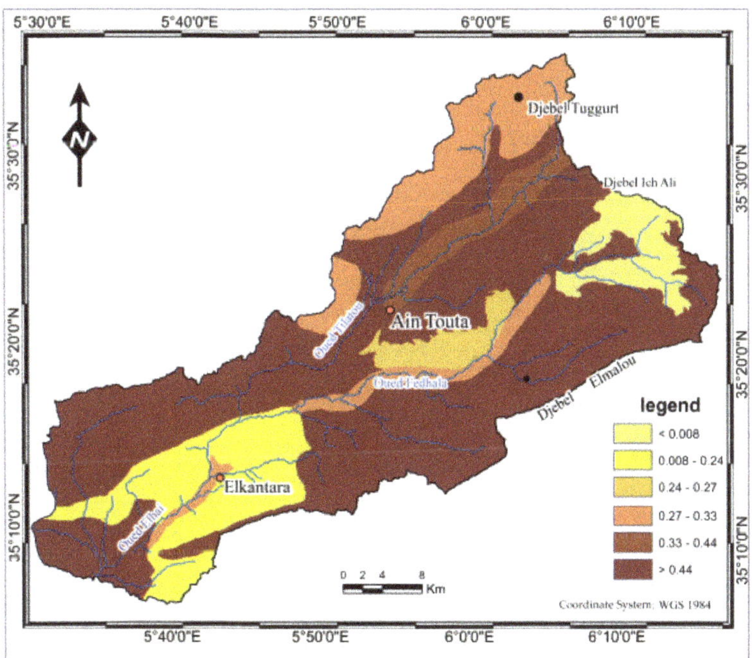

Figure 4. Soil erodibility map (K).

4.3. Topographic Factor (LS)

The topographic factor LS of [22] gives satisfactory results as to influence of length and inclination of slope in the erosion phenomenon (Figure 5). It therefore takes into account two elements: inclination (S) and slope length (L). These two factors are most often combined into a single dimensionless factor (LS). The topographic factor (LS) was calculated by [22] by the following formula:

$$LS = (L/22.13)^m \cdot (0.065 + 0.045 \cdot S + 0.065 \cdot S^2) \quad (10)$$

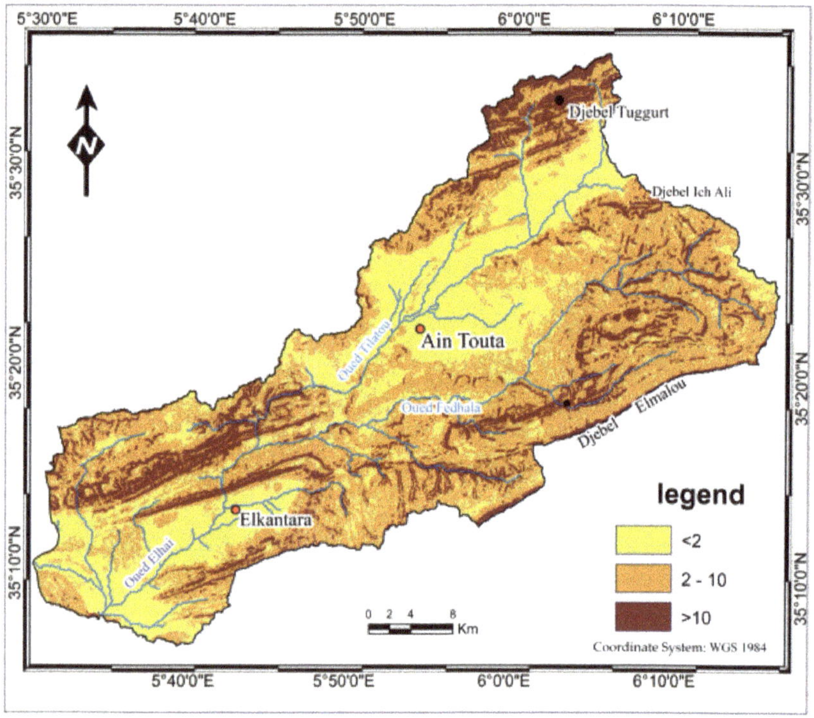

Figure 5. Topographic factor map (LS).

L is the slope length in m;

S is the incline of the slope in %;

m is a parameter such that m = 0.5 if slope is >5%; m = 0.4 if slope is 3.5 to 4.5%; m = 0.3 if slope is 1 to 3% and m = 0.2 if slope is <1%.

4.4. Land Use Factor—Plant Cover (C)

Factor C considers land use (plant cover, development and agricultural practices). In fact, the erosion more particularly affects certain types of crops, while it is less severe or simply absent for certain activities and developments. The type of vegetation cover must absolutely be considered since the damping of raindrops, the slowing of runoff and infiltration depend on it. The factor C is defined as a ratio of soil loss on land cultivated under specific conditions to corresponding soil loss on fallow land [44]. It can be calculated using Table 4 below.

Table 4. Factor C according to [45].

Vegetation Cover Type	C	Vegetation Cover Type	C
Dense reforestation	0.058	Clear Matorral, dense course	0.20
Moderately dense forest	0.13	Very degraded Matorral	0.22
Degraded Matorral, dense course	0.17	Steppes in alfa	0.32
Clear reforestation	0.18	Cereal cultivation	0.70

The land use map for the studied watershed was established by interpretation of satellite images using GoogleEarth (Figure 6). A visual digitization work was assisted by cartographic documents collected from the forest conservation service and the National Bureau of Studies for Rural Development (NBSRD). The quality of work is considered satisfied after a comparison of results obtained with observations and knowledge of field.

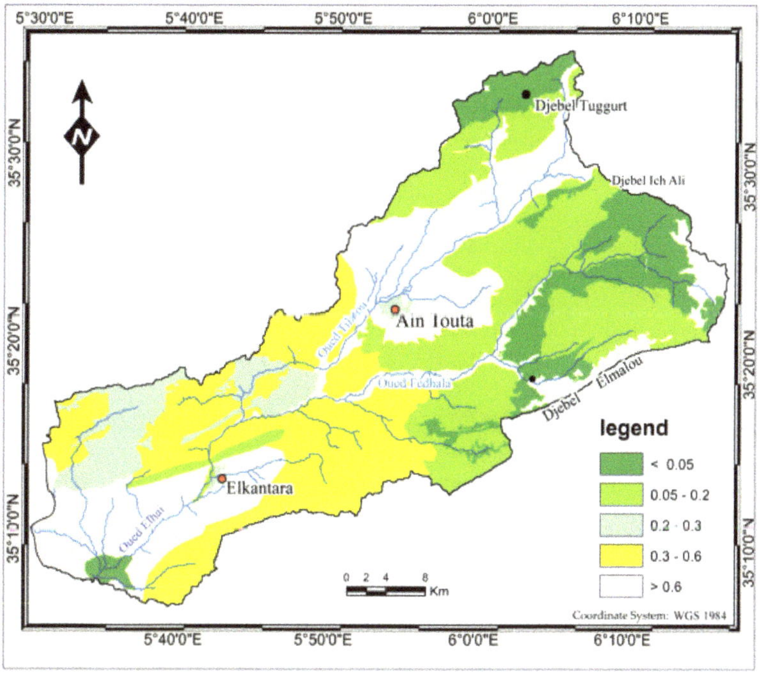

Figure 6. Land use factor map (C).

4.5. The Anti-Erosion Practical Factor (P)

Contour, alternating strip or terraced crops, bench reforestation and ridging are the most effective soil conservation practices. The values of *P* are less than or equal to 1. A value of 1 is assigned to land on which none of the above practices are used. The values of P vary according to the adopted practice and also according to the slope. Throughout the Oued el-Hai watershed, there are no anti-erosion facilities, and farmers do not use anti-erosion cultivation practices. The crops are mainly cereals, and the plowing is rarely parallel to the contour lines. There are some attempts to rehabilitate forests by reforestation, but not in benches. In this context, the value (P = 1) is assigned to the entire area of the basin.

Soil erosion is an alarming and very serious problem, especially in the study area, where different factors contribute to the rapid erosion of soil. Factors, such as steep slope of the region, very erodible soils, alarming degradation of the plant cover, and climate cause the issue of soil erosion. The study area is fed by precipitation most often in the form of violent and short thunderstorms [46]. For the estimation of soil loss in the study area, the revised universal soil loss equation (RUSLE) model was used in combination with geographic information system (GIS).

It is not possible to calculate the soil loss by using conventional methods due to the high cost, and because they are time consuming [47]. Therefore, the revised universal soil loss equation (RUSLE) model is most widely used because it is simple, easy, and requires less data and time. RUSLE is a straightforward and empirically based model that has the ability to predict the long-term average annual rate of soil erosion on slopes using data on rainfall pattern, soil type, topography, crop system and management practices. The integration of the RUSLE model with GIS data generates more accurate results. This model can easily be implemented, and required data are easily available for most countries. The use of GIS allows us to map the spatial distribution of soil erosion risk.

In the present research, a soil erosion rate map was generated for the Oued el-Hai watershed. Several data sources were used for the generation of RUSLE model input factors and were stored as raster GIS layers in the ArcGIS software. Potential soil loss is estimated from the product of factors (R, K, LS, C and P), which represents the geo-environmental scenario of the study area in the spatial analyst extension of Arc GIS software.

We analyzed each of the five main factors responsible for the erosion phenomenon in our study area. We integrated the results of these analyzes into a general model and of quantified soil losses. The crossing of data taken into account in erosive process is based on a combination in a mesh model. Each information layer is represented by a "raster" image, where the value of each cell is equal to a sensitivity level to erosion for the considered parameter. This level of sensitivity is represented by a value defined previously and also different for each parameter. All of these images constitute a multi-varied space. The data are crossed by multiplying values of the five factors R, K, LS, C and P in the "raster calculator" of ArcGIS Spatial Analyst module. The result of this multiplication is a "raster" layer giving the amount of potentially erodible soil in $t{\cdot}ha^{-1}{\cdot}yr^{-1}$ over the entire study area. The values obtained are grouped into classes for the purposes of map legibility (Figure 7).

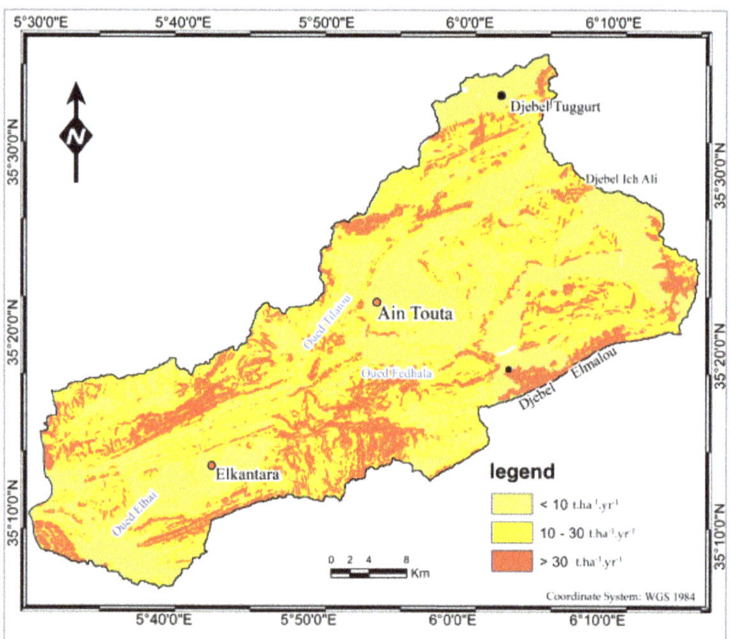

Figure 7. The erosion map of Oued el-Hai watershed presenting three main classes of soil loss rate and their geographic distribution.

The adopted classification is based on the soil's tolerance to loss. It assumes that on average, the soils can tolerate losses of up to 10 $t{\cdot}ha^{-1}{\cdot}yr^{-1}$ while allowing a high level of agricultural production. Above 30 $t{\cdot}ha^{-1}{\cdot}yr^{-1}$, the loss is high and the soils are severely degraded, which can adversely affect production. In the Oued el-Hai watershed, the results obtained are organized into three classes (see Table 5).

Table 5. Soil loss classes mainly presented in the study area.

Class	Area (km^2)	Percentage (%)
Less than 10 $t{\cdot}ha^{-1}{\cdot}yr^{-1}$	750.3	45.2
From 10 to 30 $t{\cdot}ha^{-1}{\cdot}yr^{-1}$	524.5	31.6%
Greater than 30 $t{\cdot}ha^{-1}{\cdot}yr^{-1}$	385.1	23.2%
Total	1660	100

Class 1. The soil loss rate is less than 10 $t{\cdot}ha^{-1}{\cdot}yr^{-1}$ and covers 45.2% of the total area of the watershed. It occupies the plains of Ain Touta and El Kantara because of the low slope and is also located under the dense and moderately dense forests of Jebel Tuggurt and Jebel Garouaou.

Class 2. The loss is 10 to 30 $t{\cdot}ha^{-1}{\cdot}yr^{-1}$, and it represents 31.6% of the catchment area. It concerns sectors with steep slopes and weak vegetation, as also, on the reliefs foothills, transition sectors between steep slopes and lowlands. It mainly extends on the cereal-oriented highlands.

Class 3. The loss rate is greater than 30 $t{\cdot}ha^{-1}{\cdot}yr^{-1}$, and covers 23.2% of the total area of the watershed. It concerns, in particular, the banks of watercourses and abrupt sectors devoid of plant cover (bare land).

Average loss by water erosion in aquifers for all the homogeneous units is about 16.69 $t{\cdot}ha^{-1}{\cdot}yr^{-1}$. The erosion rates differ from one area to another of the watershed, depending on the influence of different factors that control the erosion process. However, the minimum potential erosion rate values are mainly located on the plain of Ain Touta and lower than El Kantara because of the low slope. For some forests, such as Jebel Tuggurt and Jebel Garoua, despite the value of the K factor and the low value of the plant cover factor, their potential erosion rate remains high. The extreme high erosion rate values are mainly located in the mountainous areas with quite steep slopes. They are not agricultural lands, mostly bare soils with very low vegetation cover (Figure 8). There is some information on the results of USLE/RUSLE application in some regions in Algeria but they are general, and cannot be compared due to the great difference in the physical, climate and soil characteristics. Algeria is a wide country from east to west and from north to south. Concerning our study area, no previous studies have been made yet that uses this application in particular or the erosion in general. The current study is the first one on this region, which uses the RUSLE approach. Therefore, we could not compare our results.

The above study is helpful to researchers and planners for better planning to control the loss of soil in the high severity zones. The results obtained from the study are significant in recognizing and obtaining a complete understanding of risks related to the erosion of soil for the study area: not only the risks, but also figuring out the dominant and significant factors contributing to soil erosion. The factors, such as land use, land cover, and topography of the area, and their contribution to soil erosion, can be analyzed.

The management of natural resources and planning/policymaking can use the results of this study to reduce the land degradation in this region. The assessment of potential risk can be done in the regions of complex topography and high precipitation/rainfall rates. This will assist in recognizing priority areas for imposing plans and precautionary measures to resolve the conflict of erosion and other issues related to land degradation.

Figure 8. Some forms of water erosion in Oued el-Hai watershed. (**1–4**) Bad lands; (**5**) ravine on clayey soil (regressive erosion) in Ain Touta; (**6–8**) Underminingof the banks of Oued Tilatou and Oued Ain Touta, respectively. (Photos taken by authors).

5. Conclusions

The work of this investigation exposes the application results of the universal soil loss equation using a geographic information system in the Oued el-Hai watershed. The results show that the watershed is highly prone to and severely affected by soil erosion. If the rates of soil erosion continue at the same rate, it will probably cause severe land degradation. The watershed is losing an average of 16.69 $t \cdot ha^{-1} \cdot yr^{-1}$. This value corresponds to the not-tolerated erosion by the soils, which undergo an aggressive climate with low precipitation, but with stormy character. Located in time and space, they do not allow pedological changes to compensate for the soil loss. The percentage of the area under the influence of very high soil erosion (i.e., 23.2%) is an alarming indicator that this region needs to be addressed properly in the matter of soil erosion.

This serious situation is favored by the other factors of erosion, which also combine to accelerate erosion, such as steep slopes, very erodible soils and the alarming degradation of the plant cover. However, integrating the model into a GIS has many advantages. It allows to rationally manage a multitude of qualitative and quantitative data relating to the various factors of soil degradation and to establish a synthetic map of the sensitivity degrees distribution to erosion in different areas of the watershed.

Although the validity of USLE soil losses is subject to debate (it only applies to layer erosion and the estimated loss amounts do not include losses from other types of erosion (linear, solifluxion, etc.), the method provides significant help to policymakers and planners to plan erosion control interventions, and to take precautionary measures according to the obtained values of soil erosion, especially in areas where layer erosion predominates over linear erosion. It also allows to choose suitable cultivation techniques and anti-erosion methods, considering the risk nature (low, high or very high) and to monitor the impact of land use and development. The integration of GIS with the RUSLE model technique can be used in other areas of Algeria, in order to estimate and monitoring soil erosion on wider scales. In addition, the use of other methods, such as the analytic hierarchy process (AHP), might aid in ameliorating and comparing the results in our study area.

Author Contributions: Aida Bensekhria, conceiving and designing the study; Aida Bensekhria, data collection; Rabah Bouhata, methodology, and investigation; Aida Bensekhria, software and mapping; Aida Bensekhria and Rabah Bouhata, analysis and interpretation; Aida Bensekhria and Rabah Bouhata, writing—original draft preparation; Rabah Bouhata, review and editing. All authors provided critical review and final approval of the manuscript. All authors have read and agreed to the published version of the manuscript.

Funding: This research received no external funding.

Institutional Review Board Statement: Not applicable.

Informed Consent Statement: Not applicable.

Data Availability Statement: Not applicable.

Acknowledgments: The authors are deeply grateful to the unknown reviewers and the editors for their critical reading of the manuscript.

Conflicts of Interest: The authors declare no conflict of interest.

References

1. FAO; ITPS. *État des Ressources en Sols du Monde—Résumé Technique*; Organisation des Nations Unies pour L'alimentation et L'agriculture et Groupe Technique Intergouvernemental sur les Sols: Rome, Italy, 2015.
2. FAO. *GLADIS—Global Land Degradation Information System*; Food and Agriculture Organization of the United Nations: Rome, Italy, 2015.
3. Keesstra, S.D.; Bouma, J.; Wallinga, J.; Tittonell, P.; Smith, P.; Cerdà, A.; Montanarella, L.; Quinton, J.N.; Pachepsky, Y.; van der Putten, W.H.; et al. The significance of soils and soil science towards realization of the United Nations Sustainable Development Goals. *Soil* **2016**, *2*, 111–128. [CrossRef]
4. Meddi, M.; Boukhari, Y.; Morsli, B. Etude du Ruissellement Et Du Transport Solide Dans Les Monts De Beni-Chougrane, Algérie: Utilisation De La Simulation De Pluie. In Proceedings of the Symposium S1 Held during the Seventh IAHS Scientific Assembly, Foz do Iguaçu, Brazil, 4–9 April 2005; Volume 291, pp. 21–28.
5. Morsli, B.; Habi, M.; Meddi, M. Dynamique de l'érosion en zone méditerranéenne algérienne: Facteurs explicatifs de variation du ruissellement et de l'érosion sous différentes occupations du sol. *Rev. Sci. L'eau* **2013**, *26*, 89–105. [CrossRef]
6. Touaibia, B. Problématique de l'érosion et du transport solide en Algérie septentrionale. *Sécheresse* **2010**, *21*, 1–6. [CrossRef]
7. Mazour, M.; Roose, E. Influence de la couverture végétale sur le ruissellement et l'érosion des sols sur parcelles d'érosion dans des bassins versants du Nord-Ouest de l'Algérie. *Bull. Réseau Eros.* **2002**, *21*, 320–330.
8. Arar, A.; Chenchouni, H. A "simple"geomatics-based approach for assessing water erosion hazard at mountain areas. *Arab J. Geosci.* **2014**, *7*, 1–12. [CrossRef]
9. Selmi, K.; Khanchoul, K. Sediment load estimation in the Mellegue catchment, Algeria. *J. Water Land Dev.* **2016**, *31*, 129–137. [CrossRef]
10. Chebbani, R.; Djilli, K.; Roose, E. Étude des risques d'érosion dans le bassin versant de l'Isser. Algérie. *Bull. Réseau Eros.* **2009**, *19*, 85–95.

11. Djoukbala, O.; Mazour, M.; Hasbaia, M.; Benselama, O. Estimating of water erosion in semi arid regions using RUSLE equationunder GIS environment. *Environ. Earth Sci.* **2018**, *77*, 345. [CrossRef]
12. Belarbi, F.; Bouchelkia, H.; Remini, B.; Benmansour, A. Quantification and study of monthly variation of suspendedsedimentloads in Tafna basin—Algeria. *J. Water Land Dev.* **2018**, *37*, 29–38. [CrossRef]
13. Remini, B. L'envasement des barrages. *Bull. Réseau Eros.* **2000**, *20*, 165–171.
14. Demmak, A. Contribution à l'étude de l'érosion et des Transports Solides en Algérie. Ph.D. Thesis, Université de Paris VI, Paris, France, 1982; p. 323.
15. Hallouz, F.; Meddi, M.; Mahé, G.; Alirahmani, S.; Keddar, A. Modeling of discharge and sediment transport through the SWAT model in the basin of Harraza (Northwest of Algeria). *Water Sci.* **2018**, *32*, 79–88. [CrossRef]
16. Touahir, S.; Asri, A.; Remini, B.; Saad, H. Prédiction de l'érosion hydrique dans le bassin versant de l'oued Zeddine et de l'envasement du barrage OuledMellouk (Nord-Ouest algérien). *Géomorphologie* **2018**, *24*, 167–182. [CrossRef]
17. De Jong, S.M.; Paracchini, M.L.; Bertolo, F.; Folving, S.; Megier, J.; de Roo, A.P.J. Regional assessment of soil erosion using the distributed model SEMMED and remotely sensed data. *Catena* **1999**, *37*, 291–308. [CrossRef]
18. Morgan, R.P.C. A simple approach to soil loss prediction: A revised Morgan–Morgan–Finney model. *Catena* **2001**, *44*, 305–322. [CrossRef]
19. Boardman, J. Soil erosion science: Reflections on the limitations of current approaches. *Catena* **2006**, *68*, 73–86. [CrossRef]
20. Neitsch, S.L.; Arnold, J.G.; Kiniry, J.R.; Williams, J.R. *Soil and Water Assessment Tool—Theoretical Documentation Version 2009*; Texas Water Resources Institute: College Station, TX, USA, 2011.
21. Morgan, R.P.C.; Quinton, J.N.; Smith, R.E.; Govers, G.; Poesen, J.W.A.; Auerswald, K. The European soil erosion model (EUROSEM): A process-based approach for predicting soil loss from fields and small catchments. *Earth Surf. Process. Landf.* **1998**, *23*, 527–544. [CrossRef]
22. Wischmeier, W.H.; Smith, D.D. *Predicting Rainfall Erosion Losses—A Guide to Conservation Planning, Agriculture Handbook*; No. 537; USDA/Science and Education Administration 1978; US. Govt. Printing Office: Washington, DC, USA, 1978.
23. Renard, K.; Foster, G.; Weesies, G.; McCool, D.; Yoder, D. *Predicting Soil Erosion by Water: A Guide to Conservation Planning with the Revised Universal Soil Loss Equation (RUSLE). Agriculture Handbook 703*; Department of Agriculture, Agricultural Research Service: Washington, DC, USA, 1997.
24. Alewell, C.; Borrelli, P.; Meusburger, K.; Panagos, P. Using the USLE: Chances, challenges and limitations of soil erosion modeling. *Int. Soil Water Conserv. Res.* **2019**, *7*, 203–225. [CrossRef]
25. Mihi, A.; Benarfa, N.; Arar, A. Assessing and mapping water erosion-prone areas in northeastern Algeria using analytic hierarchy process, USLE/RUSLE equation, GIS, and remote sensing. *Appl. Geomat.* **2019**, *12*, 179–191. [CrossRef]
26. Radoane, M.; Vespremeanu-Stroe, A. *Landform Dynamics and Evolution in Romania*; Springer Nature: Cham, Switzerland, 2017; pp. 371–396.
27. Markose, V.J.; Jayappa, K.S. Soilloss estimation and prioritization of sub-watersheds of Kali River basin, Karnataka, India, using RUSLE and GIS. *Environ. Monit. Assess.* **2016**, *188*, 1–16. [CrossRef]
28. Benchettouh, A.; Kouri, L.; Jebari, S. Spatial estimation of soil erosion risk using RUSLE/GIS techniques and practices conservation suggested for reducing soil erosion in Wadi Mina watershed (northwest, Algeria). *Arab. J. Geosci.* **2017**, *10*, 79. [CrossRef]
29. Imamoglu, A.; Dengiz, O. Determination of soil erosion risk using RUSLE model and soil organic carbon loss in Alaca catchment (Central Black Searegion, Turkey). *Rend. Lince* **2016**, *28*, 11–23. [CrossRef]
30. Erdoğan, M.; Esbah, H.; Berberoglu, S.; Garzia, F.; Guarascio, M.; Maestas, F. Erosion risk mapping usingruslewith GIS: Case study of BüyükMenderes river basin of Turkey. *Int. J. Saf. Secur. Eng.* **2016**, *6*, 132–140. [CrossRef]
31. Bou-Imajjane, L.; Belfoul, M.A. Soil Loss Assessment in Western High Atlas of Morocco: Beni Mohand Watershed Study Case. *Appl. Environ. Soil Sci.* **2020**, *2020*, 6384176. [CrossRef]
32. Elaloui, A.; Marrakchi, C.; Fekri, A. USLE-based assessment of soil erosion by water in the watershed upstream Tessaoute (Central High Atlas, Morocco). *Model Earth Syst. Environ.* **2017**, *3*, 873–885. [CrossRef]
33. Fernández, C.; Vega, J.A. Evaluation of RUSLE and PESERA models for predicting soil erosion losses in the first year after wildfire in NW Spain. *Geoderma* **2016**, *273*, 64–72. [CrossRef]
34. Khemiri, K.; Jebari, S. Évaluation de l'érosion hydrique dans des bassins versants de la zone semi-aride tunisienne avec les modèles RUSLE et MUSLE couplés à un Système d'information géographique. *Cah. Agric.* **2021**, *30*, 7. [CrossRef]
35. Rellini, I.; Scopesi, C.; Olivari, S.; Firpo, M.; Maerker, M. Assessment of soil erosion risk in a typical Mediterranean environmentusing a high resolution RUSLE approach (Portofinopromontory, NW-Italy). *J. Maps* **2019**, *15*, 356–362. [CrossRef]
36. Miloudi, B. Evaluation et Cartographie Du Risque D'érosion Hydrique Dans Le Bassin Versant De Oued El Hai. Master's Thesis, Université Batna 2, Batna, Algérie, 2018.
37. Teshome, A.; Halefom, A.; Teshome, M.; Ahmad, I.; Taddele, Y.; Dananto, M.; Demisse, S.; Szucs, P. Soil erosion modeling using GIS and revised universal soil loss equation approach: A case study of Guna-Tana landscape, Northern Ethiopia. *Model. Earth Syst. Environ.* **2020**, *7*, 125–134. [CrossRef]
38. Xie, H.; Zhang, Y.; Wu, Z.; Lv, T. A Bibliometric Analysis on Land Degradation: Current Status, Development, and Future Directions. *Land* **2020**, *9*, 28. [CrossRef]
39. Belay, H.T.; Malede, D.A.; Geleta, F.B. Erosion Risk Potential Assessment Using GIS and RS for Soil and Water Resource Conservation Plan: The Case of Yisir Watershed, Northwestern Ethiopia. *Agric. For. Fish.* **2020**, *9*, 11. [CrossRef]

40. Chuenchum, P.; Xu, M.; Tang, W. Estimation of Soil Erosion and Sediment Yield in the Lancang–Mekong River Using the Modified Revised Universal Soil Loss Equation and GIS Techniques. *Water* **2020**, *12*, 135. [CrossRef]
41. Atoma, H.; Suryabhagavan, K.V.; Balakrishnan, M. Soil erosion assessment using RUSLE model and GIS in Hulukawatershed, Central Ethiopia. *Sustain. Water Resour. Manag.* **2020**, *6*, 12. [CrossRef]
42. El HageHassan, H.; Charbel, L.; Touchart, L. Modélisation de l'érosion hydrique à l'échelle du bassin versant du Mhaydssé. Békaa-Liban. *VertigO* **2018**. [CrossRef]
43. Arnoldus, H.M.J. An approximation of the rainfall factor in the Universal Soil Loss Equation. In *De Boodt and Gabriels: Assessment of Erosion*; FAO Land and Water Development Division; Wiley&Sons: Chichester, UK, 1980; pp. 127–132.
44. Guettouche, M.S. *Perspective De L'application De L'équation Dite "Universelle" De Perte De Sol Wischmeier En Algérie. Analyse Théorique*; Université Des Sciences et de la Technologie Houari Boumediene: Bab-Ezzouar, Algeria, 1988.
45. Achite, M.; Meddi, M. Contribution à l'étude des courbes IDF dans la région de Chlef. In Proceedings of the Conférence Nationale sur l'eau et Environnement, Biskra, Algérie, 12–13 November 2005.
46. Sadik, A.; Bouhlassa, S.; Auajjar, J.; Faleh, A.; Macaire, J. Utilisation d'un SIG pour l'évaluation et la cartographie des risques d'érosion par l'Equation universelle des pertes en sol dans le Rif oriental (Maroc): Cas du bassin versant de l'oued Boussouab, Bulletin de l'Institut Scientifique, Rabat. *Sci. Terre* **2004**, *26*, 69–79.
47. Amin, M.; Romshoo, S.A. Comparative assessment of soil erosion modeling approaches in a Himalayan watershed. *Model. Earth Syst. Environ.* **2019**, *5*, 175–192. [CrossRef]

International Journal of *Geo-Information*

Article

Comparison of Ensemble Machine Learning Methods for Soil Erosion Pin Measurements

Kieu Anh Nguyen [1], Walter Chen [1,*], Bor-Shiun Lin [2] and Uma Seeboonruang [3,*]

1 Department of Civil Engineering, National Taipei University of Technology, Taipei 10608, Taiwan; t106429401@ntut.edu.tw
2 Disaster Prevention Technology Research Center, Sinotech Engineering Consultants, Taipei 11494, Taiwan; bosch.lin@sinotech.org.tw
3 Faculty of Engineering, King Mongkut's Institute of Technology Ladkrabang, Bangkok 10520, Thailand
* Correspondence: waltchen@ntut.edu.tw (W.C.); uma.se@kmitl.ac.th (U.S.); Tel.: +886-2-27712171 (ext. 2628) (W.C.); +66-2329-8334 (U.S.)

Abstract: Although machine learning has been extensively used in various fields, it has only recently been applied to soil erosion pin modeling. To improve upon previous methods of quantifying soil erosion based on erosion pin measurements, this study explored the possible application of ensemble machine learning algorithms to the Shihmen Reservoir watershed in northern Taiwan. Three categories of ensemble methods were considered in this study: (a) Bagging, (b) boosting, and (c) stacking. The bagging method in this study refers to bagged multivariate adaptive regression splines (bagged MARS) and random forest (RF), and the boosting method includes Cubist and gradient boosting machine (GBM). Finally, the stacking method is an ensemble method that uses a meta-model to combine the predictions of base models. This study used RF and GBM as the meta-models, decision tree, linear regression, artificial neural network, and support vector machine as the base models. The dataset used in this study was sampled using stratified random sampling to achieve a 70/30 split for the training and test data, and the process was repeated three times. The performance of six ensemble methods in three categories was analyzed based on the average of three attempts. It was found that GBM performed the best among the ensemble models with the lowest root-mean-square error (RMSE = 1.72 mm/year), the highest Nash-Sutcliffe efficiency (NSE = 0.54), and the highest index of agreement (d = 0.81). This result was confirmed by the spatial comparison of the absolute differences (errors) between model predictions and observations using GBM and RF in the study area. In summary, the results show that as a group, the bagging method and the boosting method performed equally well, and the stacking method was third for the erosion pin dataset considered in this study.

Keywords: soil erosion; erosion pin; ensemble machine learning; Shihmen Reservoir watershed; bagging; boosting; stacking

Citation: Nguyen, K.A.; Chen, W.; Lin, B.-S.; Seeboonruang, U. Comparison of Ensemble Machine Learning Methods for Soil Erosion Pin Measurements. *ISPRS Int. J. Geo-Inf.* **2021**, *10*, 42. https://doi.org/10.3390/ijgi10010042

Received: 6 December 2020
Accepted: 16 January 2021
Published: 19 January 2021

Publisher's Note: MDPI stays neutral with regard to jurisdictional claims in published maps and institutional affiliations.

1. Introduction

Soil erosion is a severe global issue affecting farming and the environment in tropical and subtropical areas. In particular, soil erosion leads to environmental damage such as soil nutrient loss, pollution by sedimentation, and the increased possibility of flooding. The rate of soil erosion depends on soil characteristics, climate, slope steepness [1], land use, and protective vegetation [2]. In addition, eroded soils lose 75% to 80% of carbon content [3], which results in a deficit of terrestrial carbon budget. Anthropogenic activity is a major cause of soil erosion [4]. Soil degradation rapidly has intensified with the rising population in the 20th century [5] and beyond.

Many theoretical/empirical models can be used to study soil erosion. According to Borrelli et al. [6], 435 distinct models and model variants were used to analyze soil erosion from 1994 to 2017 in 1697 scientific articles in the Scopus database. The top five most applied models are RUSLE (Revised Universal Soil Loss Equation), USLE (Universal Soil Loss Equation), WEPP (Water Erosion Prediction Project), SWAT (Soil and Water Assessment Tool), and WaTEM/SEDEM (Water and Tillage Erosion Model and Sediment Delivery Model). More than half of these studies (58% of 1697 articles) included validation of the research results in the form of expert knowledge, measured sediment yield, measured erosion rates, or comparison with other models.

Taiwan has numerous natural hazards, and typhoons and earthquakes occur every year. The average annual precipitation is 2500 mm, which is concentrated from May to September [7]. For this reason, a large amount of soil is eroded and transported away by flowing water. In recent years, many studies have been conducted to model soil erosion in Taiwan. Traditionally, a physically or empirically based soil erosion model is needed for such studies. For example, Fan and Wu [8] developed equations to evaluate the relationship between slope steepness, soil properties, rainfall intensity, and interrill soil erosion rate in Taiwan. Lo [9] used the agricultural nonpoint source pollution model (AGNPS) to quantify soil erosion in the Bajun River basin and Tsengwen Reservoir watershed. Alternatively, Chiu et al. [10] measured the 137Cs concentrations at 60 sampling sites in the Shihmen Reservoir watershed to estimate soil erosion. Chen et al. [11] applied the USLE model to the same watershed to study the effect of the digital elevation model (DEM) on soil erosion. Finally, Liu et al. [12] followed up by applying slope units to soil erosion modeling in the same watershed.

In contrast to the traditional soil erosion modeling approach, in which a physically or empirically based model is needed to make a prediction and field data are collected to verify the model's correctness, a machine learning (ML) based approach does not require an a priori model. The field measurements (such as those of erosion pins) are used directly to formulate rules and make generalizations from the data (i.e., predictions). Although ML-based approaches have been extensively used in relevant fields such as landslides susceptibility mapping [13,14], soil thickness prediction [15], digital soil mapping [16], and biomass retrieval [17], it has only recently been applied to soil erosion pin study [18,19]. To improve upon previous methods of quantifying soil erosion, the study explored the possible application of ensemble machine learning algorithms to the Shihmen Reservoir watershed in northern Taiwan. Three categories of ensemble methods were considered in this study: (a) Bagging, (b) boosting, and (c) stacking. The results were compared among the ensemble methods in terms of three statistical indices: (1) Root-mean-square error (RMSE), (2) Nash-Sutcliffe efficiency (NSE), and (3) index of agreement (d).

2. Methods

An erosion pin is a wooden or metal rod inserted into the ground for measuring the change of the ground surface. The pin referred to in this study is made of metal with a diameter of 15.9 mm and a length of 300 mm. About 270 mm of the pin is embedded in the ground with the exposed part painted red. The erosion pin is one of the simplest and most effective methods for monitoring ground surface variation due to soil erosion and sediment deposition [20]. It has been used to monitor sheet erosion, gully erosion, landslides, and stream bank erosion [21–23]. Lin et al. [24] has documented the procedures for installing erosion pins in Taiwan. Measurements of erosion pins have been collected from various watersheds. The Shihmen Reservoir watershed data show that the average soil erosion was 90.6 t per hectare per year [12].

Ensemble machine learning is a technique that combines several base ML models (either homogeneous or heterogeneous) to make better predictions. Note that the word "model" has a different meaning in machine learning than for soil erosion. As explained earlier, a machine learning (ML) based approach does not require an a priori (soil erosion) model to work. In addition, ensemble methods gain performance over individual ML models [25]. Ensemble methods have been applied to many diverse fields such as banking [26], big data security [27], and breast cancer diagnosis [28]. In the environment and related fields, Pham et al. [29] applied several ensemble methods (AdaBoost, Bagging, Dagging,

MultiBoost, Rotation Forest, and Random Subspace) to evaluate the landslide susceptibility in the Himalayan India. The results showed that the area under the curve (AUC) of the receiver operating characteristic (ROC) curve were all higher than 0.876. Similarly, Tehrany et al. [30] used an ensemble support vector machine (SVM) and the weights-of-evidence method to conduct the flood susceptibility mapping in Terengganu of Malaysia. Their results improved flood modeling by 29%.

An ensemble method is viewed as a compound model. The purpose of such a model is to achieve better predictive performance by reducing the noise or error between observed and predicted data. Ensemble methods are usually grouped into bootstrap aggregating (bagging), boosting, and stacking methods. All three categories attempt to tune their predictions to the observations by decreasing model variance, bias, or both simultaneously. The main difference is that bagging and boosting usually work with homogeneous models, whereas stacking excels in combining heterogeneous models. Their respective approaches are described in the following sub-sections. In this study, we will select two methods from each category for analysis and comparison.

2.1. Bagging

Bagging is a technique that builds multiple homogeneous models from different subsamples of the same training dataset to obtain more accurate predictions than its individual models. It is an application of the bootstrap procedure to high-variance machine learning problems. For example, random forest (RF) is the bagging of decision trees (DT). Using CART (classification and regression trees) as an example, RF randomly samples the training dataset multiple times (with replacement) to obtain many subsamples. Then, a decision tree is built for each subsample using CART. Finally, RF issues its prediction by combing the results of all decision trees either by voting (classification) or averaging (regression). RF is a very effective machine learning tool, and it has been applied to various research problems, including soil erosion [18]. Therefore, we chose RF and bagged MARS (multivariate adaptive regression splines) in this study to compare with other ensemble algorithms.

MARS was first introduced by Friedman and Roosen [31]. It explores the relationship between the dependent and independent variables in a way very similar to least-squares regression [32]. The advantages of MARS include its computational efficiency, its ability to yield easy to interpret models, and its function to quantify the contribution of predictor variables. However, its lack of accurate prediction is one of the most significant drawbacks [32]. To address this issue, bagging was introduced to MARS to become bagged MARS to improve the classification accuracy. The bagged MARS model was implemented by the "earth" package in R, and the RF model was implemented by the "randomForest" package.

2.2. Boosting

Boosting refers to a group of algorithms that utilize weighted averages to make weak learning algorithms stronger learning algorithms. Unlike bagging that relies on each model running independently and then aggregated at the end, boosting runs sequentially by using later models to fix the prediction errors of the predecessor models in the sequence. For this study, we have selected Cubist and gradient boosting machine (GBM) for comparison with other ensemble methods.

Cubist is a prediction-oriented regression model proposed by Quinlan [33,34]. The general idea of the Cubist regression model is briefly described here. During the tree growth stage, many branches and leaves are grown. Linear regression models are added to the leaves of the tree. The Cubist method creates a series of "if-then" rules. Each rule has an associated multivariate linear model [35]. The corresponding model is used to calculate the predicted value if a set of variables satisfies the conditions of the rule. Rules are eliminated via pruning and/or combined for simplification. The main advantage of the Cubist method is the addition of multiple training committees to balance case weights. In this study, the Cubist model was implemented by the "caret" and "Cubist" packages in R.

GBM was proposed by Friedman [36,37] as a simple and highly adaptive method for machine learning [38]. It is an improved boosting algorithm for regression and classification problems. The basic theory of GBM is to produce a prediction model constructed by a group of weak learning algorithms, typically decision trees. Each tree is grown sequentially by using the information from previously grown trees [39]. In this study, the GBM model was implemented by the "gbm" package in R.

2.3. Stacking

Stacking, sometimes called stacked generalization, is an ensemble machine learning method that combines multiple heterogeneous base or component models via a meta-model. The base model is trained on the complete training data, and then the meta-model is trained on the predictions of the base models. The advantage of stacking is the ability to explore the solution space with different models in the same problem. In this study, two stacking ensembles were chosen. They are (1) RF + DT + LM + Artificial Neural Networks (ANN) + SVM, and (2) GBM + DT + LM + ANN + SVM. For implementation, the "caretEnsemble" package in R was used.

2.4. Model Assessment

In order to evaluate the performance of ensemble models, three statistical indices were used as the evaluation criteria: (1) Root-mean-square error (RMSE), (2) Nash–Sutcliffe efficiency (NSE), and (3) index of agreement (d). These statistical indices have been frequently used in many studies [40,41].

RMSE is a good indicator for evaluating the model performance for continuous variables. In this study, RMSE represents the differences between erosion pin measurements and ensemble model predictions. It can be written as follows:

$$\mathrm{RMSE} = \sqrt{\frac{\Sigma(\mathrm{P}-\mathrm{O})^2}{\mathrm{n}}} \tag{1}$$

where P and O are the predicted and observed values, respectively.

NSE defines the relative magnitude between the "noise" and "information" [42]. Its value ranges from $-\infty$ to 1. The closer the value of NSE to 1, the more efficient the model is. In the case that NSE is negative, the model is considered poor because the observed mean serves as a better prediction than the model. NSE is defined as follows:

$$\mathrm{NSE} = 1 - \frac{\Sigma(\mathrm{P}-\mathrm{O})^2}{\Sigma(\mathrm{O}-\overline{\mathrm{O}})^2} \tag{2}$$

where $\overline{\mathrm{O}}$ is the mean observed value.

Finally, the index of agreement (d) is often used to represent the model performance [43]. Its value ranges from 0 to 1, with higher values indicating better agreement between the predictions and observations. The d index is defined as follows:

$$\mathrm{d} = 1 - \frac{\Sigma(\mathrm{P}-\mathrm{O})^2}{\Sigma\left(|\mathrm{P}-\overline{\mathrm{O}}| + |\mathrm{O}-\overline{\mathrm{O}}|\right)^2} \tag{3}$$

Note that we do not use R^2 (coefficient of determination) as a statistical index in the model evaluation. As pointed out by Nguyen et al. [18,19], R^2 evaluates the fit to the regression line only. High R^2 does not mean small differences between predictions and observations. This point is illustrated in Figure 1, in which a poor model has a perfect R^2 but only mediocre (even poor) values of other statistical indices. Good predictions should fall on the 45° line instead.

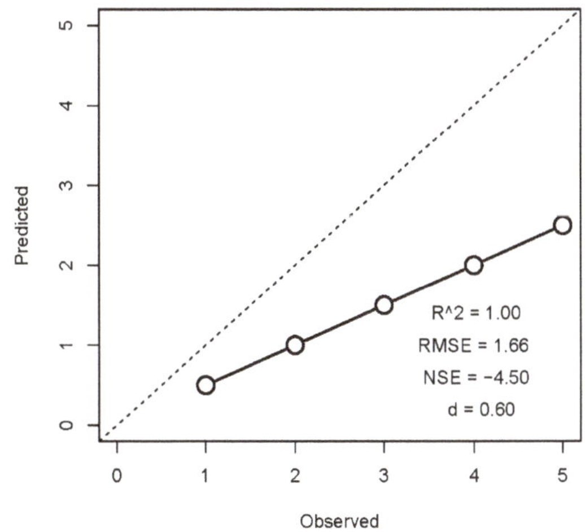

Figure 1. The comparison of R^2, root-mean-square error (RMSE), Nash–Sutcliffe efficiency (NSE), and d of a poor predictive model.

2.5. Study Area

In this study, the Shihmen Reservoir watershed in northern Taiwan was selected as the research area (Figure 2). Shihmen Reservoir watershed covers 76,340 ha of land area with a maximum elevation of 3527 m. The rainy season of the watershed coincides with the typhoon months. Therefore, heavy rainfalls are common [44]. In northern Taiwan, Shihmen Reservoir plays an essential role in providing drinking water for domestic use, irrigation for agriculture, and flood control for typhoon-related disasters [45]. Figure 2 also shows a photo of a metal erosion pin with the exposed part painted red and a picture of measurement being taken by a micrometer.

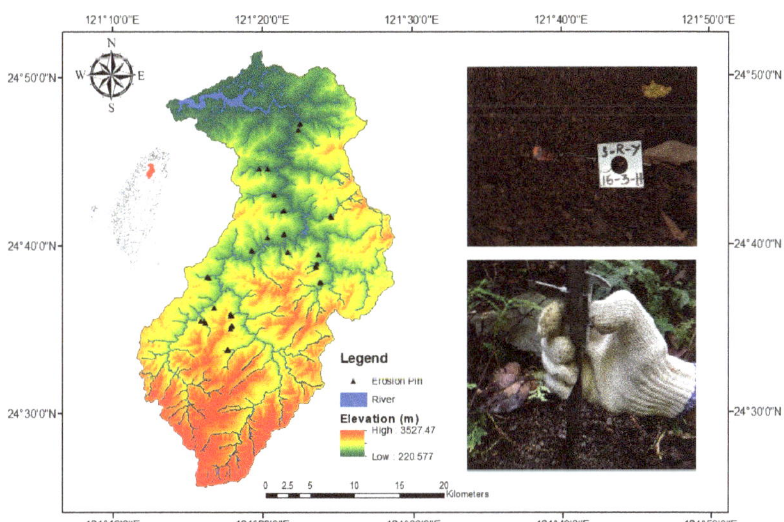

Figure 2. The study area and locations of erosion pins.

The workflow of this research is shown in Figure 3, and includes three parts: (1) Data collection, (2) data preparation, and (3) soil erosion pin analysis. During the first stage, the target variable and predictive variables were compiled into a dataset. Then, the dataset was split into training and test data using a stratified random sampling method. Finally, the training data were fed to the ensemble methods to create prediction models. The models were tested using the test data, and the statistical indices were computed.

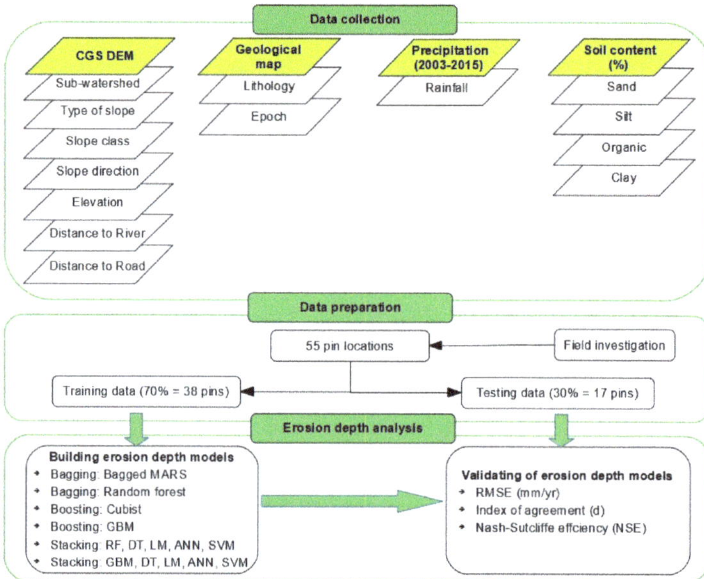

Figure 3. The soil erosion pin modeling flow chart using ensemble methods.

Fourteen attributes in four categories were used as the predictive variables, as shown in Figure 3. Five of the attributes were point data that were not available watershed-wide but only available at the erosion pins' locations. The dataset was separated into two groups using a 70/30 split, which is the common ratio used by many studies [46–48]. The entire process was repeated three times to determine the average result.

The target variable is the erosion pin measurement. A total of 550 pins were installed on 55 slopes (10 pins per slope) in the Shihmen Reservoir watershed (Figure 2). Lin et al. [24] documented the installation procedures. The erosion depth measurements were collected from 8 September 2008 to 10 October 2011. The measurements were averaged by slopes.

To derive topography-related attributes, such as elevation, sub-watershed, slope class, slope direction, distance to river, and distance to road, we used the Central Geological Survey (CGS) DEM created from an airborne LiDAR (light detection and ranging) survey, which has a spatial resolution of 10 m (Table 1). The DEM data were created in 2013 [11].

The average annual rainfall of the study area was calculated from 22 rainfall stations from 2003 to 2015. The slopes were classified into seven classes: (1) <5%, (2) 5–15%, (3) 15–30%, (4) 30–40%, (5) 40–55%, (6) 55–100%, and (7) >100%, based on the classification system of the Soil and Water Conservation Bureau. Slope direction is the aspect of a slope, which can be flat or facing north, northeast, east, southeast, south, southwest, west, or northwest. The distance to river was calculated based on the river network map. Similarly, the distance to road was found by the road network map with a scale of 1:5000. Both distances were calculated as the shortest distance between the individual erosion pin and the river or road network using ArcGIS 10.2 software.

Finally, lithology and epoch were from the geological maps of the Central Geological Survey. The scales were both 1:50,000. The soil contents were the percent of sand, silt, organic, and clay. The data were provided by Lin et al. [49].

Table 1. The predictive factors used in the modeling of soil erosion pin of the study area.

Type	Factor/Attribute	Source of Data	Scale or Resolution
CGS DEM	Sub-watershed	Central Geological Survey, Ministry of Economic Affairs of Taiwan	10 m × 10 m
	Type of slope		
	Slope class		
	Slope direction		
	Elevation		
	Distance to river		
	Distance to road		
Geological map	Lithology	Central Geological Survey	1:50,000
	Epoch		
Precipitation	Average annual rainfall	Northern Region Water Resources Office, Water Resources Agency, Ministry of Economic Affairs (2003–2015, 22 stations)	10 m × 10 m
Soil content	% sand	Lin et al. (2019)	10 m × 10 m
	% silt		
	% organic % clay		

3. Results and Discussion

The results of three types of ensemble learning (that is, bagging, boosting, and stacking) and their comparisons are presented below.

3.1. Bagging

In the bagging category, we selected bagged MARS and random forest (also used by [18]) as the representative ensemble learning methods. Three repeated samplings (groupings) of the same dataset were made to create three different 70/30 splits using the stratified random sampling method. After fitting the model, the resulting statistical metrics, including the RMSE, NSE, and d values were calculated and these are listed in Table 2. Among the three indices of bagged MARS, RMSE ranges from 0.92 to 1.83 mm/year for the training data and from 1.70 to 2.18 mm/year for the test data. In addition, NSE ($-\infty$ to 1) varies from 0.38 to 0.83 for the training data and from 0.19 to 0.60 for the test data. Finally, d (0 to 1) goes from 0.64 to 0.94 for the training data and from 0.60 to 0.85 for the test data. As expected, all three indicators are better for the training data than for the test data with no exception. The same observation can be made on the RF metrics as well.

To evaluate the relative appropriateness of bagged MARS and RF as tools for ensemble learning, we compared the RMSE values of the two ensemble methods. The results were mixed, as shown in Figure 4. RF out-performed bagged MARS two out of three times in both the training data and the test data, as shown in Figure 4a,c and was superior overall. The same conclusion was also confirmed by the lower average RMSE values of RF than bagged MARS.

If we plot the results on a Taylor diagram, we can observe further differences between RF and bagged MARS. As can be seen from Figure 4b, all three sampling (grouping) results of the RF cluster together with very similar RMSE, correlation, and standard deviation. This indicates consistent training results across different samples. By contrast, only two of the sampling results of bagged MARS cluster together. The third is very far away with a substantially smaller correlation and standard deviation and much larger RMSE than the other two. This shows that bagged MARS did not work equally well with different training data. In this case, it had difficulties fitting a model to grouping (sampling) #3. Hence, it lost to RF and fell short in the statistical comparison. Figure 4d shows a Taylor diagram comparison between the test datasets. A larger spread is observed in the data for both bagged MARS and RF. However, notice the closeness of data points of the same color. This shows that bagged MARS and RF generated similar predictions on the same test dataset. For instance, although the green triangle and green circle are very far apart in Figure 4b, they are relatively close to each other in Figure 4d.

Table 2. Statistical metrics of bagging ensemble models (bagged multivariate adaptive regression splines (MARS) and random forest (RF)).

Model	Statistical Index	Grouping 1		Grouping 2		Grouping 3		Average	
		Training	Test	Training	Test	Training	Test	Training	Test
bagged MARS	RMSE	0.92	1.93	1.00	1.70	1.83	2.18	1.25	1.94
	NSE	0.83	0.42	0.79	0.60	0.38	0.19	0.67	0.40
	d	0.94	0.78	0.93	0.85	0.64	0.60	0.84	0.74
RF	RMSE	0.93	1.68	0.89	1.77	0.97	1.79	0.93	1.75
	NSE	0.83	0.56	0.83	0.57	0.82	0.45	0.83	0.53
	d	0.94	0.82	0.93	0.85	0.93	0.72	0.93	0.80

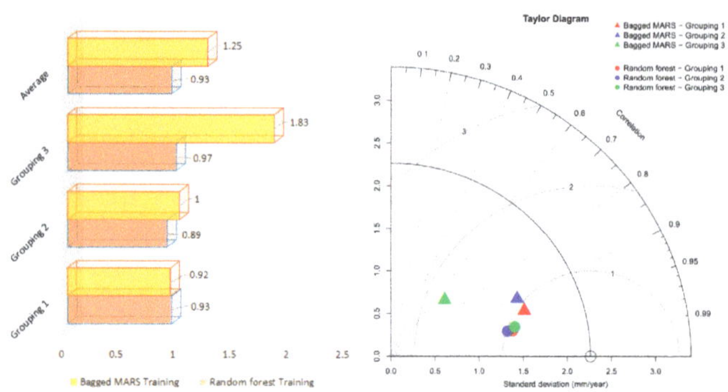

(**a**) Training data (mm/year) (**b**) Training data (mm/year)

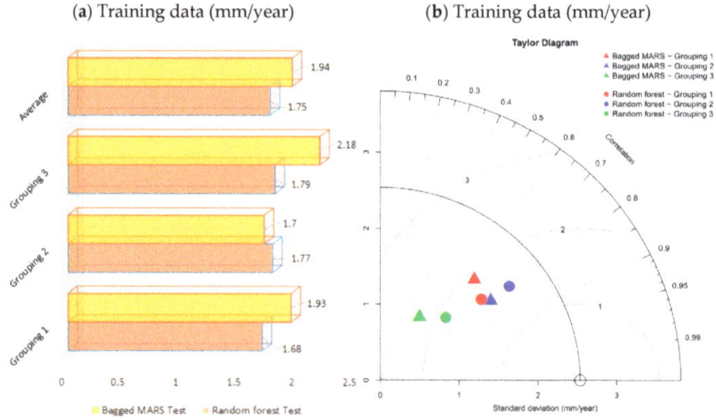

(**c**) Test data (mm/year) (**d**) Test data (mm/year)

Figure 4. The comparison of RMSE (mm/year), standard deviation, and correlation between bagged MARS and random forest.

3.2. Boosting

Two boosting ensemble methods were explored in this study. They were the GBM and Cubist models. Table 3 shows the performance metrics (RMSE, NSE, and d) under three different samplings (groupings) of the erosion pin data. For groupings #1 and #2, GBM won against Cubist in every category (lower RMSE, higher NSE, and higher d values). However, for grouping #3, the result is mixed. On the one hand, GBM lost to Cubist in all three indices concerning the training data. On the other hand, GBM beat Cubist in the test data by winning two of the three indices. The RMSE of GBM is lower than that of Cubist (1.71 vs. 1.98). Similarly, the NSE of GBM is higher than that of Cubist (0.50 vs. 0.33). However, GBM's d value is lower than Cubist's (0.75 vs. 0.78). This is one of the rare instances where NSE and d do not agree with each other. Putting everything together, we can conclude that GBM is still superior to Cubist.

Table 3. Statistical metrics of boosting ensemble models (gradient boosting machine (GBM) and Cubist).

Model	Statistical Index	Grouping 1		Grouping 2		Grouping 3		Average	
		Training	Test	Training	Test	Training	Test	Training	Test
GBM	RMSE	0.19	1.86	0.53	1.59	1.11	1.71	0.61	1.72
	NSE	0.99	0.46	0.94	0.65	0.77	0.50	0.90	0.54
	d	1.00	0.81	0.98	0.87	0.91	0.75	0.96	0.81
Cubist	RMSE	0.69	2.08	0.94	1.78	0.88	1.98	0.84	1.95
	NSE	0.91	0.33	0.81	0.56	0.86	0.33	0.86	0.41
	d	0.97	0.70	0.93	0.83	0.95	0.78	0.95	0.77

Figure 5 visually compares the average RMSE (mm/year) values between GBM and Cubist. For the training data, the RMSE of Cubist is 0.84 mm/year, inferior to the 0.61 mm/year of GBM. For the test data, a similar result is obtained. The RMSE of Cubist is 1.95 mm/year, not as good as the 1.72 mm/year of GBM. As a result, it can be confirmed that GBM is the better performing model in the category of boosting ensemble methods.

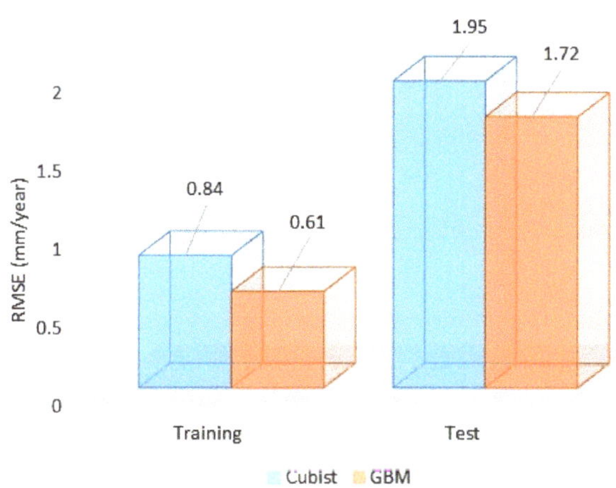

Figure 5. Comparison of the average RMSE (mm/year) values between Cubist and GBM.

3.3. Stacking

Stacking is the third and final category of ensemble methods examined in this study. The basic idea of stacking is to combine several weak models together to use their predictions as attributes in an overall meta-model. The meta-model is trained to yield better predictions than the base models (component models). Using the same six ML models studied by Nguyen et al. [18,19], we created Table 4 and classified the ML models into four categories: Tree model, neural network model, hyperplane model, and linear regression model. The average RMSE values of these models are also shown in Table 4. Based on Table 4, we picked the four weakest models (one from each category) and used them as the base models in the stacking ensemble learning. They are DT, ANN, SVM, and LM. Furthermore, RF and GBM, one from bagging and the other from boosting, were chosen as the meta-models to form two sets of stacking models: (1) RF + DT + LM + ANN + SVM and (2) GBM + DT + LM + ANN + SVM. All of the statistical indices were computed in R in this study.

Table 4. The comparison of four types of machine learning (ML) models by the average RMSE (mm/year) values.

Type of Model	Model	Average RMSE (mm/year)	
		Training	Test
Tree model	Decision tree	1.73	2.45
	Random forest	0.93	1.75
Neural network model	ANN	1.23	2.36
	ANFIS	0.01	2.05
Hyperplane model	SVM	1.43	2.61
Linear regression model	LM	1.25	3.47

The results of RF-stacking (RF + DT + LM + ANN + SVM) and GBM-stacking (GBM + DT + LM + ANN + SVM) are shown in Table 5. For the training datasets, GBM-stacking outperformed RF-stacking in groupings #1 and #3 in all three statistical indices (RMSE, NSE, and d). In grouping #2, GBM-stacking also won against RF-stacking in terms of RMSE (1.45 vs. 1.46), but lost to RF-stacking in terms of d values (0.79 vs. 0.83). This is a rare case that an inconsistency between RMSE and d is observed.

Table 5. Statistical metrics of stacking ensemble models (RF + DT + LM + ANN + SVM and GBM + DT + LM + ANN + SVM).

Model	Statistical Index	Grouping 1		Grouping 2		Grouping 3		Average	
		Training	Test	Training	Test	Training	Test	Training	Test
RF + DT + LM + ANN + SVM	RMSE	1.38	2.04	1.46	2.48	1.70	1.94	1.51	2.15
	NSE	0.63	0.35	0.55	0.15	0.46	0.35	0.55	0.29
	d	0.87	0.74	0.83	0.65	0.72	0.69	0.81	0.69
GBM + DT + LM + ANN + SVM	RMSE	1.34	2.12	1.45	2.67	1.62	1.97	1.47	2.26
	NSE	0.65	0.30	0.55	0.02	0.51	0.33	0.57	0.22
	d	0.87	0.71	0.79	0.53	0.76	0.66	0.81	0.64

As for the test datasets, the best-performing model is reversed. In all three different groupings (samplings), RF-stacking out-performed GBM-stacking in all statistical indices with no exception. In summary, although GBM-stacking performed best with the training data, it lost the test data. Since the predictive performance of a model is based on unseen test data, we conclude that RF-stacking is the better stacking model of the two. The visual comparison between RF-stacking and GBM-stacking is shown in Figure 6, where all three indices were plotted (RMSE, NSE, and d). The numbers in the figure represent the average values of three different groupings (partitioning).

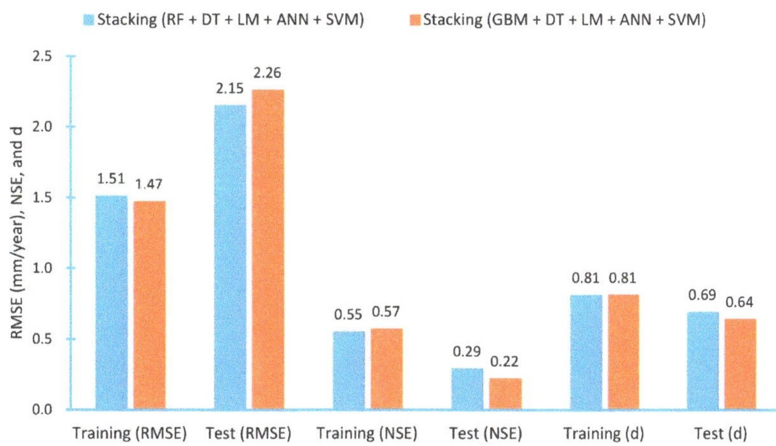

Figure 6. The comparison of RMSE (mm/year), NSE, and d between RF-stacking and GBM-stacking.

3.4. Comparison of Ensemble Models

So far, we have compared six ensemble methods in three categories and determined the best-performing model for each category. The next question is how they compare with one another in a six-way comparison. The results were compiled in Figure 7 using the RMSE values. Among the six ensemble models examined in this study, if we only consider the training data, the overall best performing model is GBM (0.61 mm/year), followed by Cubist (0.84 mm/year). Both of them belong to the boosting category. Their average d values are as high as 0.96 and 0.95, respectively. However, if we only focus on the test data, although the overall winning model is still GBM (1.72 mm/year), RF (1.75 mm/year) will replace Cubist (1.95 mm/year) as the second best-performing model. If NSE and d are considered instead of RMSE, the conclusion does not change. In those cases, GBM and RF remain the two best ensemble models. Based on Figure 7, we rank the models from the best to the worst as follows:

1. Training: GBM > Cubist > RF > bagged MARS > Stacking (GBM) > Stacking (RF)
2. Test: GBM > RF > bagged MARS > Cubist > Stacking (RF) > Stacking (GBM)

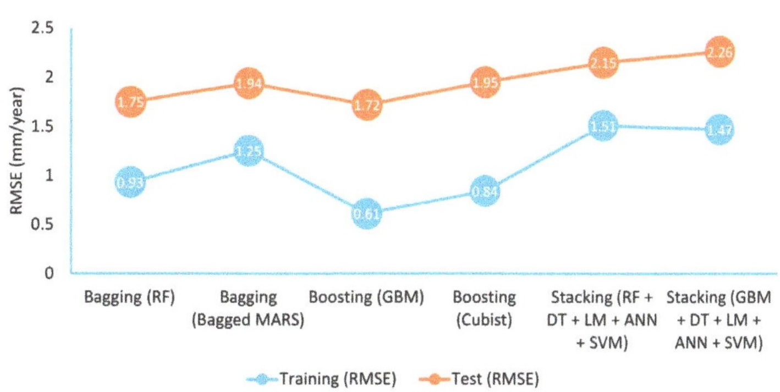

Figure 7. The comparison of ensemble models by the average RMSE (mm/year) values.

For machine learning, it is more appropriate to judge ML models' performance by the unseen test data. Therefore, for the test data, GBM (boosting) is the overall winner, followed by RF (bagging). The next best-performing models are bagged MARS (bagging) and Cubist (boosting). Clearly, the top four places were split evenly between the bagging method and the boosting method. Hence, these two types of ensemble models, bagging and boosting, rival each other in performance and are equally good in predicting soil erosion depths measured by erosion pins. By contrast, the stacking ensemble methods (RF + DT + LM + ANN + SVM and GBM + DT + LM + ANN + SVM) seem to lag behind both the bagging and boosting methods. We found this result very intriguing. Despite GBM and RF being the top-two performing models, they do not perform as well when they are used in the stacking approach to combine other weaker models. Perhaps this is because GBM and RF worked with 14 attributes directly when used individually as single models. When GBM and RF were used as meta-models in the stacking approach, they were only trained on the predictions of the base models (DT, LM, ANN, and SVM). Not being able to work with the underlying 14 attributes directly seems to have undermined the ability of GBM and RF to make better predictions.

3.5. Model Predictions and Factor Importance

Because GBM and RF are the two best ensemble models in a class of their own (RMSE = 1.72 and 1.75 mm/year, respectively), we only display their respective ensemble learning results in Figures 8 and 9. As shown in the figures, the size and color of the circles represent the absolute differences (errors) between model predictions and observations ($|Obs - Pre|$). Proportional symbols were used. Therefore, the larger the circle, the bigger the difference. Similarly, the redder the dot color, the more significant the error. The contrasting results produced by GBM and RF are evident. For GBM, it is clear from Figure 8 that most points have low error except for those in the eastern part of the study area. By contrast, Figure 9 shows that RF has large errors in both the eastern and southern parts of the study area. To better visualize the error distribution, we further mapped the spatially interpolated values (absolute errors) throughout the study area. As shown in Figure 8, the resulting watershed is mostly green (associated with low error) for GBM, except in the eastern part, where large errors due to the steep topography and undesirable slope directions bring a return to red colors. However, according to Figure 9, the watershed is only in green in the northern region for RF. The rest of the watershed is colored red or yellow.

Furthermore, we created two vertical profiles on the map, one in an east-west direction and the other in a north-south direction, to compare the model predictions and observations for both GBM and RF. As shown in Figure 8, the black line is the observation, and the blue line is the GBM prediction. Both lines show a similar trend and move in the same direction: If the observation increases so does the prediction, and if the observation decreases so does the prediction. However, it seems evident that GBM tends to under-estimate the high values of observation and over-estimate the low values of observation in both the north-south and the east-west profiles. Furthermore, if we plot the same profiles in Figure 9 and use a red line to represent the RF prediction, we can see a similar result. The RF prediction also moves in the same direction as the observation. The RF model also underestimates the high values of observation and overestimates the low values of observation in the north–south and the east–west profiles. However, the difference is that RF errors more in comparison with GBM.

In summary, both GBM and RF perform better in the watershed's northern region, where the reservoir is located. Since the watershed slopes from south to north (as shown previously in Figure 2), it implies that the models more adequately captured the erosion behavior (measured by erosion pins) at the lower elevations than at the higher elevations. The eastern part is problematic. Neither of the two models works well here. In general, GBM outperformed RF because GBM matches the available measurements better than RF in the west and south. This makes GBM the overall best model, which is consistent with the RMSE-based results.

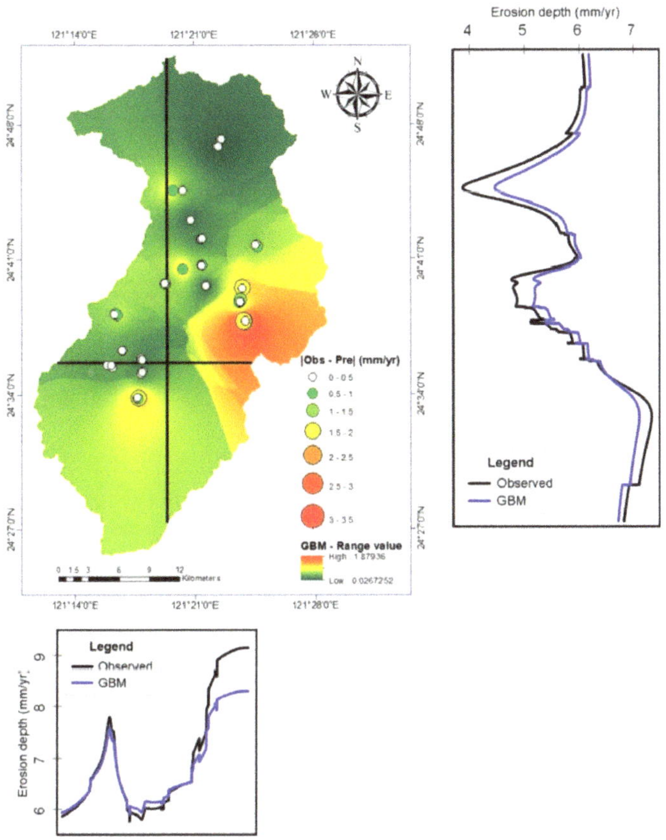

Figure 8. Map and profile of the study area showing the absolute difference between model prediction (GBM) and observation (mm/year).

Both GBM and RF generate a rank of importance for the 14 attributes used in this study. The six simulations (three of GBM and three of RF) are combined in a boxplot to illustrate the range of ranks (factor importance) as shown in Figure 10. Several points in this figure merit a closer look. First, the grey box in the figure shows the range of ranks between the first and third quartiles of the results. It is evident that each attribute has a variable range of ranks. Second, the black line in the box shows the median rank (1 being the most important and 14 being the least important). The median rank and the gray box can be used to compare the relative importance of attributes. Therefore, it can be seen from Figure 10 that A (slope direction) and B (type of slope) are the two overall most important attributes in the GBM and RF models. They consistently rank higher than the other 12 attributes. Attribute D (elevation) is interesting. It has the fourth-lowest median rank in the comparison. At the same time, it also has the longest grey box and the broadest range than any of the other attributes. This means that the elevation has changeable importance depending on the model and dataset. In some instances, the elevation is very important, whereas in others, it is not.

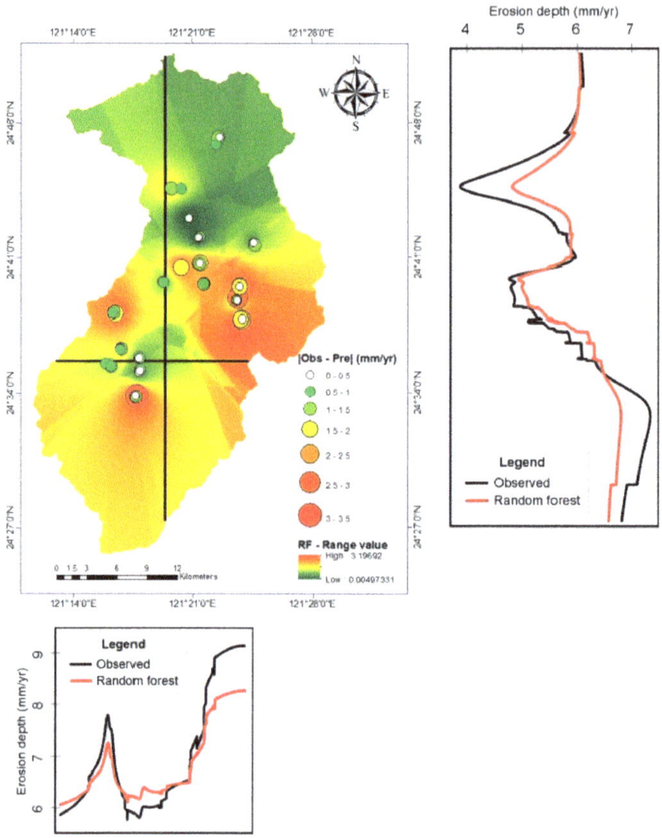

Figure 9. Map and profile of the study area showing the absolute difference between model prediction (RF) and observation (mm/year).

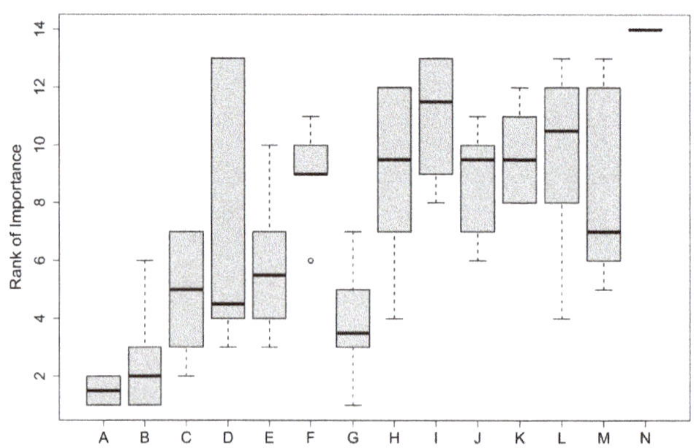

Figure 10. The relative importance of the 14 attributes used in this study. The lower the number, the higher the importance of the attribute. (A: Slope direction, B: Type of slope, C: % Organic, D: Elevation, E: Distance to road, F: % Sand, G: Subwatershed, H: % Clay, I: Slope class, J: Lithology, K: Distance to river, L: Average annual rainfall, M: % Silt, N: Epoch).

4. Conclusions

Soil erosion is a significant threat to the environment and the livelihood of the region, and should be taken seriously to mitigate the disastrous consequences. Hence, a precise spatial prediction of soil erosion is a critical need. In this study, we applied six ensemble learning methods (bagged MARS, RF, GBM, Cubist, RF-stacking, and GBM-stacking) of three categories (bagging, boosting, and stacking) to model the measurements of erosion pins in the Shihmen Reservoir watershed of Taiwan. The purpose was to improve the modeling accuracy and forecasting capability related to soil erosion. In the process of assessing the performance of the three types of ensemble approaches, we have learned how effective the ensemble methods are with unseen test data.

The study results show that the ensemble methods improve the prediction accuracy as measured by three statistical indices—RMSE, NSE, and d. Among the three categories of ensemble methods, bagging and boosting work equally well on the unseen test data. Stacking is the least favorable approach, with its RMSE trailing behind other types of ensemble algorithms. Individually speaking, we found GBM to be the best fitting model. Its RMSE, NSE, and d values are 1.72 mm/year, 0.54, and 0.81, respectively. The second-best model is RF. We used both GBM and RF to map the absolute differences (errors) between model predictions and observations ($|Obs - Pre|$) in the study area. The results show that both models perform well in the watershed's northern region (where the reservoir is located) and perform relatively poorly in the watershed's eastern part due to the steep topography and undesirable slope directions. What makes GBM superior to RF is that GBM also works well in the western and southern parts of the study area while RF does not. This conclusion is consistent with the RMSE-based results.

Finally, as an additional discovery in the study, we noticed two cases of inconsistent statistical indices during the model comparison. One of them happened when we compared GBM with Cubist (test data of grouping #3). The other occurred when we examined the performance of the two stacking models (training data of grouping #2). In both of these cases, RMSE and NSE favored one model, but d preferred the other. Therefore, potentially contradictory conclusions could be made if we only rely on one single index. This danger exemplifies the need to present multiple indices in such studies.

Author Contributions: Conceptualization, Walter Chen; data curation, Bor-Shiun Lin; formal analysis, Kieu Anh Nguyen and Walter Chen; funding acquisition, Walter Chen and Uma Seeboonruang; investigation, Kieu Anh Nguyen and Walter Chen; methodology, Walter Chen; project administration, Walter Chen and Uma Seeboonruang; resources, Walter Chen and Bor-Shiun Lin; software, Kieu Anh Nguyen; supervision, Walter Chen and Uma Seeboonruang; visualization, Kieu Anh Nguyen; writing—original draft, Kieu Anh Nguyen and Walter Chen; writing—review and editing, Walter Chen, Bor-Shiun Lin, and Uma Seeboonruang. All authors have read and agreed to the published version of the manuscript.

Funding: This study was partially supported by the National Taipei University of Technology-King Mongkut's Institute of Technology Ladkrabang Joint Research Program (Grant Numbers NTUT-KMITL-106-01, NTUT-KMITL-107-02, and NTUT-KMITL-108-01) and the Ministry of Science and Technology (Taiwan) Research Project (Grant Number MOST 109-2121-M-027-001).

Institutional Review Board Statement: Not applicable.

Informed Consent Statement: Not applicable.

Data Availability Statement: Data sharing is not applicable to this article.

Conflicts of Interest: The authors declare no conflict of interest.

References

1. Ruiz-Sinoga, J.D.; Martínez-Murillo, J.F. Hydrological response of abandoned agricultural soils along a climatological gradient on metamorphic parent material in southern Spain. *Earth Surf. Process. Landf.* **2009**, *34*, 2047–2056. [CrossRef]
2. García-Ruiz, J.M. The effects of land uses on soil erosion in Spain: A review. *Catena* **2010**, *81*, 1–11. [CrossRef]
3. Morgan, R.P.C. *Soil Erosion and Conservation*; John Wiley & Sons: Hoboken, NJ, USA, 2009.

4. Islam, R.; Jaafar, W.Z.W.; Hin, L.S.; Osman, N.; Hossain, A.; Mohd, N.S. Development of an intelligent system based on ANFIS model for predicting soil erosion. *Environ. Earth Sci.* **2018**, *77*, 186. [CrossRef]
5. Lal, R. Soil degradation by erosion. *Land Degrad. Dev.* **2001**, *12*, 519–539. [CrossRef]
6. Borrelli, P.; Alewell, C.; Alvarez, P.; Anache, J.A.A.; Baartman, J.; Ballabio, C.; Bezak, N.; Biddoccu, M.; Cerdà, A.; Chalise, D.; et al. Soil erosion modelling: A global review and statistical analysis. *EarthArxiv* **2020**. [CrossRef]
7. Yeh, S.-C.; Wang, C.-A.; Yu, H.-C. Simulation of soil erosion and nutrient impact using an integrated system dynamics model in a watershed in Taiwan. *Environ. Model. Softw.* **2006**, *21*, 937–948. [CrossRef]
8. Fan, J.-C.; Wu, M.-F. Effects of soil strength, texture, slope steepness and rainfall intensity on interrill erosion of some soils in Taiwan. In Proceedings of the 10th International Soil Conservation Organization Meeting, Purdue University, USDA-ARS National Soil Erosion Research Laboratory, W. Lafayette, IN, USA, 24–29 May 1999.
9. Lo, K.F.A. Erosion assessment of large watersheds in Taiwan. *J. Soil Water Conserv.* **1995**, *50*, 180–183.
10. Chiu, Y.-J.; Chang, K.-T.; Chen, Y.-C.; Chao, J.-H.; Lee, H.-Y. Estimation of soil erosion rates in a subtropical mountain watershed using 137Cs radionuclide. *Nat. Hazards* **2011**, *59*, 271–284. [CrossRef]
11. Chen, W.; Li, D.-H.; Yang, K.-J.; Tsai, F.; Seeboonruang, U. Identifying and comparing relatively high soil erosion sites with four DEMs. *Ecol. Eng.* **2018**, *120*, 449–463. [CrossRef]
12. Liu, Y.-H.; Li, D.-H.; Chen, W.; Lin, B.-S.; Seeboonruang, U.; Tsai, F. Soil Erosion Modeling and Comparison Using Slope Units and Grid Cells in Shihmen Reservoir Watershed in Northern Taiwan. *Water* **2018**, *10*, 1387. [CrossRef]
13. Huang, Y.; Zhao, L. Review on landslide susceptibility mapping using support vector machines. *Catena* **2018**, *165*, 520–529. [CrossRef]
14. Reichenbach, P.; Rossi, M.; Malamud, B.D.; Mihir, M.; Guzzetti, F. A review of statistically-based landslide susceptibility models. *Earth-Sci. Rev.* **2018**, *180*, 60–91. [CrossRef]
15. Lagomarsino, D.; Tofani, V.; Segoni, S.; Catani, F.; Casagli, N. A Tool for Classification and Regression Using Random Forest Methodology: Applications to Landslide Susceptibility Mapping and Soil Thickness Modeling. *Environ. Model. Assess.* **2017**, *22*, 201–214. [CrossRef]
16. Heung, B.; Ho, H.C.; Zhang, J.; Knudby, A.; Bulmer, C.E.; Schmidt, M.G. An overview and comparison of machine-learning techniques for classification purposes in digital soil mapping. *Geoderma* **2016**, *265*, 62–77. [CrossRef]
17. Ali, I.; Greifeneder, F.; Stamenkovic, J.; Neumann, M.; Notarnicola, C. Review of Machine Learning Approaches for Biomass and Soil Moisture Retrievals from Remote Sensing Data. *Remote Sens.* **2015**, *7*, 16398–16421. [CrossRef]
18. Nguyen, K.A.; Chen, W.; Lin, B.-S.; Seeboonruang, U.; Thomas, K. Predicting Sheet and Rill Erosion of Shihmen Reservoir Watershed in Taiwan Using Machine Learning. *Sustainability* **2019**, *11*, 3615. [CrossRef]
19. Nguyen, K.A.; Chen, W.; Lin, B.-S.; Seeboonruang, U. Using Machine Learning-Based Algorithms to Analyze Erosion Rates of a Watershed in Northern Taiwan. *Sustainability* **2020**, *12*, 2022. [CrossRef]
20. Haigh, M.J. The use of erosion pins in the study of slope evolution. *Br. Geomorphol. Res. Group Tech. Bull.* **1977**, *18*, 31–49.
21. Ghimire, S.K.; Higaki, D.; Bhattarai, T.P. Estimation of Soil Erosion Rates and Eroded Sediment in a Degraded Catchment of the Siwalik Hills, Nepal. *Land* **2013**, *2*, 370–391. [CrossRef]
22. Couper, P.; Stott, T.; Maddock, I. Insights into river bank erosion processes derived from analysis of negative erosion-pin recordings: Observations from three recent UK studies. *Earth Surf. Process. Landf. J. Br. Geomorphol. Res. Group* **2002**, *27*, 59–79. [CrossRef]
23. Lawler, D.M.; Couperthwaite, J.; Bull, L.J.; Harris, N.M. Bank erosion events and processes in the Upper Severn basin. *Hydrol. Earth Syst. Sci.* **1997**, *1*, 523–534. [CrossRef]
24. Lin, B.-S.; Thomas, K.; Chen, C.-K.; Ho, H.-C. Evaluation of soil erosion risk for watershed management in Shenmu watershed, central Taiwan using USLE model parameters. *Paddy Water Environ.* **2016**, *14*, 19–43. [CrossRef]
25. Dietterich, T.G. Ensemble methods in machine learning. Lecture Notes in Computer Science (including subseries Lecture Notes in Artificial Intelligence and Lecture Notes in Bioinformatics). In *International Workshop on Multiple Classifier Systems*; Springer: Berlin/Heidelberg, Germany, 2000; Volume 1857 LNCS, pp. 1–15.
26. Erdal, H.; Karahanoğlu, İ. Bagging ensemble models for bank profitability: An empirical research on Turkish development and investment banks. *Appl. Soft Comput.* **2016**, *49*, 861–867. [CrossRef]
27. Abawajy, J.; Kelarev, A.; Chowdhurry, M.U. Large Iterative Multitier Ensemble Classifiers for Security of Big Data. *IEEE Trans. Emerg. Top. Comput.* **2014**, *2*, 352–363. [CrossRef]
28. Hsieh, S.-L.; Hsieh, S.-H.; Cheng, P.-H.; Chen, C.-H.; Hsu, K.-P.; Lee, I.-S.; Wang, Z.; Lai, F. Design Ensemble Machine Learning Model for Breast Cancer Diagnosis. *J. Med. Syst.* **2012**, *36*, 2841–2847. [CrossRef]
29. Pham, B.T.; Bui, D.T.; Prakash, I.; Dholakia, M. Hybrid integration of Multilayer Perceptron Neural Networks and machine learning ensembles for landslide susceptibility assessment at Himalayan area (India) using GIS. *Catena* **2017**, *149*, 52–63. [CrossRef]
30. Tehrany, M.S.; Pradhan, B.; Jebur, M.N. Flood susceptibility mapping using a novel ensemble weights-of-evidence and support vector machine models in GIS. *J. Hydrol.* **2014**, *512*, 332–343. [CrossRef]
31. Friedman, J.H.; Roosen, C.B. An introduction to multivariate adaptive regression splines. *Stat. Methods Med. Res.* **1995**, *4*, 197–217. [CrossRef]
32. Otok, B.W.; Akbar, M.S.; Guritno, S.; Subanar, S. Ordinal Regression Model using Bootstrap Approach. *J. ILMU DASAR* **2007**, *8*, 54–67.

33. Quinlan, J.R. Learning with continuous classes. In Proceedings of the 5th Australian Joint Conference on Artificial Intelligence, Hobart, Australia, 16–18 November 1992.
34. Quinlan, J.R. Combining instance-based and model-based learning. In Proceedings of the 10th International Conference on Machine Learning, Amherst, MA, USA, 27–29 June 1993.
35. Zhou, J.; Li, E.; Wei, H.; Li, C.; Qiao, Q.; Armaghani, D.J. Random Forests and Cubist Algorithms for Predicting Shear Strengths of Rockfill Materials. *Appl. Sci.* **2019**, *9*, 1621. [CrossRef]
36. Friedman, J.H. Greedy function approximation: A gradient boosting machine. *Ann. Stat.* **2001**, *29*, 1189–1232. [CrossRef]
37. Friedman, J.H. Stochastic gradient boosting. *Comput. Stat. Data Anal.* **2002**, *38*, 367–378. [CrossRef]
38. Zhou, J.; Li, E.; Yang, S.; Wang, M.; Shi, X.; Yao, S.; Mitri, H.S. Slope stability prediction for circular mode failure using gradient boosting machine approach based on an updated database of case histories. *Saf. Sci.* **2019**, *118*, 505–518. [CrossRef]
39. Ridgeway, G. Generalized Boosted Models: A guide to the GBM package. *Update* **2007**, 1–15.
40. Acharya, G.; Cochrane, T.; Davies, T.R.H.; Bowman, E. Quantifying and modeling post-failure sediment yields from laboratory-scale soil erosion and shallow landslide experiments with silty loess. *Geomorphology* **2011**, *129*, 49–58. [CrossRef]
41. Du, S.; Zhang, J.; Deng, Z.; Li, J. A New Approach of Geological Disasters Forecasting using Meteorological Factors based on Genetic Algorithm Optimized BP Neural Network. *Elektron. Elektrotech.* **2014**, *20*, 57–62. [CrossRef]
42. Nash, J.E.; Sutcliffe, J.V. River flow forecasting through conceptual models part I—A discussion of principles. *J. Hydrol.* **1970**, *10*, 282–290. [CrossRef]
43. Willmott, C.J. On the validation of models. *Phys. Geogr.* **1981**, *2*, 184–194. [CrossRef]
44. Chen, C.-S.; Chen, Y.-L. The Rainfall Characteristics of Taiwan. *Mon. Weather Rev.* **2003**, *131*, 1323–1341. [CrossRef]
45. Chang, F.-J.; Chang, Y.-T. Adaptive neuro-fuzzy inference system for prediction of water level in reservoir. *Adv. Water Resour.* **2006**, *29*, 1–10. [CrossRef]
46. Chen, W.; Panahi, M.; Pourghasemi, H.R. Performance evaluation of GIS-based new ensemble data mining techniques of adaptive neuro-fuzzy inference system (ANFIS) with genetic algorithm (GA), differential evolution (DE), and particle swarm optimization (PSO) for landslide spatial modelling. *Catena* **2017**, *157*, 310–324. [CrossRef]
47. Rogan, J.; Franklin, J.; Stow, D.; Miller, J.; Woodcock, C.; Roberts, D. Mapping land-cover modifications over large areas: A comparison of machine learning algorithms. *Remote Sens. Environ.* **2008**, *112*, 2272–2283. [CrossRef]
48. Ramos-Pollán, R.; Guevara López, M.Á.; Oliveira, E. Introducing ROC curves as error measure functions: A new approach to train ANN-based biomedical data classifiers. Lecture Notes in Computer Science (including subseries Lecture Notes in Artificial Intelligence and Lecture Notes in Bioinformatics). In *Iberoamerican Congress on Pattern Recognition*; Springer: Berlin/Heidelberg, Germany, 2010; pp. 517–524.
49. Lin, B.-S.; Chen, C.-K.; Thomas, K.; Hsu, C.-K.; Ho, H.-C. Improvement of the K-Factor of USLE and Soil Erosion Estimation in Shihmen Reservoir Watershed. *Sustainability* **2019**, *11*, 355. [CrossRef]

International Journal of
Geo-Information

MDPI

Article

Correlation between Geochemical and Multispectral Patterns in an Area Severely Contaminated by Former Hg-As Mining

Carlos Boente [1,2], Lorena Salgado [1], Emilio Romero-Macías [2], Arturo Colina [1], Carlos A. López-Sánchez [3] and José Luis R. Gallego [1,*]

1 Environmental Biogeochemistry & Raw Materials Group and INDUROT, Campus de Mieres, University of Oviedo, C/Gonzalo Gutiérrez Quirós. S/N, 33600 Mieres, Spain; carlos.boente@dimme.uhu.es (C.B.); salgadolorena@uniovi.es (L.S.); arturo.indurot@uniovi.es (A.C.)

2 Department of Mining, Mechanical, Energetic and Civil Engineering, University of Huelva, 21819 Huelva, Spain; romaci@uhu.es

3 SMartForest Group, Department of Organisms and Systems Biology, Polytechnic School of Mieres, University of Oviedo, 36000 Mieres, Spain; lopezscarlos@uniovi.es

* Correspondence: jgallego@uniovi.es

Received: 24 October 2020; Accepted: 9 December 2020; Published: 10 December 2020

Abstract: In the context of soil pollution, plants suffer stress when exposed to extreme concentrations of potentially toxic elements (PTEs). The alterations to the plants caused by such stressors can be monitored by multispectral imagery in the form of vegetation indices, which can inform pollution management strategies. Here we combined geochemistry and remote sensing techniques to offer a preliminary soil pollution assessment of a vast abandoned spoil heap in the surroundings of La Soterrana mining site (Asturias, Spain). To study the soil distribution of the PTEs over time, twenty-seven soil samples were randomly collected downstream of and around the main spoil heap. Furthermore, the area was covered by an unmanned aerial vehicle (UAV) carrying a high-resolution multispectral camera with four bands (red, green, red-edge and near infrared). Multielement analysis revealed mercury and arsenic as principal pollutants. Two indices (from a database containing up to 55 indices) offered a proper correlation with the concentration of PTEs. These were: CARI2, presenting a Pearson Coefficient (PC) of 0.89 for concentrations >200 mg/kg of As; and NDVIg, PC of −0.67 for >40 mg/kg of Hg. The combined approach helps prediction of those areas susceptible to greatest pollution, thus reducing the costs of geochemical campaigns.

Keywords: soil pollution; mining; multispectral images; UAV; vegetation index

---

1. Introduction

Mining and industrial activities commonly release contaminants, which in turn affect the quality of environmental compartments [1,2]. In this regard, pollution has been reported not only in urban zones [3,4] but also in more rural or natural areas, where the quality of sensitive ecosystems is being seriously compromised [5]. Although industries are not usually located in rural areas, this is not true for mining operations as the ores of interest are commonly found in these areas. In the case of mining activities in hilly regions and subsequent waste disposal, residues are exposed to hard weathering and a challenging geomorphology. Therefore, pollutants can be widely dispersed, and the costs and uncertainties of geochemical campaigns may be very high [6].

Given these considerations and in the context of soil pollution studies, there is a need to find appropriate non-invasive approaches that reduce the cost and time required for these campaigns. Remote sensing techniques are widely used in natural environment research, such as studies on forest studies [7–9], relative climate change [10–12], topography [13–17], and precision agriculture [18,19], among others. Of note, in recent years, some studies have combined techniques of remote sensing and geochemistry, especially in areas affected by heavy metal pollution, either with satellites or high-altitude manned aircrafts [20–23] or unmanned aerial vehicles (UAVs) [24–28]. Remote sensing uses electromagnetic radiation to obtain information about an object or phenomenon without involving physical contact [29]. Data are gathered through measurements of electromagnetic reflectance and different frequencies [30]. Within this discipline, UAVs are one of the supports used for the development of remote sensing tools for soil applications, with most of the required information being collected by means of optical sensors, which have also been recently used in geological applications [31,32].

To date, studies addressing the electromagnetic spectrum have been based on the use of proximal or airborne sensors. However, several authors have recently used proximal aerial sensors, also known as UAVs, for this purpose, due to the advantages they present over satellite, and high-altitude manned aircrafts, such as short revist time and high spatial resolution [33]. This approach has also found applications in soil contamination research [33,34]. The direct study of contaminants in soil through remote sensing is complex; therefore, indirect techniques are used, among which vegetation as a bioindicator of soil condition stands out [30]. One of the most common ways to analyze vegetation with multispectral remote sensors is through vegetation indices, which are simple algebraic operations between the different bands of the electromagnetic spectrum [35–37]. In this regard, Shabou et al. [38] used the NDVI and MIR indices over test fields to achieve bare soil measurements from satellite images.

With respect to target pollutants, metal mining is a frequent source of potentially toxic elements (PTEs). These elements are characterized not only by their toxicity, but also by their capacity to bioaccumulate, biomagnify and persist in the food chain [39], giving rise to multiple health and environmental problems [40]. Among PTEs, mercury (Hg) is highly toxic to organisms, even at very low concentrations, especially when present in methyl-Hg form [41]. In this context, in 2005, the EU promoted an initiative to reduce the emissions of this element. Later, in 2017, the Minamata Convention, a global treaty to regulate the emissions and distribution of this toxic element, was approved. This treaty encourages countries to reduce and monitor the different sources of Hg pollution [42]. Since then, various studies have addressed the environmental impact of Hg mining [43–45].

Historically, Spain has been one of the world's largest producers of Hg, especially the Almadén district [46], the largest Hg deposit in the world. In the case of Asturias (NW Spain), numerous cinnabar deposits have been mined, the remains of which are now scattered throughout the region mainly as spoil heaps enriched in Hg-As waste. The impact of some of these mining operations has been examined in detail. As shown in the studies carried out at the mines of El Terronal [47] and Caunedo [43], and at the Muñón-Cimero complex (Soterraña), which is the focus of the present study [48,49]. Here we describe an innovative approach that combines a classical geochemical campaign and a UAV-based multispectral airborne camera to monitor the impact of Hg-As spoil heaps. The main objective of this study is a first approach to the possibility of evaluating As or Hg levels in soil polluted areas through low-cost multispectral sensors.

2. Materials and Methods

In order to summarize and synthetize the procedures followed, the different steps carried out along the study (data collection, geochemistry and remote sensing techniques, statistical analysis, etc.) are shown in Figure 1.

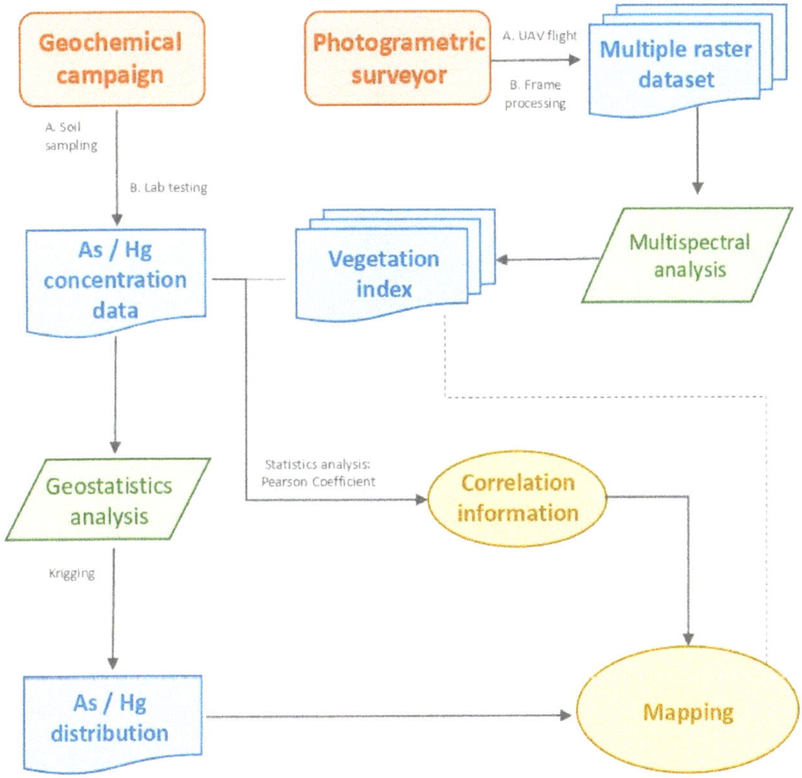

Figure 1. Research workflow.

2.1. *Context of the Study Area: Activity in La Soterraña, Geology and Vegetation*

The main activity of the Hg mountain mine of La Soterraña (Asturias, Spain. Figure 2) took place between the 1940s and the mid-1970s, the latter decade marking the crisis of the Hg sector. During its time of operation, ore mining and processing were carried out onsite. The room and pillar mining method was used on a low temperature hydrothermal epigenetic type deposit, in which the main mineral paragenesis comprised cinnabar, realgar and oropiment, with important deposits of arsenopyrite, marcasite and pyrite hosted in fractured limestones and shales [48]. Once the materials had been extracted, the ore was treated at La Soterraña mine itself, together with the material of other smaller mines located in the province. For this purpose, rotary kilns, condensers and coolers were used, as well as other equipment typical of Hg metallurgy.

The tailings generated by ore mining and processing were indiscriminately dumped in a spoil heap, which is the principal focus of pollution. At present, the mechanical dispersion of this material is the main cause of pollutant distribution in the surrounding soils, in addition to the processes of oxidation and leaching. This main spoil heap is located over layers of limestone and shale, along an average slope of 30°. There is virtually no vegetation on the heap due to the high concentrations of metals and its steep slope. Downstream of the heap, herbaceous colonies of *Centaurea nigra* and *Equisetum telmateia* predominate. These species are of special interest due to their capacity to bioaccumulate Hg and As [49,50]. In fact, *E. telmateia* and *C. nigra* showed notable Bioaccumulation Indices (BAC); $BAC_{As} = 14.39$ and $BAC_{Hg} = 76.95$, and $BAC_{As} = 2.30$ and $BAC_{Hg} = 4.84$ respectively [51]. In the surroundings of the spoil heap, there are also notable tree species, including *Betula celtibérica, Quercus robur* and *Castanea sativa*.

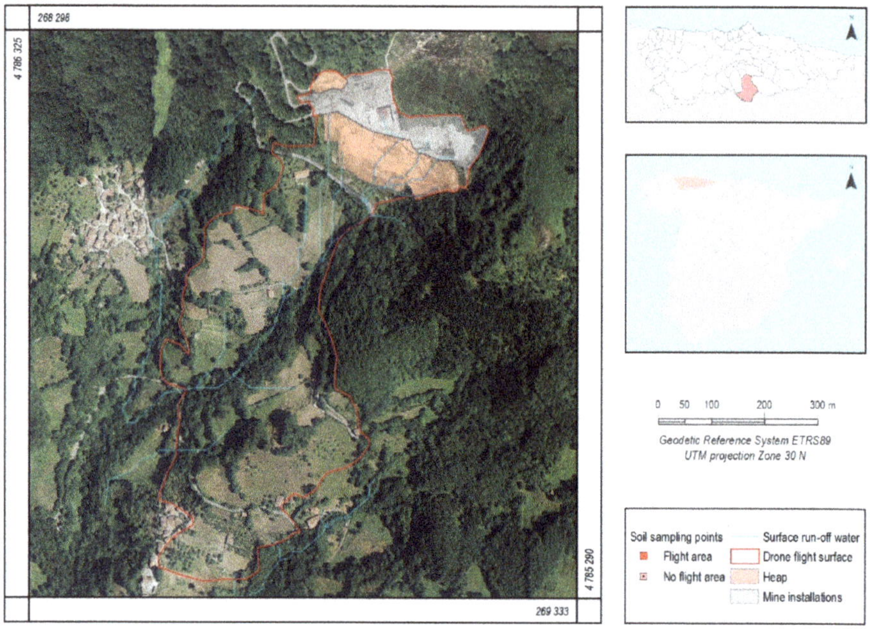

Figure 2. Location of the study area and sampling design.

2.2. UAV Flight, Sampling, Preparation and Chemical Determinations

The first step consisted of designing the coverage of UAV flights. For the purpose of this study, we selected a surface area of 25 ha downstream of the spoil heap (Figure 1), which is the area most affected by erosion and washing by surface runoff.

The UAV flight was carried out with a BlueGrass Multiespectal model UAV (Parrot SA, Paris, France) which incorporates a Parrot Sequoia plus camera. This camera includes a multispectral sensor, with four spectral bands: green, red, red edge and near infrared, and a sunlight sensor, which provides absolute reflectance measurements without the need of calibration. The use of this camera with the software PIX4D Mapper (Pix4D S.A. Prilly, Switzerland) allowed to obtain the optimal results, as reflected by other authors [52–54]. With this procedure, a reduction in the time of the photogrammetric survey and data processing is achieved. The photogrammetric survey is carried out with a transversal and longitudinal reception of 70%, due to the strong orography, and with an average height of 90 m. The date of the flight was chosen based on the vegetation conditions during the early spring of 2019 in order to obtain radiometric information of the herbaceous vegetation when younger; in addition, in this period of the year soil was not affected by extreme weather conditions such as droughts, or alterations due to agro-livestock activities such as mowing or cattle grazing.

In turn, a geochemical campaign composed of a total of 27 soil samples was designed. The samples were gathered randomly. Ten of these samples were collected within the limits of the area covered by the UAV flight (Figure 2), while the others were gathered outside this area in order to assess the degree of pollution in the surroundings for future research purposes. Each sample comprised five increments of the first 20 cm of the soil by means of a Dutch Edelman probe, forming a 1-m edge square and its central point to improve representativeness.

Sample preparation started with in-situ sieving of the soil through 20-mm mesh to remove large particles. The finer fraction was then placed in sterilized plastic bags and stored at 4 °C until being taken to the laboratory. Here, the fractions were air-dried at room temperature to prevent the evaporation of Hg and quartered to maintain sample representativeness. Finally, for each sample, 1 g was ground in an agate mortar to a particle size of < 100 μm and sent to an ISO 9002 and ISO-17025 accredited laboratory in Vancouver (BC, Canada). The concentrations of the following PTEs were determined after aqua regia extraction and ICP-MS (in parenthesis, their respective detection limits): As (0.5 ppm); Cd (0.1 ppm); Co (0.1 ppm); Cr (1 ppm); Cu (0.2 ppm); Hg (0.01 ppm); Mn (1 ppm); Mo (0.1 ppm); Ni (0.1 ppm); Pb (0.1 ppm); Sb (0.1 ppm); Tl (0.1 ppm); V (2 ppm); and Zn (1 ppm).

2.3. Statistical Analysis and Graphical Representations

The geochemical data were analyzed by means of descriptive statistics and compared with the official Risk-Based Soil Screening Levels (RBSSLs) [55]. Moreover, Pearson's Coefficient between elements was also calculated. These operations allowed us to identify the main pollutants in the study area. Additionally, the concentration of As and Hg for the whole area was mapped by means of ordinary kriging, applying log-transformation to improve normality [56]. Apart from providing an estimation of the distribution of elements in the soil, the interpolations are also useful to establish graphical comparisons between distribution and the maps showing the vegetation indices. All map representations and graphical information were treated through Geographical Information Systems (GIS), using a licensed copy of ArcMap version 10.2.1. (ESRI, Redlands, CA, USA).

2.4. Multi-Spectral Analysis Tools: Vegetation Index

The multispectral images used were obtained from a Sequoia + multispectral camera mounted on a UAV Parrot Bluegrass Fields. The camera had a spectral resolution of four bands—green, (550 nm ± 40 nm), red (660 nm ± 40 nm), red edge (735 nm ± 10 nm) and near infrared (790 ± 40 nm)—all with a high spatial resolution of 16 cm.

The first step of multispectral analysis was the achievement of a raster dataset from each band, which was carried out with software PIX4D Mapper using all photograms per each spectral resolution. The second step consisted in obtaining a new raster dataset with information derived from an algebraic operation between the bands obtained in first step. These new raster datasets are known as vegetation indices, due to their widespread use for determining vegetation status. Up to a total of 55 different vegetation indices were tested with the aim to obtain the best correlations.

The list of the 55 vegetation indices used was extracted from a remote sensing index database [57]. Initially all of them were studied, adapted to the available data and tested. Those found with the greatest correlation with the available analytical data were the Chlorophyll Absorption Ratio Index 2 (CARI-2), and the Normalized Difference Vegetation Index-green (NDVIg).

The CARI2 index was defined by Kim [58] to study leaf area index (LAI), absorbed photosynthetically active radiation, and primary production. It is described by Equation (1):

$$CARI-2 = \frac{\left|(a\cdot|Red| + |Red| + b)\right|}{\sqrt{a^2+1}} \cdot \frac{Red\ Edge}{Red} \tag{1}$$

where:

$$a = ([Red\ Edge] - [Green])/150$$
$$b = [Green] - (a \cdot [Green])$$

In contrast, NDVIg is a modification of the classical NDVI index that uses absorbance of green light (550 nm), which is more sensitive to a wide range of chlorophyll-a concentrations than red light [56]. It is defined using Equation (2):

$$NDVI_g = \frac{Near\ Infrared - Green}{Near\ Infrared + Green} \tag{2}$$

3. Results and Discussion

3.1. Geochemical Characterization

A previous study by Fernández et al. [59] reported maximum total concentrations of 23,000 mg/kg of As and 6000 mg/kg of Hg, corresponding to the mining waste of the spoil heap. As a consequence of spoil heap weathering, the soils of the area showed extremely high concentrations of both elements, as reflected in the descriptive statistics (Table 1) corresponding to the chemical analyses of the 27 soil samples taken downstream of the heap. Table 1 also includes the official Risk-Based Soil Screening Levels (RBSSLs) or threshold values in force for the use of soil [55].

Table 1. Descriptive statistics for the 27 soil samples analyzed. Concentrations expressed as mg/kg.

PTE	RBSSL	Range	Mean	Median	Typical Deviation	CV
As	40	44–9920	632.1	150.0	1876.4	296.8
Cd	2	0.2–0.9	0.4	0.5	0.2	40.3
Co	25	2.7–17	13.4	13.4	3.0	22.1
Cr	10,000	18–57	26.9	25.0	8.0	29.6
Cu	55	11–98	33.2	28.8	18.8	56.6
Hg	1	1.95–860	68.0	29.2	158.0	232.4
Mn	2135	251–1270	707.2	704.0	248.3	35.1
Mo	6	0.4–3.6	1.1	1.0	0.6	61.6
Nor	65	10–42	28.9	28.0	6.5	22.5
Pb	70	24–74	37.5	34.0	11.2	29.8
Sb	5	0.2–44	3.4	2.0	8.1	238.0
Tl	1	0.1–2	1.2	2.0	0.9	74.1
V	50	20–83	38.9	35.0	12.6	32.4
Zn	455	65–214	121.3	113.0	29.7	24.5

The differences between the mean and the median, or the high coefficients of variation (CV), strongly suggest Hg and As as the main pollutants in the area as they present irregular distributions. All the samples exceeded the RBSSL for Hg (1 mg/kg) and As (40 mg/kg) [46], values that correspond to the category "Other Uses", to which the soil of study currently belongs because it is located in agricultural areas dedicated to pasture or crop cultivation.

Furthermore, most of the samples were also above the soil-chemical background levels of the area determined in pristine areas by Loredo et al. (4.18 mg/kg for Hg and 39 mg/kg for As) [48]. Although the enrichment of these elements was consistent with the mineral paragenesis of the zone described in Section 2.1, the extremely high concentrations detected revealed a gradual degradation of the spoil heap over time due to weathering processes. These are attributable mainly to water erosion by surface and subsurface runoff, which circulates throughout the heap during the months of flooding (Figures 2 and 3). In this regard, higher concentrations of Hg and As were observed in areas close to the runoff, thereby contributing to the dispersion of these pollutants downstream.

With respect to the rest of the PTEs, Sb also appeared in an anomalous form, although to a lesser extent than Hg and As, as revealed by its range or high CV, although the similarity between average and median reveals more normality of the Sb distribution. In this regard, Sb appeared to be linked to As (metalloids usually geochemically associated) [60]. The rest of the PTEs presented stable values and were not hazardous for the environment in this case. The highest concentrations of As and Hg were recorded within the UAV flight zone and in the two samples to the north, near the mining area (Figure 3). Outside the area covered by the UAV, although the concentrations of both contaminants were substantially lower for most samples, they continued to be high. Given this observation, it would be advisable for future studies to collect more samples outside this area in order to better delimit the extent of contamination. Finally, bivariate correlations revealed high correlations between Hg and Cr (0.777), V (0.736) and Pb (0.697). In contrast, As showed an almost full correlation only with Sb (0.991). Of note, there was a low correlation between As-Hg (0.270), reflecting independence between variables.

This observation can be explained by the highly irregular distribution of As and Hg and their very different behaviors.

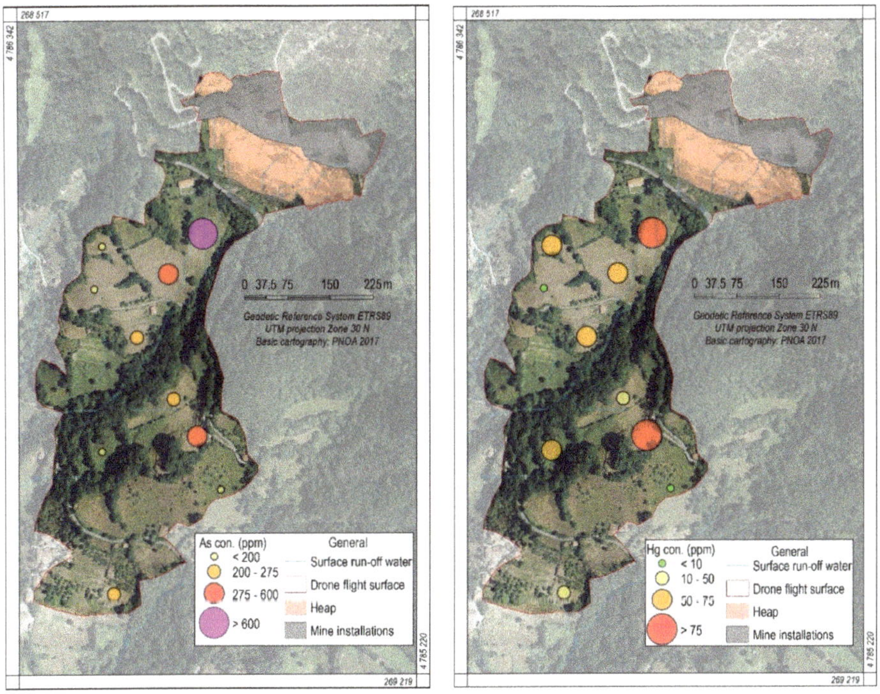

(**a**) (**b**)

Figure 3. Concentrations of: (**a**) As and (**b**) Hg, in the study area.

3.2. Mutispectral Analysis

The results presented in Section 3.1 revealed the severe contamination present within the UAV flight area, the main contaminants being Hg and As. Therefore, it is feasible that such high concentrations affect the physiology of plants in the zone. Of the 55 indices tested, those with the greatest correlation with the available analytical data were CARI-2 and NDVIg for As and Hg, respectively. These indices are shown in Figure 4.

The correlations between the concentrations of Hg and As obtained by chemical analysis and the vegetation indices were established by means of Pearson's coefficient and are shown in Figure 5. The results obtained showed that the higher the concentration, the higher the correlation, in absolute value. In the case of As, it was revealed a significant correlation starting with Pearson value's of 0.732 (corresponding to concentrations above 117 ppm) up to 0.970 (close to 400 ppm). Hg also showed a significant correlation above 40 ppm (−0.670 Pearson coefficient), up to −0.730 (70 ppm).

The CARI-2 index correlated better with geochemical values for As, maintaining a logarithmic (increasing) correlation that reached stabilization at 200 ppm (Pearson's Coef. = 0.89). The positive correlation reflects a direct relationship, that is, the higher the concentration, the greater the stress present in plants. However, when the concentration reached 200 ppm, the plants reach the limit of tolerance for the vegetation described in Section 2.1.

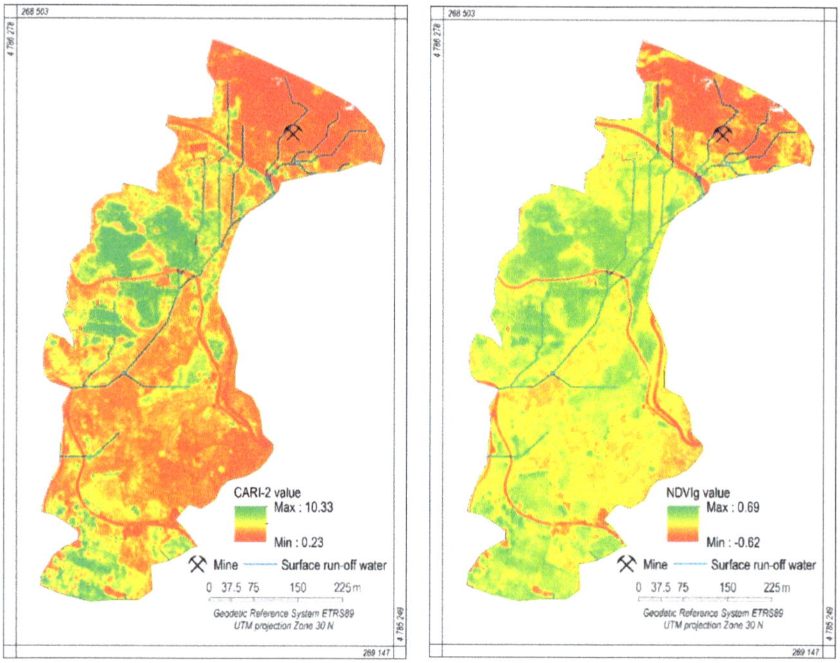

Figure 4. Results of the vegetation indices **Left**: CARI2; good correlation with As and **Right**: NDVIg; good correlation with Hg.

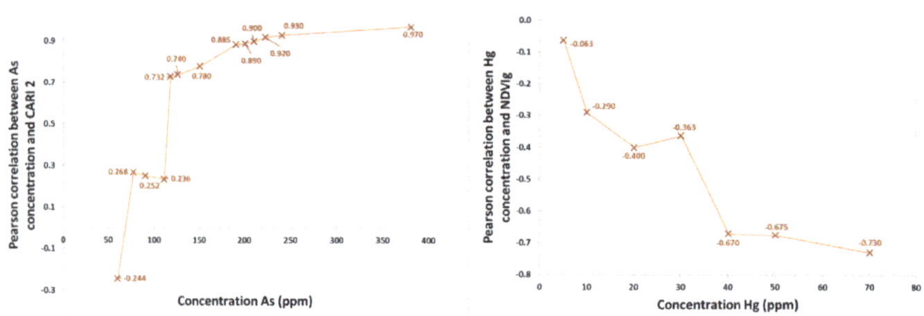

Figure 5. Pearson's coefficients between the concentration of As and CARI-2 vegetation index (**left**) and Hg and NDVIg vegetation index (**right**).

Similarly, the NDVIg index presented an optimal correlation with Hg by means of an exponential (decreasing) function whose stabilization was achieved at 40 ppm, a concentration under the industrial limit and above the limits for residential and recreational uses, but unlike As, with a negative correlation (Pearson's Coef. = −0.67). This implies that the higher the concentration, the lower the value of the NDVIg index. This finding allowed us to identify a direct alteration of the vegetation with increasing concentrations of Hg, since the definition of the index itself shows that the lower its value, the less photosynthetic activity occurs [61].

In any case, it can be said that the results of the vegetation indices presented high reliability for As and acceptable reliability for Hg at the points where the concentrations of these two pollutants were very high. All things considered, the mapping of the CARI2 and NDVIg indices (Figure 4) allows prediction of where the concentrations of As and Hg, respectively, will be highest, thus reducing the costs of sampling campaigns.

3.3. *Geostatistical Analysis*

Figure 6 shows the geostatistical interpolations obtained by ordinary kriging. As anticipated in the statistical analysis, the distribution of As and Hg was highly irregular along the UAV flight area. Extremely polluted zones near the spoil heap alternated with others downstream presenting lower concentrations of these heavy metals, although always exceeding RBSSLs. However, for academic purposes, such alternation is not necessarily negative as the differences in concentrations allowed us to test the reliability of the vegetation indices at a range of concentrations.

Our results also revealed that Hg showed less mobility than As in soils, a property that was addressed in previous studies [62–64]. This was appreciable since changes in concentration were far greater for Hg, or in other words, the bullseye was more easily identifiable. The lower mobility of Hg also contributes to explaining why Pearson's Coefficient was lower for this heavy metal. In this case, low normality hampered the estimation. Nevertheless, even under this circumstance, the correlation obtained was acceptable.

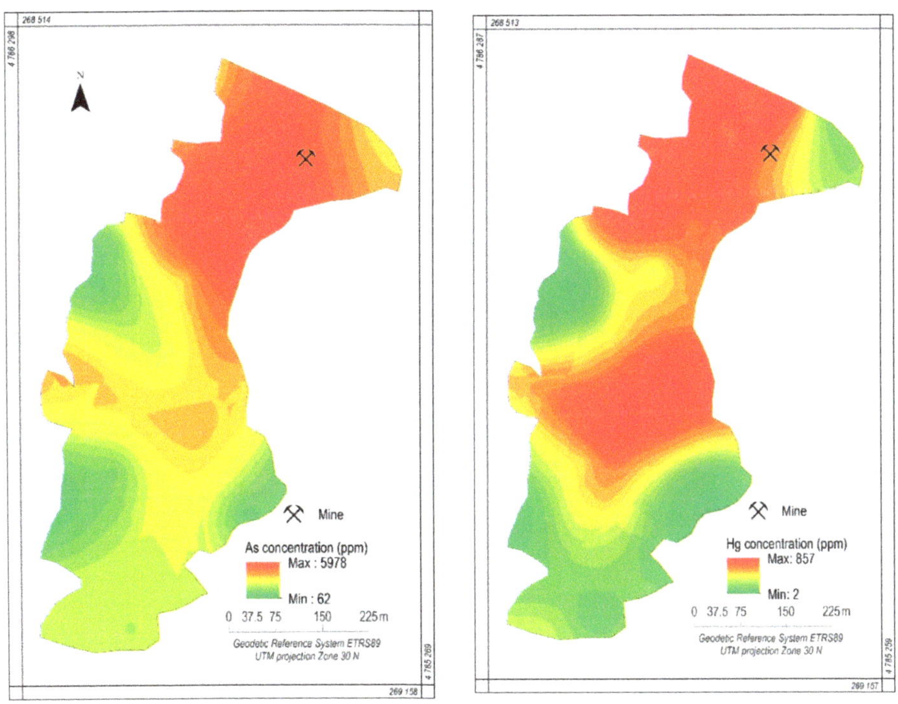

Figure 6. Kriging interpolation. **Left**: As; and **Right**: Hg.

3.4. *Synergy between Multispectral Data and Geostatistical Analysis*

Figure 7 aims to synergize the geostatistical results (Figure 6) and the multispectral analysis (Figure 4) Areas in red correspond to concentrations of As and Hg exceeding the thresholds described in the previous section (>200 mg/kg for As and >40 mg/kg for Hg). Our findings imply that, in these locations, it is possible to use multispectral techniques to predict the extent to which the pollution affects plants and subsequently predicting the degree of pollution. More specifically, around 89% of reliability was achieved for As and 67% for Hg on the basis of Pearson's Coefficient (Figure 5).

We also observed that most of the area lent itself to study; although results are more reliable for As than for Hg. This finding could be attributable to the previously mentioned higher mobility of As, which reached higher concentrations than Hg far from the spoil heap. In contrast, the areas with low concentrations of Hg and As are shown in green, where there was no correlation between the pollution levels and the CARI2/NDVIg results.

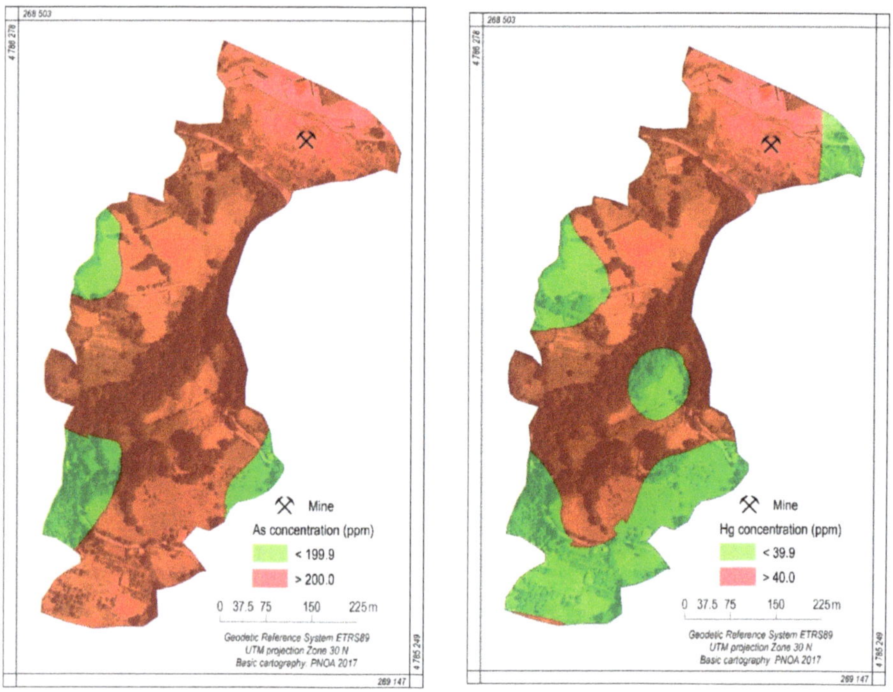

Figure 7. In red, areas that lend themselves to study by means of multispectral remote sensing techniques with a high degree of reliability. **Left**: As; **right**: Hg.

4. Conclusions

Hg-As mining spoil heaps pose an environmental risk due to their highly toxic load. Therefore, it is important to monitor their evolution in order to evaluate the dispersion of the contaminants. Here we examined the impact of weathering on the As- and Hg-rich spoil heap of the former La Soterraña mine. To this end, the synergy between classical geochemical tools and remote sensing methodologies using a UAV-based airborne multispectral camera was explored.

First, the geochemical campaign confirmed severe As and Hg pollution of soils downstream of the spoil heap, and to a lesser extent the presence of Sb, in concentrations several orders of magnitude above RBSSLs. On the basis of the high level of pollution found, we reasoned that the vegetation of the area would be affected. This notion was confirmed by remote sensing techniques involving a multispectral camera mounted on a UAV. Algebraic operations with remote sensing data, also called vegetation indices, revealed reliable results between As and the CARI-2 index, and between Hg and the NDVIg index, the former giving better results.

In this regard, our results demonstrated a very high correlation between plant stress and high concentrations of As (0.89 correlation when As concentration > 200 ppm) and Hg (0.67 correlation when Hg concentration > 40 ppm). These findings pave the way for new studies that combine classical geochemical tools and remote sensing.

The dispersion of the contaminants is attributable to the physical-chemical weathering of the spoil heap, mainly through solid waste leaching processes, which results in the distribution of As and Hg towards lower levels of the heap. However, thanks to the high correlations between the concentrations of the two contaminants and the CARI-2 and NDVIg indices, the mapping of these indices facilitates the identification of pollution hotspots in areas particularly affected by contamination, thereby reducing the costs of geochemical characterization. On the whole, this study emphasizes the potential of combining geochemical tools and remote sensing technology for soil pollution research. However, future research efforts are expected to bring about the fine tuning of both the geochemical aspects and the multispectral information.

Author Contributions: Conceptualization, Carlos Boente, Lorena Salgado & José Luis R. Gallego; Methodology, Lorena Salgado, Arturo Colina & Carlos A. López-Sánchez; Software, Arturo Colina & Lorena Salgado; Validation, José Luis R. Gallego, Arturo Colina & Emilio Romero-Macías; Formal Analysis, Carlos Boente, Lorena Salgado & Carlos A. López-Sánchez; Investigation, Carlos Boente & Lorena Salgado; Data Curation, Lorena Salgado & Carlos Boente; Writing-Original Draft Preparation, Carlos Boente, Lorena Salgado & Emilio Romero-Macías; Writing-Review & Editing, Arturo Colina & José Luis R. Gallego; Visualization, Lorena Salgado & Carlos Boente; Supervision, José Luis R. Gallego; Project Administration, José Luis R. Gallego; Funding Acquisition, José Luis R. Gallego & Emilio Romero-Macías. All authors have read and agreed to the published version of the manuscript.

Funding: This research received no external funding.

Conflicts of Interest: The authors declare no conflict of interest.

References

1. Jacobson, M.Z. Review of solutions to global warming, air pollution, and energy security. *Energy Environ. Sci.* **2009**, *2*, 148–173. [CrossRef]
2. Horvat, M.; Nolde, N.; Fajon, V.; Jereb, V.; Logar, M.; Lojen, S.; Jacimovic, R.; Falnoga, I.; Liya, Q.; Faganeli, J.; et al. Total mercury, methylmercury and selenium in mercury polluted areas in the province Guizhou, China. *Sci. Total Environ.* **2003**, *304*, 231–256. [CrossRef]
3. Wołejko, E.; Wydro, U.; Jabłońska-Trypuć, A.; Butarewicz, A.; Łoboda, T. The effect of sewage sludge fertilization on the concentration of PAHs in urban soils. *Environ. Pollut.* **2018**, *232*, 347–357. [CrossRef] [PubMed]
4. Biasioli, M.; Barbieris, R.; Ajmonemarsan, F. The influence of a large city on some soil properties and metals content. *Sci. Total Environ.* **2006**, *356*, 154–164. [CrossRef] [PubMed]
5. Li, P.; Feng, X.B.; Qiu, G.L.; Shang, L.H.; Li, Z.G. Mercury pollution in Asia: A review of the contaminated sites. *J. Hazard. Mater.* **2009**, *168*, 591–601. [CrossRef]
6. Gallini, L.; Ajmone-Marsan, F.; Scalenghe, R. The contamination legacy of a decommissioned iron smelter in the Italian Alps. *J. Geochemical Explor.* **2018**, *186*, 121–128. [CrossRef]
7. Prada, M.; Cabo, C.; Hernández-Clemente, R.; Hornero, A.; Majada, J.; Martínez-Alonso, C. Assessing Canopy Responses to Thinnings for Sweet Chestnut Coppice with Time-Series Vegetation Indices Derived from Landsat-8 and Sentinel-2 Imagery. *Remote Sens.* **2020**, *12*, 3068. [CrossRef]
8. Novo-Fernández, A.; Franks, S.; Wehenkel, C.; López-Serrano, P.M.; Molinier, M.; López-Sánchez, C.A. Landsat time series analysis for temperate forest cover change detection in the Sierra Madre Occidental, Durango, Mexico. *Int. J. Appl. Earth Obs. Geoinf.* **2018**, *73*, 230–244. [CrossRef]
9. Barrio-Anta, M.; Castedo-Dorado, F.; Cámara-Obregón, A.; López-Sánchez, C.A. Predicting current and future suitable habitat and productivity for Atlantic populations of maritime pine (Pinus pinaster Aiton) in Spain. *Ann. For. Sci.* **2020**, *77*, 1–19. [CrossRef]
10. Thomas, I.A.; Jordan, P.; Mellander, P.-E.; Fenton, O.; Shine, O.; ÓhUallacháin, D.; Creamer, R.; McDonald, N.T.; Dunlop, P.; Murphy, P.N.C. Improving the identification of hydrologically sensitive areas using LiDAR DEMs for the delineation and mitigation of critical source areas of diffuse pollution. *Sci. Total Environ.* **2016**, *556*, 276–290. [CrossRef]
11. Woodruff, L.; Cannon, W.F.; Smith, D.B.; Solano, F. The distribution of selected elements and minerals in soil of the conterminous United States. *J. Geochemical Explor.* **2015**, *154*, 49–60. [CrossRef]

12. Wang, X.; Dan, Z.; Cui, X.; Zhang, R.; Zhou, S.; Wenga, T.; Yan, B.; Chen, G.; Zhang, Q.; Zhong, L. Contamination, ecological and health risks of trace elements in soil of landfill and geothermal sites in Tibet. *Sci. Total Environ.* **2020**, *715*, 136639. [CrossRef] [PubMed]
13. Roces-Díaz, J.V.; Cabo, C.; Prendes, C.; Ordoñez, C.; Santín, C. Automatic delineation of forest patches in highly fragmented landscapes using coloured point clouds. *Forests* **2020**, *11*, 198. [CrossRef]
14. Polat, N.; Uysal, M. An Experimental Analysis of Digital Elevation Models Generated with Lidar Data and UAV Photogrammetry. *J. Indian Soc. Remote Sens.* **2018**, *46*, 1135–1142. [CrossRef]
15. Bhattacharya, A.; Arora, M.K.; Sharma, M.L. Usefulness of adaptive filtering for improved Digital Elevation Model generation. *J. Geol. Soc. India* **2013**, *82*, 153–161. [CrossRef]
16. Aguilar, F.J.; Mills, J.P.; Delgado, J.; Aguilar, M.A.; Negreiros, J.G.; Pérez, J.L. Modelling vertical error in LiDAR-derived digital elevation models. *ISPRS J. Photogramm. Remote Sens.* **2010**, *65*, 103–110. [CrossRef]
17. Reutebuch, S.E.; Mc Gaughey, R.J.; Andersen, H.E.; Carson, W.W. Accuracy of a high-resolution lidar terrain model under a conifer forest canopy. *Can. J. Remote Sens.* **2003**, *29*, 527–535. [CrossRef]
18. Khanal, S.; Fulton, J.; Shearer, S. An overview of current and potential applications of thermal remote sensing in precision agriculture. *Comput. Electron. Agric.* **2017**, *139*, 22–32. [CrossRef]
19. Mulla, D.J. Twenty five years of remote sensing in precision agriculture: Key advances and remaining knowledge gaps. *Biosyst. Eng.* **2013**, *114*, 358–371. [CrossRef]
20. Pascucci, S.; Belviso, C.; Cavalli, R.M.; Palombo, A.; Pignatti, S.; Santini, F. Using imaging spectroscopy to map red mud dust waste: The Podgorica Aluminum Complex case study. *Remote Sens. Environ.* **2012**, *123*, 139–154. [CrossRef]
21. Liu, M.; Wang, T.; Skidmore, A.K.; Liu, X. Heavy metal-induced stress in rice crops detected using multi-temporal Sentinel-2 satellite images. *Sci. Total Environ.* **2018**, *637–638*, 18–29. [CrossRef] [PubMed]
22. Zhang, Z.; Liu, M.; Liu, X.; Zhou, G. A New Vegetation Index Based on Multitemporal Sentinel-2 Images for Discriminating Heavy Metal Stress Levels in Rice. *Sensors* **2018**, *18*, 2172. [CrossRef] [PubMed]
23. Wu, Y.; Zhang, X.; Liao, Q.; Ji, J. Can Contaminant Elements in Soils Be Assessed by Remote Sensing Technology. *Soil Sci.* **2011**, *176*, 196–205. [CrossRef]
24. Tan, K.; Wang, H.; Chen, L.; Du, Q.; Du, P.; Pan, C. Estimation of the spatial distribution of heavy metal in agricultural soils using airborne hyperspectral imaging and random forest. *J. Hazard. Mater.* **2020**, *382*, 120987. [CrossRef]
25. Shi, T.; Chen, Y.; Liu, Y.; Wu, G. Visible and near-infrared reflectance spectroscopy—An alternative for monitoring soil contamination by heavy metals. *J. Hazard. Mater.* **2014**, *265*, 166–176. [CrossRef] [PubMed]
26. Kemper, T.; Sommer, S. Use of airborne hyperspectral data to estimate residual heavy metal contamination and acidification potential in the Guadiamar floodplain Andalusia, Spain after the Aznacollar mining accident. In *Remote Sensing for Environmental Monitoring, GIS Applications, and Geology IV*; Ehlers, M., Posa, F., Kaufmann, H.J., Michel, U., De Carolis, G., Eds.; International Society for Optics and Photonics: Bellingham, WA, USA, 2004; p. 224.
27. van der Meij, B.; Kooistra, L.; Suomalainen, J.; Barel, J.M.; De Deyn, G.B. Remote sensing of plant trait responses to field-based plant–soil feedback using UAV-based optical sensors. *Biogeosciences* **2017**, *14*, 733–749. [CrossRef]
28. Gholizadeh, A.; Kopačková, V. Detecting vegetation stress as a soil contamination proxy: A review of optical proximal and remote sensing techniques. *Int. J. Environ. Sci. Technol.* **2019**, *16*, 2511–2524. [CrossRef]
29. Elachi, C.; Zimmerman, P.D. Introduction to the Physics and Techniques of Remote Sensing. *Phys. Today* **1988**, *41*, 126. [CrossRef]
30. Gholizadeh, A.; Saberioon, M.; Ben-Dor, E.; Borůvka, L. Monitoring of selected soil contaminants using proximal and remote sensing techniques: Background, state-of-the-art and future perspectives. *Crit. Rev. Environ. Sci. Technol.* **2018**, *48*, 243–278. [CrossRef]
31. Erol, S.; Özögel, E.; Kuçak, R.A.; Erol, B. Utilizing Airborne LiDAR and UAV Photogrammetry Techniques in Local Geoid Model Determination and Validation. *ISPRS Int. J. Geo-Inf.* **2020**, *9*, 528. [CrossRef]
32. Balková, M.; Bajer, A.; Patočka, Z.; Mikita, T. Visual Exposure of Rock Outcrops in the Context of a Forest Disease Outbreak Simulation Based on a Canopy Height Model and Spectral Information Acquired by an Unmanned Aerial Vehicle. *ISPRS Int. J. Geo-Inf.* **2020**, *9*, 325. [CrossRef]
33. Vågen, T.-G.; Winowiecki, L.A.; Tondoh, J.E.; Desta, L.T.; Gumbricht, T. Mapping of soil properties and land degradation risk in Africa using MODIS reflectance. *Geoderma* **2016**, *263*, 216–225. [CrossRef]

34. Choe, E.; van der Meer, F.; van Ruitenbeek, F.; van der Werff, H.; de Smeth, B.; Kim, K.W. Mapping of heavy metal pollution in stream sediments using combined geochemistry, field spectroscopy, and hyperspectral remote sensing: A case study of the Rodalquilar mining area, SE Spain. *Remote Sens. Environ.* **2008**, *112*, 3222–3233. [CrossRef]
35. Ferencz, C.; Bognár, P.; Lichtenberger, J.; Hamar, D.; Tarcsai, G.; Timár, G.; Molnár, G.; Pásztor, S.; Steinbach, P.; Székely, B.; et al. Crop yield estimation by satellite remote sensing. *Int. J. Remote Sens.* **2004**, *25*, 4113–4149. [CrossRef]
36. Rouse, J.W.; Haas, R.H.; Schell, J.A.; Deering, D.W. Monitoring the vernal advancement and retrogradation (green wave effect) of natural vegetation. *Prog. Rep. RSC* **1978**, *1973*, 112.
37. Clarke, T.R.; Moran, M.S.; Barnes, E.M.; Pinter, P.J.; Qi, J. Planar domain indices: A method for measuring a quality of a single component in two-component pixels. In Proceedings of the International Geoscience and Remote Sensing Symposium (IGARSS), Sydney, Australia, 9–13 July 2001; Volume 3, pp. 1279–1281.
38. Shabou, M.; Mougenot, B.; Chabaane, Z.L.; Walter, C.; Boulet, G.; Aissa, N.B.; Zribi, M. Soil clay content mapping using a time series of Landsat TM data in semi-arid lands. *Remote Sens.* **2015**, *7*, 6059–6078. [CrossRef]
39. Reimann, C.; Filzmoser, P.; Garrett, R.G. Background and threshold: Critical comparison of methods of determination. *Sci. Total Environ.* **2005**, *346*, 1–16. [CrossRef] [PubMed]
40. He, Z.L.; Yang, X.E.; Stoffella, P.J. Trace elements in agroecosystems and impacts on the environment. *J. Trace Elem. Med. Biol.* **2005**, *19*, 125–140. [CrossRef]
41. Syversen, T.; Kaur, P. The toxicology of mercury and its compounds. *J. Trace Elem. Med. Biol.* **2012**, *26*, 215–226. [CrossRef]
42. Evers, D.C.; Keane, S.E.; Basu, N.; Buck, D. Evaluating the effectiveness of the Minamata Convention on Mercury: Principles and recommendations for next steps. *Sci. Total Environ.* **2016**, *569–570*, 888–903. [CrossRef]
43. Boente, C.; Albuquerque, M.T.D.; Gerassis, S.; Rodríguez-Valdés, E.; Gallego, J.R. A coupled multivariate statistics, geostatistical and machine-learning approach to address soil pollution in a prototypical Hg-mining site in a natural reserve. *Chemosphere* **2019**, *218*, 767–777. [CrossRef] [PubMed]
44. Kulikova, T.; Hiller, E.; Jurkovič, Ľ.; Filová, L.; Šottník, P.; Lacina, P. Total mercury, chromium, nickel and other trace chemical element contents in soils at an old cinnabar mine site (Merník, Slovakia): Anthropogenic versus natural sources of soil contamination. *Environ. Monit. Assess.* **2019**, *191*, 263. [CrossRef] [PubMed]
45. Forján, R.; Baragaño, D.; Boente, C.; Fernández-Iglesias, E.; Rodríguez-Valdes, E.; Gallego, J.R. Contribution of fluorite mining waste to mercury contamination in coastal systems. *Mar. Pollut. Bull.* **2019**, *149*, 110576. [CrossRef] [PubMed]
46. Jiménez-Moreno, M.; Barre, J.P.G.; Perrot, V.; Bérail, S.; Rodríguez Martín-Doimeadios, R.C.; Amouroux, D. Sources and fate of mercury pollution in Almadén mining district (Spain): Evidences from mercury isotopic compositions in sediments and lichens. *Chemosphere* **2016**, *147*, 430–438. [CrossRef]
47. González-Fernández, B.; Rodríguez-Valdés, E.; Boente, C.; Menéndez-Casares, E.; Fernández-Braña, A.; Gallego, J.R. Long-term ongoing impact of arsenic contamination on the environmental compartments of a former mining-metallurgy area. *Sci. Total Environ.* **2018**, *610–611*, 820–830. [CrossRef]
48. Loredo, J.; Ordonez, A.; Alvarez, R. Environmental impact of toxic metals and metalloids from the Muñón Cimero mercury-mining area (Asturias, Spain). *J. Hazard. Mater.* **2006**, *136*, 455–467. [CrossRef]
49. Matanzas, N.; Sierra, M.J.; Afif, E.; Díaz, T.E.; Gallego, J.R.; Millán, R. Geochemical study of a mining-metallurgy site polluted with As and Hg and the transfer of these contaminants to *Equisetum* sp. *J. Geochem. Explor.* **2017**, *182*, 1–9. [CrossRef]
50. Fernández, S.; Poschenrieder, C.; Marcenò, C.; Gallego, J.R.; Jiménez-Gámez, D.; Bueno, A.; Afif, E. Phytoremediation capability of native plant species living on Pb-Zn and Hg-As mining wastes in the Cantabrian range, north of Spain. *J. Geochem. Explor.* **2017**, *174*, 10–20. [CrossRef]
51. Moreno-Jiménez, E.; Peñalosa, J.M.; Manzano, R.; Carpena-Ruiz, R.O.; Gamarra, R.; Esteban, E. Heavy metals distribution in soils surrounding an abandoned mine in NW Madrid (Spain) and their transference to wild flora. *J. Hazard. Mater.* **2009**, *162*, 854–859. [CrossRef]

52. Handique, B.K.; Goswami, C.; Gupta, C.; Pandit, S.; Gogoi, S.; Jadi, R.; Jena, P.; Borah, G.; Raju, P.L.N. Hierarchical classification for assessment of horticultural crops in mixed cropping pattern using UAV-borne multi-spectral sensor. *ISPRS Int. Arch. Photogramm. Remote Sens. Spat. Inf. Sci.* **2020**, *XLIII-B3-2*, 67–74. [CrossRef]
53. Franzini, M.; Ronchetti, G.; Sona, G.; Casella, V. Geometric and Radiometric Consistency of Parrot Sequoia Multispectral Imagery for Precision Agriculture Applications. *Appl. Sci.* **2019**, *9*, 5314. [CrossRef]
54. Cubero-Castan, M.; Schneider-Zapp, K.; Bellomo, M.; Shi, D.; Rehak, M.; Strecha, C. Assessment of the Radiometric Accuracy in a Target Less Work Flow Using Pix4D Software. In Proceedings of the 2018 9th Workshop on Hyperspectral Image and Signal Processing: Evolution in Remote Sensing (WHISPERS), Amsterdam, The Netherlands, 23–26 September 2018; IEEE: Piscataway, NJ, USA, 2018; pp. 1–4.
55. BOPA Generic Reference Levels for Heavy Metals in Soils from Principality of Asturias, Spain. 2014. Available online: http://sede.612asturias.es/bopa/2014/04/21/2014e06617.pdf (accessed on 25 November 2020).
56. Oliver, M.A.; Webster, R. A tutorial guide to geostatistics: Computing and modelling variograms and kriging. *CATENA* **2014**, *113*, 56–69. [CrossRef]
57. A Database for Remote Sensing Indices. Available online: https://www.indexdatabase.de/db/i.php (accessed on 25 November 2020).
58. Kim, M.S. The Use of Narrow Spectral Bands for Improving Remote Sensing Estimations of Fractionally Absorbed Photosynthetically Active Radiation. Ph.D. Thesis, University of Maryland, College Park, MD, USA, 1994.
59. Fernández, B.; Lara, L.M.; Menéndez-Aguado, J.M.; Ayala, J.; García-González, N.; Salgado, L.; Colina, A.; Gallego, J.L.R. A multi-faceted, environmental forensic characterization of a paradigmatic brownfield polluted by hazardous waste containing Hg, As, PAHs and dioxins. *Sci. Total Environ.* **2020**, *726*, 138546. [CrossRef] [PubMed]
60. Gitelson, A.; Szilagyi, F.; Mittenzwey, K.-H. Improving quantitative remote sensing for monitoring of inland water quality. *Water Res.* **1993**, *27*, 1185–1194. [CrossRef]
61. Okkenhaug, G.; Zhu, Y.-G.; He, J.; Li, X.; Luo, L.; Mulder, J. Antimony (Sb) and Arsenic (As) in Sb Mining Impacted Paddy Soil from Xikuangshan, China: Differences in Mechanisms Controlling Soil Sequestration and Uptake in Rice. *Environ. Sci. Technol.* **2012**, *46*, 3155–3162. [CrossRef]
62. Gil-Díaz, M.; Rodríguez-Valdés, E.; Alonso, J.; Baragaño, D.; Gallego, J.R.; Lobo, M.C. Nanoremediation and long-term monitoring of brownfield soil highly polluted with As and Hg. *Sci. Total Environ.* **2019**, *675*, 165–175. [CrossRef]
63. Baragaño, D.; Boente, C.; Rodríguez-Valdés, E.; Fernández-Braña, A.; Jiménez, A.; Gallego, J.R.; González-Fernández, B. Arsenic release from pyrite ash waste over an active hydrogeological system and its effects on water quality. *Environ. Sci. Pollut. R.* **2020**, *27*, 10672–10684. [CrossRef]
64. Gallego, J.R.; Ortiz, J.E.; Sánchez-Palencia, Y.; Baragaño, D.; Borrego, A.; Torres, T. A multivariate examination of the timing and accumulation of potentially toxic elements at Las Conchas bog (NW Spain). *Environ. Pollut.* **2019**, *254*, 113048. [CrossRef]

International Journal of
Geo-Information

MDPI

Article

A Time Series Investigation to Assess Climate Change and Anthropogenic Impacts on Quantitative Land Degradation in the North Delta, Egypt

Mohamed A. E. AbdelRahman [1,*], Ahmed A. Afifi [2] and Antonio Scopa [3]

1 Division of Environmental Studies and Land Use, National Authority for Remote Sensing and Space Sciences (NARSS), Cairo 11769, Egypt
2 Soils and Water Use Department, National Research Centre, Giza 12622, Egypt; ah.afifi@nrc.sci.eg
3 Scuola di Scienze Agrarie, Forestali, Alimentari ed Ambientali (SAFE), Università degli Studi della Basilicata, Viale dell'Ateneo Lucano, 10, 85100 Potenza, Italy; antonio.scopa@unibas.it
* Correspondence: maekaoud@narss.sci.eg; Tel.: +20-1004781114

Abstract: In the current study the processes of soil deterioration over the past five decades was evaluated. Land degradation risk, status, and rate were assessed in Kafr El-Sheikh Governorate, Egypt, in 2016 using OLI and ETM (2002) remote sensing data, and soil data from 1961.A quantitative deterioration was produced based on the comparative study approach in the integrated weighted sum, weighted overlay, and fuzzy model. The parameters used were soil depth, texture, pH, EC, OM, SAR, ESP, CEC, $CaCO_3$, BD, N, P, K. The variables were based on the measurements derived from the Universal Soil Loss Equation (USLE). The results of the implemented USLE in the GIS model-builder revealed the prevalence of severe soil deterioration processes in the region, and include four main deterioration risks: water-logging, soil compaction, salinization, and alkalization. During 2002–2016, soil sealing took place on 36,297.87 ha of the study area (9.7% of the total area). Urban sprawl was one of the most noticed problems that became apparent during the fieldwork during the inventory of land resources in the investigation area. Soil sealing is one of the hidden manifestations of desertification, and it is the implicit explanation for the lost land for the agricultural production process. The study showed that the investigated soil, as a part of north Nile Delta, is a very fragile system that needs to be protected, especially under the effect of climate change in areas overloaded with population, and because of their negative effects on soil properties. According to the results of this study, it is recommended that the same approach be applied to similar agricultural semi-arid regions to help in building a database of land resources for agricultural use that will be very useful for the decision-maker to monitor changes on agricultural lands.

Keywords: climate changes; dynamic land degradation; ArcGIS model builder; remote sensing

Citation: AbdelRahman, M.A.E.; Afifi, A.A.; Scopa, A. A Time Series Investigation to Assess Climate Change and Anthropogenic Impacts on Quantitative Land Degradation in the North Delta, Egypt. *ISPRS Int. J. Geo-Inf.* **2022**, *11*, 30. https://doi.org/10.3390/ijgi11010030

Academic Editors: Dimitris Triantakonstantis and Wolfgang Kainz

Received: 31 October 2021
Accepted: 28 December 2021
Published: 31 December 2021

1. Introduction

Soil is the product of interacting effects of multiple environmental controls, combined in specific geographical/historical backgrounds to produce highly improbable (i.e., unlikely to be duplicated) outcomes. Consequently, soil degradation represents a global environmental and developmental concern [1]. Up-to-date, quantitative information is required to provide policy recommendations and to properly plan action to increase food security and food safety, stimulate economic development, ensure water security, maintain environmental integrity and conserve resources [1]. Therefore, land degradation is defined as a long-term deterioration in ecosystem services in terms of net primary productivity. The main land degradation processes are degradation of vegetative cover, wind and rainfall-induced erosion, salinization, reduction of soil organic matter content, soil crusting and compaction, and presence of toxic substances [2]. In addition, they differentiate between factors (variables) and processes (forms) of degradation in soils. Degradation processes of soils incorporate forms of biological, physical and chemical inter-actions, which influence the soil's capacity for self-regulation and its production efficiency. Whereas degradation factors (variables) of soils are components of human-induced, natural agents, and catalysts that set those forms in motion and movement, driving to changes in soil characteristics and its life-support properties. Changes in soil properties resulting from physical degradation include hydrological or physical characteristics which lead to changes in soil physical properties, which have negative effects on environmental quality, crop growth, and animal production, which in turn negatively affect farm income. Soil degradation is defined as a quality reduction of soil, whereby its actual and/or potential productivity and/or capability decrease in uses as a multi-purpose resource, as a consequence of both human-induced and natural causes [3–8]. Many studies have indicated that the most important natural processes of land degradation are fragipan and clay pan formation, laterization, and hard-setting [4]. The processes and effects on soil for increasing the production determination are soil compaction, salt accumulation and nutrient leaching, accelerated erosion, desertification, and acidification. During the last fifty years, the quick populace development in Egypt has caused an extraordinary request for nourishment and other agrarian items. Since 50% of the food needs were generated locally, much attention has been paid to expanding the production from agriculture in Egypt. This can be achieved through two main strategies, which comprise reclamation of vast desert areas to be productive land, and the intensive cultivation of productive land using high technology management. In both cases it is of paramount importance to obtain information on soil properties and the distribution of these properties in these areas. Therefore, there is a need to build a framework that can provide accurate, valuable and timely data on soil and water to decision makers and policy planners. In Egypt, the problems caused by soil deterioration are not only serious but also far-reaching in terms of the affected regions and the millions of individuals who suffer from its consequences [3,5–20] related to climate changes and also to the conditions under which the small Egyptian farmers have to live and work. Land degradation severity can be efficiently monitored from multi-temporal satellite images. Environmental sensitivity could assess the response of the environment to any change in one or more external factors. The Mediterranean Desertification and Land Use (MEDALUS) model focuses on recognizing environmentally sensitive areas (ESAs) that reveal different sensitivity to desertification through multi-factor approaches. To define the ESA index (ESAI) four qualities are evaluated: soil, climate, vegetation, and management. Through this model, more attention can be given to areas which are most vulnerable to degradation. This model is simple, robust, adaptable, and widely applicable in the Mediterranean region [21]. Three general types of ESAs to desertification can be distinguished: critical, fragile, potential ESAs—besides the non-sensitive areas. The northern part of the Nile Delta, located between the Rosetta and Damietta branches, is formed from coastal, alluvial and lacustrine deposits. Most of this area is affected by much soil degradation, i.e., salinity, alkalinity, poor drainage, wind and water erosion and compactions. On the other hand, there are many on-land reclamation processes and soil conservation performance as addition of amendments (gypsum, super phosphate, sulphur), tile drainage, sub-soiling, etc. The satellite images and Geographic Information System (GIS) are tools developed to save time and money in soil studies and to give more accurate results for soil mapping. Consequently, different remote sensing and GIS data were used in this study to investigate and to evaluate the soils of the studied area and monitor soil degradation.

The effect of climate change on soil is a complex and slow process because the soil is severely influenced not only by direct climate change (e.g., the effect of temperatures on the decay of soil organisms) [22]. Soil organic matter (OM) was subjected to decay as a result of climate change, resulting in a reduction in the stability of the soil layer and the rate of water infiltration, and in an increase in the danger of congestion, water flow, and erosion [23]. The fastest chemical or mineral changes caused by climate change might include salt loss and nutrient intake as leakage rises, and salinity in water when total water movement occurs owing to greater evaporation or less rainfall or irrigation water. The composition of the mineral clay, according to the mineralogy of the coarse particles, will usually change slightly over time. Changes in the surface characteristics of the clay particle are often slower than changes in pile formation or crystalline structure, which occur significantly more quickly. Such environmental changes have a significant impact on the physical and chemical characteristics of the soil [24–26].

Determination of the land degradation status could be considered the cornerstone of any agricultural development program, in which recognizing its various types is a necessity for taking sound scientific and practical measures to combat it and preserve the agricultural soil in a proper and maintainable manner. The current work aims to assess soil degradation in the last five decades from 1961 to 2016. This was achieved by building a database of land resources for agricultural use and building a new land degradation assessment model.

2. Materials and Methods

2.1. Study Area

The study area is located in the administrative borders of Kafr El-Sheikh Governorate (Figure 1), which is famous for agriculture, especially rice in the north part of the governorate. The entire area (3.748 km^2) is located in the north of the Nile delta and overlooks the Mediterranean Sea. It extends a distance of 100 km between the two branches of the Nile on the coast of the Mediterranean Sea, and is bordered by the Rashid River branch on the western side with a length of 85 km. It is bounded on the eastern side by the Dakahlia Governorate, and its borders from the southern side are the Gharbia Governorate. Kafr El-Sheikh Governorate is famous for the cultivation of cotton, rice, corn and wheat crops. These crops are irrigated with water from the lake behind the Zefta Barrage located on the Damietta River branch of the Nile. In the early eighties, sugar beet was introduced to the governorate, and at that time fish farming began. It is also mentioned that the two cities of Desi Bella are among the most important cities whose industries are related to agricultural products, such as the cotton industry, rice milling, and the sugar beet industry. There are also other industries in the capital, Kafr El-Sheikh.

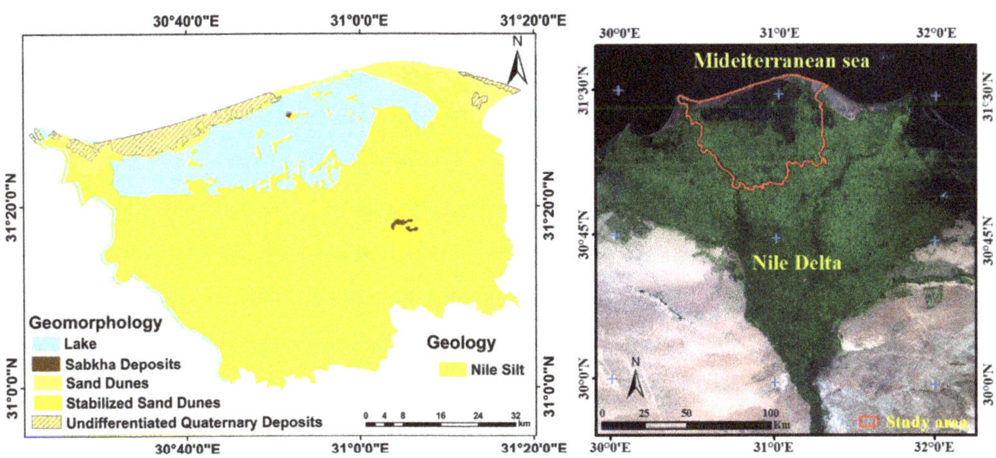

Figure 1. Geomorphology associated with geology (after [27] Conoco (1987)). Location map of the study area.

The study area is located north of the Nile Delta in Egypt. The soil of the Nile River Delta is considered one of the areas rich in nutrients, due to the silt deposits (mud) resulting from the flow of the Nile River [27]. Branches of the Nile River spread to the Mediterranean Sea. In ancient times, the frequent floods of the river led to the spread of silt in the vicinity of the Nile Delta, which led to an increase in its area east and west. It has two main branches: Al-Rashid in the west, and Damietta in the east [27]. Figure 1 shows geomorphology associated with geology: (1) Nile silt occupies the largest area with fine-grained sediment (silt and clay) deposited from suspension on a flood plain by floodwaters that cannot be contained within the stream channel. (2) Quaternary marine deposits are common on lower parts of the present land surface (such as carbon deposits and gypsum metal) and terrestrial deposits are common on higher parts of the present sea floor. (3) Sabkha deposits are salt-flat soils that are flat and very saline areas of sand or silt lying just above the watertable and often contain soft nodules and enterolithic veins of gypsum or anhydrite. (4) Sand dunes are a ridge of sand created by the wind and are found in deserts or near lakes and oceans. (5) Stabilized dunes are areas where wind moves the dunes inward from the lake. If the sand dune is stabilized by vegetation such as this dune slope, movement is halted.

2.2. Remote Sensing Data and Image Processing

Satellite images for the study area in 2002 and 2016 were used for the visual interpretation of land degradation, while supervised classification was used to extract the land use/land cover (LULC) classes. Both satellite images have a spatial resolution of 30 m and can cover the area with one scene.

The recent administrative boundary of Kafr El-Sheikh Governorate was used to subset the area from the satellite Images. Then supervised classification techniques in ENVI software were applied, using the combination of selected bands that accurately help to visually separate the targeted LULC classes. The ETM 2002 satellite imagery was processed using a band combination of 4 (NIR), 5 (MIR), and 3 (Red), whereas the Landsat 8 imagery (OLI/TIRS 2016) was processed using a band combination of 5-6-4. The distinct LULC classes in the area were detected using a supervised technique. Using this method, four LULCs were identified: agricultural land, dunes, fish farm, lake, and urban. Different training sets were assigned for each LULC class before executing the supervised classification for each image, and then confirmed using hundreds of ground truth points taken from the field in 2016 from particular places that shared the same locations in the 2002 and 2016 images. A signature file was constructed for each date based on the training samples and the separability analysis within each class. The implementation of these hundred points were used to create the classified LULC thematic maps for each assigned date using the resulted signature file (2002 and 2016).

The detail on satellites used, the kind of images, and their bandwidths are indicated in Table 1. The source of the images is from https://www.usgs.gov (accessed on 30 June 2016). ENVI software was used to process and analyze the images.

Table 1. Criteria of the satellite images used for years 2002 and 2016.

Landsat 7 Enhanced Thematic Mapper Plus (ETM+) for the Year 2002			Landsat 8 Operational Land Imager (OLI) and Thermal Infrared Sensor (TIRS) for the Year 2016		
Bands	**Wavelength (mm)**	**Resolution (m)**	**Bands**	**Wavelength (mm)**	**Resolution (m)**
Band 1—Blue	0.45–0.52	30	Band 1—Ultra Blue (coastal/aerosol)	0.435–0.451	30
Band 2—Green	0.52–0.60	30	Band 2—Blue	0.452–0.512	30
Band 3—Red	0.63–0.69	30	Band 3—Green	0.533–0.590	30
Band 4—NIR	0.77–0.90	30	Band 4—Red	0.636–0.673	30
Band 5—SWIR 1	1.55–1.75	30	Band 5—NIR	0.851–0.879	30
Band 6—Thermal	10.40–12.50	60 * (30)	Band 6—SWIR 1	1.566–1.651	30
Band 7—SWIR 2	2.09–2.35	30	Band 7—SWIR 2	2.107–2.294	30
Band 8—Panchromatic	0.52–0.90	15	Band 8—Panchromatic	0.503–0.676	15

Table 1. *Cont.*

Landsat 7 Enhanced Thematic Mapper Plus (ETM+) for the Year 2002			Landsat 8 Operational Land Imager (OLI) and Thermal Infrared Sensor (TIRS) for the Year 2016		
Bands	Wavelength (mm)	Resolution (m)	Bands	Wavelength (mm)	Resolution (m)
			Band 9—Cirrus	1.363–1.384	30
* ETM+ Band 6 is acquired at 60 m resolution, but products are resampled to 30 m pixels.			Band 10—Thermal 1	10.60–11.19	100 * (30)
			Band 11—Thermal 2	11.50–12.51	100 * (30)
			* TIRS bands are acquired at 100 m resolution, but are resampled to 30 m in delivered data product.		

2.3. Fieldwork and Laboratory Analysis

The soil survey was conducted with a semi-detailed survey by combining remote sensing data and field visits to identify different soil patterns. DEM SRTM 30 m is preprocessed to generate terrain to represent the height of the ground surface to extract the landform units used in the field visit. Longitudes and latitudes, as well as elevation, are defined in the field by using GPS "System Corporation MAGELLAN"—GPS NAV DLX-10 TM for recognizing the soil profile location within the studied area. Sixty soil profiles were generated up to depth of soil water in the study area (Figure 2). Detailed morphological descriptions were carried out following the guidelines for soil description, which include various surface characteristics, i.e., coordinates, elevation, slope, topography, landform, vegetation, parental material, land use, drainage, and different soil layers [28]. Description of soil profile layers was carried out in the field, and includes: color, texture, structure, consistency, cementation and compaction, roots, and all other features recognized in each layer. Soil samples were collected from each described layer according to the morphological features to determine the physical and chemical characteristics, i.e., soil depth, texture, pH, EC, OM, SAR, ESP, CEC, $CaCO_3$, BD, N, P, K. A total of 213 soil samples were collected from the different horizons of each profile and the appropriate laboratory analyses were executed.

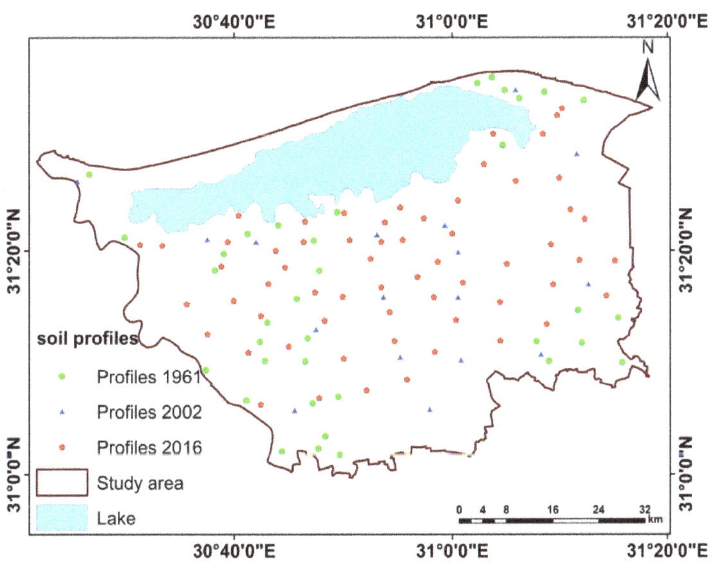

Figure 2. Location of the soil profiles in the study area.

2.4. Preparation of Soil Samples and Methods of Analysis

The soil samples collected in 2016 were air-dried, gently crushed, sieved through a 2 mm sieve and kept for analysis. Fractions of less than 2 mm were subjected to the following analyses.

Particle size distribution was analyzed after the following pretreatments: removal of carbonates using HCl, removal of organic matter using H_2O_2, and removal of soluble salts by leaching using distilled water. Sodium hexameta-phosphate (EDTA) was used as a dispersing agent, and then the pipette method was carried out to determine the % clay, % silt, and % sand [29]. Total calcium carbonate was determined by Collin's calcimeter [30]. Organic matter content was determined according to Nelson and Sommers [31]. Soil reaction (pH) was determined in the soil paste using the Beckman pH meter. The electrical conductivity (EC) of the saturated soil paste extract was carried out according to Black (1985) [32]. The saturation of soil water extract components was determined, including the carbonates and bicarbonates, by titration using phenolphthalein and bromocresol green as indicators [33]; the chlorides using Mohr's method [33]; the sulphates were calculated by the difference between the total cations and anions. Calcium and magnesium were determined by the versenate (EDTA) method using ammonium chloride as a buffer and EBT as an indicator for calcium [33]. Sodium and potassium were determined photometrically using a Perkin Elmer flame photometer [33].

2.5. Soil Classification

Soils were categorized from soil order through soil family according to the keys of Soil Taxonomy, and soil taxa were interpreted and presented according to the Soil Survey Manual [34].

2.6. Active Land Degradation Assessment

This study is based on the comparison between the data extracted from the Survey Report of the Ministry of Agriculture [35–37] and the analysis of the recent remote sensing validated by field trip missions taking place in 2016. The FAO/UNEP [38] and UNEP Staff [39] approach methodologies were applied for assessing soil degradation.

2.7. Quantitative Land Degradation Assessment

A quantitative assessment for the soil degradation method was applied after the Universal Soil Loss Equation (USLE) for four land degradation processes: salinity, alkalinity, compaction, and waterlogging. The criteria were used to determine the degree, class, and rate of different types belonging to land degradation, as shown in Tables 2 and 3.

Table 2. Criteria used to determine the degree of different types of degradation after FAO/UNEP [2].

Criteria/Degradation Type	Indicator	Unit	None1	Slight2	Moderate3	Severe4
Salinization	ECe	dS m^{-1}	<4	4–8	8–16	16–32
		Class	Non	Slight	Moderate	Severe
Alkalization	ESP	%	<10	10–15	15–30	30–50
		Class	Non	Slight	Moderate	Severe
Compaction	Bulk density	g m^{-1}	<1.2	1.2–1.4	1.4–1.6	>1.6
		Class	loose	Slight	Hard	Very hard
Waterlogging	Waterlogging	cm	>150	150–100	100–50	<50
		Class	Non	Slight	Moderate	Severe

Table 3. Criteria used to determine the classes and rates of different types of degradation after FAO/UNEP [2].

Chemical Degradation	Salinization/Increase in EC dS m^{-1} y^{-1}	Alkalinization/Increase in ESP % y^{-1}
Non to slight	<0.5	<0.5
Moderate	0.5–3	0.5–3
High	3–5	3–7
Very high	>5	>7
Physical Degradation	**Compaction (Increase in Bulk Density g $cm^{-3}y^{-1}$)**	**Waterlogging (Increase in Water Table per cm y^{-1})**
Non to slight	<0.1	<1
Moderate	0.1–0.2	1–3
High	0.2–0.3	3–5
Very high	>0.3	>5

2.8. Spatial Modeling for Land Degradation

GIS supports studies of land degradation by providing a good platform of data base storage, modeling, presentation of results and development of a user-friendly interface. Combined with GPS, the navigation of degradation areas can easily be evaluated. The spatial models were created using ArcGIS 10.5, a software from ESRI, using the Model Builder extension. Model Builder adds utility to ArcGIS; in particular, a model window provides advanced techniques for extending software capabilities, by allowing the user to design and share your models as tools and to create and implement simple workflows. Within the model document, users create models. The diagrams represent model processes, as shown afterwards in Figure 3. These processes are defined as chained model nodes depicting input data, geo-processing functions, and output or derived data. Each type of model node is represented as a distinctly shaped and colored icon. Using the Model Builder in ArcGIS, a model was developed for each of the respective land degradations in the years examined. Once the land degradation variables of each model were determined by the USLE for each type, all input shape files of the land degradation types were converted to discrete grid formats using the vector conversion functions. Grid files are composed of pixels, to which one can assign different values.

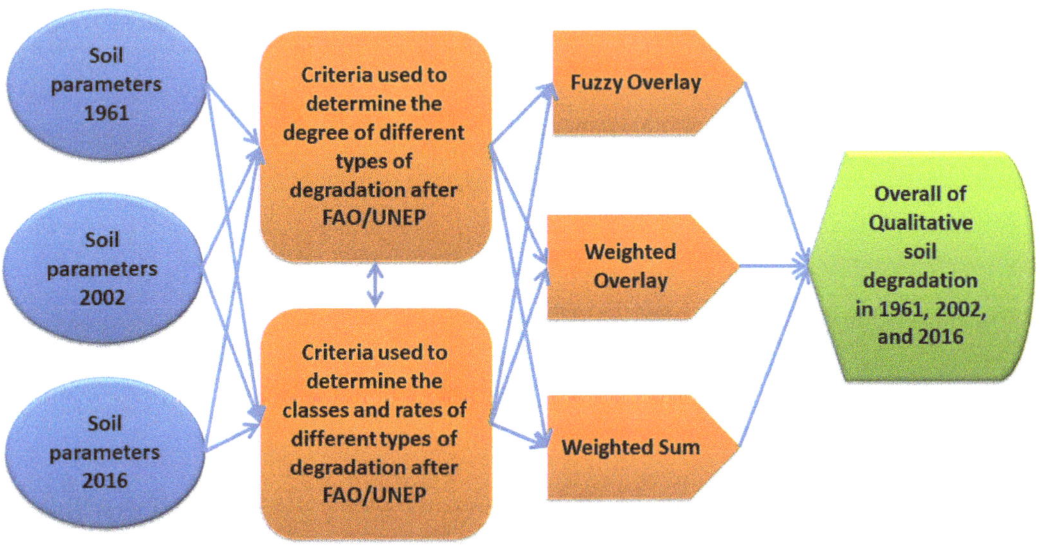

Figure 3. Model builder for overall qualitative soil degradation in 1961, 2002, and 2016.

3. Results and Discussions

3.1. LULC Classes Using Supervised Classification

The different LULC classes were extracted from the satellite images from2002 and 2016 and the results obtained are presented in Table 4 and Figures 4 and 5.

Table 4. The percentage of area coverage for LULC classes in 2002 and 2016.

LULC Classes	2002 (%)	2016 (%)
Agricultural Land/Natural Vegetation	72	68
Dunes/Sabkha/Sand Sheet	9	6
Fish Farm	6	10
Burullus Lake	8	6
Urban	5	10

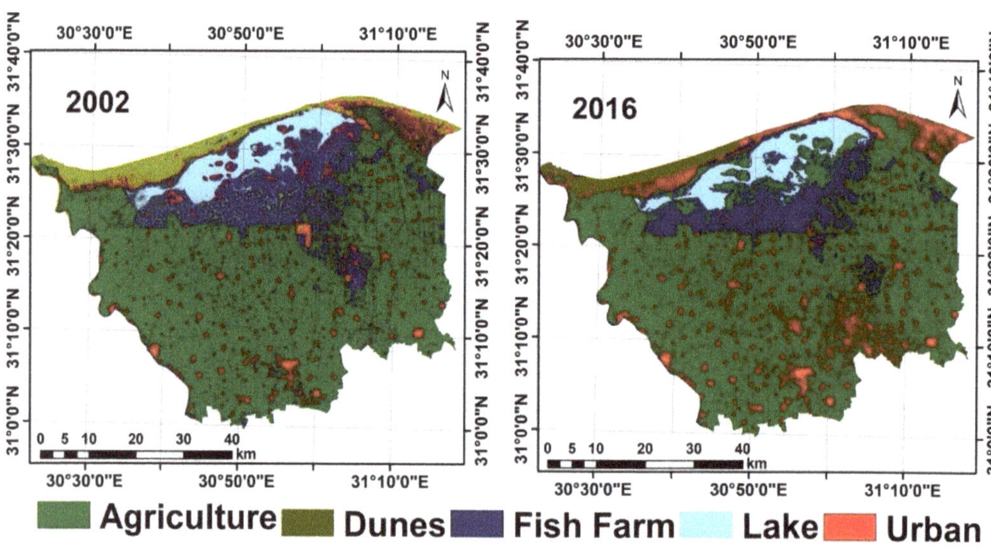

Figure 4. The land use/land cover classification maps in years of 2002 and 2016.

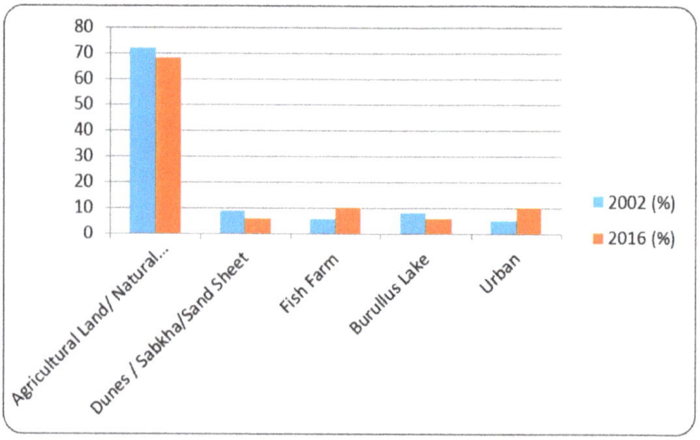

Figure 5. The percentage for each LULC class from 2002 and 2016.

During the classification processes, it was difficult to differentiate some categories of land use, such as plants in agricultural land and natural plants, so they were combined into one classification called agriculture. Similarly, it was also difficult to differentiate between sand dunes, sabkha and sand in one category, namely the Dunes. Despite the efforts of the Egyptian state to prevent construction on agricultural lands, an increase in urbanization on agricultural lands was observed in the period between 2002 and 2016 at the expense of agricultural lands. Where the capacity of agricultural land declined by 4%, all came as a result of the increase in urbanization expansion in the region.

The LULC-classified maps underwent an accuracy assessment examination using overall accuracy and the Kappa coefficient. The overall classification accuracies for the dates studied are 0.96 and 0.97 for agriculture; for dunes they are 0.90, 0.91; for fish farm, 0.95 and 0.96, and for urban, 0.92 and 0.93 for the years 2002 and 2016, respectively, which indicates a high accuracy for all classified maps. Additionally, the overall Kappa statistics are 0.95, 0.90, 0.94, and 0.92 for 2002, and 2016, respectively.

3.2. Physiographic and Soil Units in the Area

Building the physiographic map was based on an interpretation of the slope derived from the Digital elevation model (DEM SRTM 30 m) overlapped with satellite images and integrated with field work survey and analytical data. It reveals that the study area contains three main classes: flood plain, lacustrine deposits, and coastal plain. The field studies and laboratory analyses of the collected soil samples resulted in the characteristics shown in the Table 5. A significant decrease of 3% of total area from the area coverage of the Dunes, Sabkha and Sand Sheet classes, is referred to by Dunes in Figure 4. This change was due to the continuous reclamation processes over the non-cultivated areas and urban encroachment as well. The area covered by the lake water decreased by 2% between 2002 and 2016, in complete contrast to the area cover of the water of fish farms, which increased by 4% of the total area.

Table 5. Soil properties in the study area.

Characteristics	Soils of Flood Plain	Soils of Lacustrine Deposited	Soils of Coastal Plain
Texture	Clay for all profiles	Sandy clay loam and loamy sand	Sandy soils
Depth & Drainage	Moderately deep to deep soils	Moderately deep to deep soils & poor and moderately to well-drained	Very deep
Structure	Moderate-to-strong structure evolution and blocky to massive and columnar. The consistence is slightly hard when dry, slightly sticky and slightly plastic.	Moderate-to-strong structure evolution and blocky to massive and columnar. The consistence is slightly hard when dry, slightly sticky and slightly plastic.	Single grain. The consistence is hard when dry, non-sticky and non-plastic.
Horizon boundary	Diffuse smooth boundary	Diffuse smooth boundary	Clear and diffuse smooth boundary
Bulk density		1.30 and 1.47 g cm^{-3}, increases with depth	
$CaCO_3$ content	1.04 to 5.38%	0.46 to 3.97%	0.21 to 1.24%
pH	8.00 to 8.60	7.80 to 8.50	7.1 to 8.8
ECe	0.99–7.90 dS m^{-1}	1.58–79.0 dS m^{-1}	1.38–3.36 dS m^{-1}
Dominated cation	Dominated by Na^+ cation followed by Ca^{2+} or Mg^{2+} cations alternately, while K^+ cation is the least soluble component.	Dominated by Na^+ cation followed by Ca^{2+} and Mg^{2+} cations, while K^+ cation is the least soluble component.	Dominated by Na^+, Mg^{2+}, Ca^{2+}, while K^+ cation is the least soluble component.
Dominated Anion	Dominated by Cl^- ion or $SO_4{}^{2-}$ ion alternately followed by $HCO_3{}^-$. Thus, sodium chloride or sodium sulfate seems to be dominating in these soils.	Cl^- ion followed by $SO_4{}^{2-}$ ion and $HCO_3{}^-$	Cl^- ion followed by $SO_4{}^{2-}$ and by $HCO_3{}^-$

Table 5. *Cont.*

Characteristics	Soils of Flood Plain	Soils of Lacustrine Deposited	Soils of Coastal Plain
Organic matter content	0.34 to 2.78%	0.33 to 3.03%	0.52 to 2.19%
Cation Exchange Capacity	33.12 to 46.92 meq/100 g of soil	33.12 to 46.92 meq/100 g of soil	6.9 to 18.36 meq/100 g of soil
ESP	3.41 to 14.86%	6.01 to 43.85%	5.09% and 12.69%

3.3. Soil Classification of the Study Area

The pedological study of the study area produced the following soil map units. The spatial distribution of these units is shown in Figure 6 and Table 6. The highest magnitude soil unit is *VerticTorrifluvents*, occupying an area of 1812.96 km^2 (58.71%) of the southern part of the investigated area. It is bonded at the north end by batches of *TypicHaplosalids* and mostly by *AquallicSalorthids*, which surround the Lake Brullus area. The latter soil classification unit occupies an area of cultivated land equivalent to 810.24 km^2 (26.2%). *TypicHaplosalids* has a covered area of 7.91% (244.40 km^2) compared with the previous units, which have a lower frequency around the lake. While the frequencies are very slight, they were found in the rest of the soil units. The dune area was 85.67 km^2, representing 2.8% in narrow strips along the northern border of the study area. Figure 3 and Table 5 show that the order of the soil units depending on the occupation area is as follows: *VerticTorrifluvents*, *AquallicSalorthids*, *TypicHaplosalids*, *TypicQuartzipsamments*, *Sanddunes*, *TypicTorrifluvents*, *TypicTorripsamments* and *TypicSalorthids*.

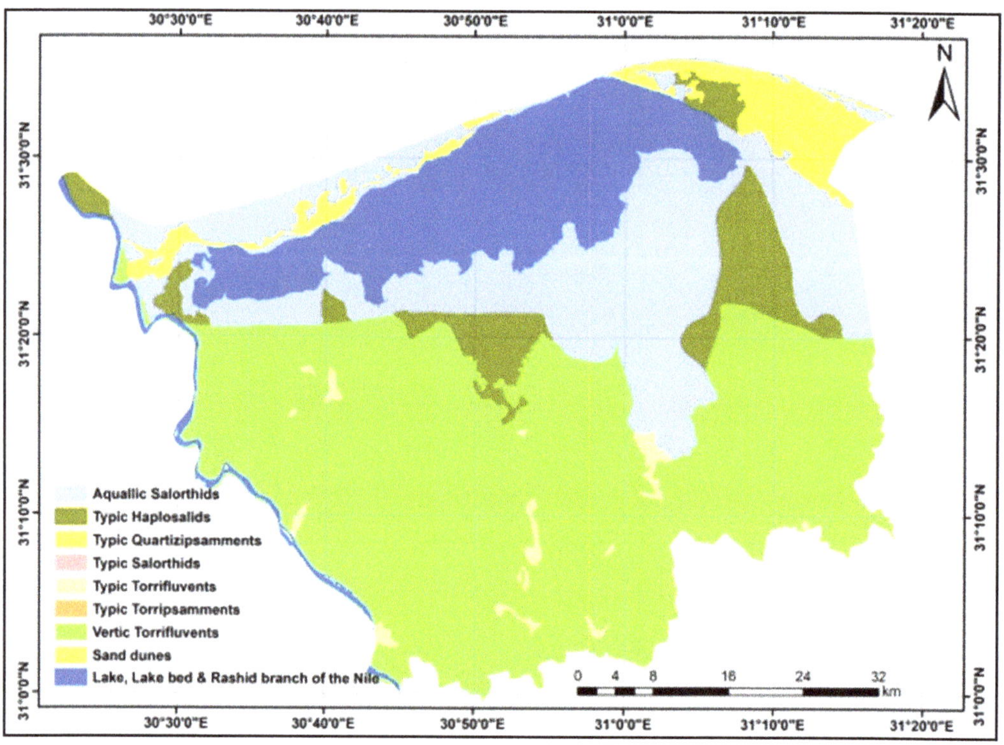

Figure 6. Soil taxonomy units of the study area, 2016.

Table 6. Areas and frequencies of soil sub-groups in Kafr El Sheikh Governorate.

Type (Sub Group)	Area (km^2)	%
AquallicSalorthids	810.24	26.24
TypicHaplosalids	244.40	7.91
TypicQuartzipsamments	94.07	3.05
TypicSalorthids	0.06	0.00
TypicTorrifluvents	40.44	1.31
TypicTorripsamments	0.13	0.00
VerticTorrifluvents	1812.96	58.71
Sand dunes	85.67	2.77
Total	3087.97	100.00

It was noted that *AquallicSalorthids*, *TypicSalorthids* and *TypicHaplosalids* were found to be the highly affected by salinity, alkalinity, compaction, and waterlogging. *VerticTorrifluvents* and *TypicTorrifluvents* were found to be highly affected by salinity, alkalinity, and compaction. *TypicQuartzipsamments* and *TypicTorripsamments* of dune area were found to be affected by salinity hazards.

3.4. Comparative Studies of Physiochemical Soil Characteristics

An assessment of soil physical and chemical properties was carried out in the studied area in1961 [35–37]. The results revealed that the soil properties had strongly changed due to land reclamation, conservation, and land management. This research focused on data available in 1961, 2002, and 2016. Since the sampling points are not at the same locations, the estimated degradation is based on changes in the spatial distribution of properties generated using the geostatistical method in ArcGIS. The quantitative comparison of the land degradation at the different dates (1961, 2002, and 2016) was carried out using spatio-temporal variance techniques.

A very important factor in crop productivity management is soil depth (Figure 7). In this study, the soil depth differed in each of the study periods, where the average depth (90–100 cm) occupied 45% of the total area in 1961, while in 2002 it reached 26% of the total area, and in 2016 it was 29% of the total area. Despite this fluctuation in the values, an increase in the shallow soil depth (≤70 cm) ratio was observed from 13%, and 18% to 24% of the total area for the years studied, respectively. It is very clear that the mismanagement of the area has caused soil deterioration in form of waterlogging. It is certain that the depth of the soil in hot areas is an influential factor in the salinity of the land. These results are in direct agreement with the results obtained in the previous studies [7,40,41].

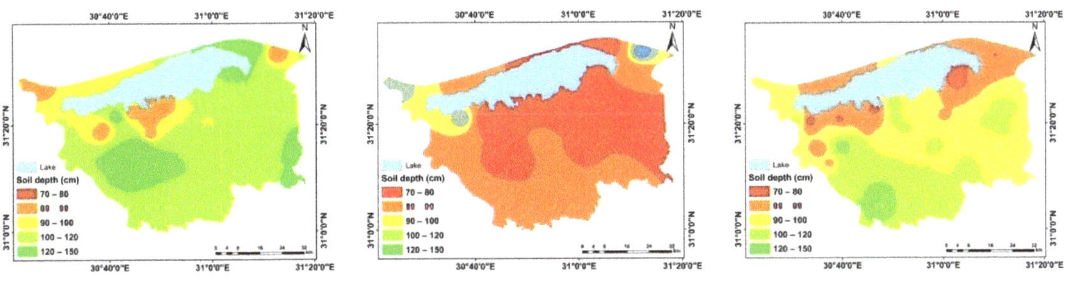

Figure 7. Soil depth maps in 1961, 2002, and 2016.

Soil salinity (Figure 8) is a limiting production hazard factor to the crops, especially those sensitive to this parameter. The degree of soil salinity hazard in the area was found to range from light to severe in all study periods. The proportion of non-saline and saline land results was reduced to 53%, 41%, and 33% in 1961, 2002, and 2016, respectively. This decrease in good lands in terms of low salinity confirms the negative impact of humans on the area and consequently the poor agricultural management.

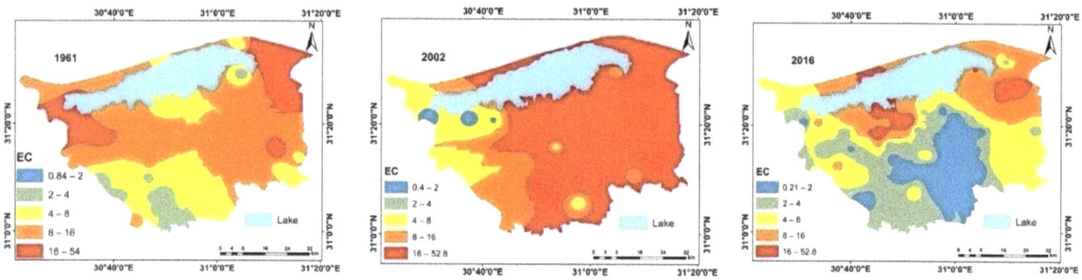

Figure 8. Soil EC maps in 1961, 2002, and 2016.

Weak soil structure leads to weak soil ability to infiltrate irrigation water through the different horizons of the sector. It is characterized as having a high pH (>8.5). In most cases, a solid clay layer is formed that may be deaf at depths ranging from 0.5 to 1 m. This thus affects the height of the depth of the groundwater within the soil sector and increases the proportion of shallow lands. The predominant presence of sodium carbonate makes these lands alkaline soils, with their unfavorable physico-chemical properties. It is clear from Figure 9 that the soil pH increased from 8 with limited alkalinity to 8.8, which is very alkaline, during the years 1961, and 2002 to 2016. The alkalinization degradation could be illustrated by the sodium adsorption Ratio (SAR) where there is a clear change in the SAR value, especially around the lake and in the eastern part of the Governorate, as seen in Figure 10.

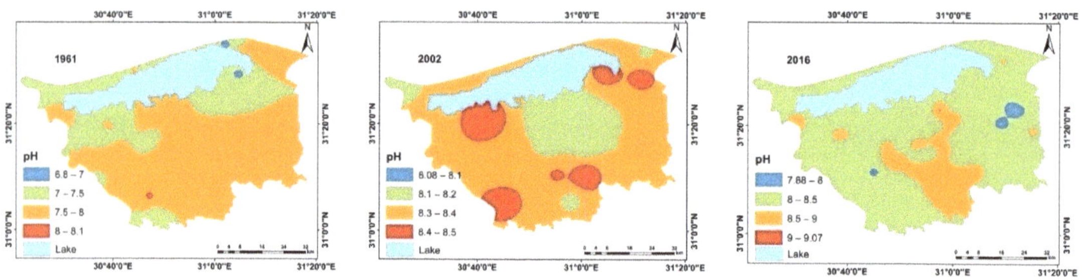

Figure 9. pH prediction maps in 1961, 2002, and 2016.

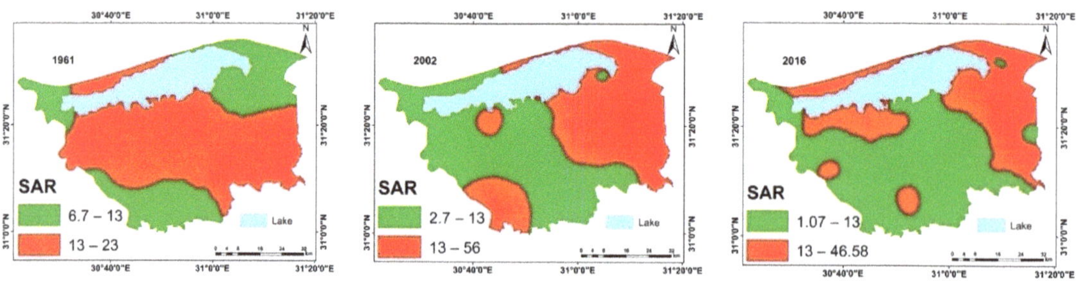

Figure 10. Prediction SAR maps in 1961, 2002, and 2016.

As seen in the Figure 11 lime content maps, all the soils studied are non-calcareous soils. Alkaline saline lands are considered among the most significant problems facing land management in the Nile Delta in Egypt. The results indicated the emergence of a remarkable improvement with the increase in the proportion of lime ins the soil on some of the soil's natural and vital properties. The values of the bulk density and of the percentage of each of the permanent wilting coefficients, as well as the number of micropores, decreased, while the percentage of field capacity, available water, total pores and water-retaining pores increased.

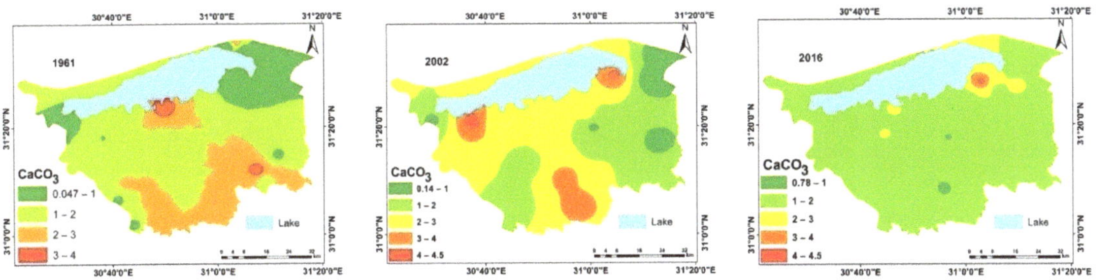

Figure 11. Lime content maps in 1961, 2002, and 2016.

There is no doubt that the most important factor by which soil quality is measured in terms of aeration, porosity, movement of water, and solute is the degree of soil compaction, which is expressed by the bulk density (Figure 12). Bulk densities above thresholds indicate impaired function and therefore affect soil quality by restricting the growth of roots; their weakness is due to the weak movement of water and air in the soil, and thus to shallow rooting of the plant and weak plant growth. This obvious negative effect of increasing the bulk density of the soil reduces the yield of crops due to the available vegetation cover, which is also important for protecting the soil from erosion. As seen in Figure 9, the soils in 2002 and 2016 are moderately compacted; most of the areas are 1.47, which, except in some parts in 2002, is high, but the texture is different—it is in the coastal dunes' area. The increase in bulk density was noted to be associated with soil compaction. Bulk density values of 1.30 to 1.47 g cm^{-3} would indicate soil compaction, but 1.3 g cm^{-3} values were also close to normal for coarse-textured soils. From Figure 9, it can be seen that the soil bulk density values have decreased gradually over the years.

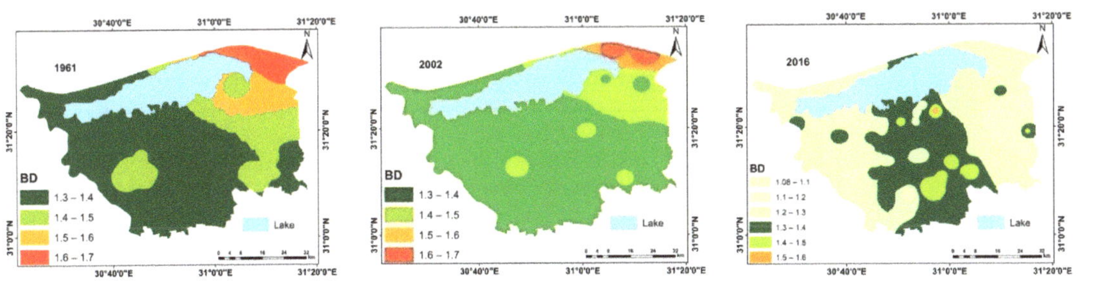

Figure 12. Soil bulk density maps in 1961, 2002, and 2016.

The importance of organic matter (OM) lies in its direct positive effects on the physical, chemical, and biological properties of the soil. It is known that the region had fish farming activity during the past three decades, which may be a determinant effect for the increase in the percentage of OM during the periods examined, as is evident from Figure 13 [42].

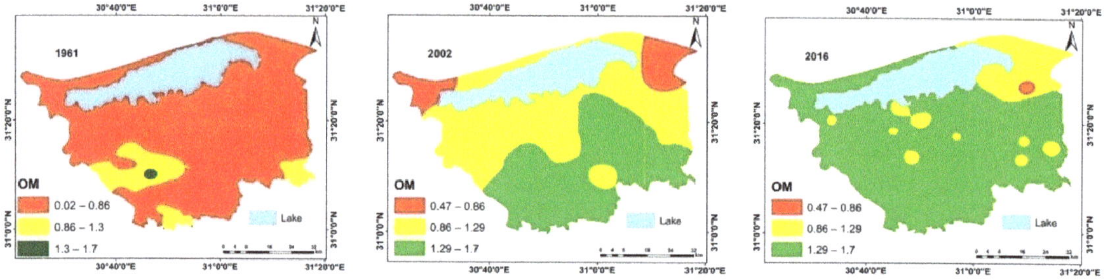

Figure 13. Organic matter content map in 1961, 2002, and 2016.

3.5. Climate Attributes

Changes in an area's average temperature and precipitation in Egypt were assessed based upon over 12 recent GCMs using a new version of MAGICC/SCENGEN coupled gas-cycle/climate modulus (MAGICC, Climate Resources Exchange International Pte. Ltd., Malacca Centre, Singapore), which drives a spatial climate change scenario generator (SCENGEN, Scorpion Computer Services). MAGICC is a simple climate model that computes the mean global surface air temperature and sea-level rise for particular emissions scenarios for greenhouse gases and sulfur dioxide [43]. The scenario for the development of irrigation conditions of the study area was derived from historical data: climatic data from 35 years (1979–2014). According to the Intergovernmental Panel of Climate Change report [44], the study area will likely be affected by increasing air temperature and decreasing precipitation systematically. The predicted scenario of climatic conditions of the study area, which is needed to study this, was worked out using results from the IPCC (Geneva, Switzerland), 2007 [44]. The annual precipitation is expected to decrease from 11 (current) to 8 mm y^{-1} in 2080. To compare the current climate with the projected climate in 2080, we chose temperature data for June, July and August, as these are the hottest months (Figure 14). Extremely hot days can be enough to reduce yield or even kill an entire crop. We also used December, January, and February as the coolest months. These conditions impose work on geneticists and plant breeders to produce new crop varieties that are better adapted to the expected climatic changes. In addition, climatic change is projected to have a significant impact on the amount of irrigation water entering the study province, which in turn will affect soil degradation and agricultural productivity.

The climate of Egypt is affected by several factors, the most important of which are the location, surface features, the general system of pressure, atmospheric depressions, and water bodies, all of which helped divide Egypt into several distinct climatic regions. The Mediterranean Sea is characterized by hot and dry summer months and moderate winters, with little rain falling on the coast. Egypt's climate can be distinguished into two climatic seasons: the hot, dry summer, which extends between May and October, and the mild, rainy winter, which extends between the months of November and April. The agricultural areas in the study area are located along the Nile Delta on the Mediterranean coast. These lands are characterized as fertile sedimentary soils formed thousands of years ago as a result of pumping of the Nile silt coming with the waters of the Nile River. These lands depend for irrigation on the waters of the Nile River, and these lands are alkaline in nature as a result of the high percentage of salt in them with the hot weather conditions.

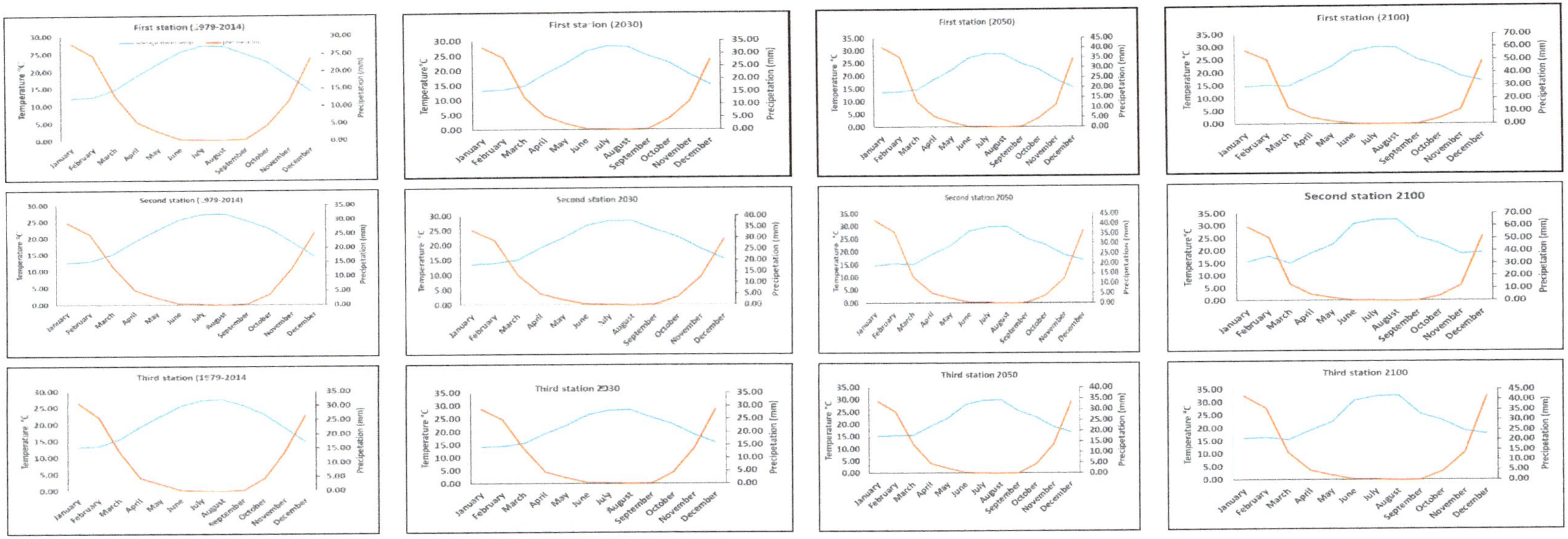

Figure 14. *Cont.*

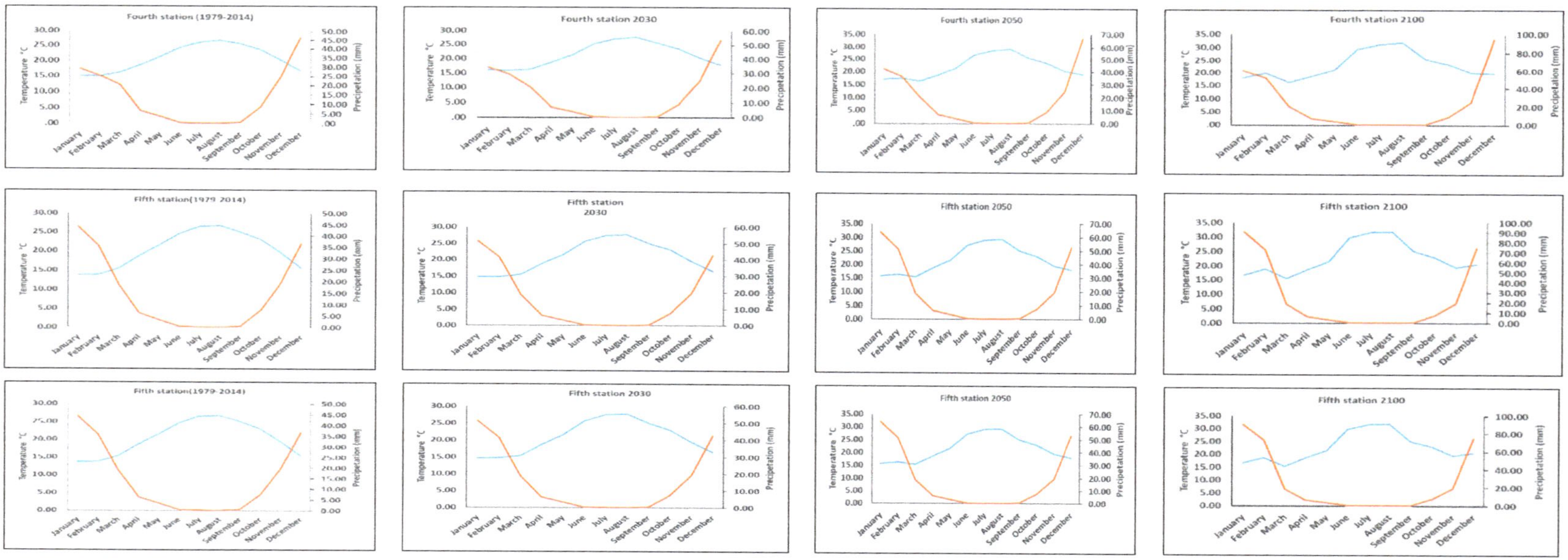

Figure 14. Climate change data.

Agriculture is concentrated in the Nile Delta, but it is exposed to multiple erosion processes as a result of surface irrigation and irrigation with water mixed with high salinity agricultural drainage water, which causes the process of flooding the lands with water and the occurrence of the phenomenon of drumming, especially with the intensive use of chemical fertilizers and pesticides. It also causes the lack of soil compensation for what has been lost of organic matter through organic fertilizers, especially with the interruption of the silt of the Nile after the construction of the High Dam. Another result is the absence of agricultural cycles and the appropriate crop composition, as well as their exposure to dredging operations with the aim of using the surface layer in the manufacture of red bricks, and their exposure to construction and crawl operations, which require modern and economical irrigation methods with cultivation of crops that are less water-intensive, that maintain the agricultural biodiversity that characterizes the region, and that reduce urban sprawl and construction at the expense of agricultural expansion in areas of the delta.

3.6. Overall of Quantitative Soil Degradation Assessment

Spatial models for overall qualitative land degradation in 1961, 2002, and 2016 were created using the Model Builder tool in ArcGIS 10.1 (spatial analyst extension), and a spatial model is represented as a flowchart in Figure 12. Land degradation variables (salinization, alkalization, compaction, lime content, and waterlogging) were exported in a raster, and each data set scored on a scale of 1 to 5 (very low, low, moderate, high, and very high), as shown in Figure 15. The data sets were then weighted according to their influence on the overall model (more weight = more influence). The model was applied to the land degradation variables to prioritize only those that were more determining.

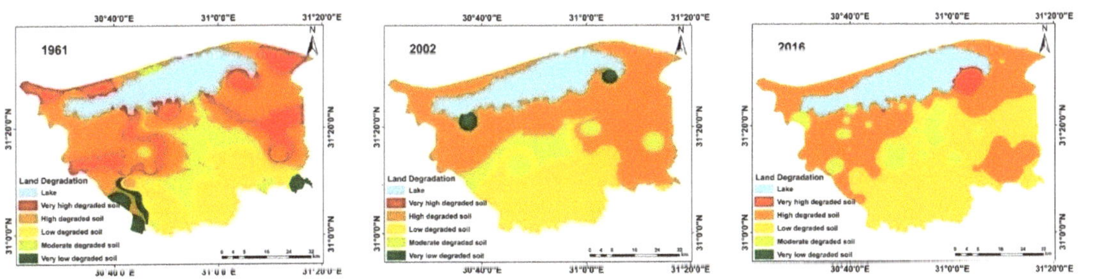

Figure 15. Overall qualitative soil degradation in 1961, 2002, and 2016.

During 2002–2016, soil sealing took place on 36,297.87 ha from the study area (9.7% of the total area) through urban sprawl and fishpond construction. It is expected that urban sprawl will increase 7% by 2020, which means that an additional 3,751 ha of productive soils will be lost from high capable soils from the study area.

As seen in Figures 16 and 17, and Table 7, the overall degradation change for the highly degraded soil increased with time from 1961 to 2016, while for the very highly degraded soil it decreased with time. However, the lowly degraded soil increased with time, and all the change was at the expense of the highly degraded soils due to the reclamation process and the soil management.

The results indicate the anthropogenic effects on the soil of the area over 35 years. This was obviously highlighted through soil sealing through the significant increase of urbanization over the highly fertile agricultural area. This urban sprawl in combination with the climate caused the negative impact on the soil. These results are in conjunction with the previous studies carried out in the region [45–52]. Accordingly, the changes of land properties and the overall land degradation processes in the area should be given more attention to apply decision support systems for better planning of land use.

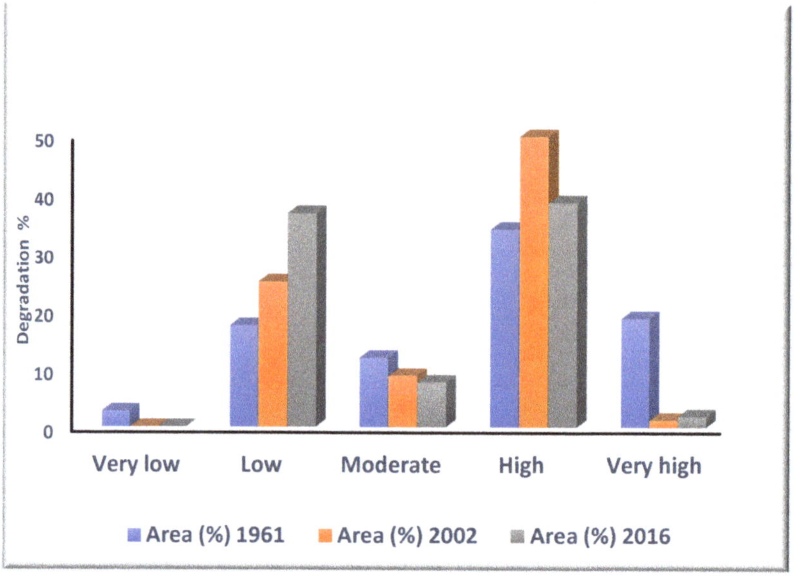

Figure 16. Overall qualitative soil degradation types in 1961, 2002, and 2016.

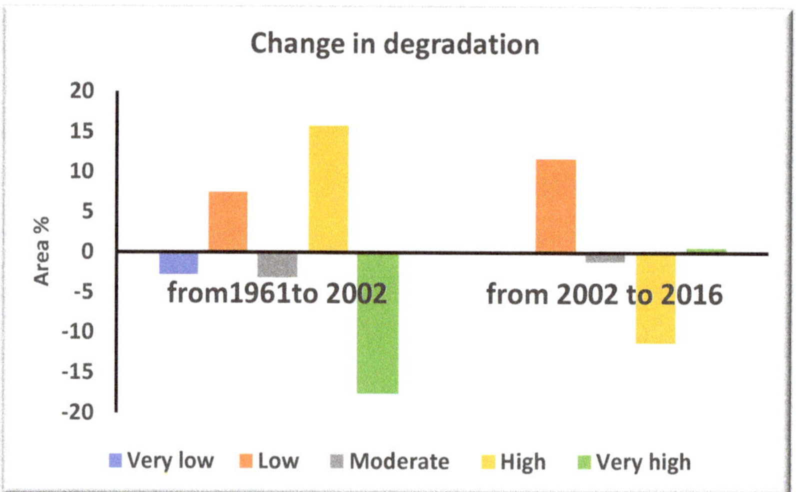

Figure 17. Overall qualitative soil degradation change in 1961, 2002, and 2016.

Table 7. Overall qualitative soil degradation change in 1961, 2002, and 2016.

Element	Area (%) 1961	Area (%) 2002	Area (%) 2016
Very lowly degraded soil	2.70	0.00	0.00
Lowly degraded soil	17.50	24.99	36.74
Moderately degraded soil	12.02	8.90	7.80
Highly degraded soil	34.06	49.88	38.62
Very highly degraded soil	18.74	1.26	1.86

4. Conclusions

The planned and frequent use of satellite-based technology to monitor soil degradation hazard tends to have economic feasibility. Timely depiction of forms of land degradation changes (salinity, water-logging) on the ground as a result of agriculture operations can help authorities and farmers to implement timely measures to mitigate the situation, and in turn can ensure that land productivity is maintained at the planned level. Furthermore, it could be a base for precision agriculture, whereby resources such as irrigation water and fertilizer are optimized.

The study needs to be implemented in all areas of the Nile Delta in order to make a future plan for agricultural use through building a land resource database for agricultural use. It is a very important, serious call for building data banks for the very limited agricultural land in the Nile valley and Delta. Urban expansion over agricultural land is a very important phenomenon and determines the yearly loss of fertile agricultural land in the Nile Valley and Delta; we therefore propose that decision-makers accurately monitor soils and that the Nile valley and Delta constitute a protectorate.

The results showed the importance of studies for inventorying land resources, directing their uses, and preparing their various maps by making use of remote sensing technologies and geographic information systems and integrating them with traditional methods. Also, studying and monitoring the deterioration and assessing its condition is an important step in rehabilitating degraded lands and improving the means and methods of sustainable development of land resources.

Thus, this study contributes toward improving land management and productivity and rationalizing its investment in order to reach sustainable development, including fertilization, the addition of soil conditioners, the selection of means and methods of water use. This also includes non-conventional water resources (saline, brackish, waste) and the rationalization of the use of all types of water in agriculture, as well as identification of irrigation needs from drainage and associated agricultural operations.

The most important recommendations of the study are:

- To develop integrated management systems for land resources and combating desertification.
- To develop land resources, choose their optimal use, and increase their productivity, by preparing land maps (soil-land use-land degradation, etc.).
- To develop and evaluate traditional agricultural operations in dry areas and encourage the exchange of knowledge and successful technologies.
- To develop non-conventional water uses to increase farm income while preserving land from degradation.

Author Contributions: Conceptualization, Mohamed A. E. AbdelRahman, Ahmed A. Afifi and Antonio Scopa; methodology, Mohamed A. E. AbdelRahman; software, Mohamed A. E. AbdelRahman and Ahmed A. Afifi; validation, Mohamed A. E. AbdelRahman, Ahmed A. Afifi and Antonio Scopa; formal analysis, Mohamed A. E. AbdelRahman and Ahmed A. Afifi; investigation, Mohamed A. E. AbdelRahman and Ahmed A. Afifi; resources, Mohamed A. E. AbdelRahman and Ahmed A. Afifi; data curation, Mohamed A. E. AbdelRahman, Ahmed A. Afifi and Antonio Scopa; writing—original draft preparation, Mohamed A. E. AbdelRahman, Ahmed A. Afifi and Antonio Scopa; writing—review and editing, Mohamed A. E. AbdelRahman, Ahmed A. Afifi and Antonio Scopa; visualization, Mohamed A. E. AbdelRahman, Ahmed A. Afifi and Antonio Scopa; supervision, Mohamed A. E. AbdelRahman, Ahmed A. Afifi and Antonio Scopa; project administration, Ahmed A. Afifi; funding acquisition, Mohamed A. E. AbdelRahman, Ahmed A. Afifi and Antonio Scopa. All authors have read and agreed to the published version of the manuscript.

Funding: This research received no external funding.

Institutional Review Board Statement: Not applicable.

Informed Consent Statement: Not applicable.

Acknowledgments: The manuscript presented a scientific participation between the scientific institutions in two countries (Egypt and Italy). The authors are grateful to scientific institutions from the two countries for their support in carrying out the work. A.A.A. would like to thank both M.A.E.A. and A.S. for their participation in the project of the National Research Center No. 11080306.

Conflicts of Interest: The authors would like to certify that they have no conflict of interest in the data collection, analyses, interpretation, writing of the manuscript, orin the decision to publish the results. The authors would also like to declare that the funding of the study has been supported by the authors' institutions.

References

1. Bai, Z.G.; Dent, D.L.; Olsson, L.; Schaepman, M.E. *Global Assessment of Land Degradation and Improvement 1: Identification by Remote Sensing*; Report 2008/01; ISRIC—World Soil Information: Wageningen, The Netherlands, 2008. [CrossRef]
2. FAO. *FAO/UNEPExpert Consultation on Methodology for Assessing Soil Degradation*; FAO: Rome, Italy, 1978.
3. AdelRahman, M.A.E. Quantification of Land Degradation Indicators Using Remote Sensing and GIS Techniques. Master's Thesis, Zagazig University, Zagazig, Egypt, 2009.
4. Belal, A.B.A.; Al-Ashri, K.M.A. GIS based land evaluation in Baharyia oasis, western desert, Egypt. *J. Soil Sci. Agric. Eng. Mansoura Univ.* **2011**, *2*, 11–24. [CrossRef]
5. AbdelRahman, M.A.E. Assessment of Land Degradation and Land Use Planning by Using Remote, Sensing and GIS Techniques in Chamarajanagar District. Ph.D. Thesis, UAS, Bangalore University, Bangalore, Karnataka, India, 2014.
6. AbdelRahman, M.A.E.; Natarajan, A.; Srinivasamurty, C.A.; Hegde, R. Estimating soil fertility status in physically degraded land using GIS and remote sensing techniques in Chamarajanagar district, Karnataka, India. *Egypt. J. Remote Sens. Space Sci.* **2016**, *19*, 95–108. [CrossRef]
7. AbdelRahman, M.A.E.; Natarajan, A.; Rajendra, H.; Prakash, S.S. Assessment of land degradation using comprehensive geostatistical approach and remote sensing data in GIS-model builder. *Egypt. J. Remote Sens. Space Sci.* **2018**, *22*, 323–334. [CrossRef]
8. AbdelRahman, M.A.E.; Metwaly, M.M.; Shalaby, A. Quantitative assessment of soil saline degradation using remote sensing indices in Siwa Oasis. *Remote Sens. Appl. Soc. Environ.* **2019**, *13*, 53–60. [CrossRef]
9. AbdelRahman, M.A.E.; Shalaby, A.; Essa, E.F. Quantitative land evaluation based on fuzzy-multi-criteria spatial model for sustainable land-use planning. *Modeling Earth Syst. Environ.* **2018**, *4*, 1341–1353. [CrossRef]
10. AbdelRahman, M.A.E.; Shalaby, A.; Mohamed, E.S. Comparison of two soil quality indices using two methods based on geographic information system. *Egypt. J. Remote Sens. Space Sci.* **2018**, *22*, 127–136. [CrossRef]
11. AbdelRahman, M.A.E.; Tahoun, S. GIS model-builder based on comprehensive geostatistical approach to assess soil quality. *Remote Sens. Appl. Soc. Environ.* **2019**, *13*, 204–214. [CrossRef]
12. AbdelRahman, M.A.E.; Natarajan, A.; Srinivasamurthy, C.A.; Hegde, R.; Prakash, S.S. Assessment of Soil Quality by Using Remote Sensing and GIS Techniques; A Case study, Chamrajanagar District, Karnataka, India. *Acta Sci. Agric.* **2018**, *2*, 5–12.
13. AbdelRahman, M.A.E.; Arafat, S.M. An Approach of Agricultural Courses for Soil Conservation Based on Crop Soil Suitability Using Geomatics. *Earth Syst. Environ.* **2020**, *4*, 273–285. [CrossRef]
14. AbdelRahman, M.A.E.; Natarajan, A.; Hegde, R. Assessment of land suitability and capability by integrating remote sensing and GIS for agriculture in Chamarajanagar district, Karnataka, India. *Egypt. J. Remote Sens. Space Sci.* **2016**, *19*, 125–141. [CrossRef]
15. AbdelRahman, M.A.E.; Tahoun, S.A.; Abdel Bary, E.A.; Arafat, S.M. Detecting Land Degradation Processes Using Geo Statistical Approach in Port Said, Egypt. *Zagazig J. Agric. Res.* **2008**, *35*, 1361–1379.
16. AbdelRahman, M.A.E.; Rehab, H.H.; Yossif, T.M.H. Soil fertility assessment for optimal agricultural use using remote sensing and GIS technologies. *Appl. Geomat.* **2021**, *13*, 1–14. [CrossRef]
17. AbdelRahman, M.A.E.; Zakarya, Y.M.; Metwaly, M.M.; Koubouris, G. Deciphering Soil Spatial Variability through Geostatistics and Interpolation Techniques. *Sustainability* **2021**, *13*, 194. [CrossRef]
18. Abuzaid, S.A.; Abdellatif, D.A.; Fadl, M.E. Modeling soil quality in Dakahlia Governorate, Egypt using GIS techniques. *Egypt. J. Remote Sens. Space Sci.* **2021**, *24*, 255–264. [CrossRef]
19. Abuzaid, A.S.; Fadl, M.E. Mapping potential risks of long-term wastewater irrigation in alluvial soils, Egypt. *Arab. J. Geosci.* **2018**, *11*, 433. [CrossRef]
20. Ali, R.R.; Abd El-Kader, A.A.; Essa, E.F.; AbdelRahman, M.A.E. Application of Remote Sensing to Determine Spatial Changes in Soil Properties and Wheat Productivity Under Salinity Stress. *Plant Arch.* **2019**, *19*, 616–621.
21. Bakr, N.; Weindorf, D.C.; Bahnassy, M.H.; El-Badawi, M.M. Multi-temporal assessment of land sensitivity to desertification in a fragile agro-ecosystem: Environmental indicators. *Ecol. Indic.* **2012**, *15*, 271–280. [CrossRef]
22. IPCC. *Climate Change 2007: Mitigation. Contribution of Working Group III to the Fourth Assessment Report of the Intergovernmental Panel on Climate Change*; Cambridge University Press: Cambridge, UK; New York, NY, USA, 2007.
23. DEFRA. *Impact of Climate Change on Soil Functions*; Final Project Report; Research and Development: London, UK, 2005.
24. Pathak, H.; Aggarwal, P.K.; Singh, S.D. *Climate Change Impact, Adaptation and Mitigation in Agriculture: Methodology for Assessment and Application*; Indian Agricultural Research Institute: New Delhi, India, 2012; pp. 1–302.
25. Varallyay, G.; Farkas, C.S. Impact of Climate Change on Soils. In *Climate Change: Environment-Risk-Society*; Harnos, Z.S., Csete, L., Eds.; SzaktudasKiadoHaz: Budapest, Hungary, 2008; pp. 91–129. (In Hungarian)

26. Brinkman, R. Resilience Against Climate Change? In *Soil Minerals, Transformations and Surface Properties, EC, pH*; Scharpenseel, H.W., Shomaker, M., Ayoub, A., Eds.; Elsevier: New York, NY, USA, 1990; pp. 51–60.
27. Conoco Geologic Map of Egypt. *Egyptian General Authority for Petroleum (UNESCO Joint Map Project), 20 Sheets, Scale 1:50, 000*; Conoco Geologic Map of Egypt: Cairo, Egypt, 1987.
28. FAO. *Guidelines for Soil Description*, 4th ed.; FAO: Rome, Italy, 2006.
29. Richards, L.A. *Diagnosis and Improvement of Saline and Alkali Soils United State Department of Agriculture Hand Book*; No. 60; United State Government Printing Office: Washington, DC, USA, 1954.
30. Nelson, R.E. Carbonate and gypsum. In *Methods of Soil Analysis, Part 2: Chemical and Microbiological Properties*; Page, A.L., Ed.; American Society of Agronomy: Madison, WI, USA, 1982; pp. 181–197. ISBN 9780891180722.
31. Nelson, D.W.; Sommers, L.E. Total carbon, organic carbon and organic matter. In *Methods of Soil Analysis. Part 2: Chemical and Microbiological Properties*; Page, A.L., Ed.; American Society of Agronomy: Madison, WI, USA, 1982; pp. 539–579. ISBN 9780891180722.
32. Black, G.A. *Methods of Soil Analysis*; American Society of Agronomy: Madison, WI, USA, 1965; Volume 15, p. 72.
33. Jackson, M.L. *Soil Chemical Analysis*; Prentice Hall of India, Pvt. Ltd.: New Delhi, India, 1973; p. 498.
34. USDA-Natural Resources Conservation Service- Soil Survey Staff. *Keys to Soil Taxonomy*, 12th ed.; USDA: Washington, DC, USA, 2014.
35. Ministry of Agriculture and Land Reclamation, Egypt. *Survey Report*; Ministry of Agriculture and Land Reclamation, Egypt: Giza, Egypt, 1961.
36. Abo-Waly, M.E.M. Pedological Studies on Different Salt Affected Soils South of the Mohite Drain at Kafr El-Sheikh Governorate (A.R.E). Master's Thesis, Soil Science Department, Faculty of Agriculture, Tanta University, Tanta, Egypt, 1982.
37. Saffan, M.M. To Know the Soil in the Shalma Region (North Nile Delta, Egypt). Ph.D. Thesis, Justus Liebig University, Giessen, Germany, 1984.
38. FAO; UNEP; UNESCO. *A Provisional Methodology for Soil Degradation Assessment*; FAO: Rome, Italy, 1979; ISBN 9251008698.
39. UNEP. *UNEP STAFF Global Assessment of Soil Degradation (GLASOD)*; UNEP, UN, GLASOG Project: Wageningen, The Netherlands, 1991.
40. Aboelsoud, H.M.; AbdelRahman, M.A.E. Rapid Field Technique for Soil Salinity Appraisal in North Nile Delta using EM38 through Some Empirical Relations. *Int. J. Plant Soil Sci.* **2017**, *14*, 1–9. [CrossRef]
41. Shalaby, A.; AbdelRahman, M.A.E.; Belal, A.A.A. GIS Based Model for Land Evaluation Mapping: A Case Study North Delta Egypt. *Egypt. J. Soil Sci.* **2017**, *57*, 339–351. [CrossRef]
42. AbdelRahman, M.A.E.; Shalaby, A.; Aboelsoud, M.H.; Moghanm, F.S. GIS spatial model based for determining actual land degradation status in Kafr El-Sheikh Governorate, North Nile. *Modeling Earth Syst. Environ.* **2018**, *4*, 359–372. [CrossRef]
43. Rao, D.L.N. *Biological Amelioration of Degraded and Contaminated Soils. Biotechnological Approach in Soil Macro Organism for Sustainable Crop Production*; Central Soil Salinity Research Institutes: Karnal, India, 1997; pp. 261–275.
44. IPCC. *Climate Change 2007: Impacts, Adaptation and Vulnerability. Contribution of Working Group II to the Fourth Assessment Report of the Intergovernmental Panel on Climate Change*; Parry, M.L., Canziani, O.F., Palutikof, J.P., van der Linden, P.J., Hanson, C.E., Eds.; Cambridge University Press: Cambridge, UK, 2007; p. 976. ISBN 978 0521 88010-7.
45. El-Ramady, H.; Abowaly, M.; Elbehiry, F.; Omara, A.; Elsakhawy, T.; Mohamed, S.A.; Belal, A.; Elbasiouny, H.; Abdalla, Z. Stressful Environments and Sustainable Soil Management: A Case Study of Kafr El-Sheikh, Egypt. *Envir. Biod. Soil Sec.* **2019**, *3*, 193–213. [CrossRef]
46. Emam, K.; Soliman, K.M. Geospatial analysis, source identification, contamination status, ecological and health risk assessment of heavy metals in agricultural soils from Qallin city, Egypt. *Stoch. Environ. Res. Risk Assess.* **2021**, 1–23. [CrossRef]
47. Elbehiry, F.; Elbasiouny, H.; El-Ramady, H.; Brevik, E.C. Mobility, distribution, and potential risk assessment of selected trace elements in soils of the Nile Delta, Egypt. *Environ. Monit. Assess.* **2019**, *191*, 713. [CrossRef] [PubMed]
48. Hendawy, E.; Belal, A.A.; Mohamed, E.S.; Elfadaly, A.; Murgante, B.; Aldosari, A.A.; Lasaponara, R. The Prediction and Assessment of the Impacts of Soil Sealing on Agricultural Land in the North Nile Delta (Egypt) Using Satellite Data and GIS Modeling. *Sustainability* **2019**, *11*, 4662. [CrossRef]
49. Abuzaid, A.S.; Bassouny, M.A. Multivariate and Spatial Analysis of Soil Quality in Kafr El-Sheikh Governorate, Egypt. *J. Soil Sci. Agric. Eng.* **2018**, *9*, 333–339. [CrossRef]
50. Elbasiouny, H. Assessment of Environmental Sensitivity to Desertification, Soil Quality and Sustainability in An Area of The North Nile Delta, Egypt. *Egypt. J. Soil. Sci.* **2018**, *58*, 399–415. [CrossRef]
51. Moghanm, F.S.; Belal, A.B. Assessment and Mapping of Environmentally Sensitive Areas to Desertification Using Gis in An Area of The North Delta Region of Egypt. *Egypt. J. Soil Sci.* **2018**, *58*, 325–335.
52. Bakr, N.; Afifi, A.A. Quantifying land use/land cover change and its potential impact on rice production in the Northern Nile Delta, Egypt. *Remote Sens. Appl. Soc. Environ.* **2019**, *13*, 348–360. [CrossRef]

International Journal of *Geo-Information*

MDPI

Article

New Insight on Soil Loss Estimation in the Northwestern Region of the Zagros Fold and Thrust Belt

Arsalan Ahmed Othman [1,*], Ahmed K. Obaid [2], Diary Ali Mohammed Amin Al-Manmi [3], Ahmed F. Al-Maamar [4], Syed E. Hasan [5], Veraldo Liesenberg [6], Ahmed T. Shihab [4] and Younus I. Al-Saady [4]

1 Iraq Geological Survey, Sulaymaniyah Office, Sulaymaniyah 46001, Iraq
2 Department of Geology, University of Baghdad, Al-Jadiryah Street, Baghdad 10071, Iraq; ahmedobaid@uobaghdad.edu.iq
3 Department of Geology, College of Science, University of Sulaimani, Sulaymaniyah 46001, Iraq; diary.amin@univsul.edu.iq
4 Iraq Geological Survey, Al-Andalus Square, Baghdad 10068, Iraq; ahmed.f@geosurviraq.iq (A.F.A.-M.); ahmedshihab@geosurviraq.iq (A.T.S.); younusalsaady@geosurviraq.iq (Y.I.A.-S.)
5 Department of Earth & Environmental Sciences, University of Missouri, Kansas City, MO 64110-2499, USA; hasans@umkc.edu
6 Department of Forest Engineering, Santa Catarina State University (UDESC), Lages, SC 88520-000, Brazil; veraldo.liesenberg@udesc.br
* Correspondence: arsalan.aljaf@geosurviraq.iq; Tel.: +964-773-055-2201

Abstract: Soil loss is one of the most important causes of land degradation. It is an inevitable environmental and socio-economic problem that exists in many physiographic regions of the world, which, besides other impacts, has a direct bearing on agricultural productivity. A reliable estimate of soil loss is critical for designing and implementing any mitigation measures. We applied the widely used Revised Universal Soil Loss Equation (RUSLE) in the Khabur River Basin (KhRB) within the NW part of the Zagros Fold and Thrust Belt (ZFTB). The areas such as the NW Zagros range, characterized by rugged topography, steep slope, high rainfall, and sparse vegetation, are most susceptible to soil erosion. We used the Digital Elevation Model (DEM) of the Shuttle Radar Topography Mission (SRTM), Tropical Rainfall Measuring Mission (TRMM), Harmonized World Soil Database (HWSD), and Landsat imagery to estimate annual soil loss using the RUSLE model. In addition, we estimated sediment yield (SY) at sub-basin scale, in the KhRB where a number of dams are planned, and where basic studies on soil erosion are lacking. Estimation of SY will be useful in mitigation of excessive sedimentation affecting dam performance and watershed management in this region. We determined the average annual soil loss and the SY in the KhRB to be 11.16 $t.ha^{-1}.y^{-1}$ and 57.79 $t.ha^{-1}.y^{-1}$, respectively. The rainfall and runoff erosivity (R factor), slope length (L factor), and slope steepness (S factor), are the three main factors controlling soil loss in the region. This is the first study to determine soil loss at the sub-basin scale along with identifying suitable locations for check dams to trap the sediment before it enters downstream reservoirs. The study provides valuable input data for design of the dams to prevent excessive siltation. This study also aims at offering a new approach in relating potential soil erosion to the actual erosion and hypsometric integrals.

Keywords: RUSLE; soil loss; siltation; Zagros; sub-basin; hypsometric integral

Citation: Othman, A.A.; Obaid, A.K.; Al-Manmi, D.A.M.A.; Al-Maamar, A.F.; Hasan, S.E.; Liesenberg, V.; Shihab, A.T.; Al-Saady, Y.I. New Insight on Soil Loss Estimation in the Northwestern Region of the Zagros Fold and Thrust Belt. *ISPRS Int. J. Geo-Inf.* **2021**, *10*, 59. https://doi.org/10.3390/ijgi10020059

Received: 5 December 2020
Accepted: 28 January 2021
Published: 1 February 2021

1. Introduction

Soil loss by water erosion is one of the most influential factors in land degradation, which cause significant environmental and socio-economic consequences that are critical to human welfare [1,2]. The evolution of the earth's landscape results from interplay among three main factors, which are lithology, climate, and tectonics [3,4]. These factors control, directly or indirectly, the erosional rate and soil losses [5]. Geomorphological processes, including erosion, continually modify the landscape, reduce area relief, and generate a vast amount of sediment that directly impact agricultural, ecological, and water resources.

Based on their lithological characteristics, different rock units, exposed in the Zagros region, offer varying levels of resistance to erosional forces, thereby exhibiting different erosional patterns [3]. In terms of climate, the northwestern part of the Zagros region is characterized by a heavy rainfall season preceded by a long dry season [6,7]. Collectively, the lithological and meteorological features sculpt the land, creating distinctive features, such as rugged topography with steep slopes [8], that occur in the northwestern Zagros [9]. Tectonics has also been a dominant factor in modifying the landscape of the Zagros Fold and Thrust Belt (ZFTB). For instance, high topography and steep slopes to the northeast of the Dezful and Kirkuk embayments result from intense thrusting caused by low strain at the two embayments [9,10]. This intense thrusting and the resulting steep slopes cause rivers to incise deep topography, yielding large volume of sediments, which can potentially affect the safety of the downstream dams and shorten reservoir life.

Hypsometry—the relation between forms of drainage basins (catchments) and their slopes [11]—is a function of river action, which, in turn, is controlled by lithology, climate, and tectonics. An individual basin can be represented by different intervals of elevation that enclose different areas. These intervals of elevations and areas form the hypsometric curve, which can be integrated into the Hypsometric Integral (HI) [12]. The amount of residual terrain above the lowest horizontal plane of a basin is the HI of a basin and can be used as a proxy for the erosional stage and landforms development [12,13]. HI has been used in several studies to express the relation between uplift and erosion and it gives reliable results on the degree of dissection to describe the erosional cycle through the youthful, mature, and old stages [14]. This allows the determination of active and inactive erosional regions.

Soil erosion removes the most fertile topsoil containing rich organic matter and nutrients, adversely impacting soil productivity [15,16]. An estimated 84% of land degradation worldwide results from erosion [17,18]. Due to the global nature of the problem and its direct impact on agricultural productivity, availability of a reliable and cost-effective method for estimating soil loss is of critical importance. Many models have been proposed to quantify soil loss that can be grouped into: (1) physical models [19,20], which are based on principles of water erosion; (2) conceptual models, that define significant human and natural factors, which affect the processes of erosion and sedimentation [21]; and (3) empirical models, which identify the different interacting erosion variables. Of these, the empirical models are easy to apply and require less computational time [22,23].

The popular Universal Soil Loss Equation (USLE) model was developed to estimate the amount of soil or sediment (mass of soil/sediment loss per unit area over time) reaching the outlet of a basin undergoing water erosion [24]. The USLE was purposed to foresee the mean rates of long-term erosion but not for prediction of soil loss events [22] where there is high fluctuation of soil losses from individual events [25] and such fluctuation cannot not be accurately included in the USLE model. This limitation has been removed in the Revised Universal Soil Loss Equation (RUSLE) by taking into consideration the rainfall energy [15]. Recent research, integrating RUSLE in the Geographic Information System (GIS) and remote sensing techniques, aided by the ever-increasing power and efficiency of computers [26], enables estimating soil loss for large areas quickly and at a reasonable cost [27–38] in different climatic zones [39].

The RUSLE model is widely used for soil loss estimation in many countries of the world due to its flexibility, direct application at field-scale, and value in establishing conservation measures and policies [40]. It is designed to estimate the average annual soil loss (A) based on five factors [41]; these factors are the rainfall erosivity factor (R), the slope length and steepness factor (LS), the conservation practice factor (P), the soil erodibility factor (K), the cover management factor (C). [42] stated that soil erosion can be classified into two types: potential erosion (PE) and actual erosion (AE). The R, K, and LS factors were examined as PE, whereas the C and P factors, in addition to R, K, LS factors, were considered as actual soil erosion [42]. We used the RUSLE model to estimate AE.

Soil erosion is the primary process for sediment delivery to rivers and streams. The amount of sediments exported at the basin outlet over a given period of time constitutes the sediment yield (SY) [43]. A barrier in the flow path, such as a dam, causes sediment deposition in the dam's reservoir in the form of bedload and suspended load [44]. Gradual accumulation of sediments leads to reduced reservoir storage capacity that lowers its useful life [45–47]. Mosul Dam, downstream of the present study area, is a good example of excessive reservoir siltation problem [7,48].

Several investigators have used the RUSLE model to estimate soil loss in neighboring countries (i.e., Iran [49–55] and Turkey [56,57]), and have applied RUSLE to determine the Sediment Delivery Ratio (SDR) [55,58]. However, RUSLE and SY models have not been used in the Kurdistan Region, north of Iraq, for soil loss estimation. Thus, our study serves to fill this critical knowledge gap and provides a reliable estimate of soil loss and SY in this part of the Zagros Range. In addition, our study also investigated the relationship between PE and HI at a basin scale.

This study also presents a new approach to relate PE—represented by the R, K, and LS factors—with rate of AE. It offers a useful tool to predict the effectiveness of soil loss control measures in a virgin area with no, or minimum human intervention. The resulting data should be valuable in implementing erosion control measures, such as ground cover management (C) and terracing and/or contour farming (P).

2. Materials and Methods

2.1. Study Area

The study area is located at the distal northwestern part of the ZFTB between 36°55′28″–37°47′01″ N latitudes and 42°34′44″–43°29′5″ E longitudes, along the border between Iraq and Turkey. It encompasses an area of ~3500 km^2 and represents the Khabur River Basin (KhRB). The area is characterized by rugged topography and high relief to the north, due to the presence of the High Zagros Fault (HZF), and subdued topography and low relief in the southern part, near Dohuk city (Figure 1). The area is characterized by its remoteness and difficulty of access, frequent landslides and lacks useful in situ hydrological and pedological information. The area encompasses the Khabur River that is the source of life-sustaining water to the region. The Zakho meteorological station data indicate significant seasonal fluctuation in the region, characterized by long, dry summers and short, rainy winter conditions. Most of the precipitation (586 mm) occurs between October and May, with a maximum monthly average of 134.3 mm occurring in January. The monthly average temperature and evaporation occurs in July, 33.96 °C, and 354.7 mm, respectively, and the minimum temperature of 4.17 °C was recorded in January (Figure 2).

A number of dams have been planned in the KhRB and, like the existing Mosul dam with a high rate of reservoir siltation at 78,000 t.ha^{-1}.y^{-1} [59]. Excessive soil erosion in the region would create a major problem for the proposed dams, as pointed out by [48]. The rate of soil loss determined by this study will, therefore, enable the decision makers to incorporate suitable soil conservation measures to mitigate reservoir siltation problem, and also to optimize agricultural productivity.

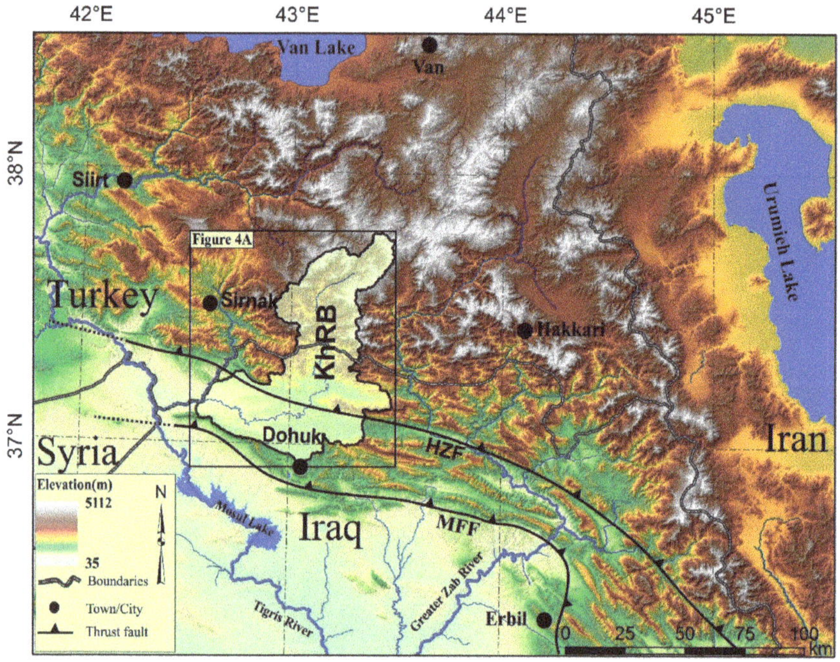

Figure 1. Shaded relief map showing location of the study area in Zagros range.

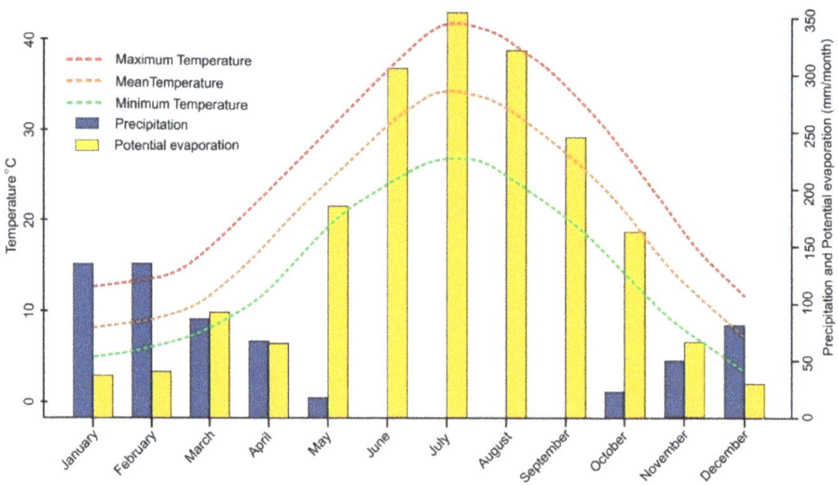

Figure 2. Monthly data of rainfall, evaporation, and temperature, collected at the Zakho meteorological station for the period 2002–2010; data modified from [48].

2.2. Data and Software

We used four scenes of the mosaicked Digital Elevation Model (DEM), with 30 m spatial resolution (1 arc-second), gathered by Shuttle Radar Topography Mission (SRTM) [60]. DEM was used to extract the drainage network and calculate the slope and runoff flow accumulations. Due to the lack of in situ meteorological data, we employed Tropical

Rainfall Measuring Mission (TRMM) data [61] to produce the rainfall map. The TRMM (3B43-V7) is a raster structure with 0.25° × 0.25° spatial resolution that collects monthly precipitation data [62]. The Harmonized World Soil Database (HWSD) [63] was used to determine soil erodibility, or the K factor. It has a cell size of 30 arc-second (~1 km). We combined seasonal impacts on vegetation using four Landsat Operational Land Imager (OLI) scenes. These scenes in Level 1T format, acquired in 2019, were used in this study. Details are listed in Table 1. The spectral images are of good quality, geo-referenced, and orthorectified with a 30 m spatial resolution, and are available without any charge, from the Earth Resources Observation and Science (EROS) center in Sioux Falls, South Dakota, U.S.A. Only scenes with excellent quality and less than 12.5 % cloud cover were chosen for this study. The region of interest is positioned in zone 38 north and the path and row of the image are 170 and 34 respectively. All input data were reprojected to the Universal Transverse Mercator (UTM) projection/WGS 1984 datum, within zone 38N using the nearest neighbor resampling method to fit with the OLI scenes.

Table 1. Landsat OLI raw data.

Date	ID	Quality	Cloud Cover
23 June 2019	LC08_L1TP_170034_20190623_20200827_02_T1	9	0.55%
26 August 2019	LC08_L1TP_170034_20190826_20200826_02_T1	9	0.02%
27 October 2019	LC08_L1TP_170034_20191013_20200825_02_T1	9	0.05%
16 December 2019	LC08_L1TP_170034_20191216_20201023_02_T1	9	12.40%

Combining the powerful capabilities of remote sensing and GIS, and using multiple datasets from different sources [64], we applied the RUSLE model to estimate soil loss in the study area. We used the MATLAB based TecDEM 2.2 [65,66] software to extract the drainage network. Specific functions within the ArcGIS 10.3 [67], such as slope, runoff flow accumulation, interpolation, raster conversion, and raster calculator, were utilized for base map preparation. In addition, we used Environment for Visualizing Images (ENVI) software for Landsat imageries preparation and processing that included implementing Boolean operation, surface reflectance conversion, and Normalized Differential Vegetation Index (NDVI) determination. R-studio script was used for statistical analysis.

2.3. Hypsometric Integral (HI)

The HI is a useful metric to identify different stages of landscape development in its geomorphological erosion cycle [12,68,69]. We used HI to determine the stage of landscape development (erosional stages) to quantify the relationship between HI and amount of soil loss. The mature (highly eroded) areas have low HI values, while the youthful (slightly eroded) areas show high HI values [12,14,69–72]. According to [73], the HI of a given area can be estimated using Equation (1):

$$HI = \frac{H_{mean} - H_{min}}{H_{max} - H_{min}} \tag{1}$$

where H_{max}, H_{min}, and H_{mean} are the maximum, minimum, and mean elevations, respectively.

An HI value close to 1 means that the terrain removal is in the lowest stage (uplift > erosion) and the land surface is in its youthful stage, while low HI values (close to 0) refers to high amount of terrain removal (erosion > uplift) and the land surface is in its mature stage. This dimensionless parameter is very useful for comparing the erosional stage of different basins regardless of their size (areas).

2.4. RUSLE Model Description

The RUSLE model, proposed by [41], can been used to estimate annual soil loss as shown in Equation (2).

$$AE = R \,.\, K \,.\, LS \,.\, C \,.\, P \tag{2}$$

where,

AE is the actual erosion (yearly rate of the soil loss (t.ha^{-1}.y^{-1}));
R rainfall and runoff erosivity factor (MJ.mm.ha^{-1}.h^{-1}.y^{-1});
K, soil erodibility factor (t.ha.h.ha^{-1}.MJ^{-1}.mm^{-1});
LS, the slope length and slope steepness factor;
C, the cover management factor;
P, the support practice factor;
LS, C, and P factors are dimensionless.

The following sections explain the method of calculating RUSLE factors.

2.4.1. Rainfall and Runoff Erosivity (R factor)

The R factor expresses the effect of precipitation on soil erosion [74,75], as soil particle detachment is directly related to precipitation [76]. Estimation of R factor requires the availability of accurate and continuous precipitation data [22,31].

Since the study area lacks reliable meteorological data, we used the monthly TRMM (3B43-V7) to calculate the yearly average precipitation (mm), due to its suitable temporal and spatial resolution [30], and because of its extensive application for assessing the R factor, as compared to other satellite data. The TRMM data covers 15 years between September 2002 and August 2017. The TRMM pixels have been changed in vectors (points) format, and these points were interpolated employing an Inverse Distance Weighting (IDW) method, to produce a raster format with a spatial resolution of 30 m.

Numerous approaches have been used to calculate the R factor by employing long-term precipitation data. [77] found that Equation (3) can be used to calculate the R factor, which is directly related to the average annual rainfall.

$$R = 81.5 + 0.38\ P_A \tag{3}$$

where R is the rainfall erosivity factor in MJ.mm.ha^{-1}.h^{-1}.y^{-1} and P_A is the average annual precipitation in mm.

Precipitation in the KhRB amounts to 562.85 mm·yr^{-1} in the south, increasing to 785 mm·yr^{-1} to the north. The suitability of applying TRMM data in the study area was evaluated by comparing it with the observed precipitation dataset of Zakho meteorological station. Sixty-two monthly precipitation measurements, covering the period from September 2002 to December 2007 were compared with the TRMM data (Figure 3). In this figure, a strong direct relationship was found, the coefficient of determination (R^2) being 0.853, at a significant p-value of <0.05. The slope and intercept were 0.9124 and 9.5485, respectively.

2.4.2. Soil Erodibility (K) Factor

The K–factor representing soil vulnerability to erosion, is affected by rainfall and runoff, [30,78]. Soil susceptibility to erosion is affected by its physical and chemical characteristics, such as its texture, mineralogy, and structure. Five soil parameters (i.e., texture, organic matter content (OM), coarse fragments, structure, and permeability) are the main properties that influence the K–factor. With the soil OM being inversely proportional to soil erodibility, such that any increase in soil OM decreases its vulnerability to detachment [36]. For this study, dataset of the HWSD was used to obtain information about soil texture and soil organic carbon (OC). We used Tables 2 and 3 to estimate the soil structure class and the soil permeability class, respectively [22,79]. We estimated the K–factor of the soil types in the study area using Equations (4) and (5) [22,79].

$$K = \left[\frac{2.1\ \times 10^{-4}\ M^{1.14}\ (12 - OM) + 3.25(s-2) + 2.5(p-3)}{100}\right] * 0.1317 \tag{4}$$

$$M = (m_{silt} + m_{vfs}) * (100 - m_c) \tag{5}$$

m_c [%] is clay fraction (<0.002 mm);

m_{silt} [%] silt fraction (0.002–0.05 mm);
m_{vfs} [%] very fine sand fraction (0.05–0.1 mm);
OM [%] the organic matter;
s is the soil structure class (s = 1: very fine granular, s = 2: fine granular, s = 3, medium or coarse granular, s = 4: blocky, platy or massive; Table 2);
p is the permeability class (p = 1: very rapid, ... , p = 6: very slow; Table 3).

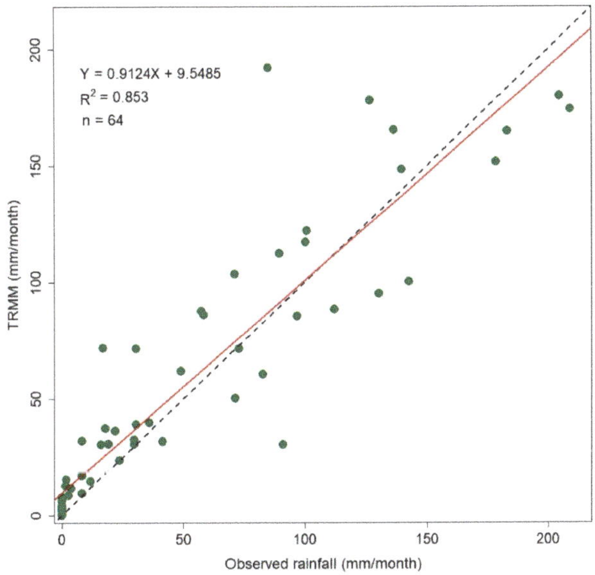

Figure 3. Linear relationship between precipitation data collected at the meteorological station of Zakho and the corresponding cell of the TRMM data for the period September 2002–December 2007 [48].

Table 2. Classes of soil structure derived from the European Soil Database.

Structure Class (s)	Soil Database
1 (very fine granular: 1–2 mm)	G (good)
2 (fine granular: 2–5 mm)	N (normal)
3 (medium or coarse granular: 5–10 mm)	P (poor)
4 (blocky, platy or massive: >10 mm)	H (peaty top soil)

Table 3. Soil permeability classes and saturated hydraulic conductivity ranges estimated from major soil textural classes.

Permeability Class (p)	Texture	Saturated Hydraulic Conductivity, mm h^{-1}
1 (fast and very fast)	Sand	>61.0
2 (moderate fast)	Loamy sand, sandy loam	20.3–61.0
3 (moderate)	Loam, silty loam	5.1–20.3
4 (moderate low)	Sandy clay loam, clay loam	2.0–5.1
5 (slow)	Silty clay loam, sand clay	1.0–2.0
6 (very slow)	Silty clay, clay	>1.0

Estimating soil OM from the organic carbon (OC) is a simple procedure but requires the use of a conversion factor [80]. However, there is no agreement on the conversion factor [81] to use for converting soil OC to OM [82]. According to [80,82], a more reliable value of *OM* can be obtained by using the conversion factor of 1.724, which is based on the assumption that soil *OM* contains 58% carbon, as shown in Equation (6).

$$OM = 1.724 * OC \tag{6}$$

2.4.3. Slope Length and Slope-Steepness Factor (LS factor)

The LS factor which is a function of slope length (*L*) and slope steepness (S), reflects local topography's influence on soil erosion. Numerous approaches have been proposed to evaluate the LS factor (e.g., [22,23,36,83–86]. We have estimated the *L* factor using Equations (7)–(9) proposed by Moore and Burch (1986).

The L factor can be computed by using a slope (in percent) and runoff flow accumulation. We divided the slope map into four classes, which are <1%, 1–3%, 3–5%, and ≥5%.

$$L = \left(\frac{\lambda}{22.12848}\right)^{m} \tag{7}$$

$$\lambda = \mathrm{FA} * \mathrm{Ps} \tag{8}$$

$$\lambda = \mathrm{FA} * \mathrm{Ps} \tag{9}$$

where, λ is the horizontal projection of slope length (m), m is a constant, based on the value of slope; FA is runoff flow accumulation, Ps is the cell size of DEM, and θ is the slope angle in percent.

We estimated the S factor by dividing the slope (in percent) into three classes: <9%, 9–18%, and ≥18% (Equation (10). Equation (11) was applied to evaluate the *LS* factor.

$$S = \begin{cases} 10.8.sin\theta + 0.03 & \theta < 9\% \\ 16.8.sin\theta - 0.05 & 9\% \leq \theta < 18\% \\ 21.9.sin\theta - 0.96 & \theta \geq 18\% \end{cases} \tag{10}$$

$$LS = L * S \tag{11}$$

where S is the slope steepness factor, θ is the slope angle in percent, and LS represents the slope length and slope steepness factors [87].

2.4.4. Cover and Management Factor (C factor)

As a dimensionless and influential factor used in the RUSLE model, the C factor refers to the amount of soil erosion related to the degree of vegetation/ground cover. Evaluation of vegetation cover is considered the most significant aspect of estimating the C factor [88].

We first performed the radiometric corrections to compute top of atmosphere reflectance (ρ) by using (Equation (12)).

$$\rho = \pi * L_{\lambda} * D^{2} * \left(E_{sun,\lambda} * cos\,(\theta)\right)^{-1} \tag{12}$$

where ρ is reflectance (at sensor), L_{λ} represents at-sensor radiance, D is sun–earth distance in astronomical units, θ is the solar zenith angle, and $E_{sun,\lambda}$ is the solar spectral irradiance. Next step involved atmospheric corrections that were performed using the Fast Line-of-Sight Atmospheric Analysis of Spectral Hypercubes (FLAASH) model. The FLAASH [89] model is based on a MODTRAN4 model to decrease the molecular and particulate scattering, absorption, and adjacency effects to retrieve at-surface reflectance values (ρ) [90–92]. The surface reflectance (ρ) of the OLI imageries is used to compute *NDVI* following Equation (13) [93], which, in turn, was used to compute the C factor using Equation (14) [94–96].

$$NDVI = \frac{\rho NIR - \rho R}{\rho NIR + \rho R} \tag{13}$$

where ρNIR and ρR are the reflectance of band 5 (851 nm–879 nm) and band 4 (636 nm–673 nm) of the Landsat-8 OLI surface reflectance.

$$C = Exp\left[-\alpha * \frac{NDVI}{(\beta - NDVI)}\right] \tag{14}$$

where α, β are constants, which were deduced from [94,95] to be 2 and 1, respectively. These constants determine the shape of the NDVI–C curve. We omitted the bare rock in the soil loss assessment, affirmed by relatively low values of $NDVI$ [32].

2.4.5. Support Practice Factor Related to Slope Direction (P)

The P factor describes soil loss ratio by a specific soil conservation or support practice to the corresponding loss up– and down–slope [23,97]. Lack of in situ data for conservation practices within the Zagros region prompted us to apply Equation (15) for estimating the P factor. We modified Equation (15) as suggested by Wener [33,37,98] and used the modified equation (Equation (16)) to calculate P.

$$P = 0.2 + 0.008 * \theta \tag{15}$$

$$P = 0.2 + 0.03 * \theta \tag{16}$$

where, P is the support practice factor, θ is the slope angle in percent. By applying Equation (16), the P factor of flat pixels (θ = 0) was found to be 0.2 (i.e., P = 0.2 + 0.008 * 0), and for the highest slope angle pixels (θ = 100) it was 1 (i.e., P = 0.2 + 0.008 * 100).

2.5. Sediment Yield (SY)

We estimated the SY in the KhRB following [99]. First, we measured the areas of the sub-basins within the KhRB. Then, we calculated the SDR (Equation (17)). Finally, we estimated the SY amount for each sub-basin following Equation (18) [99]. Results of SY were used to estimate the amount of sediment transported by streams and rivers to the downstream area to determine the check dam locations.

$$SDR = 1.817 \times {A_b}^{-0.132} \tag{17}$$

$$SY = AE \times SDR \tag{18}$$

SDR is sediment delivery ratio;
A_b is area of a basin (km^2);
AE is the actual erosion (yearly rate of soil loss (t.ha^{-1}.y^{-1}));
SY is the sediment yield (t.ha^{-1}.y^{-1}).

2.6. Work Procedure

We applied the RUSLE model to estimate soil loss for the KhRB. The KhRB was subdivided into a number of sub-basins (Figure 4A). We estimated the mean of the AE, the PE, and the mean of the HI, for each sub-basin (Figure 4B). Afterwards, we tested the relationship between AE and PE for them.

We classified various parameters in the RUSLE model into five classes, using the natural breaks technique, except the K factor that was classified into three classes based on the three soil types present in the KhRB. These classes are: very high, high, moderate, low, and very low. The natural break classification is best applied to familiar data ranges. It reduces the variance within classes and maximizes the variance between classes [72,100]. This characteristic makes the natural breaks useful when RUSLE and its factors histogram show distinct breaks.

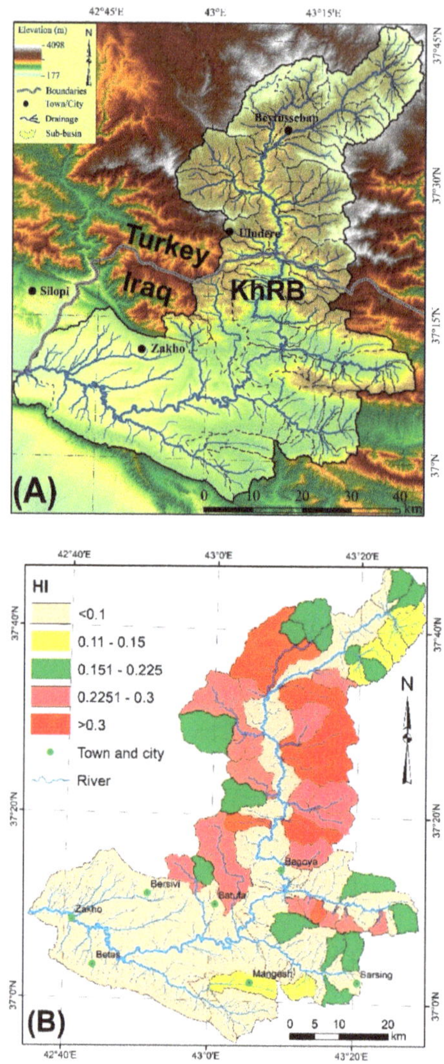

Figure 4. Map showing (**A**) the basins tested for estimating soil loss, and (**B**) the mean value of HI for each sub-basin within the study area.

3. Results

3.1. The Hypsometric Integral (HI)

Figure 4B displays the HI for the study area. The HI values range from 0.36 to 0.05; low values (<0.30) were obtained for the central and lower parts of the KhRB, indicating plain terrain with occasional slope. The HI values above 0.30 fall in the part of basin located inside Turkey. The topography of these areas possesses high relief, which might be controlled by the active thrust fault, HZF (Figure 1).

3.2. The R Factor

The value of R factor in the study area varied from ~296 to ~393 $MJ.mm.ha^{-1}.h^{-1}.y^{-1}$, with an average of ~354 $MJ.mm.ha^{-1}.h^{-1}.y^{-1}$. The R factor map of KhRB, presented in

(Figure 5A), has been sliced into five groups, to show relatively lower R factor values in the south of the KhRB compared to the mountainous terrain in the north.

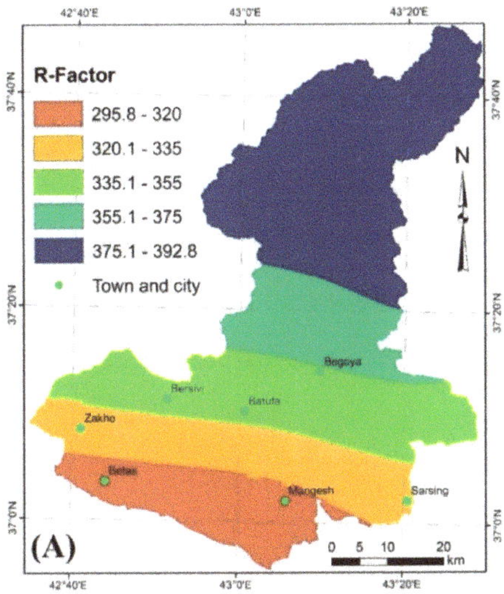

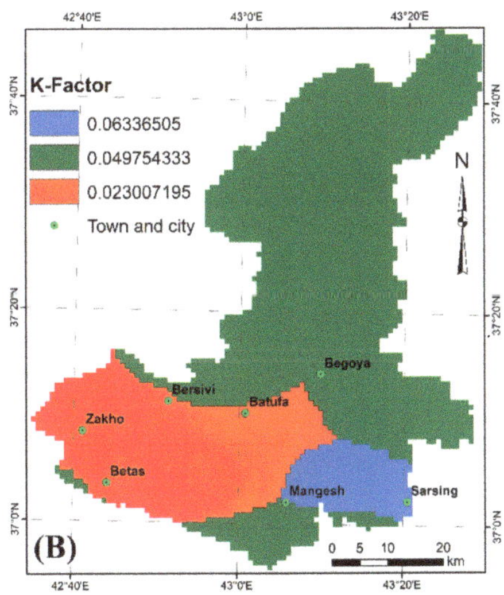

Figure 5. Maps showing the distribution of the R factor (**A**), and the K factor (**B**) within KhRB.

3.3. *K Factor*

The K factor data presented in Figure 5B show that the minimum and maximum soil erodibility is ~0.023 t·ha·h·ha^{-1}·MJ^{-1}.mm^{-1} in the southwestern part of the KhRB and ~0.063 t·ha·h·ha^{-1}·MJ^{-1}.mm^{-1} in the southeastern parts of the KhRB respectively. Major part of the basin has an intermediate to high K value (~0.0498) occurring in the central and northern part of the basin. Three types of soil occur in the KhRB: lithosols, calcic xerosols, and chromic vertisols. The textural description and the K–factor for each soil type is shown in Table 4 and Figure 5B. Variations in the K–factor values are possibly due to variations in the soil types' physical characteristics.

Table 4. K factor, type, texture, and organic material content of the three soil types.

Type of Soil	Texture Class	Sand%	Silt%	Clay%	OC%	OM	K Factor
Lithosols	Loam	43	34	23	1.4	2.4136	0.049754
Calcic Xerosols	Clay loam	40	37	23	0.56	0.96544	0.063365
Chromic Vertisols	Clay	16	29	55	0.75	1.293	0.023007

3.4. The LS Factor

The highest slope angle in the KhRB is 77°, the average being 17.5°. Nearly all high slope areas are located in the central and northern parts of the KhRB. Due to the difference in runoff flow accumulation, and slope distribution, the LS factor values range from 0 to 1619.6 with the average and standard deviation 12.05 and 12.74, respectively (Figure 6A). The KhRB, generally, is characterized by moderate LS factor values: It is <10 in the lower part of the KhRB, but the middle and upper parts of the basin show relatively higher LS factor values, which indicate the possibility of greater soil erosion and consequently accelerated land degradation.

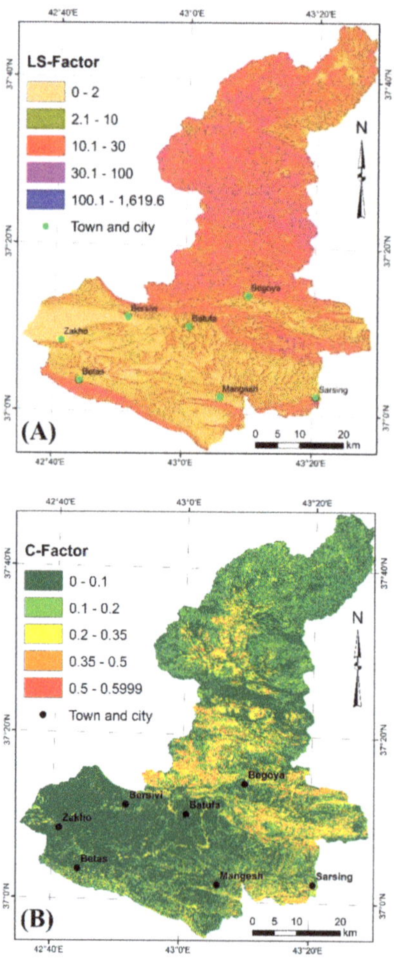

Figure 6. Maps showing distribution of the: (**A**) LS factor, and (**B**) C factor.

3.5. The C Factor

The C factor ranges from 0 to 0.5999 (Figure 6B). Less than 16% of the KhRB is covered with vegetation, and most of the barren land occurs in the upstream and downstream parts of the basin (Figure 7A) where C factor values are low (dark green and green colors). Presence of vegetation increases soil particles' adhesion and binding, making it less prone to erosion. The moderate, high, and very high (>0.2) values of the C factor (green, orange, and red colors) are mainly restricted to the midstream part of the KhRB. Four types of vegetative areas exist in the KhRB (Figure 7B). Open shrublands (dominated by woody perennials (1–2 m height), 10–60% cover), savannas (tree cover 10–30% (canopy > 2 m)), grasslands (dominated by herbaceous annuals (<2 m)), and croplands (at least 60% of area is cultivated cropland) [101].

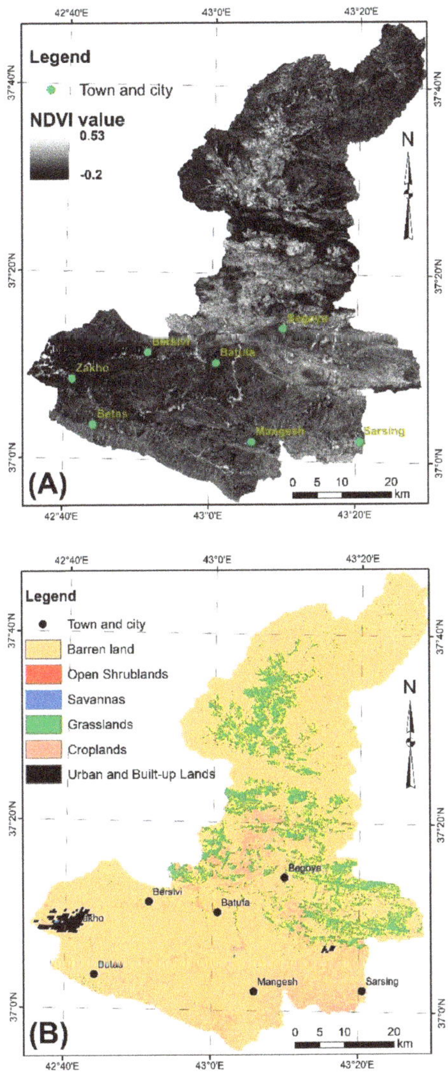

Figure 7. Maps showing (**A**) MODIS land cover (MOD12Q1) in 2019, and (**B**) the NDVI extracted from Landsat OLI, acquired on 26 August 2019.

3.6. The P Factor

Values of the P factor within the KhRB range between 0 and 1 (Figure 8A). Low values correspond to pixels with good conservation practice, while the high values denote areas of poor conservation practice [31]. P factor value depends on slope length and steepness, and farming method [102]. Therefore, an increase in the value of P factor is consistent with the rise in the LS factor value (Figures 6A and 8A). High and very high (0.35–1) values of P factor are mainly distributed in the middle part of the KhRB, while the low and very low values (0–0.22) are mainly distributed downstream of the KhRB (Figure 8A).

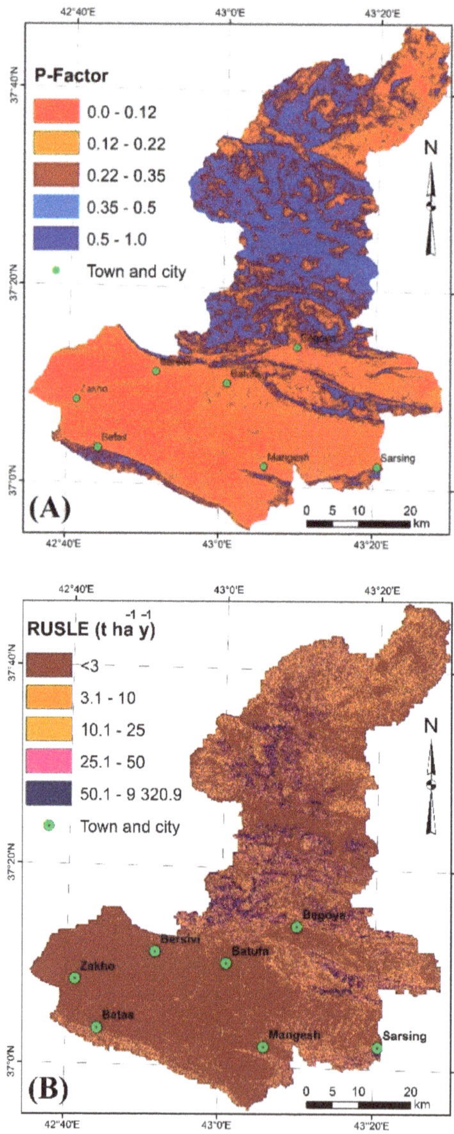

Figure 8. Maps showing distribution of the: (**A**) P factor, and (**B**) soil loss (t.ha^{-1}.y^{-1}) in the KhRB.

3.7. Estimation of Soil Loss

We integrated the RUSLE model's parameters within the GIS environment to evaluate the average annual soil loss in the KhRB and estimate it to be 11.16 t.ha^{-1}.y^{-1} (Figure 8B). The minimum and maximum yearly soil losses are 0 t.ha^{-1}.y^{-1} and 3617.6 t.ha^{-1}.y^{-1}, respectively. Total annual soil loss for the whole study area is estimated at 31,237,283.358 t.y^{-1}. The low and very low range (<10 t.ha^{-1}.y^{-1}) of the yearly soil loss values mainly occur in the northeastern and downstream reaches of the KhRB, while the moderate, high and very high range (>10 t.ha^{-1}.y^{-1}) occurs in the midstream and upstream parts of the KhRB (Figure 8B). Very high soil losses (>50 t.ha^{-1}.y^{-1}) occur only in 6.19% of the KhRB.

3.8. Relationship of PE to RUSLE and the HI

Our study revealed a strong direct correlation between: (a) PE and soil loss, and (b) PE and HI. It is well recognized that heavy rain, and steeper slopes contribute to a large volume of runoff with high erosive energy and greater amount of soil loss. Sixty-one sub-basins included in the KhRB, comprise areas ranging between 3.5 and 974.6 km^2 (average: 57.5 km^2). Analyses of annual soil loss within the KhRB at the sub-basin scale reveal a moderate linear relationship between the mean of the PE and mean of soil loss. This direct relationship is clearly displayed in Figure 9, where the statistical correlation, the *p*-value, the t-test, the slope, and the intercept of the data were >0.511, 9.572×10^{-11}, 7.8565, 0.02979, and 3.56121, respectively. Using the same procedure, we found a moderate linear relationship between the mean of HI and mean of soil loss (Figure 10), where the statistical correlation, the *p*-value, the t-test, the slope, and the intercept of the data were 0.522, 3.315×10^{-10}, 7.6743, 0.0003911, and 0.0921817, respectively.

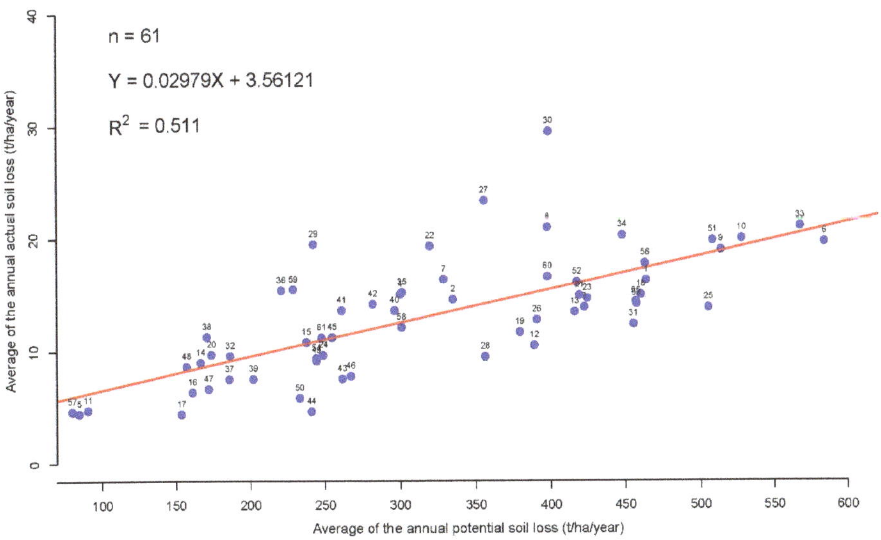

Figure 9. Relationship between average potential soil loss (i.e., R*K*LS product) and the actual soil loss (R*K*LS*P*C) in the KhRB, in individual subbasins.

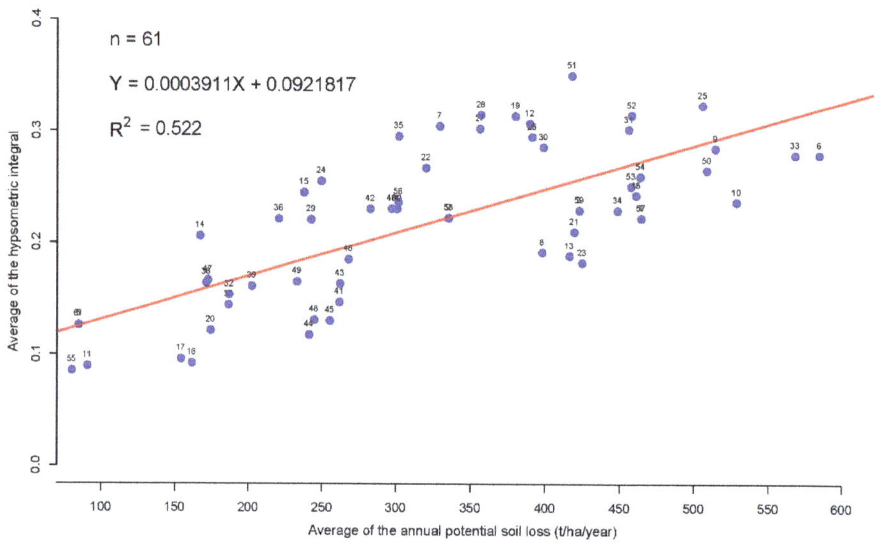

Figure 10. Relationship between average of potential erosion (i.e., R*K*LS product) and the average of the HI in the KhRB, in individual subbasins.

3.9. Estimation the Sediment Yield

In addition to SDR (Figure 11A), we also calculated the average annual quantity of SY for the 61 sub-basins included in the KhRB. The SDR values range between 0.619 and 1.373, respectively. Estimate of the yearly average of SY in the KhRB is 57.79 t.y^{-1}. The minimum and maximum yearly SY are 6.9 t.y^{-1} and 165.12 t.y^{-1}, respectively. Total annual SY for the whole study area is estimated at 21,817.68 t.y^{-1}. The low and very low range (<10 t.ha^{-1}.y^{-1}) of annual SY values mainly occur in the northeastern and downstream reaches of the KhRB, while the moderate, high and very high range (>10 t.ha^{-1}.y^{-1}) occur in the midstream part of the KhRB (Figure 11B).

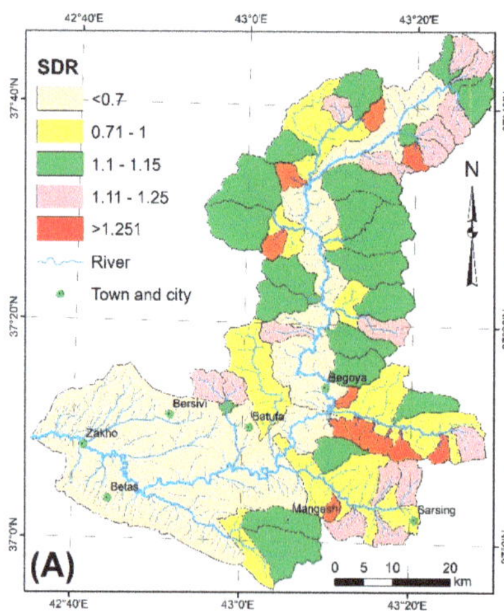

Figure 11. *Cont.*

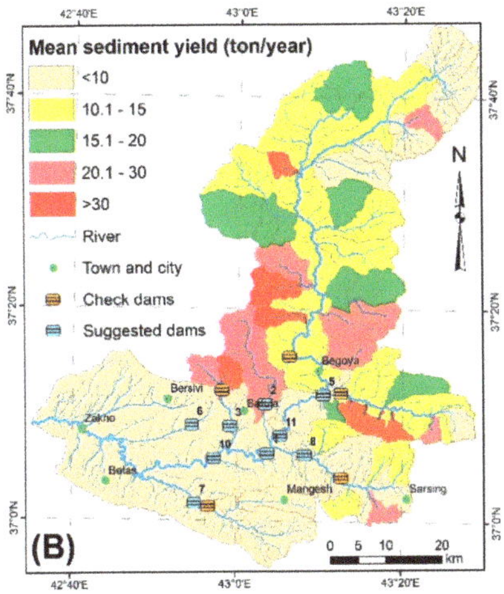

Figure 11. Map showing the distribution of (**A**) sediment delivery ratio (SDR), and (**B**) the average sediment yield within the sub-basins of the KhRB. Location of dam sites recommended in an early study by [48] along with proposed check dams are indicated in (**B**).

4. Discussion

It is well established that soil loss is controlled by the five factors of the RUSLE model (i.e., R, K, LS, C, and P). Change in one of these factors leads to a change in the quantity of soil loss. As can be seen in Equation (3), the R factor is directly related to the amount of precipitation [77]. Moreover, in the KhRB, precipitation is affected by elevation (Figure 1). The soil loss is a function of K factor, rainfall, and runoff [30,78]. The higher the K factor, the greater is the amount of soil loss and vice versa [103]. The same is true for the LS factor, where the amount of soil loss increases with increasing LS factor and vice versa [85,87]. Our results show moderate to high soil loss in areas with high R factor values (Figure 5A), which is consistent with relatively steeper slope zones (Figures 6A and 8B). Soil loss is a function of the products of all five factors in RUSLE, of which R, K, LS are natural factors and cannot be changed, P and C are factors that can be altered by human intervention. This is precisely what is done to minimize soil loss in agricultural land [104,105]. Therefore, by taking measures to reduce values of P and C factors by contour ploughing, no-till seeding, and terracing, etc. the overall soil loss can be reduced.

The C factor is relatively lower in the southern part of the KhRB (Figure 6B) because of sparse vegetation, somewhat gentler slopes, and lower rainfall amounts–all of which contribute to lower rate of soil loss (Figure 8B). On the other hand, the central and the northwestern parts of the KhRB, with lower values of the C factor, due to presence of vegetation, exhibit a relatively higher amount of soil loss due to relatively steeper slope and higher amounts of rainfall. Dynamics of the erosional process are greatly affected by the LS factor that controls runoff flow. Our results agree with the findings reported by [106] in which the altitude variance and topographic relief control the LS factor (Figure 6A); and higher value of P factor (Figure 8A) directly correlates with a higher rate of soil loss (Figure 8B). Interestingly, both of which are related to the steeper slopes (Figure 6A).

This study also concurs with [8], which stated that the main factor controlling soil loss "in the NW part of the ZFTB" is the LS factor (Figure 6A). The next important factor that affects soil loss is the R factor (Figure 5A). There is a narrow range in the values of the K factor (~0.023–0.063 $t \cdot ha \cdot h \cdot ha^{-1} \cdot MJ^{-1}.mm^{-1}$), C factor (0.2–0.87), P factor (0–1), and R factor in the sub-basins. The LS factor for ~99% of the pixels in the sub-basins gave values of <166, except some that show an extremely high value of about 1500.

From the above discussion of the various RUSLE parameters, it can be stated that slope is the main proxy that controls soil loss in the region, where precipitation is strongly affected by the slope. Regional tectonism, evidenced by the presence of HZF in the study area (Figure 1), is believed to have contributed to development of steeper slopes.

Although the RUSLE model is widely used to quantify soil loss and may be more accurate than other methods [107], it has not been applied at pixel scale to estimate PE in the study area. Our study revealed direct relationship between potential soil loss and actual soil loss at the sub-basin scale, but a with moderate-strong value of 0.51 for the coefficient of determination, R^2 (Figure 9). This may be due to the difference in precipitation, soil types, and vegetation density in the various sub-basins.

Overall, this study shows that soil loss in the study area increases with an increase in slope angle, rainfall, and lack of vegetation, and is more pronounced in regions of steeper slopes. This study agrees with [108], which reported that different slopes initiate different rates of soil loss, and progressively higher soil loss occurs on steeper slopes. Our study indicates that the direct relationship between PE and HI can be formed in quantitative relationships (Figure 10). This relation confirms the finding of [14,71] that the sub-basins in youthful stage are more active than mature stage. This allows the determination of active and inactive erosional regions, which will be helpful to select the best sites for check dams.

Further studies should be carried out to ensure safety of existing dams in Iraq. Regarding potential siltation problems in the proposed dams in Iraq, research should be undertaken to determine annual volume of soil that will enter the reservoirs to ensure safety and longevity of the planned structures.

Estimation of the Dam Siltation in the KhRB

Estimation of soil loss at sub-basin scale provided reliable data to find suitable sites for locating main dams and the check dams. We applied RUSLE with GIS and RS at the sub-basin scale for the first time in the KhRB to estimate soil loss. Since the planned dams are located in the sub-basins (Figure 11B, Table 5), they would experience excessive reservoir siltation due to high rate of soil loss. To minimize this problem, we have recommended check dams to trap the sediments upstream of the main dams.

Table 5. Features of the proposed dam (modified after [48]). The dam sites (4 and 9) are excluded because they are outside of the study area.

Dam No.	Dam Width (m)	Lake Area (km^2)	Volume (m^3)	Basin Area (km^2)	Soil Loss (m^3)	X-Coordinate	Y-Coordinate	N_v
1	1109.9	8.61	226,654,026	2420.5	336,547,265.4	43.061066	37.10736	8
2	509.6	1.31	37,026,508	103.8	18,905,137.1	43.05568	37.18427	1
3	850.0	2.35	53,640,739	74.3	7,383,939.5	42.98572	37.14923	4
5	667.0	5.34	421,706,403	266.9	21,822,711.4	43.208654	37.20219	2
6	443.0	0.46	5,769,178	26.3	458,963.8	42.907832	37.15043	0
7	817.0	4.39	191,423,581	219.9	4,950,075.4	42.914868	37.02736	1
8	1367.0	14.86	1,182,091,212	276.9	13,890,242.6	43.136919	37.10535	0
10	372.0	1. 60	55,823,171	35.6	301,711.7	42.953957	37.09876	0
11	737.0	6.04	199,256,701	1963.5	313,212,044.7	43.08669	37.13489	3

One of the common and simple ways to mitigate excessive siltation is construction of small check dams. [48] had recommended suitable locations for dams in the KhRB (Figure 11B and Table 5) based on lithology, tectonic zones, distance to active fault, distance to the lineaments, soil type, land cover, elevation, slope gradient, precipitation, stream width, curve number, distance to road, distance to towns and cities, and distance to villages [109–111]. These dam sites might be influenced by soil loss from the upstream areas leading to excessive reservoir siltation. Drawing upon our present research, we propose construction of five check dams to trap sediments transported from the upstream area of the KhRB (Figure 11). These dams will prevent ~1757 $t.y^{-1}$ of sediment, eroded from the upper reaches of the KhRB, from entering the main dams. The suggested check dams will prolong the lives of proposed dams (Figure 11B), and also increase the useful life of the Mosul dam to the southwest of the KhRB.

5. Conclusions

The classical RUSLE model has been integrated with remote sensing and GIS to estimate soil loss and sediment yield in the KhRB. Results show the spatial distribution and rate of average annual soil loss and sediment yield within selected basins that ranges from 0 $t.ha^{-1}.y^{-1}$ to 3617.6 $t.ha^{-1}.y^{-1}$, and from 6.9 $t.y^{-1}$ to 165.12 $t.y^{-1}$, respectively. The northwestern part of KhRB experiences low rate of soil loss (A) and SY. These rates are mainly influenced by topography, characterized by steep slopes, and high rainfall. Our study also demonstrates that LS and R are the most dominant factors controlling soil loss in the selected area. Analysis of the potential and actual soil erosion for KhRB shows a direct relationship. We conclude that RUSLE can serve as an effective model for estimating sediment yield, and for mitigation its adverse effects on multi-purpose river valley development projects for water supply, agriculture, etc. We recommend locating check dams in the upstream reaches of KhRB to minimize reservoir siltation and ensure safety of the main dams. Our study showed a correlation between potential soil erosion and HI. We found that HI values at the sub-basin scale can be used to identify active and inactive erosional regions. Thus, HI could serve as a useful geomorphic index for locating check dams in basins with high rate of soil loss. Further studies should be performed to find out if the HI can be used for estimating actual soil loss at specific locations.

Author Contributions: Arsalan Ahmed Othman outlined the research, prepared and processed the data, accomplished the study, and wrote the manuscript. Ahmed K. Obaid processed the data and supported the writing, analysis, and discussion. Diary Ali Mohammed Amin Al-Manmi, Ahmed F. Al-Maamar, Ahmed T. Shihab, and Younus I. Al-Saady supported the analysis and RUSLE model. Veraldo Liesenberg and Syed E. Hasan supported the writing and discussion. All authors have checked and revised the manuscript. All authors have read and agreed to the published version of the manuscript.

Funding: This research received no external funding. Veraldo Liesenberg is supported by FAPESC (2017TR1762) and CNPq (313887/2018-7).

Institutional Review Board Statement: Not applicable.

Informed Consent Statement: Not applicable.

Data Availability Statement: Data available on request due to privacy restrictions.

Acknowledgments: We thank USGS for providing Landsat and DEM data. We are grateful to the General Survey Authority and Ministry of Planning, Iraq for providing other useful data used in our study. We would like to thank the Food and Agriculture Organization (FAO) of the United Nations, Chinese Academy of Sciences (CAS), International Institute for Applied Systems Analysis (IIASA), International Soil Reference and Information Centre (ISRIC), and Joint Research Centre of the European Commission (JRC) for supplying Harmonized world soil database (HWSD). We are grateful to the U.S. National Aeronautics and Space Administration (NASA) and the Japan Aerospace Exploration Agency (JAXA) for providing Tropical Rainfall Measuring Mission (TRMM) data.

Conflicts of Interest: The authors declare no conflict of interest.

References

1. García-Ruiz, J.M.; Beguería, S.; Nadal-Romero, E.; González-Hidalgo, J.C.; Lana-Renault, N.; Sanjuán, Y. A meta-analysis of soil erosion rates across the world. *Geomorphology* **2015**, *239*, 160–173. [CrossRef]
2. Pimentel, D.; Terhune, E.C.; Dyson-Hudson, R.; Rochereau, S.; Samis, R.; Smith, E.A.; Denman, D.; Reifschneider, D.; Shepard, M. Land degradation: Effects on food and energy resources. *Science* **1976**, *194*, 149–155. [CrossRef] [PubMed]
3. Othman, A.A. *Improving Remote Sensing Based Landslide Investigations and Lithological Mapping Using Morphological Features: A Case Study in NE-Iraq and W-Iran*; Technische Universität Bergakademie: Freiberg, Germany, 2015.
4. Kuzucuoğlu, C.; Çiner, A.; Kazancı, N. (Eds.) The Geomorphological Regions of Turkey BT. In *Landscapes and Landforms of Turkey*; Springer International Publishing: Cham, Switzerland, 2019; pp. 41–178, ISBN 978-3-030-03515-0.
5. Carretier, S.; Regard, V.; Vassallo, R.; Aguilar, G.; Martinod, J.; Riquelme, R.; Pepin, E.; Charrier, R.; Hérail, G.; Farías, M.; et al. Slope and climate variability control of erosion in the Andes of central Chile. *Geology* **2013**, *41*, 195–198. [CrossRef]
6. Al-Saady, Y.; Al-Suhail, Q.; Al-Tawash, B.; Othman, A. Drainage network extraction and morphometric analysis using remote sensing and GIS mapping techniques (Lesser Zab River Basin, Iraq and Iran). *Environ. Earth Sci.* **2016**, *75*. [CrossRef]

7. Othman, A.A.; Al-Maamar, A.F.; Al-Manmi, D.A.M.; Liesenberg, V.; Hasan, S.; Al-Saady, Y.I.; Shihab, A.T.; Khwedim, K. Application of DInSAR-PSI Technology for Deformation Monitoring of the Mosul Dam, Iraq. *Remote Sens.* **2019**, *11*, 2632. [CrossRef]
8. Thornes, J.B. Modelling Soil Erosion by Grazing: Recent Developments and New Approaches. *Geogr. Res.* **2007**, *45*, 13–26. [CrossRef]
9. Obaid, A.K.; Allen, M.B. Landscape expressions of tectonics in the Zagros fold-and-thrust belt. *Tectonophysics* **2019**. [CrossRef]
10. Allen, M.B.; Talebian, M. Structural variation along the Zagros and the nature of the Dezful Embayment. *Geol. Mag.* **2011**, *148*, 911–924. [CrossRef]
11. Langbein, W.B. *Topographic Characteristics of Drainage Basins*; US Department of Interior: Tulsa, OK, USA, 1947.
12. Strahler, A.N. Hypsometric (area-altitude) analysis of erosional topography. *Geol. Soc. Am. Bull.* **1952**, *63*, 1117–1142. [CrossRef]
13. Schumm, S.A. Evolution of drainage systems and slopes in badlands at Perth Amboy, New Jersey. *Geol. Soc. Am. Bull.* **1956**, *67*, 597–646. [CrossRef]
14. Keller, E.A.; Pinter, N. *Active Tectonics: Earthquakes, Uplift, and Landscape*; Prentice Hall: Upper Saddle River, NJ, USA, 2002.
15. Li, L.; Du, S.; Wu, L.; Liu, G. An overview of soil loss tolerance. *Catena* **2009**, *78*, 93–99. [CrossRef]
16. Montanarella, L.; Pennock, D.; McKenzie, N.; Badraoui, M.; Chude, V.; Baptista, I.; Mamo, T.; Yemefack, M.; Aulakh, M.; Yagi, K.; et al. World's soils are under threat. *Soil* **2016**, *2*. [CrossRef]
17. Blanco-Canqui, H.; Lal, R. *Principles of Soil Conservation and Management*; Springer: Dordrecht, The Netherlands, 2008; ISBN 9781402087097.
18. Toy, T.J.; Foster, G.R.; Renard, K.G. *Soil Erosion: Processes, Prediction, Measurement, and Control*; Wiley: Hoboken, NJ, USA, 2002; ISBN 9780471383697.
19. Laflen, J.M.; Lane, L.J.; Foster, G.R. WEPP: A new generation of erosion prediction technology. *J. Soil Water Conserv.* **1991**, *46*, 34–38.
20. Yu, B.; Rose, C.W.; Ciesiolka, C.A.A.; Coughlan, K.J.; Fentie, B. Toward a framework for runoff and soil loss prediction using GUEST technology. *Soil Res.* **1997**, *35*, 1191–1212. [CrossRef]
21. Viney, N.R.; Sivapalan, M. A conceptual model of sediment transport: Application to the Avon River Basin in Western Australia. *Hydrol. Process.* **1999**, *13*, 727–743. [CrossRef]
22. Wischmeier, W.H.; Smith, D.D. *Predicting Rainfall Erosion Losses: A Guide to Conservation Planning*; Science and Education Administration, US Department of Agriculture: Washington, DC, USA, 1978.
23. Renard, K.G.; Foster, G.R.; Weesies, G.A.; McCool, D.K.; Yoder, D.C. *Predicting Soil Erosion by Water: A Guide to Conservation Planning with the Revised Universal Soil Loss Equation (RUSLE)*; United States Department of Agriculture: Washington, DC, USA, 1997; Volume 703.
24. Nearing, M.A.; Yin, S.; Borrelli, P.; Polyakov, V.O. Rainfall erosivity: An historical review. *Catena* **2017**, *157*, 357–362. [CrossRef]
25. Nearing, M.A.; Govers, G.; Norton, L.D. Variability in Soil Erosion Data from Replicated Plots. *Soil Sci. Soc. Am. J.* **1999**, *63*, 1829–1835. [CrossRef]
26. Boardman, J. Soil erosion science: Reflections on the limitations of current approaches. *Catena* **2006**, *68*, 73–86. [CrossRef]
27. Qiwei, C.; Kangning, X.; Rong, Z. Assessment on erosion risk based on GIS in typical Karst region of Southwest China. *Eur. J. Remote Sens.* **2020**, 1–12. [CrossRef]
28. Gianinetto, M.; Aiello, M.; Polinelli, F.; Frassy, F.; Rulli, M.C.; Ravazzani, G.; Bocchiola, D.; Chiarelli, D.D.; Soncini, A.; Vezzoli, R. D-RUSLE: A dynamic model to estimate potential soil erosion with satellite time series in the Italian Alps. *Eur. J. Remote Sens.* **2019**, *52*, 34–53. [CrossRef]
29. Saygın, S.D.; Ozcan, A.U.; Basaran, M.; Timur, O.B.; Dolarslan, M.; Yılman, F.E.; Erpul, G. The combined RUSLE/SDR approach integrated with GIS and geostatistics to estimate annual sediment flux rates in the semi-arid catchment, Turkey. *Environ. Earth Sci.* **2014**, *71*, 1605–1618. [CrossRef]
30. Phinzi, K.; Ngetar, N.S. The assessment of water-borne erosion at catchment level using GIS-based RUSLE and remote sensing: A review. *Int. Soil Water Conserv. Res.* **2019**, *7*, 27–46. [CrossRef]
31. Ganasri, B.P.; Ramesh, H. Assessment of soil erosion by RUSLE model using remote sensing and GIS—A case study of Nethravathi Basin. *Geosci. Front.* **2016**, *7*, 953–961. [CrossRef]
32. Thomas, J.; Joseph, S.; Thrivikramji, K.P. Assessment of soil erosion in a tropical mountain river basin of the southern Western Ghats, India using RUSLE and GIS. *Geosci. Front.* **2018**, *9*, 893–906. [CrossRef]
33. Terranova, O.; Antronico, L.; Coscarelli, R.; Iaquinta, P. Soil erosion risk scenarios in the Mediterranean environment using RUSLE and GIS: An application model for Calabria (southern Italy). *Geomorphology* **2009**, *112*, 228–245. [CrossRef]
34. Nasir, N.; Selvakumar, R. Influence of land use changes on spatial erosion pattern, a time series analysis using RUSLE and GIS: The cases of Ambuliyar sub-basin, India. *Acta Geophys.* **2018**, *66*, 1121–1130. [CrossRef]
35. Farhan, Y.; Zregat, D.D.; Farhan, I. Spatial Estimation of Soil Erosion Risk Using RUSLE Approach, RS, and GIS Techniques: A Case Study of Kufranja Watershed, Northern Jordan. *J. Water Resour. Prot.* **2013**, *5*, 1247–1261. [CrossRef]
36. Zerihun, M.; Mohammedyasin, M.S.; Sewnet, D.; Adem, A.A.; Lakew, M. Assessment of soil erosion using RUSLE, GIS and remote sensing in NW Ethiopia. *Geoderma Reg.* **2018**, *12*, 83–90. [CrossRef]
37. Fu, B.J.; Zhao, W.W.; Chen, L.D.; Zhang, Q.J.; Lü, Y.H.; Gulinck, H.; Poesen, J. Assessment of soil erosion at large watershed scale using RUSLE and GIS: A case study in the Loess Plateau of China. *Land Degrad. Dev.* **2005**, *16*, 73–85. [CrossRef]

38. Bircher, P.; Liniger, H.P.; Prasuhn, V. Comparing different multiple flow algorithms to calculate RUSLE factors of slope length (L) and slope steepness (S) in Switzerland. *Geomorphology* **2019**, *346*, 106850. [CrossRef]
39. Ghosal, K.; Das Bhattacharya, S. A Review of RUSLE Model. *J. Indian Soc. Remote Sens.* **2020**, *48*, 689–707. [CrossRef]
40. Chen, T.; Niu, R.; Li, P.; Zhang, L.; Du, B. Regional soil erosion risk mapping using RUSLE, GIS, and remote sensing: A case study in Miyun Watershed, North China. *Environ. Earth Sci.* **2011**, *63*, 533–541. [CrossRef]
41. Renard, K.G.; Foster, G.R.; Weesies, G.A.; Porter, J.P. RUSLE: Revised universal soil loss equation. *J. Soil Water Conserv.* **1991**, *46*, 30–33.
42. Chuenchum, P.; Xu, M.; Tang, W. Estimation of Soil Erosion and Sediment Yield in the Lancang–Mekong River Using the Modified Revised Universal Soil Loss Equation and GIS Techniques. *Water* **2020**, *12*, 135. [CrossRef]
43. Samad, N.; Hamid, M.; Muhammad, C.; Saleem, M.; Hamid, Q.; Babar, U.; Tariq, H.; Farid, M.S. Sediment yield assessment and identification of check dam sites for Rawal Dam catchment. *Arab. J. Geosci.* **2016**, *9*, 1–14. [CrossRef]
44. Dutta, S. Soil erosion, sediment yield and sedimentation of reservoir: A review. *Model. Earth Syst. Environ.* **2016**, *2*, 1–18. [CrossRef]
45. Bagherzadeh, A.; Daneshvar, M.R.M. Evaluation of sediment yield and soil loss by the MPSIAC model using GIS at Golestan watershed, northeast of Iran. *Arab. J. Geosci.* **2013**, *6*, 3349–3362. [CrossRef]
46. Zhou, Q.; Yang, S.; Zhao, C.; Cai, M.; Ya, L. A soil erosion assessment of the upper mekong river in yunnan province, China. *Mt. Res. Dev.* **2014**, *34*, 36–47. [CrossRef]
47. Eroğlu, H.; Çakir, G.; Sivrikaya, F.; Akay, A.E. Using high resolution images and elevation data in classifying erosion risks of bare soil areas in the Hatila Valley Natural Protected Area, Turkey. *Stoch. Environ. Res. Risk Assess.* **2010**, *24*, 699–704. [CrossRef]
48. Othman, A.A.; Al-Maamar, A.F.; Al-Manmi, D.A.M.; Veraldo, L.; Hasan, S.E.; Obaid, A.K.; Al-Quraishi, A.M.F. GIS-based modeling for selection of dam sites in the Kurdistan Region, Iraq. *ISPRS Int. J. Geo-Inf.* **2020**, *9*, 244. [CrossRef]
49. Mehri, A.; Salmanmahiny, A.; Tabrizi, A.R.M.; Mirkarimi, S.H.; Sadoddin, A. Investigation of likely effects of land use planning on reduction of soil erosion rate in river basins: Case study of the Gharesoo River Basin. *Catena* **2018**, *167*, 116–129. [CrossRef]
50. Ahmadi Mirghaed, F.; Souri, B.; Mohammadzadeh, M.; Salmanmahiny, A.; Mirkarimi, S.H. Evaluation of the relationship between soil erosion and landscape metrics across Gorgan Watershed in northern Iran. *Environ. Monit. Assess.* **2018**, *190*. [CrossRef] [PubMed]
51. Sheikh, V.; Kornejady, A.; Ownegh, M. Application of the coupled TOPSIS–Mahalanobis distance for multi-hazard-based management of the target districts of the Golestan Province, Iran. *Nat. Hazards* **2019**, *96*, 1335–1365. [CrossRef]
52. Zakeri, E.; Mousavi, S.A.; Karimzadeh, H. Scenario-based modelling of soil conservation function by rangeland vegetation cover in northeastern Iran. *Environ. Earth Sci.* **2020**, *79*. [CrossRef]
53. Zare, M.; Nazari Samani, A.A.; Mohammady, M.; Teimurian, T.; Bazrafshan, J. Simulation of soil erosion under the influence of climate change scenarios. *Environ. Earth Sci.* **2016**, *75*. [CrossRef]
54. Arekhi, S.; Niazi, Y.; Kalteh, A.M. Soil erosion and sediment yield modeling using RS and GIS techniques: A. case study, Iran. *Arab. J. Geosci.* **2012**, *5*, 285–296. [CrossRef]
55. Ebrahimzadeh, S.; Motagh, M.; Mahboub, V.; Mirdar Harijani, F. An improved RUSLE/SDR model for the evaluation of soil erosion. *Environ. Earth Sci.* **2018**, *77*, 1–7. [CrossRef]
56. Demirci, A.; Karaburun, A. Estimation of soil erosion using RUSLE in a GIS framework: A case study in the Buyukcekmece Lake watershed, northwest Turkey. *Environ. Earth Sci.* **2012**, *66*, 903–913. [CrossRef]
57. Imamoglu, A.; Dengiz, O. Determination of soil erosion risk using RUSLE model and soil organic carbon loss in Alaca catchment (Central Black Sea region, Turkey). *Rend. Lincei* **2017**, *28*, 11–23. [CrossRef]
58. Özalp, M.; Yüksel, E.E.; Yıldırımer, S. Subdividing large mountainous watersheds into smaller hydrological units to predict soil loss and sediment yield using the GeoWEPP model. *Polish J. Environ. Stud.* **2017**, *26*, 2135–2146. [CrossRef]
59. Issa, I. Sedimentological and Hydrological Investigation of Mosul Dam Reservoir. Ph.D. Thesis, Lulea University of Technology, Luleå, Sweden, 2015.
60. Rabus, B.; Eineder, M.; Roth, A.; Bamler, R. The shuttle radar topography mission—A new class of digital elevation models acquired by spaceborne radar. *ISPRS J. Photogramm. Remote Sens.* **2003**, *57*, 241–262. [CrossRef]
61. GSFC DAAC. Tropical Rainfall Measurement Mission Project (TRMM;3B43 V7). Available online: http://disc.gsfc.nasa.gov/datacollection/TRMM_3B42_daily_V6.shtml (accessed on 10 April 2020).
62. Kummerow, C.; Barnes, W.; Kozu, T.; Shiue, J.; Simpson, J. The Tropical Rainfall Measuring Mission (TRMM) sensor package. *J. Atmos. Ocean. Technol.* **1998**, *15*, 809–817. [CrossRef]
63. Nachtergaele, F.; Van Velthuizen, H.; Van Engelen, V.; Fischer, G.; Jones, A.; Montanarella, L.; Petri, M.; Prieler, S.; Teixeira, E.; Shi, X. *Harmonized World Soil Database (Version 1.2)*; FAO: Rome, Italy; IIASA: Laxenburg, Austria, 2012; pp. 1–50.
64. Akintorinwa, O.J.; Okoro, O. V Combine electrical resistivity method and multi-criteria GIS-based modeling for landfill site selection in the Southwestern Nigeria. *Environ. Earth Sci.* **2019**, *78*. [CrossRef]
65. Shahzad, F.; Gloaguen, R. TecDEM: A MATLAB based toolbox for tectonic geomorphology, Part 2: Surface dynamics and basin analysis. *Comput. Geosci.* **2011**, *37*, 261–271. [CrossRef]
66. Shahzad, F.; Gloaguen, R. TecDEM: A MATLAB based toolbox for tectonic geomorphology, Part 1: Drainage network preprocessing and stream profile analysis. *Comput. Geosci.* **2011**, *37*, 250–260. [CrossRef]
67. ESRI (Environmental Systems Research Institute). *ArcGIS Desktop: Release 10.3*; ESRI: Redlands, CA, USA, 2015.

68. Zhang, L.; Liang, S.; Yang, X.; Gan, W. Landscape evolution of the Eastern Himalayan Syntaxis based on basin hypsometry and modern crustal deformation. *Geomorphology* **2020**, *355*, 107085. [CrossRef]
69. Pérez-Peña, J.V.; Azañón, J.M.; Azor, A. CalHypso: An ArcGIS extension to calculate hypsometric curves and their statistical moments. Applications to drainage basin analysis in SE Spain. *Comput. Geosci.* **2009**, *35*, 1214–1223. [CrossRef]
70. Obaid, A.K.; Allen, M.B. Landscape maturity, fold growth sequence and structural style in the Kirkuk Embayment of the Zagros, northern Iraq. *Tectonophysics* **2017**, *717*, 27–40. [CrossRef]
71. Othman, A.A.; Gloaguen, R. River Courses Affected by Landslides and Implications for Hazard Assessment: A High Resolution Remote Sensing Case Study in NE Iraq–W Iran. *Remote Sens.* **2013**, *5*, 1024–1044. [CrossRef]
72. Othman, A.A.; Gloaguen, R.; Andreani, L.; Rahnama, M. Improving landslide susceptibility mapping using morphometric features in the Mawat area, Kurdistan Region, NE Iraq: Comparison of different statistical models. *Geomorphology* **2018**, *319*, 147–160. [CrossRef]
73. Pike, R.J.; Wilson, S.E. Elevation-relief ratio, hypsometric integral, and geomorphic area-altitude analysis. *Bull. Geol. Soc. Am.* **1971**, *82*, 1079–1084. [CrossRef]
74. Zeng, C.; Wang, S.; Bai, X.; Li, Y.; Tian, Y.; Li, Y.; Wu, L.; Luo, G. Soil erosion evolution and spatial correlation analysis in a typical karst geomorphology using RUSLE with GIS. *Solid Earth* **2017**, *8*, 721–736. [CrossRef]
75. Kayet, N.; Pathak, K.; Chakrabarty, A.; Sahoo, S. Evaluation of soil loss estimation using the RUSLE model and SCS-CN method in hillslope mining areas. *Int. Soil Water Conserv. Res.* **2018**, *6*, 31–42. [CrossRef]
76. Ma, X.; He, Y.; Xu, J.; Noordwijk, M.; Lu, X. Spatial and temporal variation in rainfall erosivity in a Himalayan watershed. *Catena* **2014**, *121*, 248–259. [CrossRef]
77. Karan, S.K.; Ghosh, S.; Samadder, S.R. Identification of spatially distributed hotspots for soil loss and erosion potential in mining areas of Upper Damodar Basin—India. *Catena* **2019**, *182*, 104144. [CrossRef]
78. Thomas, J. (Ed.) Frontispiece BT. In *Nineteenth-Century Illustration and the Digital: Studies in Word and Image*; Springer International Publishing: Cham, Switzerland, 2017; pp. 1–14, ISBN 978-3-319-58148-4.
79. Panagos, P.; Meusburger, K.; Ballabio, C.; Borrelli, P.; Alewell, C. Soil erodibility in Europe: A high-resolution dataset based on LUCAS. *Sci. Total Environ.* **2014**, *479*, 189–200. [CrossRef]
80. Kerven, G.L.; Menzies, N.W.; Geyer, M.D. Analytical methods and quality assurance. *Commun. Soil Sci. Plant Anal.* **2000**, *31*, 1935–1939. [CrossRef]
81. Pribyl, D.W. A critical review of the conventional SOC to SOM conversion factor. *Geoderma* **2010**, *156*, 75–83. [CrossRef]
82. Waksman, S.A.; Stevens, K.R. A critical study of the methods for determining the nature and abundance of soil organic matter. *Soil Sci.* **1930**, *30*, 97–116. [CrossRef]
83. Desmet, P.J.J.; Govers, G. A GIS procedure for automatically calculating the USLE LS factor on topographically complex landscape units. *J. Soil Water Conserv.* **1996**, *51*, 427–433.
84. Mitasova, H.; Hofierka, J.; Zlocha, M.; Iverson, L.R. Modelling topographic potential for erosion and deposition using GIS. *Int. J. Geogr. Inf. Syst.* **1996**, *10*, 629–641. [CrossRef]
85. Moore, I.D.; Burch, G.J. Physical Basis of the Length-slope Factor in the Universal Soil Loss Equation1. *Soil Sci. Soc. Am. J.* **1986**, *50*, 1294–1298. [CrossRef]
86. Wilson, J.P.; Gallant, J.C. *Terrain Analysis: Principles and Applications*; Wiley: Hoboken, NJ, USA, 2000; Volume 4, p. 479.
87. Li, Y.; Qi, S.; Liang, B.; Ma, J.; Cheng, B.; Ma, C.; Qiu, Y.; Chen, Q. Dangerous degree forecast of soil loss on highway slopes in mountainous areas of the Yunnan–Guizhou Plateau (China) using the Revised Universal Soil Loss Equation. *Nat. Hazards Earth Syst. Sci.* **2019**, *19*, 757–774. [CrossRef]
88. Pal, S.C.; Chakrabortty, R. Simulating the impact of climate change on soil erosion in sub-tropical monsoon dominated watershed based on RUSLE, SCS runoff and MIROC5 climatic model. *Adv. Space Res.* **2019**, *64*, 352–377. [CrossRef]
89. Abrams, M.; Hook, S. *ASTER User Handbook. Version 2*; Jet Propulsion Laboratory/California Institute of Technology: Pasadena, CA, USA, 2002.
90. Felde, G.W.; Anderson, G.P.; Cooley, T.W.; Matthew, M.W.; Adler-Golden, S.M.; Berk, A.; Lee, J. Analysis of Hyperion data with the FLAASH atmospheric correction algorithm. In *2003 IEEE International Geoscience and Remote Sensing Symposium*; IEEE International: New York, NY, USA, 2003; Volume 1, pp. 90–92.
91. Liesenberg, V.; Gloaguen, R. Evaluating SAR polarization modes at L-band for forest classification purposes in Eastern Amazon, Brazil. *Int. J. Appl. Earth Obs. Geoinf.* **2013**, *21*, 122–135. [CrossRef]
92. Othman, A.A.; Gloaguen, R. Integration of spectral, spatial and morphometric data into lithological mapping: A comparison of different Machine Learning Algorithms in the Kurdistan Region, NE Iraq. *J. Asian Earth Sci.* **2017**. [CrossRef]
93. Rouse, J.W.; Haas, R.H.; Schelle, J.A.; Deering, D.W. *Monitoring the Vernal Advancement or Retrogradation of Natural Vegetation*; NASA/GSFC: Greenbelt, MD, USA, 1974.
94. Chandramani, L.; Oinam, B. Soil Loss Assessment in Imphal River Watershed, Manipur, North-East India: A Spatio-Temporal Approach BT. In *Advances in Water Resources Engineering and Management*; Al Khaddar, R., Singh, R.K., Dutta, S., Kumari, M., Eds.; Springer: Singapore, 2020; pp. 81–101.
95. Durigon, V.L.; Carvalho, D.F.; Antunes, M.A.H.; Oliveira, P.T.S.; Fernandes, M.M. NDVI time series for monitoring RUSLE cover management factor in a tropical watershed. *Int. J. Remote Sens.* **2014**, *35*, 441–453. [CrossRef]

96. Nyesheja, E.M.; Chen, X.; El-Tantawi, A.M.; Karamage, F.; Mupenzi, C.; Nsengiyumva, J.B. Soil erosion assessment using RUSLE model in the Congo Nile Ridge region of Rwanda. *Phys. Geogr.* **2019**, *40*, 339–360. [CrossRef]
97. Prasannakumar, V.; Vijith, H.; Abinod, S.; Geetha, N. Estimation of soil erosion risk within a small mountainous sub-watershed in Kerala, India, using Revised Universal Soil Loss Equation (RUSLE) and geo-information technology. *Geosci. Front.* **2012**, *3*, 209–215. [CrossRef]
98. Lufafa, A.; Tenywa, M.M.; Isabirye, M.; Majaliwa, M.J.G.; Woomer, P.L. Prediction of soil erosion in a Lake Victoria basin catchment using a GIS-based Universal Soil Loss model. *Agric. Syst.* **2003**, *76*, 883–894. [CrossRef]
99. Sharda, V.N.; Ojasvi, P.R. A revised soil erosion budget for India: Role of reservoir sedimentation and land-use protection measures. *Earth Surf. Process. Landforms* **2016**, *41*, 2007–2023. [CrossRef]
100. Jenks, G.F. The Data Model Concept in Statistical Mapping. *Int. Yearb. Cartogr.* **1967**, *7*, 186–190.
101. Sulla-Menashe, D.; Friedl, M.A. *User Guide to Collection 6 MODIS Land Cover (MCD12Q1 and MCD12C1) Product*; NASA EOSDIS Land Processes DAAC: Sioux Falls, SD, USA, 2018.
102. Bouguerra, H.; Bouanani, A.; Khanchoul, K.; Derdous, O.; Tachi, S.E. Mapping erosion prone areas in the Bouhamdane watershed (Algeria) using the Revised Universal Soil Loss Equation through GIS. *J. Water L. Dev.* **2017**, *32*, 13–23. [CrossRef]
103. Kumar, S.; Kushwaha, S.P.S. Modelling soil erosion risk based on RUSLE-3D using GIS in a Shivalik sub-watershed. *J. Earth Syst. Sci.* **2013**, *122*, 389–398. [CrossRef]
104. Renard, K.G.; Yoder, D.C.; Lightle, D.T.; Dabney, S.M. Universal Soil Loss Equation and Revised Universal Soil Loss Equation. *Handb. Eros. Model.* **2011**, 135–167. [CrossRef]
105. Toy, T.J.; Foster, G.R.; Renard, K.G. RUSLE for mining, construction and reclamation lands. *J. Soil Water Conserv.* **1999**, *54*, 462–467.
106. Olorunfemi, I.E.; Komolafe, A.A.; Fasinmirin, J.T.; Olufayo, A.A.; Akande, S.O. A GIS-based assessment of the potential soil erosion and flood hazard zones in Ekiti State, Southwestern Nigeria using integrated RUSLE and HAND models. *Catena* **2020**, *194*, 104725. [CrossRef]
107. Alexakis, D.D.; Hadjimitsis, D.G.; Agapiou, A. Integrated use of remote sensing, GIS and precipitation data for the assessment of soil erosion rate in the catchment area of "Yialias" in Cyprus. *Atmos. Res.* **2013**, *131*, 108–124. [CrossRef]
108. Assouline, S.; Ben-Hur, M. Effects of rainfall intensity and slope gradient on the dynamics of interrill erosion during soil surface sealing. *Catena* **2006**, *66*, 211–220. [CrossRef]
109. Al-Ruzouq, R.; Shanableh, A.; Yilmaz, A.; Idris, A.; Mukherjee, S.; Khalil, M.; Barakat, M.; Gibril, M.B.A. Dam Site Suitability Mapping and Analysis Using an Integrated GIS and Machine Learning Approach. *Water* **2019**, *11*, 1880. [CrossRef]
110. Mohammed, A.; Pradhan, B.; Mahmood, Q. Dam site suitability assessment at the Greater Zab River in northern Iraq using remote sensing data and GIS. *J. Hydrol.* **2019**, *574*, 964–979. [CrossRef]
111. Ramakrishnan, D.; Bandyopadhyay, A.; Kusuma, K.N. SCS-CN and GIS-based approach for identifying potential water harvesting sites in the Kali Watershed, Mahi River Basin, India. *J. Earth Syst. Sci.* **2009**, *118*, 355–368. [CrossRef]

International Journal of
Geo-Information

MDPI

Article

The Soil Nutrient Digital Mapping for Precision Agriculture Cases in the Trans-Ural Steppe Zone of Russia Using Topographic Attributes

Azamat Suleymanov [1,2], Evgeny Abakumov [1], Ruslan Suleymanov [2,3], Ilyusya Gabbasova [2] and Mikhail Komissarov [2,*]

1 Department of Applied Ecology, Saint Petersburg State University, 199034 Saint Petersburg, Russia; filpip@yandex.ru (A.S.); e_abakumov@mail.ru (E.A.)
2 Laboratory of Soil Science, Ufa Institute of Biology UFRC, Russian Academy of Sciences, 450054 Ufa, Russia; soils@mail.ru (R.S.); gimib@mail.ru (I.G.)
3 Department of Geodesy, Cartography and Geographic Information Systems, Bashkir State University, 450076 Ufa, Russia
* Correspondence: mkomissarov@list.ru

Abstract: Topographic features of territory have a significant impact on the spatial distribution of soil properties. This research is focused on digital soil mapping (DSM) of main agrochemical soil properties—values of soil organic carbon (SOC), nitrogen, potassium, calcium, magnesium, sodium, phosphorus, pH, and thickness of the humus-accumulative (AB) horizon of arable lands in the Trans-Ural steppe zone (Republic of Bashkortostan, Russia). The methods of multiple linear regression (MLR) and support vector machine (SVM) were used for the prediction of soil nutrients spatial distribution and variation. We used 17 topographic indices calculated using the SRTM (Shuttle Radar Topography Mission) digital elevation model. Results showed that SVM is the best method in predicting the spatial variation of all soil agrochemical properties with comparison to MLR. According to the coefficient of determination R^2, the best predictive models were obtained for content of nitrogen (R^2 = 0.74), SOC (R^2 = 0.66), and potassium (R^2 = 0.62). In our study, elevation, slope, and MMRTF (multiresolution ridge top flatness) index are the most important variables. The developed methodology can be used to study the spatial distribution of soil nutrients and large-scale mapping in similar landscapes.

Keywords: agrochemical properties; digital soil mapping; SVM; MLR; topographic variables

Citation: Suleymanov, A.; Abakumov, E.; Suleymanov, R.; Gabbasova, I.; Komissarov, M. The Soil Nutrient Digital Mapping for Precision Agriculture Cases in the Trans-Ural Steppe Zone of Russia Using Topographic Attributes. *ISPRS Int. J. Geo-Inf.* **2021**, *10*, 243. https://doi.org/10.3390/ijgi10040243

Academic Editors: Panagiotis Tziachris, Dimitris Triantakonstantis and Wolfgang Kainz

Received: 16 February 2021
Accepted: 4 April 2021
Published: 7 April 2021

1. Introduction

Nutrients are essential for soil fertility and stable plant growth [1]. Studying, modeling, and mapping the spatial distribution of soil properties is an important task for effective farming and sustainable land management [2]. Most Russian large-scale soil maps were created prior to the late 1980s [3], which is almost the same time as the start of economic problems in the USSR. Thus, the elaboration of new cartographic data and updating of existing cartographic data is necessary, especially due to the active agricultural use and transformation of climatic conditions. Nevertheless, field soil surveys and further large-scale mapping (including updating old maps) are expensive and time-consuming processes.

In recent decades, digital soil mapping (DSM) methods have been actively used to study and map soils and their properties. DSM is based on statistical, geostatistical, and mathematical methods to identify relationships between soil properties and soil formation factors, called environmental covariates [4]. DSM methods are more cost-effective and allow the creation of maps with greater accuracy and higher spatial resolution [5]. These methods are especially relevant to digital mapping of soil nutrients, as laboratory analyses of soil nutrients are costly and time-consuming.

Relief (as well as parent material, organisms, climate, and time) is one of the main factors of soil formation and has a great influence on the distribution of soil properties [6–8], especially in arid and semiarid regions [9]. Moreover, some authors conclude that topography is the most useful factor in soil studying and predicting [4,10]. Relief attributes (such as those presented in Table 1) provide a good opportunity to map soil properties and soil classes [11–17].

Table 1. The list of topographic attributes used for calculations and modeling.

№	Topographic Attributes	Acronym
1	Elevation	El
2	Aspect	As
3	Slope	Sl
4	Profile curvature	PrCu
5	Plan curvature	PlCu
6	Flow accumulation	FlAc
7	Analytical hillshading	AnHil
8	Channel network base level	CNBL
9	Channel network distance	CND
10	Convergence index	CI
11	LS-Factor	LS-F
12	Topographic wetness index	TWI
13	Valley depth	VD
14	Closed depressions	CD
15	Relative slope position	RSP
16	Multiresolution ridge top flatness	MRRTF
17	Multiresolution valley bottom flatness	MRVBF

Information about relief as a major or additional factor was successfully applied in different studies such as the migration of radiocesium in Japan [18], analysis of shallow landslides after an extreme rainstorm in Italy [19], and proliferation of fungal communities in a boreal alpine ecosystem [20]. In the studied region (Republic of Bashkortostan), the topography data were used in mapping subtypes of Chernozems soils and heavy metals [21], in the study of erosion processes [22], and in preparation of arable land for irrigation reclamation [23].

Machine learning (ML) techniques are actively used for prediction and mapping of soil properties. Various techniques including multiple linear regression (MLR) and support vector machine (SVM) have been successfully applied to the interrelation of soil properties and covariates [24–26]. MLR is a machine learning algorithm applied to regress a target variable. MLR is a least-squares model in which a targeted soil property is predicted from selected explanatory topographic variables [27]. This method establishes a linear relationship between the predicted elements and the covariates. For a better understanding of these relationships, MLR is presented below in Equation (1) [28]:

$$y = a + \sum_{i-1}^{n} b_i \times x_i \pm \varepsilon_i \tag{1}$$

where n is the number of predictors, y—response variables, x_i—explanatory variables or predictors, a—intercept (constant term), b_i—partial regression coefficients, and ε_i—residuals.

SVM is one of the basic supervised approaches for classification and regression analysis [29,30]. SVM using binary classification defines the closest points between two classes in feature space and draws an optimal dividing line between them [31,32]. This hyperplane, from the maximized margins, is used as a criterion for subsequent classification. Several investigators have reported that the SVM approach provides lower prediction errors compared with other methods [33–35].

As relief is one of the main factors of soil formation, we hypothesized that the ML techniques using topography information can predict the spatial distribution of most soil properties with good accuracy.

The objectives of this study are the following: (i) digital mapping of soil nutrients using relief attributes in arable land at the Trans-Ural steppe zone (Republic of Bashkortostan, Russia), (ii) model performance evaluation using MLR and SVM algorithms, and (iii) creating digital maps and determining the spatial distribution of soil nutrients. We assume that the developed models can be used for further large-scale digital mapping of agricultural lands in other regions of the world with similar conditions.

2. Materials and Methods

The study site is located in the Trans-Ural steppe zone, near the southern border of the Ural Mountains (Republic of Bashkortostan, Russia) (Figure 1). The study area consists of 1400 ha and is characterized mostly by gentle (1–3°) slopes up to 8° with various exposures. The elevation of the studied field ranges from 460 m in the northwestern part of the site to 377 m above sea level in the southeastern part. The state of soil cover at studied plot is characterized as arable land (the plowing by a tractor with a turnover of the soil layer to a 10–15 cm depth repeated every year in spring). Wheat (*Triticum aestivum*) is predominantly cultivated on the plot. The reconnaissance survey of the study area showed that in the southern part of territory, water erosion processes (sheet and rill erosion) are occurring.

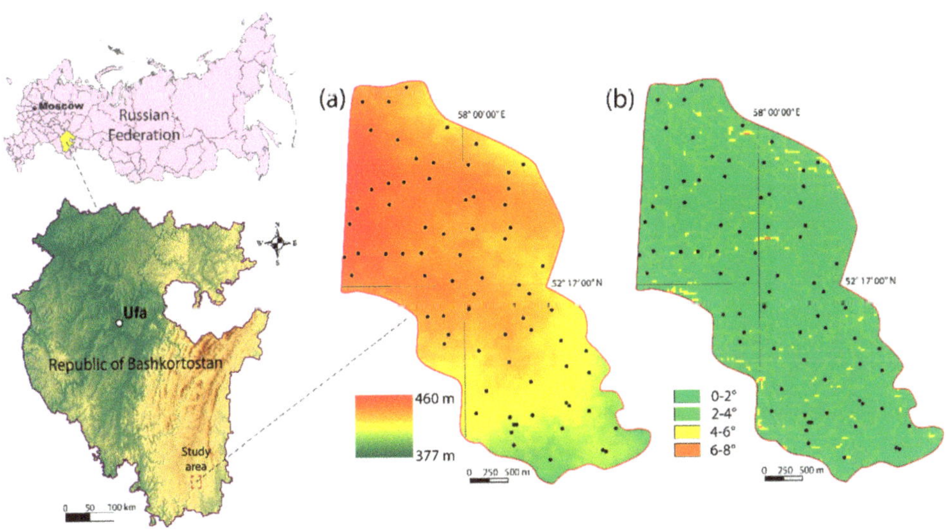

Figure 1. Location of the study area and places of soil sampling (marked as black points on (**a**) and (**b**)): (**a**)—elevation map in meters above sea level; (**b**)—slope map in degrees.

The climate of the study area is characterized as arid and slightly arid, or as Dfa by the Köppen climate classification. The average annual air temperature is 1.4 °C. The average annual amount of precipitation is 379 mm. The soil cover is presented by Haplic Chernozems [36]. The parent materials are eluvial–deluvial carbonate clay, heavy loams, and eluvium of sandy schists.

In the studied arable plot, soils were sampled from the upper horizons (0–10 cm) in representative places. The scheme of sample points location is shown on the map (Figure 1). We used a total of 76 topsoil samples, collected in July of 2020. Sample preparation was carried out according to the standard procedure: samples were air-dried, then sieved through a 1 mm sieve for further laboratory trials.

The chemical analyses were carried out using standard laboratory methods reported in [37–39]. The soil organic carbon (SOC) content, pH, alkaline hydrolyzable nitrogen (N), available phosphorus P_2O_5 (P) and K_2O (K), and exchangeable cations of Ca^{2+}, Mg^{2+}, and Na^+ were determined.

Topographic (or terrain) attributes were calculated using the digital elevation model (DEM) NASA's Shuttle Radar Topography Mission (SRTM), with a resolution of 30 m (https://www2.jpl.nasa.gov/srtm, accessed on 12 October 2020). Calculation of topographic attributes was performed in the SAGA GIS. The terrain attributes obtained from DEM are present in Table 1.

MLR and SVM techniques were used for spatial modeling of soil agrochemical properties. For choosing the final MLR model and to prevent overfitting, we used minimizing Mallows's C_p [40] and the stepwise backward selection method.

The SVM approach was conducted using a radial basis kernel. The tuning method was used for choosing the best parameters in the SVM model (C and gamma). These values were defined by the following intervals: $0.1 < C < 1.0$ (step size is 0.1) and $0.2 < \gamma < 0.5$ (step size is 0.05), according to Mattera and Haykin [41] and Cherkassky and Ma [42]. The root-mean-square error of prediction (RMSE) and coefficient of determination (R^2) values were used for determining final accuracy prediction (Equations (2) and (3)). These values were evaluated by the 10-fold cross-validation approach [43]. Randomly selected, 80% of the total samples were used for model training data, and then the remaining 20% were used for model validation and accuracy assessment. The analyzed data set was randomly divided into ten equally sized folds. Then, nine of these folds were used to calibrate the model and predict the values for the remaining fold. This procedure was repeated 10 times, each time setting a different fold aside [44]. The advantage of this method is that it performs reliably and is unbiased on smaller data sets [45].

$$RMSE = \sqrt{\frac{\sum_{i=0}^{n}(O_i - P_i)^2}{n}} \tag{2}$$

$$R^2 = \left(\frac{\sum_{i=0}^{n}(O_i - O_{avg}) \times (P_i - P_{avg})}{\sqrt{\sum_{i=0}^{n}(O_i - O_{avg})^2 \times (P_i - P_{avg})^2}} \right)^2 \tag{3}$$

where O_i and P_i are observed and predicted values of soil properties, O_{avg} and P_{avg} are the average values, and n is the number of samples.

The model with the lowest RMSE and highest R^2 values was considered to be the most applicable [46]. The R^2 was determined by the following classification [47]: models with $R^2 < 0.4$ show a poor or very low level of predictive ability; values of $0.5 < R^2 < 0.7$ indicate models with an average level of forecasting; models with $R^2 > 0.7$ are highly predictive.

Statistical data processing and estimates of model accuracy were performed using the "leaps", "e1071", and "caret" packages in R 4.0.2 [48] and RStudio (version 1.3.1073) [49].

3. Results and Discussion

The studied soil in the 0–10 cm depth has a low and medium SOC content and a neutral and slightly alkaline reaction. The content of NPK is sufficient and commensurate for this region, as well as the values of exchangeable cations [50]. According to the thickness of the humus-accumulative (AB) horizon, the soils are characterized as an eroded, uneroded, and somewhere with sediment deposition. The general statistics parameters of soil properties—mean, minimum, maximum, standard deviation (SD), and coefficient of variation (CV)—are shown in Table 2.

Table 2. Statistics description of soil properties in the 0–10 cm depth. SOC: soil organic carbon; AB: humus-accumulative.

Parameter	SOC, %	The Thickness of AB Horizon, (cm)	pH (H_2O)	N Alkaline Hydrolyzable, mg kg^{-1}	Exchangeable Cations			Available	
					Ca^{2+}	Mg^{2+}	Na^{+}	P_2O_5	K_2O
					cmol$_{(+)}$ kg^{-1}			mg kg^{-1}	
				$n = 76$					
Mean	3.7	44.1	6.8	132.6	31.8	9.7	0.3	1.8	220.2
Min	1.8	18	6.4	65	15	5	0.04	0.4	123
Max	5.6	70	8.0	189	47	15	1.7	4.6	326
SD	0.9	9.3	0.3	26.6	6.2	2.2	0.3	0.8	52.7
CV (%)	24.3	20.9	4.4	20.1	19.5	22.7	113.3	44.4	23.9

Figure 2 shows the correlation graph between soil properties and topographic attributes. According to this diagram, the SOC values correlate with N ($R = 0.66$), with the thickness of the humus-accumulative horizon ($R = 0.59$), K ($R = 0.52$), and Ca ($R = 0.48$). The relationship was also established between the elements N and K ($R = 0.57$), Mg and Ca ($R = 0.47$), Ca and N ($R = 0.53$), and between the thickness of the humus horizon with N ($R = 0.55$).

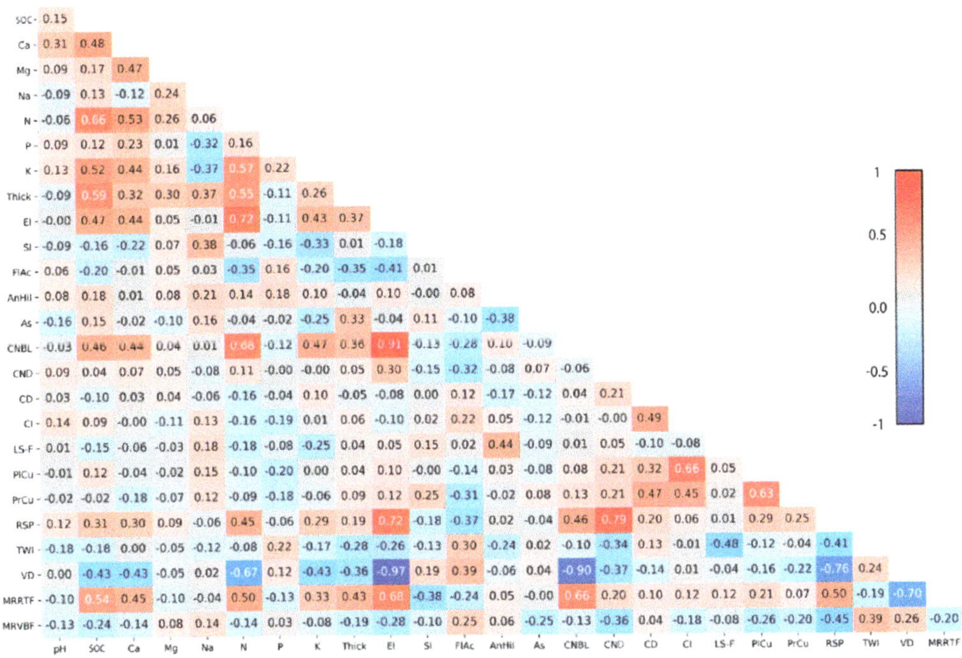

Figure 2. The correlograms of soil nutrients at topographic variables.

Among the topographic attributes and elements of the studied site, the strongest correlation was found for N. This element shows a close correlation with elevation ($R = 0.72$), channel network base level (CNBL) ($R = 0.68$), valley depth (VD) ($R = -0.67$), and multiresolution ridge top flatness (MRRTF) ($R = 0.50$). The SOC content has a relationship with the following variables: MRRTF ($R = 0.54$), elevation ($R = 0.47$), CNBL ($R = 0.46$), and VD ($R = -0.43$).

Ca is positively correlated with MRRTF values ($R = 0.45$), elevation and CNBL ($R = 0.44$). The CNBL and elevation values also show relationships to the element K ($R = 0.47$ and

$R = 0.43$, respectively). The thickness of the humus horizon shows some relationship only with MRRTF ($R = 0.43$). Elements Mg, Na, and P and pH do not correlate with calculated terrain attributes.

The final results of modeling using MLR and SVM methods are presented in Table 3. According to the R^2 and RMSE values, SVM showed a better performance for all predicting properties compared to the MLR method. The SVM method has the best ability to predict N ($R^2 = 0.74$, RMSE = 14.23). For example, Zhou [25] also developed an SVM model for soil total nitrogen prediction in northwestern China. The authors explained 50% of the variation using data about land use, land cover, climate, topography, and remote-sensing-derived variables.

Table 3. Comparison of the approaches used to predict soil properties. MLR: multiple linear regression; SVM: support vector machine.

Soil Parametr	Number of Variables	Variables for Modeling	MLR		SVM		SVM Parameters	
			R^2	RMSE	R^2	RMSE	C	Gamma
SOC	5	AnHil, As, LS-F, TWI, MRRTF	0.52	0.60	0.66	0.53	1	0.2
Na	5	Sl, AnHil, As, PlCu, MRVBF	0.35	0.21	0.49	0.20	1	0.2
Ca	4	El, CD, PrCu, MRRTF	0.31	5.84	0.43	5.33	1	0.2
N	5	El, Sl, AnHil, LS-F, PrCu	0.66	15.31	0.74	14.23	1	0.2
P	2	AnHil, TWI	0.11	1.64	0.16	0.73	0.1	0.2
Mg	2	VD, MRRTF	0.04	10.05	0.20	2.09	1	0.5
K	8	El, CNBL, Sl, As, CND, CD, LS-F, TWI	0.57	35.51	0.62	32.34	1	0.2
pH	1	TWI	0.03	6.61	0.02	0.30	0.1	0.2
Thickness of AB	3	FlAc, As, MRRTF	0.35	47.03	0.52	6.81	1	0.2

MLR and SVM showed similar performance in predicting K ($R^2 = 0.57$, RMSE = 35.51 vs. $R^2 = 0.62$, RMSE = 32.34, respectively). For example, Gopp [15] obtained an MLR model with an R^2 value of 0.45 for predicted spatial variation of exchangeable potassium on the arable plot in Russia using topography variables.

Good performance for predicting spatial distribution was obtained by SOC ($R^2 = 0.66$, RMSE = 0.53) using the SVM approach. This model uses seven topographic variables (AnHil, As, LS-F, TWI, MRRTF). Silatsa [51] concluded that terrain attributes are important variables for predicting SOC content. Similarly, Mahmoudzadeh [35] found that 58% of SOC distribution was described by topographic and climate variables using the SVM model in Iran.

The SVM technique showed almost similar coefficients of determination for the Na ($R^2 = 0.49$, RMSE = 0.20) and depth of humus horizon ($R^2 = 0.52$, RMSE = 6.81). The SVM method performed significantly better than the MLR in RMSE value for depth of humus horizon (6.8 vs. 47.03, respectively).

The SVM model explained 43% of the Ca variation ($R^2 = 0.43$, RMSE = 5.33). The above-mentioned study [15], Gopp obtained a model with a 0.67 coefficient of determination for exchangeable Ca.

Low coefficients of determination ($R^2 < 0.25$) and, consequently, no predictive capabilities were found for pH ($R^2 = 0.01$), P ($R^2 = 0.11$), and Mg ($R^2 = 0.04$). These elements are not predicted with satisfactory accuracy by the methods used. Previous results also obtained poor models for predicting the spatial distribution of pH using machine learning approaches [52,53].

The relative importance of the variables used for predicting soil elements is shown in Figure 3. The MMRTF index is a key attribute for predicting the SOC and thickness of the humus horizon variation (56% and 47%, respectively). MMRTF is a basic variable in Ca modeling (38%). This topographic attribute is a multiscale flatness index that identifies sloping and flat areas and characterizes the direction and strength of the distribution of matter in the landscape [54]. For example, an MMRTF attribute was successfully used in models to predict SOC, K, pH, and clay using regression kriging approaches [55].

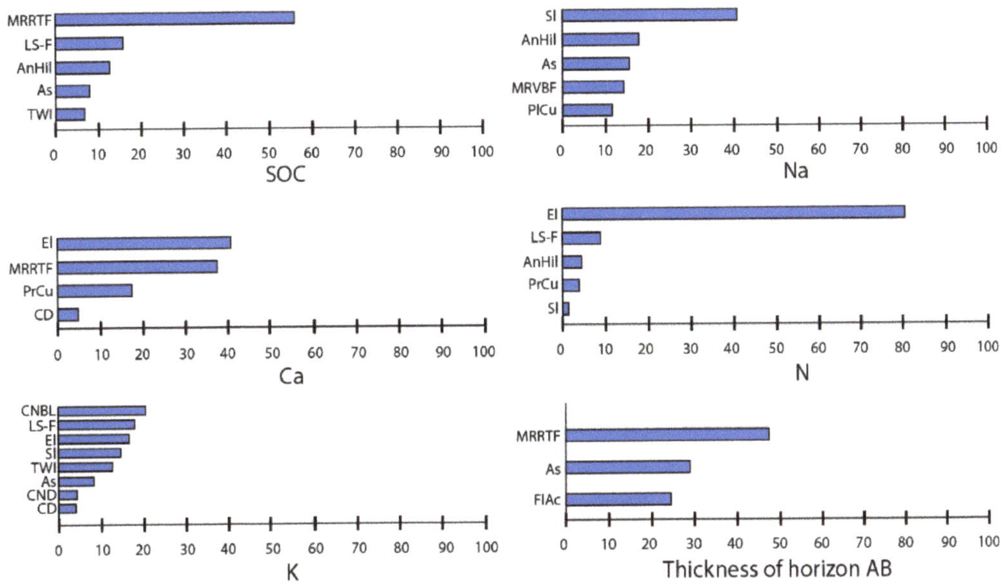

Figure 3. Relative importance (%) of topographic attributes using the SVM prediction model (for R^2 models ≥ 0.25).

Terrain elevation is the most important variable for explaining the variability of N and Ca in the study area (81% and 41%, respectively). Wang [56] reported that elevation is the most important variable affecting the spatial distribution of N. In some other digital soil mapping studies, elevation was also noted as the most important covariate for predicting soil elements [51,53,57–59].

The slope is the main covariate in the model for Na in our research (41%). The most effective variables are CNBL (21%), LS-Factor (18%), and elevation (17%). The LS-Factor attribute is the second most important parameter in the SOC and N models (16% and 9%, respectively). The LS-Factor is a combination of slope length and slope angle. This attribute is one of the five factors of the universal soil loss equation (USLE) and describing the influence of topography on soil erosion risk.

Figure 4 shows the frequency of use of all 17 topographic attributes. The most frequently used topographic attribute in modeling is aspect (slope exposition). It is used four times in various regression equations and is involved in the prediction of almost all elements (except Ca and Mg). The topographic attribute analytical hillshading (AnHil) is used three times, which is closely related to aspect; AnHil raster image is a relief with a shadow, created based on the angle of the light source and the shadow. Sahabiev [55] used a southwest position as a separate topographic parameter in the regression analysis for the prediction model for humus content, N, pH, and physical clay (the sum of soil fractions less 0.01 mm, %). The elevation, LS-Factor, and MRRTF variables were also used three times each in the prediction. These variables are closely related to each other and successfully determine areas where accumulation of soil nutrients is occurring.

Only models with a coefficient of determination $R^2 \geq 0.25$ were used to create the maps. The spatial variation of studied soil properties using the SVM predicted models is shown in Figure 5. The spatial variation of SOC, Ca, N, K, and thickness of humus horizon correlate with elevation and are characterized by higher values in the top areas of the western part and the central part of the studied plot. Furthermore, these properties gradually decrease towards the southern part of the field, where runoff processes occur. This south area sufficiently demonstrates the zones of accumulation of Ca, Na, and SOC content due to runoff and sediment deposition from elevated areas. The studied region is characterized by active wind erosion processes [60]. Thus, increased values of these nutrients are also found along roads and forest belts, designed to protect against wind erosion. The spatial distribution of Na has a different character compared to other mapped elements. The Na accumulation zones are concentrated in the southwestern part.

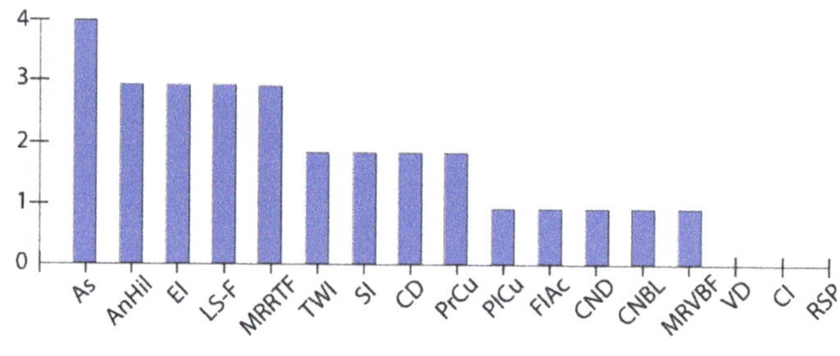

Figure 4. Frequency (%) of using topographic indices (for R^2 models ≥ 0.25).

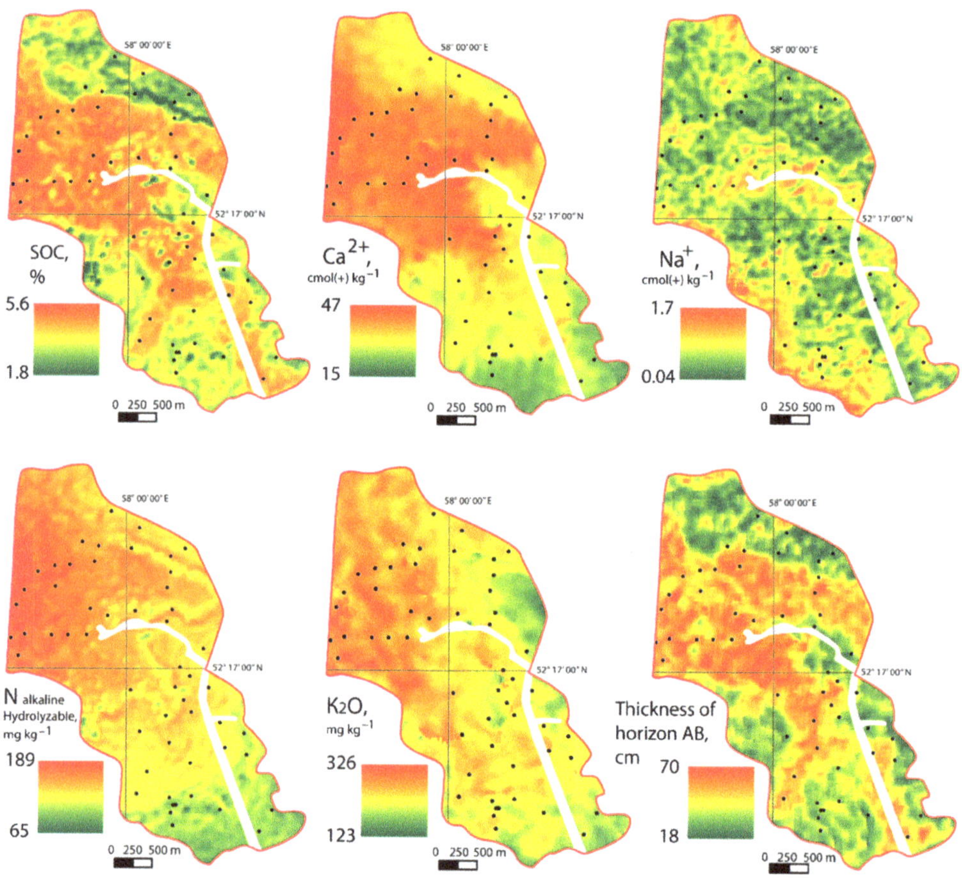

Figure 5. Spatial distribution of soil agrochemical properties using SVM method (for R^2 models ≥ 0.25). Here, and in Figure 6, areas with vegetation and roads are masked by white color.

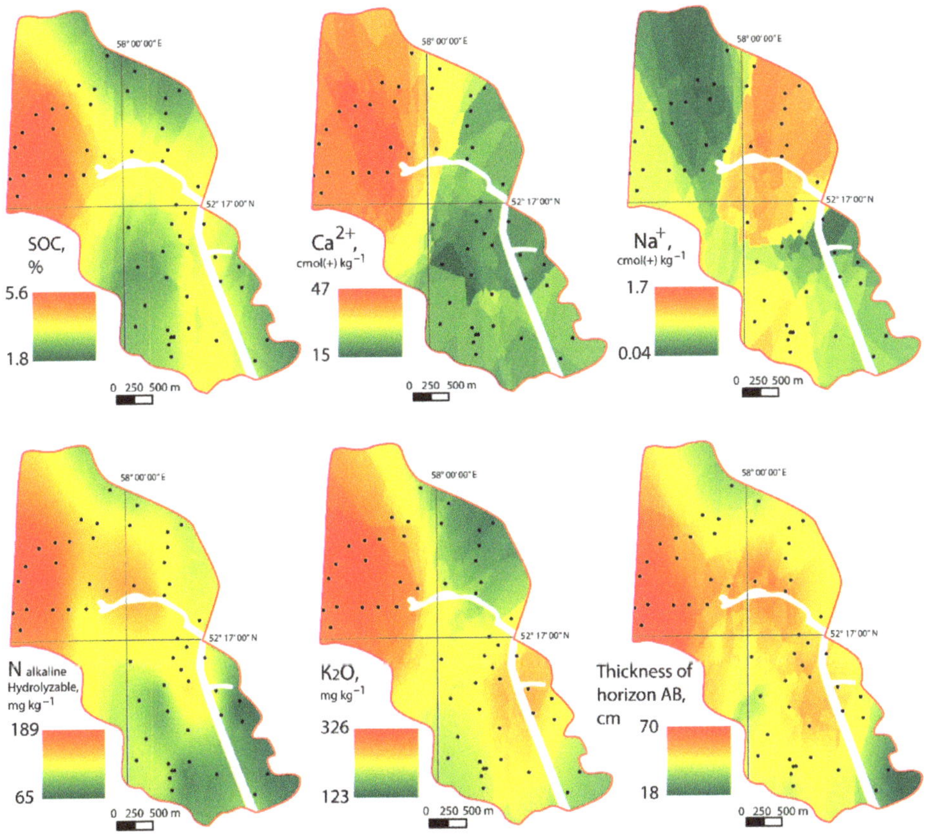

Figure 6. Spatial distribution of soil agrochemical properties using an ordinary kriging method (for R^2 models > 0.25).

For validation and assessment of the ML techniques accuracy in soil mapping (prediction of soil elements spatial distribution), we compared the results with ordinary kriging approach. The maps based and created using a kriging approach (Figure 6) have a similar distribution of studied soil parameters as obtained by the SVM method. The highest values of SOC, Ca, N, K, and thickness of humus horizon are also located in the western part on elevated areas. Thus, the accuracies obtained for the ML techniques in this study are promising and useful in digital soil mapping.

It should be noted that the soil properties could changing over the time. Thus, it is necessary to continue the present study for comparing and validating the results. At the same time, we assumed that the error in digital soil mapping with use of the ML techniques will be absent or minor. For example, the research [61] conducted earlier in the studied region showed that some properties of soils (namely, the thickness of humus-accumulative horizon, SOC, P, and N) changed insignificantly (less than 5–10%) from 1975 to 2011. Basically, those changes were associated with erosion processes and mechanical impact by tillage operations.

4. Conclusions

The monitoring of agricultural land conditions is a necessary task, especially due to intensive anthropogenic pressures and climate transformation. Large-scale mapping of soil properties is necessary for rational land management and agriculture reclamation measures.

According to the coefficient of determination (R^2) and RMSE values, the SVM method was superior to MLR in predicting all studied agrochemical soil properties. The SVM model was obtained to explain the spatial variation of N content and showed the best results (R^2 = 0.74, RMSE = 14.23). This model is characterized by highly predictive ability. The medium level of prediction was obtained for modeling of SOC, Na, K, and thickness of the humus-accumulative (AB) horizon. The most important variables for predicting the spatial distribution of soil properties were elevation, slope, and MRRTF index.

Our results showed the possibility of using topography variables with an SVM approach to predict the main soil agrochemical properties in the Trans-Ural steppe zone (Republic of Bashkortostan, Russia). The use of such ML techniques can reduce soil sampling efforts, and therefore reduce soil mapping costs and keep time/human resources. We assumed that the accuracies obtained in this study are promising for current/future local scale digital soil mapping efforts as well as for the other areas in the world with similar soil/climate conditions. We propose that the elaborated SVM models should be improved by adding other covariates, analyzing other soil formation factors, increasing the number and period of soil sampling, and using other ML techniques to obtain the best results for soil mapping.

Author Contributions: Azamat Suleymanov and Evgeny Abakumov interpreted the data and prepared the manuscript. Azamat Suleymanov, Ruslan Suleymanov, Ilyusya Gabbasova and Mikhail Komissarov performed sample collection and laboratory analyses. All authors have read and agreed to the published version of the manuscript.

Funding: The article was made with support of the Ministry of Science and Higher Education of the Russian Federation in accordance with agreement № 075-15-2020-922 date 16.11.2020 on providing a grant in the form of subsidies from the Federal budget of Russian Federation. The grant was provided for state support for the creation and development of a world-class scientific center "Agrotechnologies for the Future".

Institutional Review Board Statement: Not applicable.

Informed Consent Statement: Not applicable.

Acknowledgments: We thank the special issue guest Editors (Dimitris Triantakonstantis and Panagiotis Tziachris) and the anonymous reviewers for their careful reading of our manuscript and their many insightful comments and suggestions.

Conflicts of Interest: The authors declare no conflict of interest.

References

1. Dong, W.; Wu, T.; Luo, J.; Sun, Y.; Xia, L. Land Parcel-Based Digital Soil Mapping of Soil Nutrient Properties in an Alluvial-Diluvia Plain Agricultural Area in China. *Geoderma* **2019**, *340*, 234–248. [CrossRef]
2. Ma, Y.; Minasny, B.; Wu, C. Mapping Key Soil Properties to Support Agricultural Production in Eastern China. *Geoderma Reg.* **2017**, *10*, 144–153. [CrossRef]
3. Savin, I.Y.; Ovechkin, S.V. On the Updating of Medium-Scale Soil Maps. *Eurasian Soil Sci.* **2014**, *47*, 987–994. [CrossRef]
4. McBratney, A.B.; Mendonça Santos, M.L.; Minasny, B. On Digital Soil Mapping. *Geoderma* **2003**, *117*, 3–52. [CrossRef]
5. Arrouays, D.; Grundy, M.; Hartemink, A.E.; Hempel, J.; Heuvelink, G.B.M.; Hong, S.Y.; Lagacherie, P.; Lelyk, G.; Mcbratney, A.B.; Mckenzie, N.J.; et al. GlobalSoilMap: Toward a Fine-Resolution Global Grid of Soil Properties. *Adv. Agron.* **2014**, *125*, 93–134. [CrossRef]
6. Dokuchaev, V.V. *Our Steppes—At One Time and Now*; Yevdokimoff Press: St. Petersburg, Russia, 1892. (In Russian)
7. Jenny, H. *Factors of Soil Formation, A System of Quantitative Pedology*; McGraw-Hill: New York, NY, USA, 1941.
8. Jenny, H. Derivation of State Factor Equations of Soils and Ecosystems. *Soil Sci. Soc. Am. J.* **1961**, *25*, 385–388. [CrossRef]
9. Mehnatkesh, A.; Ayoubi, S.; Jalalian, A.; Sahrawat, K.L. Relationships between Soil Depth and Terrain Attributes in a Semi Arid Hilly Region in Western Iran. *J. Mt. Sci.* **2013**, *10*, 163–172. [CrossRef]
10. McKenzie, N.J.; Gessler, P.E.; Ryan, P.J.; O'Connell, D.A. The role of terrain analysis in soil mapping. In *Terrain Analysis: Principles and Applications*; Wilson, J.P., Gallant, J.C., Eds.; John Wiley and Sons: New York, NY, USA, 2000; pp. 245–265.
11. Moore, I.D.; Gessler, P.E.; Nielsen, G.A.; Peterson, G.A. Soil Attribute Prediction Using Terrain Analysis. *Soil Sci. Soc. Am. J.* **1993**, *57*, 443–452. [CrossRef]
12. Odeha, I.O.A.; McBratney, A.B.; Chittleborough, D.J. Spatial Prediction of Soil Properties from Landform Attributes Derived from a Digital Elevation Model. *Geoderma* **1994**, *63*, 197–214. [CrossRef]

13. Florinsky, I.V.; Eilers, R.G.; Manning, G.R.; Fuller, L.G. Prediction of Soil Properties by Digital Terrain Modelling. *Environ. Model. Softw.* **2002**, *17*, 295–311. [CrossRef]
14. Mulder, V.L.; de Bruin, S.; Schaepman, M.E.; Mayr, T.R. The Use of Remote Sensing in Soil and Terrain Mapping—A Review. *Geoderma* **2011**, *162*, 1–19. [CrossRef]
15. Gopp, N.V.; Nechaeva, T.V.; Savenkov, O.A.; Smirnova, N.V.; Smirnov, V.V. The Methods of Geomorphometry and Digital Soil Mapping for Assessing Spatial Variability in the Properties of Agrogray Soils on a Slope. *Eurasian Soil Sci.* **2017**, *50*, 20–29. [CrossRef]
16. Takoutsing, B.; Weber, J.C.; Martín, J.A.R.; Shepherd, K.; Aynekulu, E.; Sila, A. An Assessment of the Variation of Soil Properties with Landscape Attributes in the Highlands of Cameroon. *Land Degrad. Dev.* **2018**, *29*, 2496–2505. [CrossRef]
17. Sena, N.C.; Veloso, G.V.; Fernandes-Filho, E.I.; Francelino, M.R.; Schaefer, C.E.G.R. Analysis of Terrain Attributes in Different Spatial Resolutions for Digital Soil Mapping Application in Southeastern Brazil. *Geoderma Reg.* **2020**, *21*, e00268. [CrossRef]
18. Komissarov, M.; Ogura, S. Siltation and Radiocesium Pollution of Small Lakes in Different Catchment Types Far from the Fukushima Daiichi Nuclear Power Plant Accident Site. *Int. Soil Water Conserv. Res.* **2020**, *8*, 56–65. [CrossRef]
19. Bartelletti, C.; Giannecchini, R.; Avanzi, G.D.; Galanti, Y.; Mazzali, A. The Influence of Geological–Morphological and Land Use Settings on Shallow Landslides in the Pogliaschina T. Basin (Northern Apennines, Italy). *J. Maps* **2017**, *13*, 142–152. [CrossRef]
20. Chai, Y.; Jiang, S.; Guo, W.; Qin, M.; Pan, J.; Bahadur, A.; Shi, G.; Luo, J.; Jin, Z.; Liu, Y.; et al. The Effect of Slope Aspect on the Phylogenetic Structure of Arbuscular Mycorrhizal Fungal Communities in an Alpine Ecosystem. *Soil Biol. Biochem.* **2018**, *126*, 103–113. [CrossRef]
21. Gabbasova, I.M.; Suleymanov, R.R.; Khabirov, I.K.; Komissarov, M.A.; Garipov, T.T.; Sidorova, L.V.; Nazyrova, F.I. Multiple Assessment of the Soil Cover in the Area of Natural Monuments Tra-Tau and Yurak-Tau Monadnocks under Conditions of Technogenic Loads. *Eurasian Soil Sci.* **2014**, *47*, 35–46. [CrossRef]
22. Suleymanov, R.; Yaparov, I.; Saifullin, I.; Vildanov, I.; Shirokikh, P.; Suleymanov, A.; Komissarov, M.; Liebelt, P.; Nigmatullin, A.; Khamidullin, R. The Current State of Abandoned Lands in the Northern Forest-Steppe Zone at the Republic of Bashkortostan (Southern Ural, Russia). *Span. J. Soil Sci.* **2020**, *10*, 29–44. [CrossRef]
23. Suleymanov, A.R. Geomorphometric and geoinformation approach to meliorative evaluation of the territory. In *Climate Change Impacts on Hydrological Processes and Sediment Dynamics: Measurement, Modelling and Management*; Chalov, S., Golosov, V., Li, R., Tsyplenkov, A., Eds.; Springer Proceedings in Earth and Environmental Sciences; Springer International Publishing: Cham, Switzerland, 2019; pp. 72–75. [CrossRef]
24. Forkuor, G.; Hounkpatin, O.K.L.; Welp, G.; Thiel, M. High Resolution Mapping of Soil Properties Using Remote Sensing Variables in South-Western Burkina Faso: A Comparison of Machine Learning and Multiple Linear Regression Models. *PLoS ONE* **2017**, *12*, e0170478. [CrossRef]
25. Zhou, T.; Geng, Y.; Chen, J.; Sun, C.; Haase, D.; Lausch, A. Mapping of Soil Total Nitrogen Content in the Middle Reaches of the Heihe River Basin in China Using Multi-Source Remote Sensing-Derived Variables. *Remote Sens.* **2019**, *11*, 2934. [CrossRef]
26. Kingsley, J.; Lawani, S.; Esther, A.; Ndiye, K.; Sunday, O.; Penížek, V. Predictive Mapping of Soil Properties for Precision Agriculture Using Geographic Information System (GIS) Based Geostatistics Models. *Mod. Appl. Sci.* **2019**, *13*, 60. [CrossRef]
27. John, K.; Abraham Isong, I.; Michael Kebonye, N.; Okon Ayito, E.; Chapman Agyeman, P.; Marcus Afu, S. Using Machine Learning Algorithms to Estimate Soil Organic Carbon Variability with Environmental Variables and Soil Nutrient Indicators in an Alluvial Soil. *Land* **2020**, *9*, 487. [CrossRef]
28. Tranmer, M.; Murphy, J.; Elliot, M.; Pampaka, M. *Multiple Linear Regression*, 2nd ed.; Cathie Marsh Institute Working Paper; Cathie Marsh Institute: Manchester, UK, 2020.
29. Cortes, C.; Vapnik, V. Support-Vector Networks. *Mach. Learn.* **1995**, *20*, 273–297. [CrossRef]
30. Shmilovici, A. Support vector machines. In *Data Mining and Knowledge Discovery Handbook*; Maimon, O., Rokach, L., Eds.; Springer: Boston, MA, USA, 2009; pp. 231–247. [CrossRef]
31. Witten, I.H.; Frank, E. *Data Mining: Practical Machine Learning Tools and Techniques*, 2nd ed.; Elsevier: San Francisco, CA, USA, 2005; p. 525.
32. Heung, B.; Ho, H.C.; Zhang, J.; Knudby, A.; Bulmer, C.E.; Schmidt, M.G. An Overview and Comparison of Machine-Learning Techniques for Classification Purposes in Digital Soil Mapping. *Geoderma* **2016**, *265*, 62–77. [CrossRef]
33. Guan, X.; Wang, S.; Gao, Z.; Lv, Y. Dynamic Prediction of Soil Salinization in an Irrigation District Based on the Support Vector Machine. *Math. Comput. Model.* **2013**, *58*, 719–724. [CrossRef]
34. Parveen, N.; Zaidi, S.; Danish, M. Support Vector Regression Model for Predicting the Sorption Capacity of Lead (II). *Perspect. Sci.* **2016**, *8*, 629–631. [CrossRef]
35. Mahmoudzadeh, H.; Matinfar, H.R.; Taghizadeh-Mehrjardi, R.; Kerry, R. Spatial Prediction of Soil Organic Carbon Using Machine Learning Techniques in Western Iran. *Geoderma Reg.* **2020**, *21*, e00260. [CrossRef]
36. IUSS Working Group WRB. World reference base for soil resources. In *International Soil Classification System for Naming Soils and Creating Legends for Soil Maps*; World Soil Resources Reports No. 106; FAO: Rome, Italy, 2014.
37. Jones, J.B., Jr. (Ed.) *Soil Analysis Handbook of Reference Methods*, 1st ed.; CRC Press: Boca Raton, FL, USA, 1999; p. 264.
38. Pansu, M.; Gautheyrou, J. *Handbook of Soil Analysis: Mineralogical, Organic and Inorganic Methods*; Springer: Berlin, Germany, 2006; p. 993.
39. Carter, M.R.; Gregorich, E.G. (Eds.) *Soil Sampling and Methods of Analysis*, 2nd ed.; CRC Press: Boca Raton, FL, USA, 2008; p. 1264.

40. Mallows, C.L. Some Comments on Cp. *Technometrics* **2000**, *42*, 87–94. [CrossRef]
41. Mattera, D.; Haykin, S. Support vector machines for dynamic reconstruction of a chaotic system. In *Advances in Kernel Methods: Support Vector Learning*; MIT Press: Cambridge, MA, USA, 1999; pp. 211–241.
42. Cherkassky, V.; Ma, Y. Practical selection of SVM parameters and noise estimation for SVM regression. *Neural Netw.* **2004**, *17*, 113–126. [CrossRef]
43. Hengl, T.; Heuvelink, G.B.M.; Kempen, B.; Leenaars, J.G.B.; Walsh, M.G.; Shepherd, K.D.; Sila, A.; MacMillan, R.A.; de Jesus, J.M.; Tamene, L.; et al. Mapping Soil Properties of Africa at 250 m Resolution: Random Forests Significantly Improve Current Predictions. *PLoS ONE* **2015**, *10*, e0125814. [CrossRef] [PubMed]
44. Heuvelink, G.B.M.; Angelini, M.E.; Poggio, L.; Bai, Z.; Batjes, N.H.; van den Bosch, R.; Bossio, D.; Estella, S.; Lehmann, J.; Olmedo, G.F.; et al. Machine Learning in Space and Time for Modelling Soil Organic Carbon Change. *Eur. J. Soil Sci.* **2020**, 1–17. [CrossRef]
45. Zeraatpisheh, M.; Ayoubi, S.; Jafari, A.; Tajik, S.; Finke, P. Digital Mapping of Soil Properties Using Multiple Machine Learning in a Semi-Arid Region, Central Iran. *Geoderma* **2019**, *338*, 445–452. [CrossRef]
46. Jaber, S.M.; Lant, C.L.; Al-Qinna, M.I. Estimating Spatial Variations in Soil Organic Carbon Using Satellite Hyperspectral Data and Map Algebra. *Int. J. Remote Sens.* **2011**, *32*, 5077–5103. [CrossRef]
47. Vaudour, E.; Gomez, C.; Fouad, Y.; Lagacherie, P. Sentinel-2 Image Capacities to Predict Common Topsoil Properties of Temperate and Mediterranean Agroecosystems. *Remote Sens. Environ.* **2019**, *223*, 21–33. [CrossRef]
48. R Development Core Team. *R: A Language and Environment for Statistical Computing*; R Foundation for Statistical Computing: Vienna, Austria, 2015. Available online: http://www.Rproject.org (accessed on 15 November 2020).
49. RStudio. *Integrated Development Environment for R*; RStudio: Boston, MA, USA, 2015. Available online: http://www.r-studio.com (accessed on 15 November 2020).
50. Gabbasova, I.M.; Garipov, T.T.; Komissarov, M.A.; Suleimanov, R.R.; Suyundukov, Y.T.; Khasanova, R.F.; Sidorova, L.V.; Komissarov, A.V.; Suleimanov, A.R.; Nazyrova, F.I. The Impact of Fires on the Properties of Steppe Soils in the Trans-Ural Region. *Eurasian Soil Sci.* **2019**, *52*, 1598–1607. [CrossRef]
51. Silatsa, F.B.T.; Yemefack, M.; Tabi, F.O.; Heuvelink, G.B.M.; Leenaars, J.G.B. Assessing Countrywide Soil Organic Carbon Stock Using Hybrid Machine Learning Modelling and Legacy Soil Data in Cameroon. *Geoderma* **2020**, *367*, 114260. [CrossRef]
52. Sun, X.-L.; Wang, Y.; Wang, H.-L.; Zhang, C.; Wang, Z.-L. Digital Soil Mapping Based on Empirical Mode Decomposition Components of Environmental Covariates. *Eur. J. Soil Sci.* **2019**, *70*, 1109–1127. [CrossRef]
53. Mosleh, Z.; Salehi, M.H.; Jafari, A.; Borujeni, I.E.; Mehnatkesh, A. The Effectiveness of Digital Soil Mapping to Predict Soil Properties over Low-Relief Areas. *Environ. Monit. Assess.* **2016**, *188*, 195. [CrossRef]
54. Gallant, J.C.; Dowling, T.I. A Multiresolution Index of Valley Bottom Flatness for Mapping Depositional Areas. *Water Resour. Res.* **2003**, *39*. [CrossRef]
55. Sahabiev, I.A.; Ryazanov, S.S.; Kolcova, T.G.; Grigoryan, B.R. Selection of a Geostatistical Method to Interpolate Soil Properties of the State Crop Testing Fields Using Attributes of a Digital Terrain Model. *Eurasian Soil Sci.* **2018**, *51*, 255–267. [CrossRef]
56. Wang, S.; Zhuang, Q.; Wang, Q.; Jin, X.; Han, C. Mapping Stocks of Soil Organic Carbon and Soil Total Nitrogen in Liaoning Province of China. *Geoderma* **2017**, *305*, 250–263. [CrossRef]
57. Adhikari, K.; Hartemink, A.E.; Minasny, B.; Kheir, R.B.; Greve, M.B.; Greve, M.H. Digital Mapping of Soil Organic Carbon Contents and Stocks in Denmark. *PLoS ONE* **2014**, *9*, e105519. [CrossRef] [PubMed]
58. Assami, T.; Hamdi-A ssa, B. Digital Mapping of Soil Classes in Algeria—A Comparison of Methods. *Geoderma Reg.* **2019**, *16*, e00215. [CrossRef]
59. Mondejar, J.P.; Tongco, A.F. Estimating Topsoil Texture Fractions by Digital Soil Mapping—A Response to the Long Outdated Soil Map in the Philippines. *Sustain. Environ. Res.* **2019**, *29*, 31. [CrossRef]
60. Khaziev, F.K. Soils of Bashkortostan. In *Ecologic-Genetic and Agroproductive Characterization*; Gilem: Ufa, Russia, 1995; Volume 1, p. 385. (In Russian)
61. Gabbasova, I.M.; Suleimanov, R.R.; Khabirov, I.K.; Komissarov, M.A.; Fruehauf, M.; Liebelt, P.; Garipov, T.T.; Sidorova, L.V.; Khaziev, F.K. Temporal changes of eroded soils depending on their agricultural use in the southern Cis-Ural region. *Eurasian Soil Sci.* **2016**, *49*, 1204–1210. [CrossRef]

MDPI
St. Alban-Anlage 66
4052 Basel
Switzerland
Tel. +41 61 683 77 34
Fax +41 61 302 89 18
www.mdpi.com

ISPRS International Journal of Geo-Information Editorial Office
E-mail: ijgi@mdpi.com
www.mdpi.com/journal/ijgi

MDPI